Numerical Methods for Engineers and Scientists

An Introduction with Applications Using MATLAB®

THE WILEY BICENTENNIAL—KNOWLEDGE FOR GENERATIONS

$\mathcal{E}$ach generation has its unique needs and aspirations. When Charles Wiley first opened his small printing shop in lower Manhattan in 1807, it was a generation of boundless potential searching for an identity. And we were there, helping to define a new American literary tradition. Over half a century later, in the midst of the Second Industrial Revolution, it was a generation focused on building the future. Once again, we were there, supplying the critical scientific, technical, and engineering knowledge that helped frame the world. Throughout the 20th Century, and into the new millennium, nations began to reach out beyond their own borders and a new international community was born. Wiley was there, expanding its operations around the world to enable a global exchange of ideas, opinions, and know-how.

For 200 years, Wiley has been an integral part of each generation's journey, enabling the flow of information and understanding necessary to meet their needs and fulfill their aspirations. Today, bold new technologies are changing the way we live and learn. Wiley will be there, providing you the must-have knowledge you need to imagine new worlds, new possibilities, and new opportunities.

Generations come and go, but you can always count on Wiley to provide you the knowledge you need, when and where you need it!

WILLIAM J. PESCE
PRESIDENT AND CHIEF EXECUTIVE OFFICER

PETER BOOTH WILEY
CHAIRMAN OF THE BOARD

Numerical Methods
for Engineers and Scientists

An Introduction with
Applications Using MATLAB®

Amos Gilat
Vish Subramaniam

Department of Mechanical Engineering
The Ohio State University

BICENTENNIAL
BICENTENNIAL
1807
WILEY
2007
BICENTENNIAL
BICENTENNIAL

ACQUISITIONS EDITOR	Joseph Hayton
EDITORIAL ASSISTANT	Sandra Kim
SENIOR PRODUCTION EDITOR	Lisa Wojcik
COVER DESIGNER	Michael St. Martine
COVER PHOTO	Dr. Kelly Carney
Wiley 200th Anniversary logo designed by:	Richard J. Pacifico

This book was set in Times New Roman and printed and bound by RR Donnelley, Willard. The cover was printed by Phoenix Color Corp.
This book is printed on acid free paper. ∞

The image on the cover shows a numerical simulation of ice impacting a surface of the space shuttle. Courtesy of Dr. Kelly Carney, NASA Glenn Research Center, Cleveland, Ohio.

To order books or for customer service please, call 1-800-CALL WILEY (225-5945).

ISBN-13 9780471734406

Printed in the United States of America

10 9 8 7 6 5 4 3 2

To Yaela, Taly, and Edan

To Radha, Priya, and Sonya

Preface

This textbook is intended for a first course in numerical methods for students in engineering and science, typically taught in the second year of college. The book covers the fundamentals of numerical methods from an applied point of view. It explains the basic ideas behind the various methods and shows their usefulness for solving problems in engineering and science.

In the past a numerical methods course was essentially mathematical, emphasizing numerical analysis and theory. More recently, due to the availability of powerful desktop computers and computing software that is both affordable and powerful, the content and nature of a first course in numerical methods for engineering and science students are changing. The emphasis is shifting more and more toward applications and toward implementing numerical methods with ready to use tools. In a typical course, students still learn the fundamentals of numerical methods. In addition, however, they learn computer programming (or improve their programming skills if they have already been introduced to programming), and use advanced software as a tool for solving problems. MATLAB is a good example of such software. It can be used by students to write their own programs, and can be used as a tool for solving problems using its built-in functions. One of the objectives of a course in numerical methods is to prepare students in science and engineering for future courses in their areas of specialization (and their future careers) where they will have to use computers for solving problems.

Main objectives of the book

> To teach the fundamentals of numerical methods, with emphasis on the most essential methods.

> To provide students with the opportunity to enhance their programming skills using the MATLAB environment to implement algorithms.

> To teach the use of MATLAB as a tool (using its built-in functions) for solving problems in science and engineering, and for checking the results of any programs students write themselves.

Features/pedagogy of the book

- This book is written in simple, clear, and direct language. Frequently, bullets and a list of steps, rather than lengthy text, are used to list facts and details of a specific subject.

- Numerous illustrations are used for explaining the principles of the numerical methods.

- Many of the examples and end-of-chapter problems involve realistic problems in science and engineering.

- MATLAB is integrated within the text and in the examples. A light red background is used when MATLAB syntax is displayed.

- Annotating comments that explain the commands are posted alongside the MATLAB syntax.

- MATLAB's built-in functions that are associated with the numerical methods are presented in detail.

- The homework problems at the end of the chapters are divided into three groups:

(*a*) ***Problems to be solved by hand:*** Problems related to improving understanding of numerical methods. In these problems the students are asked to answer questions related to the fundamentals of numerical methods, and to carry out a few steps of the numerical methods by hand.

(*b*) ***Problems to be programmed in MATLAB:*** Problems designed to provide the opportunity to improve programming skills. In these problems students are asked to use MATLAB to write computer programs (script files, and user-defined functions) implementing various numerical methods.

(*c*) ***Problems in math, science, and engineering:*** Problems in science and engineering that have to be solved by using numerical methods. The objective is to train the students to use numerical methods for solving problems they can expect to see in future courses or in practice. Students are expected to use the programs that are presented in the book, programs that they write, and the built-in functions in MATLAB.

Organization of the book

Chapter 1: The first chapter gives a general introduction to numerical methods and to the way that computers store numbers and carry out numerical operations. It also includes a section on errors in numerical solutions and a section on computers and programming.

Chapter 2: The second chapter presents a review of fundamental mathematical concepts that are used in the following chapters that cover the numerical methods. It is intended to be used as a reminder, or a refresher, of concepts that the students are assumed (expected) to be familiar with from their first and second-year mathematics courses. Since many of these topics are associated with various numerical methods, we feel that it is better to have the mathematical background gathered in one chapter (and easier to find when needed) rather than be dispersed throughout the book. Several of the topics that are covered in Chapter 2 and that are essential in the explanation of a numerical method are repeated in other chapters where the numerical methods are presented. Most instructors will probably choose not to cover Chapter 2 as one unit in the class, but will mention a topic when needed and refer the students to the chapter.

Chapters 3 through 9: These seven chapters present the various numerical methods in an order that is typically followed in a first course on numerical methods. These chapters follow the format explained next.

Organization of a typical chapter

An itemized list of the topics that are covered in the chapter is displayed below the title of the chapter. The list is divided into *core* and *complementary* topics. The *core topics* are the most essential topics related to the subject of the chapter. The *complementary topics* include more advanced topics. Obviously, a division of topics related to one subject into core and complementary is subjective. The intent is to help instructors in the design of their course when there is not enough time to cover all the topics. In practicality, the division can be ignored in courses where all the topics are covered.

The first section of the chapter provides a general background with illustrative examples of situations in the sciences and engineering where the methods described in the chapter are used. This section also explains the basic ideas behind the specific class of numerical methods that are described in the chapter. The following sections cover the core topics of the chapter. Next, a special section discusses the built-in functions in MATLAB that implement the numerical methods described in the chapter, and how they may be used to solve problems. The later sections of the chapter cover the complementary topics.

The order of topics

It is probably impossible to write a text book where all the topics follow an order that is agreed upon by all instructors. In the present book the main subjects are in an order that is typical in a first course in numerical methods. Chapter 3 covers solution of nonlinear equations. It mostly deals with the solution of a single equation, which is a simple application of numerical methods. The chapter also includes, as a complementary topic, a section on the solution of a system of nonlinear equations. Chapter 4 deals with the solution of a system of linear equations. A complementary topic in this chapter deals with eigenvalue problems.

Chapter 5 covers curve fitting and interpolation, and Chapters 6 and 7 cover differentiation and integration, respectively. Finally, solution of ordinary differential equations (ODE) is presented in the last two chapters. Chapters 8 deals with the solution of initial-value problems (first-order, systems, and higher-order) and Chapter 9 consider boundary-value problems.

The order of some of the topics is dictated by the subjects themselves. For example, differentiation and integration need to be covered before ordinary differential equations. It is possible, however, to cover the other subjects in different order than is in the book. The various chapters and sections in the book are written in a self-contained manner that make it easy for the instructor to cover the subjects in a different order, if desired.

MATLAB programs

This book contains many MATLAB programs. The programs are clearly identified as user-defined functions, or as script files. All the programs are listed in Appendix B. The programs, or the scripts, are written is a simple way that is easy to follow. The emphasis of these programs is on the basics and on how to program an algorithm of a specific numerical method. Obviously, the programs are not general, and do not cover all possible circumstances when executed. The programs are not written from the perspective of being shortest, fastest, or most efficient. Rather, they are written such that they are easy to follow. It is assumed that most of the students only have limited understanding of MATLAB and programming, and presenting MATLAB in this manner will advance their computing skills. More advanced users of MATLAB are encouraged to write more sophisticated and efficient programs and scripts, and compare their performance with the ones in the book.

First edition promise

A. Text Accuracy

The manuscript was reviewed by more than 20 college engineering professors. Two dedicated accuracy checkers (in addition to the authors) independently checked all worked examples and homework problems. A copy editor reviewed the manuscript for correct grammar and punctuation. The final text was sent to the printer as 'camera ready' files from the authors, eliminating any potential errors that might be introduced during composition.

B. Solutions Accuracy

Fully worked solutions for the end of chapter problems were written by the authors. Two accuracy checkers independently checked and verified all final solutions.

C. Reliability

The manuscript has been class tested by over 200 students in three separate courses in order to ensure reliability, readability, and student usefulness. Student feedback from these class tests was incorporated into the final version of the text.

D. Time saving support material (available on the instructor companion site at www.wiley.com/college/gilat)

(*a*) for faculty who have adopted the text for use in their course, a fully worked solution manual, triple checked for accuracy.

(*b*) suggested course syllabi with suggested assignments to help quickly integrate the text into your course.

(*c*) conversion guides from other major numerical methods titles to show where each section of your current text is covered in this new text, helping you quickly convert from old to new.

(*d*) electronic versions of all the figures and tables from the text, for creating lecture slides and quizzes/exams based on images from the book.

(*e*) m-files of all the programs in the text.

Many people have assisted during the preparation of this book. We would like to thank the reviewers for the many comments and suggestions they have made.

Lawrence K. Agbezuge, *Rochester Institute of Technology*
David Alciatore, *Colorado State University*
Salame Amr, *Virginia State University*
John R. Cotton, *Virginia Polytechnic Institute and State University*
David Dux, *Purdue University*
Venkat Ganesan, *University of Texas-Austin*
Michael R. Gustafson II, *Duke University*
Alain Kassab, *University of Central Florida*
Tribikram Kundu, *University of Arizona*
Ronald A. Mann, *University of Louisville*
Peter O. Orono, *Indiana University Purdue University Indianapolis*
Charles Ritz, *California State Polytechnic University-Pomona*
Douglas E. Smith, *University of Missouri-Columbia*
Anatoliy Swishchuk, *University of Calgary*
Ronald F. Taylor, *Wright State University*
Brian Vick, *Virginia Polytechnic Institute and State University*

We would also like to thank Joseph Hayton, acquisition editor, Lisa Wojcik, production editor, Sandra Kim, editorial assistant, and Harry Nolan, design director, all from Wiley. Special thanks to Professor Subramaniam's daughters, Sonya and Priya, for typing early drafts of some chapters and for proofreading them.

Our intention was to write a book that is useful to students and instructors alike. We would appreciate any comments that will help to improve future editions.

Amos Gilat (gilat.1@osu.edu)
Vish Subramaniam (subramaniam.1@osu.edu)
Columbus, Ohio
February, 2007

Brief Table of Contents

Contents

Chapter 6 *Numerical Differentiation* 233

Chapter 7 *Numerical Integration* 267

Chapter 9 *Ordinary Differential Equations: Boundary-Value Problems* 387

Appendix A *Introductory MATLAB*   421

Appendix B *MATLAB Programs*   451

Chapter 1

Introduction

Core Topics

Representation of numbers on a computer (1.2).

Errors in numerical solutions, round-off errors and truncation errors (1.3).

Computers and programming (1.4).

1.1 BACKGROUND

Numerical methods are mathematical techniques used for solving mathematical problems that cannot be solved or are difficult to solve analytically. An analytical solution is an exact answer in the form of a mathematical expression in terms of the variables associated with the problem that is being solved. A numerical solution is an approximate numerical value (a number) for the solution. Although numerical solutions are an approximation, they can be very accurate. In many numerical methods, the calculations are executed in an iterative manner until a desired accuracy is achieved.

For example, Fig. 1-1 shows a block of mass m being pulled by a force F applied at an angle θ. By applying equations of equilibrium, the relationship between the force and the angle is given by:

$$F = \frac{\mu m g}{\cos\theta + \mu\sin\theta} \tag{1.1}$$

where μ is the friction coefficient and g is the acceleration due to gravity. For a given value of F, the angle that is required for moving the block can be determined by solving Eq. (1.1) for θ. Equation (1.1), however, cannot be solved analytically for θ. Using numerical methods, an approximate solution can be determined for specified accuracy. This means that when the numerical solution for θ is substituted back in Eq. (1.1) the value of F that is obtained from the expression on the right-hand side is not exactly equal to the given value of F, but is very close.

Numerical techniques for solving mathematical problems were developed and used hundreds and even thousands of years ago. Implementation of the numerical techniques was difficult since the calculations had to be carried out by hand or by use of simple mechanical

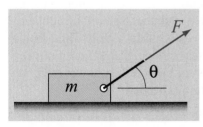

Figure 1-1: Motion of a block on a surface with friction.

1

computing devices, which limited the number of calculations that could be carried out, as well as their speed and accuracy. Today numerical methods are used with fast electronic digital computers that make it possible to execute many tedious and repetitive calculations that produce accurate (even though not exact) solutions in a very short time.

Solving a problem in science and engineering

The process of solving a problem in science and engineering is influenced by the tools (mathematical methods) that are available for solving the problem. The process can be divided into the following steps:

Problem statement

The problem statement defines the problem. It gives a description of the problem, lists the variables that are involved, and identifies the constraints in the form of boundary and/or the initial conditions.

Formulation of the solution

Formulation of the solution consists of the model (physical law or laws) that is used to represent the problem and the derivation of the governing equations that need to be solved. Examples of such laws are Newton's laws, conservations of mass, and the laws of thermodynamics. The models that are used (chosen) to solve the problem need to be consistent with the methods that are subsequently used for solving the equations. If analytical methods are expected to be used for the solution, the governing equations must be of a type that can be solved analytically. If needed, the formulation has to be simplified, such that the equations could be solved analytically. If numerical methods are used for the solution, the models and the equations can be more complicated. Even then, however, some limitations might exist. For example, if the formulation is such that a numerical solution requires a long computing time, the formulation might have to be simplified such that a solution is obtained in a reasonable time. An example is weather forecasting. The problem that is solved is large, and the numerical models that are used are very complicated. The numerical simulation of the weather, however, cannot outlast the period over which forecasting is needed.

Programming (of numerical solution)

If the problem is solved numerically, the numerical method that is used for the solution has to be selected. For every type of mathematical problem there are several (or many) numerical techniques that can be used. The techniques differ in accuracy, length of calculations, and difficulty in programming. Once a numerical method is selected, it is implemented in a computer program. The implementation consists of an *algorithm*, which is a detailed plan that describes how to carry out the numerical method, and a computer program, which is a list of commands that allows the computer to execute the algorithm to find the solution.

Interpretation of the solution

Since numerical solutions are an approximation (errors are addressed in Section 1.4), and since the computer program that executes the numerical method might have errors (or bugs), a numerical solution needs to be examined closely. This can be done in several ways, depending on the problem. For example, if the numerical method is used for solving a nonlinear algebraic equation, the validity of the solution can be verified by substituting the solution back in the equation. In more complicated problems, like a solution of a differential equation, the numerical solution can be compared with a known solution of a similar problem, or the problem can be solved several times using different boundary (or initial) conditions, and different numerical methods, and examining the subsequent differences in the solutions.

An illustration of the first two steps in the solution process of a problem is shown in Example 1-1.

Example 1-1: Problem formulation

Consider the following problem statement:
A pendulum of mass m is attached to a rigid rod of length L, as shown in the figure. The pendulum is displaced from the vertical position such that the angle between the rod and the x axis is θ_0, and then the pendulum is released from rest. Formulate the problem for determining the angle θ as a function of time, t, once the pendulum is released. In the formulation include a damping force that is proportional to the velocity of the pendulum.

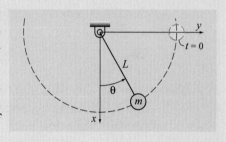

Formulate the solution for two cases:
(a) $\theta_0 = 5°$ and (b) $\theta_0 = 90°$.

SOLUTION

Physical law

The physical law that is used for solving the problem is Newton's second law of mechanics, according to which, as the pendulum swings back and forth, the sum of the forces that are acting on the mass is equal to the mass times its acceleration.

$$\Sigma \bar{F} = m\bar{a} \qquad (1.2)$$

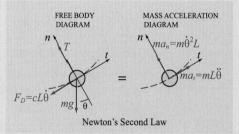

This can be visualized by drawing a free body diagram and a mass acceleration diagram, which are shown on the right. The constant c is the damping coefficient. It should be pointed out that the mass of the rod is neglected in the present solution.

Governing equation

The governing equation is derived by applying Newton's second law in the tangential direction:

$$\Sigma F_t = -cL\frac{d\theta}{dt} - mg\sin\theta = mL\frac{d^2\theta}{dt^2} \tag{1.3}$$

Equation (1.3), which is a second-order, nonlinear, ordinary differential equation, can be written in the form:

$$mL\frac{d^2\theta}{dt^2} - cL\frac{d\theta}{dt} - mg\sin\theta = 0 \tag{1.4}$$

The initial conditions are that when the motion of the pendulum starts ($t = 0$) the pendulum is at angle θ_0 and its velocity is zero (released from rest):

$$\theta(0) = \theta_0 \quad \text{and} \quad \frac{d\theta}{dt}\bigg|_{t=0} = 0 \tag{1.5}$$

Method of solution

Equation (1.4) is a nonlinear equation and cannot be solved analytically. However, in part (*a*) the initial displacement of the pendulum is $\theta_0 = 5°$, and once the pendulum is released the angle as the pendulum oscillates will be less than $5°$. For this case Eq. (1.4) can be linearized by assuming that $\sin\theta \approx \theta$. With this approximation, the equation that has to be solved is linear and can be solved analytically:

$$mL\frac{d^2\theta}{dt^2} - cL\frac{d\theta}{dt} - mg\,\theta = 0 \tag{1.6}$$

with the initial conditions Eq. (1.5).

In part (*b*), the initial displacement of the pendulum is $\theta_0 = 90°$ and the equation has to be solved numerically. An actual numerical solution for this problem is shown in Example 8-8.

1.2 REPRESENTATION OF NUMBERS ON A COMPUTER

Decimal and binary representation

Numbers can be represented in various forms. The familiar decimal system (base 10) uses ten digits 0, 1, ..., 9. A number is written by a sequence of digits that correspond to multiples of powers of 10. As shown in Fig. 1-2, the first digit to the left of the decimal point corre-

$$10^4 \quad 10^3 \quad 10^2 \quad 10^1 \quad 10^0 \quad 10^{-1} \quad 10^{-2} \quad 10^{-3} \quad 10^{-4}$$

$$\downarrow \quad \downarrow \quad \downarrow \quad \downarrow \quad \downarrow \quad \downarrow \quad \downarrow \quad \downarrow \quad \downarrow$$

$$6 \quad 0 \quad 7 \quad 2 \quad 4 \,.\, 3 \quad 1 \quad 2 \quad 5$$

$$6 \times 10^4 + 0 \times 10^3 + 7 \times 10^2 + 2 \times 10^1 + 4 \times 10^0 + 3 \times 10^{-1} + 1 \times 10^{-2} + 2 \times 10^{-3} + 5 \times 10^{-4} = 60,724.3125$$

Figure 1-2: Representation of the number 60,724.3125 in the decimal system (base 10).

Base 10	Base 2			
	2^3	2^2	2^1	2^0
1	0	0	0	1
2	0	0	1	0
3	0	0	1	1
4	0	1	0	0
5	0	1	0	1
6	0	1	1	0
7	0	1	1	1
8	1	0	0	0
9	1	0	0	1
10	1	0	1	0

Figure 1-3: Representation of numbers in decimal and binary forms.

sponds to 10^0. The digit next to it on the left corresponds to 10^1, the next digit to the left to 10^2, and so on. In the same way, the first digit to the right of the decimal point corresponds to 10^{-1}, the next digit to the right to 10^{-2}, and so on.

In general, however, a number can be represented using other bases. A form that can be easily implemented in computers is the binary (base 2) system. In the binary system, a number is represented by using the two digits 0 and 1. A number is then written as a sequence of zeros and ones that correspond to multiples of powers of 2. The first digit to the left of the decimal point corresponds to 2^0. The digit next to it on the left corresponds to 2^1, the next digit to the left to 2^2, and so on. In the same way, the first digit to the right of the decimal point corresponds to 2^{-1}, the next digit to the right to 2^{-2}, and so on. The first ten digits 1, 2, 3, ..., 10 in base 10 and their representation in base 2 are shown in Fig. 1-3. The representation of the number 19.625 in the binary system is shown in Fig. 1-4.

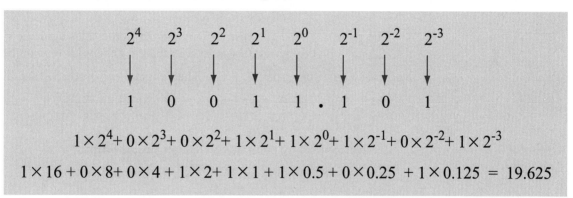

$$2^4 \quad 2^3 \quad 2^2 \quad 2^1 \quad 2^0 \quad 2^{-1} \quad 2^{-2} \quad 2^{-3}$$
$$\downarrow \quad \downarrow \quad \downarrow \quad \downarrow \quad \downarrow \quad \downarrow \quad \downarrow \quad \downarrow$$
$$1 \quad 0 \quad 0 \quad 1 \quad 1 \,.\, 1 \quad 0 \quad 1$$

$$1\times2^4+0\times2^3+0\times2^2+1\times2^1+1\times2^0+1\times2^{-1}+0\times2^{-2}+1\times2^{-3}$$

$$1\times16+0\times8+0\times4+1\times2+1\times1+1\times0.5+0\times0.25+1\times0.125 = 19.625$$

Figure 1-4: Representation of the number 19.625 in the binary system (base 2).

Another example is shown in Fig. 1-5, where the number 60,724.3125 is written in binary form.

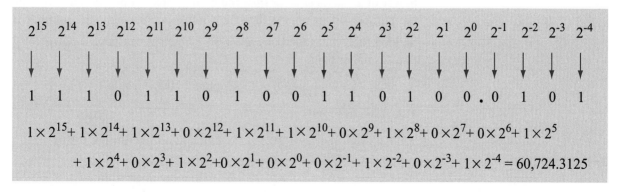

$$2^{15} \ 2^{14} \ 2^{13} \ 2^{12} \ 2^{11} \ 2^{10} \ 2^9 \ 2^8 \ 2^7 \ 2^6 \ 2^5 \ 2^4 \ 2^3 \ 2^2 \ 2^1 \ 2^0 \ 2^{-1} \ 2^{-2} \ 2^{-3} \ 2^{-4}$$
$$\downarrow \downarrow \downarrow \downarrow \downarrow \downarrow \downarrow \downarrow \downarrow \downarrow \downarrow \downarrow \downarrow \downarrow \downarrow \downarrow \downarrow \downarrow \downarrow \downarrow$$
$$1 \ 1 \ 1 \ 0 \ 1 \ 1 \ 0 \ 1 \ 0 \ 0 \ 1 \ 1 \ 0 \ 1 \ 0 \ 0 \,.\, 0 \ 1 \ 0 \ 1$$

$$1\times2^{15}+1\times2^{14}+1\times2^{13}+0\times2^{12}+1\times2^{11}+1\times2^{10}+0\times2^9+1\times2^8+0\times2^7+0\times2^6+1\times2^5$$

$$+1\times2^4+0\times2^3+1\times2^2+0\times2^1+0\times2^0+0\times2^{-1}+1\times2^{-2}+0\times2^{-3}+1\times2^{-4} = 60,724.3125$$

Figure 1-5: Representation of the number 60,724.3125 in the binary system (base 2).

Computers store and process numbers in binary (base 2) form. Each binary digit (one or zero) is called a *bit* (for binary digit). Binary arithmetic is used by computers because modern transistors can be used as extremely fast switches. Therefore, a network of these may be used to represent strings of numbers with the "1" referring to the switch being in the "on" position and "0" referring to the "off" position. Various operations are then performed on these sequences of ones and zeros.

Floating point representation

To accommodate large and small numbers, real numbers are written in floating point representation. Decimal floating point representation (also called scientific notation) has the form:

$$d.dddddd \times 10^p \tag{1.7}$$

One digit is written to the left of the decimal point, and the rest of the significant digits are written to the right of the decimal point. The number *0.dddddd* is called the **mantissa**. Two examples are:

$$6519.23 \quad \text{is written as} \quad 6.51923 \times 10^3$$

$$0.00000391 \quad \text{is written as} \quad 3.91 \times 10^{-6}$$

The power of 10, p, represents the number's order of magnitude, provided the preceding number is smaller than 5. Otherwise the number is said to be of the order of $p+1$. Thus the number 3.91×10^{-6} is of the order of 10^{-6}, $O(10^{-6})$, and the number 6.51923×10^3 is of the order of 10^4 (written as $O(10^4)$).

Binary floating point representation has the form:

$$1.bbbbbb \times 2^{bbb} \qquad (b \text{ is a decimal digit}) \tag{1.8}$$

In this form, the mantissa is $.bbbbbb$, and the power of 2 is called the **exponent**. Both, the mantissa and the exponent are written in a binary form. The form in Eq. (1.8) is obtained by normalizing the number (when it is written in the decimal form) with respect to the largest power of 2 that is smaller than the number itself. For example, to write the number 50 in binary floating point representation, the number is divided (and multiplied) by $2^5 = 32$ (which is the largest power of 2 that is smaller than 50):

$$50 = \frac{50}{2^5} \times 2^5 = 1.5625 \times 2^5 \text{ which in binary form is } 1.1001 \times 2^{101}$$

Two more examples are:

$$1344 = \frac{1344}{2^{10}} \times 2^{10} = 1.3125 \times 2^{10} \text{ which in binary form is } 1.0101 \times 2^{1001}$$

$$0.3125 = \frac{0.3125}{2^{-2}} \times 2^{-5} = 1.25 \times 2^{-2} \text{ which in binary form is } 1.01 \times 2^{-10}$$

Storing a number in computer memory

Once in binary floating point representation, the number is stored in the computer. The computer stores the values of the exponent and the mantissa separately, while the leading 1 in front of the decimal point is not stored. As already mentioned, a bit is a binary digit. The memory in the computer is organized in **bytes**, where each byte is 8 bits. According to the IEEE[1]-754 standard, computers store numbers and carry out calculations in **single-precision**[2] or in **double-precision**.[3] In single-precision, the numbers are stored in a string of 32 bits (4 bytes), and in double-precision in a string of 64 bits (8 bytes). In both cases the first bit stores the sign (0 corresponds to + and 1 corresponds to −) of the number. The next 8 bits in single precision (11 bits in double precision) are used for storing the exponent. The following 23 bits in single precision (52 bits in double precision) are used for storing the mantissa. This is illustrated for double precision in Fig. 1-6.

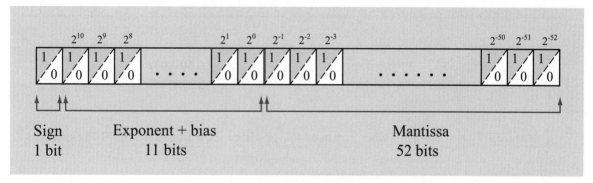

Figure 1-6: Storing in double precision a number written in binary floating point representation.

The value of the mantissa is entered, as is, in a binary form. The value of the exponent is entered with a bias. A bias means that a constant is added to the value of the exponent. The bias is introduced in order to avoid using one of the bits for the sign of the exponent (since the exponent can be positive or negative). In binary notation the largest number that can be written with 11 bits is 2047 (when all 11 digits are 1). The bias that is used is 1023, which means that if, for example, the

1. IEEE stands for the Institute of Electrical and Electronics Engineers.

2. **Precision** refers to the number of significant digits of a real number that can be stored on a computer. For example, the number 1/3 = 0.333333..., can only be represented on a computer in a chopped or rounded form with a finite number of binary digits, since the amount of memory where these bits are held is finite. The more digits to the right-hand side of the decimal point that are stored, the more **precise** is the representation of the real number on the computer.

3. This is somewhat of a misnomer. The precision in a double-precision number is not really doubled compared to a single-precision number. Rather, the "double" in double precision refers to the fact that twice as many binary digits (64 versus 32) are used to represent a real number than in the case of a single-precision representation.

exponent is 4, then the value that is stored is $4 + 1023 = 1027$. Thus the smallest exponent that can be stored by the computer is -1023 (which will be stored as 0, and the largest is 1024 (which will be stored as 2047). In single precision, 8 bits are allocated to the value of the exponent and the bias is 127.

As an example, consider storing of the number 22.5 in double precision according to the IEEE-754 standard. First, the number is normalized, $\frac{22.5}{2^4}2^4 = 1.40625 \times 2^4$. In double precision, the exponent with the bias is $4 + 1023 = 1027$, which is stored in binary form as 10000000011. The mantissa is 0.40625, which is stored in binary form as .01101000....000. The storage of the number is illustrated in Fig. 1-7.

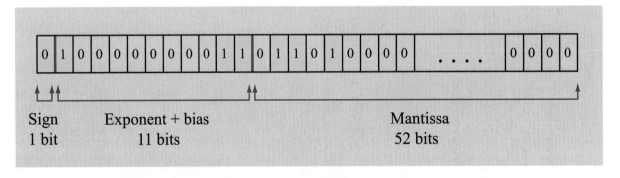

Figure 1-7: Storing the number 22.5 in double precision according to the IEEE-754 standard.

Additional notes

- The smallest positive number that can be expressed in double precision is:

$$2^{-1023} \approx 1.1 \times 10^{-308}$$

  This means that there is a (small) gap between zero and the smallest number that can be stored on the computer. Attempts to define a number in this gap causes an ***underflow*** error. (In the same way, the closest negative number to zero is -1.1×10^{-308}.)

- The largest positive number that can be expressed in double precision is approximately:

$$2^{1024} \approx 1.8 \times 10^{308}$$

  Attempts to define a larger number causes ***overflow*** error. (The same applies to numbers smaller than -2^{1024}.)

The range of numbers that can be represented in double precision is shown in Fig. 1-8.

- Since a finite number of bits is used, not every number can be accu-

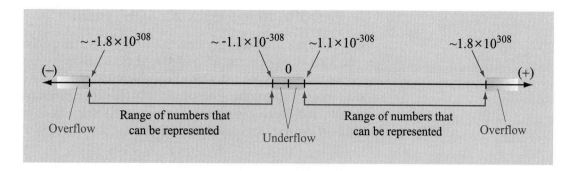

Figure 1-8: Range of numbers that can be represented in double precision.

rately written in binary form. In other words, only a finite number of exact values in decimal format can be stored in binary form. For example, the number 0.1 cannot be represented exactly in finite binary format when single precision is used. To be written in binary floating point representation 0.1 is normalized: $0.1 = 1.6 \times 2^{-4}$. The exponent -4 (with a bias) can be stored exactly, but the mantissa, 0.6 cannot be written exactly in a binary format that uses 23 bits. In addition, irrational numbers cannot be represented exactly in any format. This means that in many cases exact values are approximated. The errors that are introduced are small in one step, but when many operations are executed, the errors can grow to such an extent that the final answer is affected. These errors, as well as other errors, are discussed in the next section.

- The interval between numbers that can be represented depends on their magnitude. In double precision, the smallest value of the mantissa that can be stored is $2^{-52} \approx 2.22 \times 10^{-16}$. This is also the smallest possible difference in the mantissa between two numbers. The magnitude of the real number that is associated with this mantissa, however, depends on the exponent. For numbers of the order of 1 the smallest difference between two numbers that can be represented in double precision is then 2.22×10^{-16}. This value is also defined as the ***machine epsilon*** in double precision. In MATLAB this value is assigned to the predefined variable eps. As shown below, when the name of the variable eps is typed (Command Window) the assigned value is displayed.

```
>> eps
ans =
    2.220446049250313e-016
```

1.3 ERRORS IN NUMERICAL SOLUTIONS

Numerical solutions can be very accurate but in general are not exact. Two kinds of errors are introduced when numerical methods are used for solving a problem. One kind, which was mentioned in the previous section, occurs because of the way that digital computers store numbers and execute numerical operations. These errors are labeled ***round-off errors***. The second kind of errors is introduced by the numerical method that is used for the solution. These errors are labeled ***truncation errors***. Numerical methods use approximations for solving problems. The errors introduced by the approximations are the truncation errors. Together, the two errors constitute the ***total error*** of the numerical solution, which is the difference (can be defined in various ways) between the true (exact) solution (which is usually unknown) and the approximate numerical solution. Round-off, truncation, and total errors are discussed in the following three subsections.

1.3.1 Round-Off Errors

Numbers are represented on a computer by a finite number of bits (see Section 1.2). Consequently, real numbers that have a mantissa longer than the number of bits that are available for representing them have to be shortened. This requirement applies to irrational numbers that have to be represented in a finite form in any system, to finite numbers that are too long, and to finite numbers in decimal form that cannot be represented exactly in binary form. A number can be shortened either by ***chopping*** off, or discarding, the extra digits or by ***rounding***. In chopping, the digits in the mantissa beyond the length that can be stored are simply left out. In rounding, the last digit that is stored is rounded.

As a simple illustration, consider the number 2/3. (For simplicity, decimal format is used in the illustration. In the computer, chopping and rounding are done in the binary format.) In decimal form with four significant digits, 2/3 can be written as 0.6666 or as 0.6667. In the former instance the actual number has been chopped off, while in the latter instance the actual number has been rounded. Either way, such chopping and rounding of real numbers lead to errors in numerical computations, especially when many operations are performed. This type of numerical error (regardless of whether it is due to chopping or rounding) is known as ***round-off error***. Example 1-2 shows the difference between chopping and rounding.

Example 1-2: Round-off errors

Consider the two nearly equal numbers $p = 9890.9$ and $q = 9887.1$. Use decimal floating point representation (scientific notation) with three significant digits in the mantissa to calculate the difference between the two numbers, $(p - q)$. Do the calculation first by using chopping and then by using rounding.

SOLUTION

In decimal floating point representation, the two numbers are:

$p = 9.8909 \times 10^3$ and $q = 9.8871 \times 10^3$

If only three significant digits are allowed in the mantissa, the numbers have to be shortened. If chopping is used, the numbers become:

$p = 9.890 \times 10^3$ and $q = 9.887 \times 10^3$

Using these values in the subtraction gives:

$q = 9.890 \times 10^3 - 9.887 \times 10^3 = 0.003 \times 10^3 = 3$

If rounding is used, the numbers become:

$p = 9.891 \times 10^3$ and $q = 9.887 \times 10^3$ (q is the same as before)

Using these values in the subtraction gives:

$q = 9.891 \times 10^3 - 9.887 \times 10^3 = 0.004 \times 10^3 = 4$

The true (exact) difference between the numbers is 3.8. These results show that in the present problem rounding gives a value closer to the true answer.

The magnitude of round-off errors depends on the magnitude of the numbers that are involved since, as explained in the previous section, the interval between the numbers that can be represented on a computer depends on their magnitude. Round-off errors are likely to occur when the numbers that are involved in the calculations differ significantly in their magnitude and when two numbers that are nearly identical are subtracted from each other.

For example, consider the quadratic equation:

$$x^2 - 100.0001x + 0.01 = 0 \tag{1.9}$$

for which the exact solutions are $x_1 = 100$ and $x_2 = 0.0001$. The solutions can be calculated with the quadratic formula:

$$x_1 = \frac{-b + \sqrt{b^2 - 4ac}}{2a} \text{ and } x_2 = \frac{-b - \sqrt{b^2 - 4ac}}{2a} \tag{1.10}$$

Using MATLAB (Command Window) to calculate x_1 and x_2 gives:

```
>> format long
>> a = 1; b = -100.0001; c = 0.01;
>> RootDis = sqrt(b^2 - 4*a*c)
RootDis =
  99.99990000000000
```

```
>> x1 = (-b + RootDis)/(2*a)
x1 =
   100
>> x2 = (-b - RootDis)/(2*a)
x2 =
   1.000000000033197e-004
```

The value that is calculated by MATLAB for x_2 is not exact due to round-off errors. The round-off error occurs in the numerator in the expression for x_2. Since b is negative, the numerator involves subtraction of two numbers that are nearly equal.

In many cases, the form of the mathematical expressions that contain subtraction of two quantities that are nearly equal can be changed to a different form that is less likely to cause round-off errors. In the expression for x_2 in Eq. (1.10), this can be done by multiplying the expression by $(-b + \sqrt{b^2 - 4ac})/(-b + \sqrt{b^2 - 4ac})$:

$$x_2 = \frac{-b - \sqrt{b^2 - 4ac}}{2a} \frac{(-b + \sqrt{b^2 - 4ac})}{(-b + \sqrt{b^2 - 4ac})} = \frac{2c}{-b + \sqrt{b^2 - 4ac}} \qquad (1.11)$$

Using Eq. (1.11) in MATLAB to calculate the value of x_2 gives:

```
>> x2Mod = (2*c)/(-b+RootDis)
x2Mod =
   1.000000000000000e-004
```

Now the calculated value for x_2 is without an error. Another example of round-off errors is shown in Example 1-3.

Example 1-3: Round-off errors

Consider the function:

$$f(x) = x(\sqrt{x} - \sqrt{x - 1}) \qquad (1.12)$$

(a) Use MATLAB to calculate the value of $f(x)$ for the following three values of x: $x = 10$, $x = 1000$, and $x = 100000$.

(b) Use the decimal format with six significant digits to calculate $f(x)$ for the values of x in part (a). Compare the results with the values in part (a).

(c) Change the form of $f(x)$ by multiplying it by $\dfrac{\sqrt{x} + \sqrt{x - 1}}{\sqrt{x} + \sqrt{x - 1}}$. Using the new form with numbers in decimal format with six significant digits, calculate the value of $f(x)$ for the three values of x. Compare the results with the values in part (a).

SOLUTION

(*a*)

```
>> format long g
>> x = [10  1000  100000];
>> Fx = x.*(sqrt(x) - sqrt(x-1))
Fx =
         1.6227766016838       15.8153431255776       158.114278298171
```

(*b*) Using decimal format with six significant digits in Eq. (1.12) gives the following values for $f(x)$:

$$f(10) = 10(\sqrt{10} - \sqrt{10-1}) = 10(3.16228 - 3) = 1.62280$$

This value agrees with the value from part (*a*), when the latter is rounded to six significant digits.

$$f(1000) = 1000(\sqrt{1000} - \sqrt{1000-1}) = 1000(31.6228 - 31.6070) = 15.8$$

When rounded to six significant digits, the value in part (*a*) is 15.8153.

$$f(100000) = 100000(\sqrt{100000} - \sqrt{100000-1}) = 100000(316.228 - 316.226) = 200$$

When rounded to six significant digits, the value in part (*a*) is 158.114.

The results show that the rounding error due to the use of six significant digits increases as x increases and the relative difference between $\sqrt{x}$ and $\sqrt{x-1}$ decreases.

(*c*) Multiplying the right-hand side of Eq. (1.12) by $\dfrac{\sqrt{x} + \sqrt{x-1}}{\sqrt{x} + \sqrt{x-1}}$ gives:

$$f(x) = x(\sqrt{x} - \sqrt{x-1})\frac{\sqrt{x} + \sqrt{x-1}}{\sqrt{x} + \sqrt{x-1}} = \frac{x[x - (x-1)]}{\sqrt{x} + \sqrt{x-1}} = \frac{x}{\sqrt{x} + \sqrt{x-1}} \tag{1.13}$$

Calculating $f(x)$ using Eq. (1.13) for $x = 10, x = 1000$, and $x = 100000$ gives:

$$f(10) = \frac{10}{\sqrt{10} + \sqrt{10-1}} = \frac{10}{3.16228 + 3} = 1.62278$$

$$f(1000) = \frac{1000}{\sqrt{1000} + \sqrt{1000-1}} = \frac{1000}{31.6228 + 31.6070} = 15.8153$$

$$f(100000x) = \frac{100000}{\sqrt{100000} + \sqrt{100000-1}} = \frac{1000}{316.228 + 316.226} = 158.114$$

Now the values of $f(x)$ are the same as in part (*a*).

1.3.2 Truncation Errors

Truncation errors occur when the numerical methods used for solving a mathematical problem use an approximate mathematical procedure. A simple example is the numerical evaluation of $\sin(x)$, which can be done by using Taylor's series expansion (Taylor's series are reviewed in Chapter 2):

$$\sin(x) = x - \frac{x^3}{3!} + \frac{x^5}{5!} - \frac{x^7}{7!} + \frac{x^9}{9!} - \frac{x^{11}}{11!} + \ldots \tag{1.14}$$

The value of $\sin\left(\frac{\pi}{6}\right)$ can be determined exactly with Eq. (1.14) if an infinite number of terms are used. The value can be approximated by using only a finite number of terms. The difference between the true (exact) value and an approximate value is the truncation error, denoted by E^{TR}. For example, if only the first term is used:

$$\sin\left(\frac{\pi}{6}\right) = \frac{\pi}{6} = 0.5235988 \quad E^{TR} = 0.5 - 0.5235988 = -0.0235988$$

If two terms of the Taylor's series are used:

$$\sin\left(\frac{\pi}{6}\right) = \frac{\pi}{6} - \frac{(\pi/6)^3}{3!} = 0.4996742 \qquad E^{TR} = 0.5 - 0.4996742 = 0.0003258$$

Another example of truncation error that is probably familiar to the reader is the approximate calculation of derivatives. The value of the derivative of a function $f(x)$ at a point x_1 can be approximated by the expression:

$$\left.\frac{df(x)}{dx}\right|_{x = x_1} = \frac{f(x_2) - f(x_1)}{x_2 - x_1} \tag{1.15}$$

where x_2 is a point near x_1. The difference between the value of the true derivative and the value that is calculated with Eq. (1.15) is called a ***truncation error***.

The truncation error is dependent on the specific numerical method or algorithm used to solve a problem. Details on truncation errors are discussed throughout the book as various numerical methods are presented. The truncation error is independent of round-off error; it exists even when the mathematical operations themselves are exact.

1.3.3 Total Error

Numerical solution is an approximation. It always includes round-off errors and depending on the numerical method, can also include truncation errors. Together, the round-off and truncation errors yield the total numerical error that is included in the numerical solution. This total error, also called the true error, is the difference between the true (exact) solution and the numerical solution:

$$TrueError = TrueSolution - NumericalSolution \tag{1.16}$$

The absolute value of the ratio between the true error and the true solution is called the true relative error:

$$TrueRelativeError = \left|\frac{TrueSolution - NumericalSolution}{TrueSolution}\right| \tag{1.17}$$

This quantity which is non-dimensional and scale-independent indicates how large the error is relative to the true solution.

The true error and the true relative error in Eqs. (1.16) and (1.17) cannot actually be determined in problems that require numerical methods for their solution since the true solution is not known. These error quantities can be useful for evaluating the accuracy of different numerical methods. This is done by using the numerical method for solving problems that can be solved analytically and evaluating the true errors.

Since the true errors cannot, in most cases, be calculated, other means are used for estimating the accuracy of a numerical solution. This depends on the specific method and is discussed in more detail in later chapters. In some methods the numerical error can be bounded, while in others an estimate of the order of magnitude of the error is determined. In practical applications numerical solutions can also be compared to experimental results, but it is important to remember that experimental data have errors and uncertainties as well.

1.4 COMPUTERS AND PROGRAMMING

As mentioned earlier in Section 1.1, the fundamentals of numerical methods for solving mathematical problems that cannot be solved analytically were developed and used many years ago. The introduction of modern digital computers provided a means for applying these methods more accurately, and to problems requiring a large number of repetitive calculations. A computer can store a large quantity of numbers and can execute mathematical operations with these numbers very quickly. To carry out the calculations required for implementing a specific numerical method, the computer has to be provided with a set of instructions, called a computer program. Since binary format is used in the mathematical operations and for storing numbers, the instructions have to be in this form, and require the use of what is called machine language. In the early days of computers, computer programs were written in low-level computer languages (a language called assembler). Programming in this way was tedious and prone to errors because it had to be very detailed, and it was done in a form much different from the form used in everyday mathematics.

Later on, operating systems were introduced. Operating systems may be viewed as interfaces or layers enabling easier contact and communication between human users and the machine language of the computer. The instructions written in the language of the operating system are converted by the system to machine language commands that are executed by the computer. Examples of operating systems are Unix (written in the programming language called C) developed by Bell Laboratories in the 1970s and DOS (Disk Operating System) used by Microsoft Inc. Although operating systems simplify communication with the computer, they are still relatively difficult to use, require long codes, and are not written for the special needs of engineers and scientists.

Computer programs used by scientists and engineers are often written in programming languages that operate on top of the operating system. These higher-level computer languages are easier to use and enable the engineer or scientist to concentrate on problem solving rather than on tedious programming. Computer languages that are often used in science and engineering are Fortran, C, and C++. In general, for the same task, computer programs that are written in high-level computer languages are shorter (require less commands) than programs written in lower level languages. This book uses MATLAB, which is a high-level language for technical computing. For example, multiplication of two matrices in MATLAB is denoted by the regular multiplication operation, while other languages require the writing of a loop with several lines of code.

Algorithm

When a computer is used for obtaining a numerical solution to a problem, the program carries out the operations associated with the specific numerical method that is used. Some of the numerical methods are simple to implement, but sometimes the numerical procedures are complicated and difficult to program.

Before a numerical method is programmed, it is helpful to plan out all the steps that have to be followed in order to implement the numerical method successfully. Such a plan is called an **algorithm**, which is defined as step-by-step instructions on how to carry out the solution. Algorithms can be written in various levels of detail. Since the focus of this book is on numerical methods, the term *algorithm* is used here only in the context of instructions for implementing the numerical methods.

As a simple example, consider an algorithm for the solution of the quadratic equation:

$$ax^2 + bx + c = 0 \tag{1.18}$$

for which the solution in the case of real roots is given by the quadratic formula:

$$x_1 = \frac{-b + \sqrt{b^2 - 4ac}}{2a} \qquad x_2 = \frac{-b - \sqrt{b^2 - 4ac}}{2a} \tag{1.19}$$

Algorithm for solving for the real roots of a quadratic equation

Given are the three constants of the quadratic equation a, b, and c.

1. Calculate the value of the discriminant $D = b^2 - 4ac$.

2. If $D \geq 0$, calculate the two roots using Eq. (1.19).

3. If $D = 0$, calculate the root: $x = \dfrac{-b}{2a}$, and display the message: "The equation has a single root."

4. If $D < 0$, display the message: "The equation has no real roots."

Once the algorithm is devised, it can be implemented in a computer program.

Computer programs

A computer program (code) is a set (list) of commands (operations) that are to be executed by the computer. Different programming languages use different syntax for the commands, but in general, commands can be grouped into several categories:

- Commands for input and output of data. These commands are used for importing data into the computer, displaying on the monitor, or storing numerical results in files.
- Commands for defining variables.
- Commands that execute mathematical operations. These include the standard operations (addition, multiplication, power, etc.) and commands that calculate values of commonly used functions (trigonometric, exponential, logarithmic, etc.).
- Commands that control the order in which commands are executed and enable the computer to choose different groups of commands to be executed under different circumstances. These commands are typically associated with conditional statements that provide the means for making decisions as to which commands to execute in which order. Many languages have "if-else" commands for this purpose, but many other commands for this purpose exist.
- Commands that enable the computer to repeat sections of the program. In many languages these are called loops. These commands are very useful in the programming of numerical methods, since many methods use iterations for obtaining accurate solutions.
- Commands that create figures and graphical displays of the results.

A computer program can be written as one long list of commands, but typically it is divided into smaller well-defined parts (subprograms). The parts are self-contained programs that perform part of the overall operations that have to be carried out. With this approach the various parts can be written and tested independently. In many computer languages, the subprograms are called subroutines and functions.

As already mentioned, in this book numerical methods are implemented by using MATLAB, which is a relatively new language for technical computing. MATLAB is powerful and easy to use. It includes many built-in functions that are very useful for solving problems in science and engineering.

It is assumed that the reader of this book has at least some knowledge of MATLAB and programming. For those who do not, an introduction to MATLAB is presented in the appendix. It includes a section on conditional statements and loops, which are the basic building blocks of programming. For a more comprehensive introduction to programming, the reader is referred to books on computer programming. To help the reader follow the MATLAB programs listed in this book, comments and explanations are posted next to the program listings.

1.5 PROBLEMS

Problems to be solved by hand

Solve the following problems by hand. When needed, use a calculator or write a MATLAB script file to carry out the calculations.

1.1   Convert the binary number 1011101 to decimal format.

1.2   Convert the binary number 11000101.101 to decimal format.

1.3   Convert the binary number 10010101110001.01110101 to decimal format.

1.4   Write the number 81 in the following forms (in part (*c*) follow the IEEE-754 standard):
(*a*) Binary form.   (*b*) Base 2 floating point representation.   (*c*)   32 bit single-precision string.

1.5   Write the number 66.25 in the following forms (in part (*c*) follow the IEEE-754 standard):
(*a*) Binary form.   (*b*) Base 2 floating point representation.   (*c*)   32 bit single-precision string.

1.6   Write the number -0.625 in the following forms (in part (*c*) follow the IEEE-754 standard):
(*a*) Binary form.   (*b*) Base 2 floating point representation.   (*c*)   32 bit single-precision string.

1.7   Write the number 0.533203125 in the following forms (in part (*c*) follow the IEEE-754 standard):
(*a*) Binary form.   (*b*) Base 2 floating point representation.   (*c*)   32 bit single-precision string.

1.8   Write the number 256.1875 in the following forms (in part (*c*) follow the IEEE-754 standard):
(*a*) Binary form.   (*b*) Base 2 floating point representation.   (*c*)   64 bit double-precision string.

1.9   Write the number -30952 in the following forms (in part (*c*) follow the IEEE-754 standard):
(*a*) Binary form.   (*b*) Base 2 floating point representation.   (*c*)   64 bit double-precision string.

1.10   Write the number 0.33203125 in the following forms (in part (*c*) follow the IEEE-754 standard):
(*a*) Binary form.   (*b*) Base 2 floating point representation.   (*c*)   64 bit double-precision string.

1.11   Write the number 0.001220703125 in the following forms (in part (*c*) follow the IEEE-754 standard):
(*a*) Binary form.   (*b*) Base 2 floating point representation.   (*c*)   64 bit double-precision string.

1.12   Write the number 0.2 in binary form with sufficient number of digits so that the true relative error is less than 0.005.

1.13   Consider the function $f(x) = \dfrac{1 - \cos(x)}{\sin(x)}$.
(*a*) Use the decimal format with six significant digits (apply rounding) to calculate (using a calculator) $f(x)$ for $x = 0.007$.
(*b*) Use MATLAB (use `format long`) to calculate the value of $f(x)$ and the true relative error, due to rounding, in the value of $f(x)$ that was obtained in part (*a*).
(*c*) Multiply $f(x)$ by $\dfrac{1 + \cos(x)}{1 + \cos(x)}$ to obtain a form of $f(x)$ that is less prone to rounding errors. With the

new form, use the decimal format with six significant digits (apply rounding) to calculate (using a calculator) $f(x)$ for $x = 0.007$. Compare the value with the values in parts (a) and (b).

1.14 Consider the function $f(x) = \dfrac{\sqrt{9+x}-3}{x}$.

(a) Use the decimal format with six significant digits (apply rounding) to calculate (using a calculator) $f(x)$ for $x = 0.005$.

(b) Use MATLAB (use `format long`) to calculate the value of $f(x)$ and the true relative error due to rounding in the value of $f(x)$ that was obtained in part (a).

(c) Multiply $f(x)$ by $\dfrac{\sqrt{9+x}+3}{\sqrt{9+x}+3}$ to obtain a form of $f(x)$ that is less prone to rounding errors. With the new form, use the decimal format with six significant digits (apply rounding) to calculate (using a calculator) $f(x)$ for $x = 0.005$. Compare the value with the values in parts (a) and (b).

1.15 Consider the function $f(x) = \dfrac{e^x - 1}{x}$.

(a) Use the decimal format with five significant digits (apply rounding) to calculate (using a calculator) $f(x)$ for $x = 0.00275$.

(b) Use MATLAB (use `format long`) to calculate the value of $f(x)$ and the true relative error due to rounding in the value of $f(x)$ that was obtained in part (a).

1.16 The Taylor series expansion of $\cos(x)$ is given by:

$$\cos(x) = 1 - \frac{x^2}{2!} + \frac{x^4}{4!} - \frac{x^6}{6!} + \frac{x^8}{8!} - \frac{x^{10}}{10!} + \dots \tag{1.20}$$

Use the first three terms in Eq. (1.20) to calculate the value of $\cos\left(\dfrac{\pi}{3}\right)$. Calculate the truncation error. Use the decimal format with six significant digits (apply rounding).

1.17 Taylor series expansion of the function $f(x) = e^x$ is:

$$f(x) = e^x = 1 + x + \frac{x^2}{2!} + \frac{x^3}{3!} + \frac{x^4}{4!} + \frac{x^5}{5!} + \dots \tag{1.21}$$

Use Eq. (1.21) to calculate the value of e^{-2} for the following cases. In each case also calculate the true relative error. (Use MATLAB with `format long` to calculate the true value of e^{-2}.) Use decimal numbers with six significant numbers (apply rounding).
(a) Use the first four terms. (b) Use the first six terms. (c) Use the first eight terms.

1.18 Use the first seven terms in Eq. (1.21) to calculate an estimated value of e. Do the calculation with MATLAB (use `format long` to display the numbers). Determine the true relative error. For the exact value of e use in MATLAB `exp(1)`.

1.19 Develop an algorithm to determine whether or not a given integer is a prime number.

1.20 Develop an algorithm for adding all prime numbers between 0 and a given number.

1.21 Develop an algorithm for converting integers given in decimal form to binary form.

Problems to be programmed in MATLAB
Solve the following problems using the MATLAB environment. Do not use MATLAB's built-in functions for changing the form of numbers.

1.22 Write a MATLAB program in a script file that implements the algorithm developed in Problem 1.19. The program should start by assigning a value to a variable x. When the program is executed, a message should be displayed that states whether or not the value assigned to x is a prime number. Execute the program with $x = 79$, $x = 126$, and $x = 367$.

1.23 Write a user-defined MATLAB function that implements the algorithm developed in Problem 1.20. Name the function sp = sumprime(int), where the input argument int is a number larger than 1, and the output argument sp is the sum of all the prime numbers that are smaller than int. Use the function to calculate the sum of all the prime numbers between 0 and 30.

1.24 Write a user-defined MATLAB function that converts integers to binary form. Name the function b = integerTObina(d), where the input argument d is the integer to be converted and the output argument b is a vector with 1s and 0s that represent the binary number. The largest number that could be converted with the function should be a binary number with 20 1s. If a larger number is entered as d, the function should display an error message. Use the function to convert the numbers 81, 30952, and 1500000.

1.25 Write a user-defined MATLAB function that converts real numbers in decimal form to binary form. Name the function b = deciTObina(d), where the input argument d is the number to be converted and the output argument b is a 30-element-long vector with 1s and 0s that represent the binary number. The first 15 elements of b store the digits to the left of the decimal point, and the last 15 elements of b store the digits to the right of the decimal point. If more than 15 positions are required in the binary form for the digits to the right of the decimal point, the digits should be chopped. If the number that is entered as d is larger than can be stored in b, the function should display an error message. Use the function in the Command Window to convert the numbers 85.321, 0.00671, and 3006.42.

1.26 The value of π can be calculated with the series:

$$\pi = 4\sum_{n=1}^{\infty}(-1)^{n-1}\frac{1}{2n-1} = 4\left(1 - \frac{1}{3} + \frac{1}{5} - \frac{1}{7} + \frac{1}{9} - \frac{1}{11} + \ldots\right) \tag{1.22}$$

Write a MATLAB program in a script file that calculates the value of π by using *n* terms of the series and calculates the corresponding true relative error. (For the true value of π, use the predefined MATLAB variable pi.) Use the program to calculate π and the true relative error for:
(*a*) $n = 10$ (*b*) $n = 20$ (*c*) $n = 40$

Chapter 2

Mathematical Background

Core Topics

Concepts from calculus (2.2).

Vectors (2.3).

Matrices and linear algebra (2.4).

Ordinary differential equations (ODEs) (2.5).

Functions of two or more independent variables, partial differentiation (2.6)

Taylor series (2.7)

2.1 BACKGROUND

The present book focuses on the numerical methods used for calculating approximated solutions to problems that cannot be solved (or are difficult to solve) analytically. It is assumed that the reader has knowledge of calculus, linear algebra, and differential equations. This chapter has two objectives. One is to present, as a reference, some background useful for analysis of numerical methods. The second objective is to review some fundamental concepts and terms from calculus that are useful in the derivation of the numerical methods themselves. The chapter also defines many of the mathematical terms and the notation used in the rest of the book.

It should be emphasized that the topics that are presented in this section are meant as a review, and so are not covered thoroughly. This section is by no means a substitute for a text in calculus where the subjects are covered more rigorously. For each topic, the objective is to have a basic definition with a short explanation. This will serve as a reminder of concepts with which the reader is assumed to be familiar. If needed, the reader may use this information as the basis for seeking for a reference text on linear algebra and calculus that has a more rigorous exposition of the topic.

Some of the topics explained in this chapter are repeated in other chapters where the numerical methods are presented. A typical chapter starts with a section that explains the analytical background and the reasons the corresponding numerical method is needed. This repetition is necessary in order to have a complete presentation of the numerical method.

This chapter is organized as follows. Some fundamental concepts from pre-calculus and calculus are covered in Section 2.2. Vectors, as well as the ideas of linear independence and linear dependence are discussed in Section 2.3. This is followed by a review of matrices and introductory linear algebra in Section 2.4. Ordinary differential equations are briefly discussed in Section 2.5. Functions with two or more independent variables and partial differentiation are reviewed in Section 2.6. Finally, Taylor series expansion for a function of a single variable and for a function of two variables is described in Section 2.7.

2.2 CONCEPTS FROM PRE-CALCULUS AND CALCULUS

Function

A function written as $y = f(x)$ associates a unique number y (***dependent variable***) with each value of x (***independent variable***) (Fig. 2-1). The span of values that x can have from its minimum to its maximum value is called the ***domain***, and the span of the corresponding values of y is called the ***range***. The domain and range of the variables are also called intervals. When the interval includes the endpoints (the first and last values of the variable), then it is called a ***closed interval***; when the endpoints are not included, the interval is called an ***open interval***. If the endpoints of the interval of x are a and b, then the closed interval of x is written as $[a, b]$, and the open interval as (a, b). A function can have more than one independent variable. For example, the function $T = f(x, y, z)$ has three independent variables, where one unique number T is associated with each set of values x, y, and z.

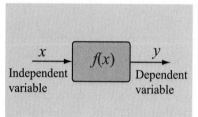

Figure 2-1: A function.

Limit of a function

If a function $f(x)$ comes arbitrarily close to a single number L as x approaches a number a from either the right side or the left side, then the limit of $f(x)$ is said to approach L as x approaches a. Symbolically the limit is expressed by:

$$\lim_{x \to a} f(x) = f(a) = L \qquad (2.1)$$

The formal definition states that if $f(x)$ is a function defined on an open interval containing a and L is a real number, then for each number $\varepsilon > 0$, there exists a number $\delta > 0$ such that if $0 < |x - a| < \delta$ then $|f(x) - L| < \varepsilon$. Since δ can be chosen to be arbitrarily small, $f(x)$ can be made to approach the limit L as closely as desired. Equation (2.1) by the way it is written implies that the limit exists. This is not always the case and sometimes functions do not have limits at certain points. However, those that do have limits cannot have two different limits as $x \to a$. In other words, if the limit of a function exists, then it is unique.

Continuity of a function

A function $f(x)$ is said to be **continuous** at $x = a$ if the following three conditions are satisfied:

(1) $f(a)$ exists, (2) $\lim_{x \to a} f(x)$ exists, and (3) $\lim_{x \to a} f(x) = f(a)$

A function is continuous on an open interval (a, b) if it is continuous at each point in the interval. A function that is continuous on the entire real axis $(-\infty, \infty)$ is said to be **everywhere continuous**.

Numerically, continuity means that small variations in the independent variable give small variations in the dependent variable.

Intermediate value theorem

The intermediate value theorem is a useful theorem about the behavior of a function in a closed interval. Formally, it states that if $f(x)$ is continuous on the closed interval $[a, b]$ and M is any number between $f(a)$ and $f(b)$, then there exists at least one number c in $[a, b]$ such that $f(c) = M$ (Fig. 2-2). Note that this theorem tells you that at least one c exists, but it does not provide a method for finding this value of c. Such a theorem is called an existence theorem. The intermediate value theorem implies that the graph of a continuous function cannot have a vertical jump.

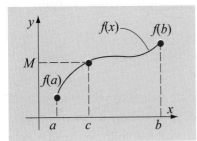

Figure 2-2: Intermediate value theorem.

Derivatives of a function

The ordinary derivative, first derivative, or simply, derivative of a function $y = f(x)$ at a point $x = a$ in the domain of f is denoted by $\frac{dy}{dx}$, y', $\frac{df}{dx}$, or $f'(a)$, and is defined as:

$$\frac{dy}{dx}\bigg|_{x = a} = f'(a) = \lim_{x \to a} \frac{f(x) - f(a)}{x - a} \tag{2.2}$$

Equation (2.2) defines an ordinary derivative because the function $f(x)$ is a function of a single independent variable. Note that the quantity $\frac{f(x) - f(a)}{x - a}$ represents the **slope** of the secant line connecting the points $(a, f(a))$ and $(x, f(x))$. In the limit as $x \to a$, it can be seen from Fig. 2-3 that the limit is the tangent line at the point $(a, f(a))$. Therefore, the derivative of the function $f(x)$ at the point $x = a$ is the slope of the tangent to the curve $y = f(x)$ at that point. A function must be continuous before it can be differentiable. A function that is continuous and differentiable over a certain interval is said to be **smooth**.

There are two important ways to interpret the first derivative of a function. One is as mentioned before, as the slope of the tangent to the curve described by $y = f(x)$ at a point. This first interpretation is especially useful in finding the maximum or minimum of the curve $y = f(x)$

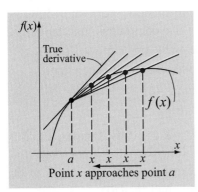

Point x approaches point a

Figure 2-3: Derivative of a function.

since the slope (and hence the first derivative) must be zero at those points. The second interpretation of the derivative is as the rate of change of the function $y = f(x)$ with respect to x. In other words, $\dfrac{dy}{dx}$ represents how fast y changes as x is changed.

Higher order derivatives may be obtained by successive application of the definition (2.2) to each derivative. In other words, the second derivative $\dfrac{d^2y}{dx^2}$ is obtained by differentiating the first derivative once, that is, $\dfrac{d\left(\dfrac{dy}{dx}\right)}{dx}$. Similarly, the third derivative is the first derivative of the second derivative or $\dfrac{d^3y}{dx^3} = \dfrac{d}{dx}\left(\dfrac{d^2y}{dx^2}\right)$, and so on.

Chain rule for ordinary differentiation

The chain rule is useful for differentiating functions whose arguments themselves are functions. For instance, if $y = f(u)$, where $u = g(x)$, then $y = f(g(x))$ is also differentiable and the following chain rule holds:

$$\frac{dy}{dx} = \left(\frac{dy}{du}\right)\left(\frac{du}{dx}\right) \tag{2.3}$$

Mean value theorem for derivatives

The mean value theorem is very useful in numerical analysis when finding bounds for the order of magnitude of numerical error for different methods. Formally, it states that if $f(x)$ is a continuous function on the closed interval $[a, b]$ and differentiable on the open interval (a, b), then there exists a number c within the interval, $c \in (a, b)$, such that:

$$f'(c) = \left.\frac{dy}{dx}\right|_{x = c} = \frac{f(b) - f(a)}{b - a} \tag{2.4}$$

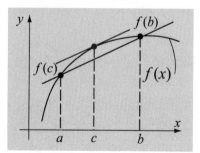

Figure 2-4: Mean value theorem for derivatives.

Simply stated, the mean value theorem for derivatives states, as illustrated in Fig. 2-4, that within the interval there exists a point c such that the value of the derivative of $f(x)$ is exactly equal to the slope of the secant line joining the endpoints $(a, f(a))$, and $(b, f(b))$.

Integral of a function

There are two types of integrals of a function of one variable, the indefinite integral and the definite integral. The **indefinite integral** is the opposite or inverse of the derivative and is therefore also referred to as the **antiderivative**. For instance, if $g(x)$ is the derivative of $f(x)$, that is, if $g(x) = \dfrac{df(x)}{dx}$, then the antiderivative or indefinite integral of $g(x)$ is $f(x)$ and is written as:

$$f(x) = \int g(x)dx \tag{2.5}$$

A **definite integral**, I, is denoted by:

$$I = \int_a^b f(x)dx \tag{2.6}$$

The definite integral is a number, and is defined on a closed interval $[a, b]$. a and b are the lower and upper limits of integration, respectively, and the function $f(x)$ that is written next to the integral sign $\int$ is called the integrand. A definite integral can be defined by using the Riemann sum. Consider a function $f(x)$ that is defined and continuous on $[a, b]$. The domain can be divided into n subintervals defined by $\Delta x_i = x_{i+1} - x_i$ where $i = 1, ..., n$. The Riemann sum for $f(x)$ on $[a, b]$ is defined as $\sum_{i=1}^{n} f(c_i)\Delta x_i$ where c_i is a number in the subinterval $[x_i, x_{i+1}]$. A definite integral $\int_a^b f(x)dx$ is defined as the limit of the Riemann sum when the length of all the subintervals of $[a, b]$ approaches zero:

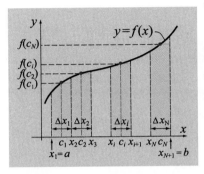

Figure 2-5: Definite integral using a Riemann sum.

$$I = \int_a^b f(x)dx = \lim_{\Delta x_i \to 0} \sum_{i=1}^{n} f(c_i)\Delta x_i \tag{2.7}$$

It can be seen from Fig. 2-5 that the value of $\lim_{\Delta x_i \to 0} \sum_{i=1}^{n} f(c_i)\Delta x_i$ equals the area under the curve specified by $y = f(x)$. This interpretation of the integral as the area under the curve is useful in developing approximate methods for numerically integrating functions (see Chapter 7).

Fundamental theorem of calculus

The connection between differentiation and integration is expressed by the fundamental theorem of calculus, which states that if a function $f(x)$ is continuous over the closed interval $[a, b]$ and $F(x)$ is an antiderivative of $f(x)$ over $[a, b]$, then

$$\int_a^b f(x)dx = F(b) - F(a) \tag{2.8}$$

Mean value theorem for integrals

One way of interpreting the definite integral of a monotonically increasing function $f(x)$ over an interval $[a, b]$ is as the area under the curve $y = f(x)$. It can be shown that the area under the curve is bounded by the area of the lower rectangle $f(a)(b-a)$ and the area of the upper rectangle $f(b)(b-a)$. The mean value theorem for integrals states that

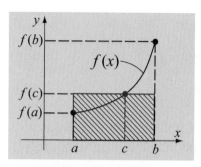

Figure 2-6: Mean value theorem for an integral.

somewhere between these two rectangles there exists a rectangle whose area is exactly equal to the area under the curve (Fig. 2-6). Formally, the theorem states that if $f(x)$ is continuous over the closed interval $[a, b]$, then there exists a number $c \in [a, b]$ such that:

$$\int_a^b f(x)dx = f(c)(b-a) \tag{2.9}$$

Average value of a function

The value $f(c)$ appearing in the mean value theorem for integrals, Eq. (2.9), is called the *average value* of the function $f(x)$ over the interval $[a, b]$. Thus, the average value of $f(x)$ over $[a, b]$ is denoted $\langle f \rangle$ and defined as:

$$\langle f \rangle = \frac{1}{(b-a)} \int_a^b f(x)dx \tag{2.10}$$

Second fundamental theorem of calculus

This theorem allows one to evaluate the derivative of a definite integral. Formally, it states that if $f(x)$ is continuous over an open interval containing the number a, then for every x in the interval,

$$\frac{d}{dx}\left[\int_a^x f(\xi)d\xi \right] = f(x) \tag{2.11}$$

where ξ is a dummy variable representing the coordinate along the interval.

2.3 VECTORS

Vectors are quantities (mathematical or physical) that have two attributes: magnitude and direction. In contrast, objects with a single attribute, such as magnitude, are called *scalars*. Examples of scalars are mass, length, and volume. Examples of vectors are force, momentum, and acceleration. One way to denote a quantity that is a vector is by writing a letter with a small arrow (or a short line) above the letter, $\vec{V}$. (In many books names of vectors are written in bold type.) The magnitude of the vector is denoted by the letter itself, V, or is written as $|V|$. A vector is usually defined with respect to a coordinate system. Once a coordinate system has been chosen, a vector may be represented graphically in such a space by a directed line segment (i.e., line with an arrow) (Fig. 2-7). Projections of the vector onto each of the coordinate axes define the components of the vector. If $V_x \hat{i}$ is the x component (i.e., projection of the vector on the x axis), $V_y \hat{j}$ is the y component, and $V_z \hat{k}$ is the z component, then the vector $\vec{V}$ can be written as:

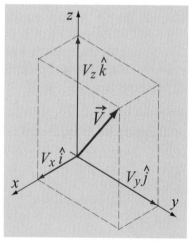

Figure 2-7: Vector V with components V_x, V_y, and V_z.

$$\vec{V} = V_x \hat{i} + V_y \hat{j} + V_z \hat{k} \tag{2.12}$$

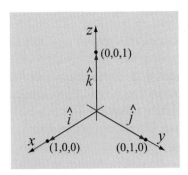

Figure 2-8: Unit vectors.

where $\hat{i}$, $\hat{j}$, and $\hat{k}$ are the unit vectors in the x, y, and z directions, respectively. Unit vectors, shown in Fig. 2-8, are vectors that have a specific direction and a magnitude of 1. In addition, a vector may be written by listing the magnitudes of its components in a row or a column:

$$\vec{V} = \begin{bmatrix} V_x & V_y & V_z \end{bmatrix} \quad \text{or} \quad \vec{V} = \begin{bmatrix} V_x \\ V_y \\ V_z \end{bmatrix} \tag{2.13}$$

Sometimes a vector is denoted by V_i, where $i = x, y, z$ or $i = 1, 2, 3$.

The magnitude of a vector in a three-dimensional Cartesian space is its length, and is determined by:

$$|\vec{V}| = \sqrt{V_x^2 + V_y^2 + V_z^2} \tag{2.14}$$

The direction of the vector can be specified by the unit vector (a unit vector in the direction of the vector). The unit vector, written as $\hat{V}$, is obtained by dividing Eq. (2.12) by the magnitude (length) given by Eq. (2.14):

$$\hat{V} = \frac{\vec{V}}{|\vec{V}|} = \frac{V_x\hat{i} + V_y\hat{j} + V_z\hat{k}}{\sqrt{V_x^2 + V_y^2 + V_z^2}} = l\hat{i} + m\hat{j} + n\hat{k} \tag{2.15}$$

where $l = \dfrac{V_x}{\sqrt{V_x^2 + V_y^2 + V_z^2}}$, $m = \dfrac{V_y}{\sqrt{V_x^2 + V_y^2 + V_z^2}}$, and $n = \dfrac{V_z}{\sqrt{V_x^2 + V_y^2 + V_z^2}}$

are called direction cosines and are equal to the cosine of the angles between the vector and the x, y, and z coordinate axes, respectively.

In physical situations, vectors are restricted to at most three dimensions. The idea of vectors, however, is generalized in mathematics (linear algebra) to dimensions beyond three. A vector then is a list (or a set) of n numbers (elements or components) written in a row or a column, where the name of the vector is written inside brackets:

$$[V] = \begin{bmatrix} V_1 & V_2 & \dots & V_n \end{bmatrix} \quad \text{or} \quad [V] = \begin{bmatrix} V_1 \\ V_2 \\ \dots \\ V_n \end{bmatrix} \tag{2.16}$$

An element of a vector is referred to as V_i where the subscript i denotes the position of the element in the row or the column. When the components are written as a row, the vector is referred to as a ***row vector***, and when written as a column, the vector is called a ***column vector***. In Eq. (2.16) the row vector is called a $(1 \times n)$ vector, indicating that it has 1 row and n columns. The column vector in Eq. (2.16) is called a $(n \times 1)$ vector.

2.3.1 Operations with Vectors

Two vectors are equal if they are of the same type (row or column) and all the elements that are in the same position are equal to each other. Some of the regular mathematical operations are defined for vectors where as others are not. For example, vectors can be added, subtracted, and multiplied in certain ways but cannot be divided. There are also operations that are unique to vectors. Basic operations are summarized next.

Addition and subtraction of two vectors

Two vectors can be added or subtracted only if they are of the same type (i.e., both row vectors or both column vectors) and of the same size (i.e., the same number of components or elements). Given two (row or column) vectors $\vec{V} = [V_i] = [V_1, ..., V_n]$ and $\vec{U} = [U_i] = [U_1, ..., U_n]$, the sum of the two vectors is:

$$\vec{V} + \vec{U} = [V_i + U_i] = \left[V_1 + U_1, V_2 + U_2, ..., V_n + U_n\right] \qquad (2.17)$$

Similarly, for subtraction:

$$\vec{V} - \vec{U} = [V_i - U_i] = \left[V_1 - U_1, V_2 - U_2, ..., V_n - U_n\right] \qquad (2.18)$$

Multiplication of a vector by a scalar

When a vector is multiplied by a scalar, each element is multiplied by the scalar. Given a vector $\vec{V} = [V_i] = \left[V_1, V_2, ..., V_n\right]$ and scalar α, the two can be multiplied to yield:

$$\alpha\vec{V} = \alpha[V_i] = \left[\alpha V_1, \alpha V_2, ..., \alpha V_n\right] \qquad (2.19)$$

A similar property holds for the case where $\vec{V}$ is a column vector.

Transpose of a vector

The transpose operation turns a row vector into a column vector and vice versa. For example, if $\vec{V} = \left[V_1, V_2, ..., V_n\right]$ is a $(1 \times n)$ row vector, then the transpose of $\vec{V}$, written as $\vec{V}^T$, is the following $(n \times 1)$ column vector:

$$\vec{V}^T = \begin{bmatrix} V_1 \\ V_2 \\ ... \\ V_n \end{bmatrix} \qquad (2.20)$$

Multiplication of two vectors

There are different ways to multiply two vectors. Two of the ways that produce physically meaningful results are the dot product and cross product. The dot product results in a scalar quantity, while the cross product results in a vector quantity.

Dot or scalar product of two vectors

The dot product of two vectors $\vec{V} = [V_i]$ and $\vec{U} = [U_i]$ is defined as:

$$\vec{V} \bullet \vec{U} = [V_i][U_i] = V_1 U_1 + V_2 U_2 + \ldots V_n U_n \qquad (2.21)$$

The result of such a multiplication is a number or scalar. Sometimes the dot product is written in the short-hand form:

$$\vec{V} \bullet \vec{U} = V_i U_i \qquad (2.22)$$

where the repeated subscripts imply summation over all possible values of that subscript, that is, $V_i U_i = \sum_{i=1}^{n} V_i U_i$.

The dot product can be given a geometric interpretation when the two vectors are drawn in a coordinate system. It can be shown from simple geometry and trigonometry that:

$$\vec{V} \bullet \vec{U} = |\vec{V}||\vec{U}| \cos \theta \qquad (2.23)$$

where $|\vec{V}|$ and $|\vec{U}|$ are the magnitudes of the vectors (see Eq. (2.14)) and θ is the angle formed by the two vectors.

Cross or vector product of two vectors

A cross, or a vector product, of two vectors is another vector. For two vectors $\vec{V} = V_x \hat{i} + V_y \hat{j} + V_z \hat{k}$ and $\vec{U} = U_x \hat{i} + U_y \hat{j} + U_z \hat{k}$, defined in a three-dimensional Cartesian coordinate system, the cross product $\vec{W} = \vec{V} \otimes \vec{U}$ is defined by:

$$\vec{W} = \vec{V} \otimes \vec{U} = (V_y U_z - V_z U_y)\hat{i} + (V_z U_x - V_x U_z)\hat{j} + (V_x U_y - V_y U_{1x})\hat{k} \qquad (2.24)$$

As illustrated in Fig. 2-9, the vector $\vec{W}$ is perpendicular to the plane that is formed by $\vec{V}$ and $\vec{U}$, and the magnitude of $\vec{W}$ is given by:

$$|\vec{W}| = |\vec{V}||\vec{U}| \sin \theta \qquad (2.25)$$

where θ is the angle formed by the two vectors $\vec{V}$ and $\vec{U}$.

Linear dependence and linear independence of a set of vectors

A set of vectors $\vec{V}_1$, $\vec{V}_2$,..., $\vec{V}_n$ is said to be **linearly independent** if

$$\alpha_1 \vec{V}_1 + \alpha_2 \vec{V}_2 + \ldots + \alpha_n \vec{V}_n = 0 \qquad (2.26)$$

is satisfied if and only if $\alpha_1 = \alpha_2 = \ldots = \alpha_n = 0$. Otherwise, the vectors are said to be **linearly dependent**. In other words, if **any** of the numbers $\alpha_1, \alpha_2, \ldots, \alpha_n$ is not identically zero, then the set of vectors is linearly dependent. As an example, consider the column vectors

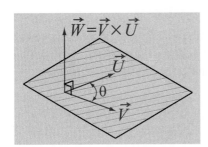

Figure 2-9: Cross product.

$\vec{V} = \begin{bmatrix} 1 \\ 0 \end{bmatrix}$, $\vec{U} = \begin{bmatrix} 0 \\ 1 \end{bmatrix}$, and $\vec{W} = \begin{bmatrix} 2 \\ 3 \end{bmatrix}$. By inspection, it can be seen that $\vec{V}$ and $\vec{U}$ are linearly independent. However because $2\vec{V} + 3\vec{U} - \vec{W} = 0$, $\vec{W}$ is linearly dependent on $\vec{V}$ and $\vec{U}$. Equation (2.26) is called a ***linear combination*** of vectors. A vector is therefore linearly dependent on a set of other vectors if it can be expressed as a linear combination of these other vectors.

Triangle inequality

The addition of two vectors, $\vec{V}$ and $\vec{U}$, can be represented geometrically (Fig 2-10) by a parallelogram whose two sides are the vectors that are being added, and the resulting sum, $(\vec{V} + \vec{U})$ is the main diagonal, as shown in Fig. (2-10). The triangle inequality refers to the fact that the sum of the lengths of two sides of a triangle is always larger than or at least equal to the length of the third side. It is written as:

$$\left| \vec{V} + \vec{U} \right| \leq \left| \vec{V} \right| + \left| \vec{U} \right| \tag{2.27}$$

This property is useful for matrices as well.

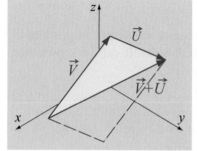

Figure 2-10: Triangle inequality.

2.4 MATRICES AND LINEAR ALGEBRA

A ***matrix*** is a rectangular array of numbers. The ***size*** of a matrix refers to the number of rows and columns it contains. An $(m \times n)$ matrix ("m by n matrix") has m rows and n columns:

$$[a] = \begin{bmatrix} a_{11} & a_{12} & \dots & a_{1n} \\ a_{21} & a_{22} & \dots & a_{2n} \\ \dots & \dots & \dots & \dots \\ a_{n1} & a_{n2} & \dots & a_{mn} \end{bmatrix} \tag{2.28}$$

The name of a matrix is written with brackets. An element (or entry) of a matrix is referred to as a_{ij} where the subscripts i and j denote the number of the row and the number of the column where the element is positioned.

Matrices are useful in the analysis of systems of linear equations and in other applications. Matrices can be added, subtracted, multiplied, and used in mathematical operations that are special for matrices.

Relationship between matrices and vectors

There is a close relationship between matrices and vectors. The matrix may be thought of as being composed of row vectors, or alternatively, column vectors. On the other hand, a vector is a special case of a matrix. A row vector is simply a matrix with one row and several columns, and a column vector is simply a matrix with several rows and one column.

2.4.1 Operations with Matrices

Mathematical operations performed with matrices fall in the general area of mathematics known as **linear algebra**. As with vectors, only certain mathematical operations are defined for matrices. These operations include multiplication by a scalar, addition, subtraction, and multiplication. As with vectors, division is not an allowed operation. Two matrices are equal if they are of the same size and all the elements that are in the same position in both matrices are equal.

Multiplication by a scalar

If $[a] = [a_{ij}]$ is a matrix and α is a scalar, then $\alpha[a] = [\alpha a_{ij}]$ is obtained by multiplying every element or entry of the matrix by the number α.

Addition and subtraction of two matrices

Two matrices can be added or subtracted only if they are of the same size. The matrix $[a]$ and the matrix $[b]$ (both $(n \times m)$) are added (or subtracted) by adding (or subtracting) the corresponding elements of the two matrices. The result $[c]$ is a matrix of the same size where:

$$[c_{ij}] = [a_{ij}] + [b_{ij}] \qquad (2.29)$$

$$\begin{bmatrix} 2 & -1 \\ 5 & 3 \\ 6 & 1 \end{bmatrix} + \begin{bmatrix} 1 & 3 \\ -5 & 2 \\ 3 & 7 \end{bmatrix} = \begin{bmatrix} 3 & 2 \\ 0 & 5 \\ 9 & 8 \end{bmatrix}$$

Figure 2-11: Addition of matrices.

for addition, as illustrated in Fig. 2-11, and:

$$[c_{ij}] = [a_{ij}] - [b_{ij}] \qquad (2.30)$$

for subtraction.

Transpose of a matrix

The transpose operation of a matrix rearranges the matrix such that the rows are switched into columns (or vice versa, the columns are switched into rows) (Fig. 2-12). In other words, the position (row number, column number) of each element in the matrix is switched around. The transpose of $[a]$ is written as $[a]^T$. For example, the element $[a_{12}]$ becomes $[a_{21}^T]$, and so on. In general:

$$\begin{bmatrix} 2 & -1 & 0 \\ 5 & 3 & 1 \\ 6 & 1 & -4 \\ 7 & -2 & 9 \end{bmatrix}^T = \begin{bmatrix} 2 & 5 & 6 & 7 \\ -1 & 3 & 1 & -2 \\ 0 & 1 & -4 & 9 \end{bmatrix}$$

Figure 2-12: Transpose of a matrix.

$$[a]^T = [a_{ij}^T] = [a_{ji}] \qquad (2.31)$$

Thus, the transpose of a (3×4) matrix such as $[a] = \begin{bmatrix} a_{11} & a_{12} & a_{13} & a_{14} \\ a_{21} & a_{22} & a_{23} & a_{24} \\ a_{31} & a_{32} & a_{33} & a_{34} \end{bmatrix}$

is the (4×3) matrix: $[a]^T = \begin{bmatrix} a_{11} & a_{21} & a_{31} \\ a_{12} & a_{22} & a_{32} \\ a_{13} & a_{23} & a_{33} \\ a_{14} & a_{24} & a_{34} \end{bmatrix}$.

Multiplication of matrices

The multiplication $[c] = [a][b]$ of a matrix $[a]$ times a matrix $[b]$ is defined only when the number of columns of matrix $[a]$ is equal to the number of rows of the matrix $[b]$. There are no restrictions on the number of rows of $[a]$ or the number of columns of $[b]$. The result of the multiplication is a matrix $[c]$ that has the same number of rows as $[a]$ and the same number of columns as $[b]$. So, if matrix $[a]$ is $(m \times q)$ and matrix $[b]$ is $(q \times n)$, then the matrix $[c]$ is $(m \times n)$ (Fig. 2-13). For example, as shown in Eq. (2.32), if $[a]$ is (3×4) and $[b]$ is (4×2), then $[c]$ is (3×2).

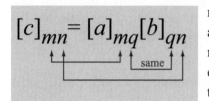

Figure 2-13: Multiplication of matrices.

$$\begin{bmatrix} c_{11} & c_{12} \\ c_{21} & c_{22} \\ c_{31} & c_{32} \end{bmatrix} = \begin{bmatrix} a_{11} & a_{12} & a_{13} & a_{14} \\ a_{21} & a_{22} & a_{23} & a_{24} \\ a_{31} & a_{32} & a_{33} & a_{34} \end{bmatrix} \begin{bmatrix} b_{11} & b_{12} \\ b_{21} & b_{22} \\ b_{31} & b_{32} \\ b_{41} & b_{42} \end{bmatrix} \tag{2.32}$$

The elements of the matrix $[c]$ are calculated by multiplying rows of $[a]$ by columns of $[b]$. Starting with the first row, the value of the element c_{11} is obtained by multiplying the first row of $[a]$ by the first column of $[b]$ in the following manner:

$$c_{11} = a_{11}b_{11} + a_{12}b_{21} + a_{13}b_{31} + a_{14}b_{41} \tag{2.33}$$

The value of the element c_{12} is obtained by multiplying the first row of $[a]$ by the second column of $[b]$:

$$c_{12} = a_{11}b_{12} + a_{12}b_{22} + a_{13}b_{32} + a_{14}b_{42} \tag{2.34}$$

In the second row of $[c]$, the value of the element c_{21} is obtained by multiplying the second row of $[a]$ by the first column of $[b]$:

$$c_{21} = a_{21}b_{11} + a_{22}b_{21} + a_{23}b_{31} + a_{24}b_{41} \tag{2.35}$$

The multiplication procedure continues until the value of the element c_{32} is calculated. In general, the multiplication rule is given by:

$$c_{ij} = a_{ik}b_{kj} = \sum_{k=1}^{q} a_{ik}b_{kj} \tag{2.36}$$

A numerical example of multiplication is shown in Fig. 2-14.

$$\begin{bmatrix} 2 & -1 \\ 8 & 3 \\ 6 & 7 \end{bmatrix} \begin{bmatrix} 4 & 9 & 1 & -3 \\ -5 & 2 & 4 & 6 \end{bmatrix} = \begin{bmatrix} (2 \cdot 4 + -1 \cdot -5) & (2 \cdot 9 + -1 \cdot 2) & (2 \cdot 1 + -1 \cdot 4) & (2 \cdot -3 + -1 \cdot 6) \\ (8 \cdot 4 + 3 \cdot -5) & (8 \cdot 9 + 3 \cdot 2) & (8 \cdot 1 + 3 \cdot 4) & (8 \cdot -3 + 3 \cdot 6) \\ (6 \cdot 4 + 7 \cdot -5) & (6 \cdot 9 + 7 \cdot 2) & (6 \cdot 1 + 7 \cdot 4) & (6 \cdot -3 + 7 \cdot 6) \end{bmatrix}$$

$$= \begin{bmatrix} 13 & 16 & -2 & -12 \\ 17 & 78 & 20 & -6 \\ -11 & 68 & 34 & 24 \end{bmatrix}$$

Figure 2-14: Numerical example of multiplication of matrices.

2.4.2 Special Matrices

Matrices with special structures or properties arise when numerical methods are used for solving problems. The following is a list of such matrices, with a short description of each.

Square matrix

A matrix that has the same number of columns as rows is called a square matrix. In such matrices, entries or elements along the diagonal of the matrix, a_{ii}, i.e. a_{11}, a_{22}, and so on, are known as the *diagonal* elements and all other entries are the *off-diagonal* elements. In a square matrix, the entries (or elements) above the diagonal, that is $[a_{ij}]$ for $j > i$, are called the *superdiagonal entries* or *above-diagonal entries*. The entries below the diagonal, that is, $[a_{ij}]$ for $i > j$, are called the *subdiagonal entries* or *below-diagonal entries*.

Diagonal matrix

A square matrix with diagonal elements that are nonzero, and off-diagonal elements that are all zeros is called a *diagonal matrix* and is denoted by $[D]$.

Upper triangular matrix

A square matrix whose subdiagonal entries are all zero is called an *upper triangular matrix* and is denoted by $[U]$.

Lower triangular matrix

A square matrix whose superdiagonal entries are all zero is called a *lower triangular matrix* and is denoted by $[L]$.

Identity matrix

The *identity matrix* $[I]$ is a square matrix whose diagonal elements are all 1s and whose off-diagonal entries are all 0s. The identity matrix is the analog of the number 1 for matrices. Any matrix that is multiplied by the identity matrix remains unchanged:

$$[a][I] = [a] \tag{2.37}$$

Zero matrix

The *zero matrix* is a matrix whose entries are all zero.

Symmetric matrix

A *symmetric matrix* is a square matrix in which $[a_{ij}] = [a_{ji}]$. For a symmetric matrix the transpose of the matrix is equal to the matrix itself:

$$[a]^T = [a] \qquad (2.38)$$

2.4.3 Inverse of a Matrix

Division is an operation that is not defined for matrices. However, an operation that is defined and serves an equivalent purpose is the *inverse* of a matrix. A square matrix $[a]$ is invertible provided there exists a square matrix $[b]$ of the same size such that $[a][b] = [I]$, where $[I]$ is the identity matrix. The matrix $[b]$ is called the *inverse* of $[a]$ and written as $[a]^{-1}$. Thus:

$$[a][a]^{-1} = [a]^{-1}[a] = [I] \qquad (2.39)$$

Example 2-1 illustrates the property expressed by Eq. (2.39).

Example 2-1: Inverse of a matrix.

Show that the matrix $[b] = \begin{bmatrix} 0.1 & 0.2 & 0 \\ 0.4 & 0.1 & 0.2 \\ 0.2 & 0.1 & 0.8 \end{bmatrix}$ is the inverse of the matrix $[a] = \begin{bmatrix} -1.2 & 3.2 & -0.8 \\ 5.6 & -1.6 & 0.4 \\ -0.4 & -0.6 & 1.4 \end{bmatrix}$.

SOLUTION

To show that the matrix $[b]$ is the inverse of the matrix $[a]$, the two matrices are multiplied.

$$[a][b] = \begin{bmatrix} -1.2 & 3.2 & -0.8 \\ 5.6 & -1.6 & 0.4 \\ -0.4 & -0.6 & 1.4 \end{bmatrix} \begin{bmatrix} 0.1 & 0.2 & 0 \\ 0.4 & 0.1 & 0.2 \\ 0.2 & 0.1 & 0.8 \end{bmatrix} =$$

$$\begin{bmatrix} (1.2 \cdot 0.1 + 3.2 \cdot 0.4 + -0.8 \cdot 0.2) & (1.2 \cdot 0.2 + 3.2 \cdot 0.1 + -0.8 \cdot 0.1) & (1.2 \cdot 0 + 3.2 \cdot 0.2 + -0.8 \cdot 0.8) \\ (5.6 \cdot 0.1 + -1.6 \cdot 0.4 + 0.4 \cdot 0.2) & (5.6 \cdot 0.2 + -1.6 \cdot 0.1 + 0.4 \cdot 0.1) & (5.6 \cdot 0 + -1.6 \cdot 0.2 + 0.4 \cdot 0.8) \\ (-0.4 \cdot 0.1 + -0.6 \cdot 0.4 + 1.4 \cdot 0.2) & (-0.4 \cdot 0.2 + -0.6 \cdot 0.1 + 1.4 \cdot 0.1) & (-0.4 \cdot 0 + -0.6 \cdot 0.2 + 1.4 \cdot 0.8) \end{bmatrix} =$$

$$= \begin{bmatrix} 1 & 0 & 0 \\ 0 & 1 & 0 \\ 0 & 0 & 1 \end{bmatrix}$$

2.4.4 Properties of Matrices

The following are general properties of matrices:

- $[a] + [b] = [b] + [a]$

- $([a] + [b]) + [c] = [a] + ([b] + [c])$

- $\alpha([a] + [b]) = \alpha[a] + \alpha[b]$, where α is a scalar

- $(\alpha + \beta)[a] = \alpha[a] + \beta[a]$, where α and β are scalars

The properties above apply to subtraction as well.

- If $[a]$ and $[b]$ are square matrices, then in general $[a][b] \neq [b][a]$ (unless one is the inverse of the other). If either $[a]$ or $[b]$ is not square, and the product $[a][b]$ exists, then the product $[b][a]$ is not defined and does not exist. In other words, when matrices are involved, the order of multiplication is important.

- $([a] + [b])[c] = [a][c] + [b][c]$, with the order of multiplication being important.

- $[a]([b] + [c]) = [a][b] + [a][c]$.

- $\alpha([a][b]) = (\alpha[a])[b] = [a](\alpha[b])$, where α is a scalar.

- If $[a]$ and $[b]$ are matrices for which $[a][b]$ is defined and exists, then $([a][b])^T = [b]^T[a]^T$. Note that the order of multiplication is changed.

- For any matrix $[a]$, $([a]^T)^T = [a]$.

- For an invertible matrix $[a]$, $([a]^{-1})^{-1} = [a]$.

- If $[a]$ and $[b]$ are two square, invertible matrices of the same size. then $([a][b])^{-1} = [b]^{-1}[a]^{-1}$.

2.4.5 Determinant of a Matrix

The determinant that is defined for square matrices is a useful quantity that features prominently in finding the inverse of a matrix and provides useful information regarding whether or not solutions exist for a set of simultaneous equations. The determinant of a matrix is often difficult to compute if the size of a matrix is larger than (3×3) or (4×4).

The determinant is a number. It is the sum of all possible products formed by taking one element from each row and each column and attaching the proper sign. The proper sign of each term is found by writing the individual terms in each product and counting the number of interchanges necessary to put the subscripts into the order $1, 2, ..., n$. If the number of such required interchanges is even, then the sign is $+$ and if the number of interchanges is odd, the sign is $-$. Formally, the determinant of a matrix $[a]_{n \times n}$ is denoted by $\det(a)$ or $|a|$ and is defined as:

$$\det(A) = |A| = \sum_j (-1)^k a_{1,j_1} a_{2,j_2} \ldots a_{n,j_n} \qquad (2.40)$$

where the sum is taken over all $n!$ permutations of degree n and k is the number of interchanges required to put the second subscripts in the order $1, 2, 3, \ldots, n$. Use of Eq. (2.40) is illustrated for $n = 1$, $n = 2$, and $n = 3$.

For $n = 1$ the matrix is (1×1), $[a] = \begin{bmatrix} a_{11} \end{bmatrix}$ and the determinant is:

$$\det(a) = a_{11}$$

For $n = 2$ the matrix is (2×2), $[a] = \begin{bmatrix} a_{11} & a_{12} \\ a_{21} & a_{22} \end{bmatrix}$ and the determinant is:

$$\det(a) = (-1)^0 a_{11} a_{22} + (-1)^1 a_{12} a_{21} = a_{11} a_{22} - a_{12} a_{21}$$

For $n = 3$ the matrix is (3×3), $[a] = \begin{bmatrix} a_{11} & a_{12} & a_{13} \\ a_{21} & a_{22} & a_{23} \\ a_{31} & a_{32} & a_{33} \end{bmatrix}$ and the determi-

nant is:

$$\det(A) = (-1)^0 a_{11} a_{22} a_{33} + (-1)^1 a_{11} a_{23} a_{32} + (-1)^1 a_{12} a_{21} a_{33}$$

$$+ (-1)^2 a_{12} a_{23} a_{31} + (-1)^2 a_{13} a_{21} a_{32} + (-1)^3 a_{13} a_{22} a_{31}$$

$$= a_{11}(a_{22} a_{33} - a_{23} a_{32}) - a_{12}(a_{21} a_{33} - a_{23} a_{31}) + a_{13}(a_{21} a_{32} - a_{22} a_{31})$$

It can be seen that evaluation of a determinant for large matrices is impractical both by hand and by computer because of the large number of operations required to consider the $n!$ permutations.

2.4.6 Cramer's Rule and Solution of a System of Simultaneous Linear Equations

A set of n simultaneous linear equations with n unknowns $x_1, x_2, \ldots, x_n$ is given by:

$$a_{11}x_1 + a_{12}x_2 + \ldots + a_{1n}x_n = b_1$$
$$a_{21}x_1 + a_{22}x_2 + \ldots + a_{2n}x_n = b_2$$
$$\ldots + \ldots + \ldots + \ldots = \ldots \qquad (2.41)$$
$$a_{n1}x_1 + a_{n2}x_2 + \ldots + a_{nn}x_n = b_n$$

The system can be written compactly by using matrices:

$$\begin{bmatrix} a_{11} & a_{12} & \ldots & a_{1n} \\ a_{21} & a_{22} & \ldots & a_{2n} \\ \ldots & \ldots & \ldots & \ldots \\ a_{n1} & a_{n2} & \ldots & a_{nn} \end{bmatrix} \begin{bmatrix} x_1 \\ x_2 \\ \ldots \\ x_n \end{bmatrix} = \begin{bmatrix} b_1 \\ b_2 \\ \ldots \\ b_n \end{bmatrix} \qquad (2.42)$$

Equation (2.42) can also be written as:

$$[a][x] = [b] \qquad (2.43)$$

where $[a]$ is the matrix of coefficients, $[x]$ is the vector of n unknowns, and $[b]$ is the vector containing the right-hand sides of each equation. Cramer's rule states that the solution to Eq. (2.41), if it exists, is given by:

$$x_j = \frac{\det(a'_j)}{\det(a)} \text{ for } j = 1, 2, ..., n \qquad (2.44)$$

where a'_j is the matrix formed by replacing the jth column of the matrix $[a]$ with the column vector $[b]$ containing the right-hand sides of the original system (2.42). It is apparent from Eq. (2.44) that solutions to (2.42) can exist only if $\det(a) \neq 0$. The only way that $\det(a)$ can be zero is if two or more columns or rows of $[a]$ are either identical or one or more columns (or rows) of $[a]$ are linearly dependent on other columns (or rows).

Example 2-2: Solving a system of linear equations using Cramer's rule.

Find the solution of the following system of equations using Cramer's rule.

$$2x + 3y - z = 5$$
$$4x + 4y - 3z = 3 \qquad (2.45)$$
$$-2x + 3y - z = 1$$

SOLUTION

Step 1: Write the system of equations in a matrix form $[a][x] = [b]$.

$$\begin{bmatrix} 2 & 3 & -1 \\ 4 & 4 & -3 \\ -2 & 3 & -1 \end{bmatrix} \begin{bmatrix} x \\ y \\ z \end{bmatrix} = \begin{bmatrix} 5 \\ 3 \\ 1 \end{bmatrix} \qquad (2.46)$$

Step 2: Calculate the determinant of the matrix of coefficients.

$$\det(A) = 2[(4 \times -1) - (-3 \times 3)] - 3[(4 \times -1) - (-3 \times -2)] - 1[(4 \times 3) - (4 \times -2)]$$
$$= 2(5) - 3(-10) - 1(20) = 10 + 30 - 20 = 20$$

Step 3: Apply Eq. (2.44) to find x, y, and z. To find x, the modified matrix a'_x is created by replacing its first column with $[b]$.

$$x = \frac{\det\left(\begin{bmatrix} 5 & 3 & -1 \\ 3 & 4 & -3 \\ 1 & 3 & -1 \end{bmatrix}\right)}{20} = \frac{(5 \cdot 5) - (3 \cdot 0) - (1 \cdot 5)}{20} = 1$$

In the same way, to find y, the modified matrix a'_y is created by replacing its second column with $[b]$.

$$y = \frac{\det \left(\begin{bmatrix} 2 & 5 & -1 \\ 4 & 3 & -3 \\ -2 & 1 & -1 \end{bmatrix} \right)}{20} = \frac{(20 \cdot 0) - (5 \cdot -10) - (1 \cdot 10)}{20} = 2$$

Finally, to determine the value of z, the modified matrix a'_z is created by replacing its third column with $[b]$.

$$z = \frac{\det \left(\begin{bmatrix} 2 & 3 & 5 \\ 4 & 4 & 3 \\ -2 & 3 & 1 \end{bmatrix} \right)}{20} = \frac{(2 \cdot -5) - (3 \cdot 10) - (5 \cdot 20)}{20} = 3$$

To check the answer, the matrix of coefficients $[a]$ is multiplied by the solution:

$$\begin{bmatrix} 2 & 3 & -1 \\ 4 & 4 & -3 \\ -2 & 3 & -1 \end{bmatrix} \begin{bmatrix} 1 \\ 2 \\ 3 \end{bmatrix} = \begin{bmatrix} 2+6-3 \\ 4+8-9 \\ -2+6-3 \end{bmatrix} = \begin{bmatrix} 5 \\ 3 \\ 1 \end{bmatrix}$$

The right-hand side is equal to $[b]$, which confirms that the solution is correct.

2.4.7 Norms

In Section 2.3, vectors were identified as having a magnitude usually specified by Eq. (2.14). From Euclidean geometry, this magnitude can be seen to be a measure of the length of a vector (not to be confused with the size or number of elements it contains). The magnitude of the vector is useful in comparing vectors so that one may determine that one vector is larger than another. Such an equivalent measure for the "magnitude" of a matrix is also useful in comparing different matrices; it is called the *Norm* and denoted as $\|[a]\|$. There is no unique way to measure the "magnitude" or norm of a matrix. Several definitions of norms are presented in Section 4.9. The norm basically assigns a real number to a matrix (or vector).

A norm must satisfy certain properties since it is a quantity for a matrix that is analogous to the magnitude or length of a vector. These are:

(1) $\|[a]\| \geq 0$ and $\|[a]\| = 0$ if and only if $[a] = [0]$ (i.e., if $[a]$ is the zero matrix).

(2) For all numbers α, $\|\alpha[a]\| = |\alpha| \|[a]\|$.

(3) For any two matrices (or vectors) $[a]$ and $[b]$, the following must be satisfied: $\|[a] + [b]\| \leq \|[a]\| + \|[b]\|$.

Condition (1) states that the "magnitude" of a matrix or vector as measured by the norm must be a positive quantity just as any length that is used to measure the magnitude of a vector. Condition (2) states that for matrices too, just like vectors, $\|[a]\|$ and $\|[-a]\|$ would have the same "magnitude." This is easy to see in the case of vectors, since the length of the vector does not change simply because its direction is reversed. Condition (3) is just the triangle inequality and is easily visualized with Euclidean geometry for vectors. The various vector and matrix norms are discussed further in Section 4.10.

2.5 ORDINARY DIFFERENTIAL EQUATIONS (ODE)

An ordinary differential equation (ODE) is an equation that contains one dependent variable, one independent variable, and ordinary derivatives of the dependent variable. If x is the independent variable and y is the dependent variable, an ODE has terms that contain x, y, $\dfrac{dy}{dx}$, $\dfrac{d^2y}{dx^2}$, ..., $\dfrac{d^ny}{dx^n}$. ODEs can be linear or nonlinear. An ODE is linear if its dependence on y and its derivatives is linear. Any **linear** ODE can be written in the following **standard** or **canonical form**:

$$a_{n+1}(x)\frac{d^ny}{dx^n} + a_n(x)\frac{d^{n-1}y}{dx^{n-1}} + ... + a_3(x)\frac{d^2y}{dx^2} + a_2(x)\frac{dy}{dx} + a_1(x)y = r(x) \qquad (2.47)$$

Note that the coefficients in Eq. (2.47) are all functions only of the independent variable x. Examples of linear ODEs are:

$$\frac{dy}{dx} = 10x$$

$$c\frac{dx}{dt} + kx = -m\frac{d^2x}{dt^2}$$

where m, k, and c are constants.

Homogeneous / nonhomogeneous ODE

An ODE can be homogeneous or nonhomogeneous. When written in the standard form (Eq. (2.47)), the ODE is **homogeneous** if on the right-hand side $r(x) = 0$. Otherwise, if $r(x) \neq 0$, then the ODE is said to be **nonhomogeneous**.

Order of an ODE

The order of an ODE is determined by the order of the highest derivative that appears in the equation. The order of an ODE can convey important information. When an ODE is solved, arbitrary constants or integration constants appear in the solution. The number of such constants that must be determined is equal to the order of the ODE. For example, the solution to a second-order ODE has two undetermined constants. This means that two constraints must be specified in order to determine these two undetermined constants. When the independent variable is position and the constraints are specified at two different positions, the constraints are called **boundary conditions**. When the independent variable is time and the constraints are specified at a single instant of time, the constraints are called **initial conditions**.

Nonlinear ODE

An ODE is ***nonlinear*** if the coefficients in Eq. (2.47) are functions of y or its derivatives, if the right-hand side r is itself a nonlinear function of y, or if the linear term $a_1(x)y$ is replaced with a nonlinear function of y. The following ODEs are all examples of nonlinear ODEs:

$$\frac{d^2y}{dt^2} + \sin y = 4$$

$$y\frac{d^2y}{dt^2} + 3y = 8$$

$$\left(\frac{dy}{dt}\right)\frac{d^2y}{dt^2} + y = 9$$

$$\frac{d^2y}{dt^2} + 8y = \tan y$$

Analytical solutions to some important linear ODEs

Certain first and second-order linear ODEs recur in many branches of science and engineering. Because of their pervasiveness, the solutions to these ODEs are given here as a reminder. The ***general solution*** to a linear, nonhomogeneous ODE is the sum of the ***homogeneous solution*** and the ***particular solution***. The homogeneous solution is the solution to the homogeneous ODE (i.e., ODE with $r(x)$ in Eq. (2.47) set to zero), and the ***particular solution*** is a solution that when substituted into the ODE satisfies the right-hand side (i.e., yields $r(x)$). It is only after the general solution is obtained that the constraints (i.e., boundary or initial conditions) must be substituted to solve for the undetermined constants.

General solution to a nonhomogeneous linear first-order ODE

The general solution to a nonhomogeneous, linear, first-order ODE of the form:

$$\frac{dy}{dx} + P(x)y = Q(x) \tag{2.48}$$

is obtained by multiplying both sides of the equation by the following integrating factor:

$$\mu(x) = e^{\int P(x)dx} \tag{2.49}$$

When this is done, Eq. (2.48) can be written in the following integrable form:

$$\frac{d}{dx}(y\mu) = Q(x)\mu(x) \tag{2.50}$$

Since $P(x)$ and $Q(x)$ are known functions Equation (2.50) can be integrated by multiplying both sides by dx and integrating:

$$y(x)\mu(x) = \int Q(x)\mu(x)dx + C_1 \tag{2.51}$$

Dividing through by $\mu(x)$ gives:

$$y(x) = \frac{1}{\mu(x)}\int Q(x)\mu(x)dx + \frac{C_1}{\mu(x)} \tag{2.52}$$

The integration constant C_1 must be determined from a constraint, which is problem-dependent.

General solution to a homogeneous second-order linear ODE with constant coefficients

A homogeneous, second-order linear ODE with constant coefficients can be written in the form:

$$\frac{d^2y}{dx^2} + b\frac{dy}{dx} + cy = 0 \tag{2.53}$$

where b and c are constants. The general solution to this equation is obtained by substituting $y = e^{sx}$. The resulting equation is called the characteristic equation:

$$s^2 + bs + c = 0 \tag{2.54}$$

The solution to Eq. (2.54) is obtained from the quadratic formula:

$$s = \frac{-b \pm \sqrt{b^2 - 4c}}{2} \tag{2.55}$$

The general solution to Eq. (2.53) is therefore:

$$y(x) = e^{-bx/2}\left[C_1 e^{\frac{x}{2}\sqrt{b^2-4c}} + C_2 e^{-\frac{x}{2}\sqrt{b^2-4c}}\right] \tag{2.56}$$

where C_1 and C_2 are integration constants that are determined from the problem-dependent constraints (i.e., boundary or initial conditions).

There are two important special cases in Eq. (2.56). In the first, the discriminant is positive, that is, $b^2 > 4c$, and the solution remains as shown in Eq. (2.56). In the second case the discriminant is negative, that is, $b^2 < 4c$, and the solution becomes:

$$y(x) = e^{-bx/2}\left[C_1 e^{\frac{i}{2}x\sqrt{b^2-4c}} + C_2 e^{-\frac{i}{2}x\sqrt{b^2-4c}}\right] \tag{2.57}$$

where $i = \sqrt{-1}$. By using Euler's formula, $e^{iz} = \cos(z) + i\sin(z)$, Eq. (2.57) can be written in another, perhaps more familiar, form:

$$\begin{aligned}y(x) = e^{-bx/2}&\left[C_1\cos\left(x\frac{\sqrt{4c-b^2}}{2}\right) + C_1 i\sin\left(x\frac{\sqrt{4c-b^2}}{2}\right)\right.\\&\left.+ C_2\cos\left(x\frac{\sqrt{4c-b^2}}{2}\right) - C_2 i\sin\left(x\frac{\sqrt{4c-b^2}}{2}\right)\right]\end{aligned} \tag{2.58}$$

which can be combined to yield:

$$y(x) = e^{-bx/2}\left[(C_1 + C_2)\cos\left(x\frac{\sqrt{4c - b^2}}{2}\right) + (C_1 - C_2)i\sin\left(x\frac{\sqrt{4c - b^2}}{2}\right)\right] \quad (2.59)$$

Since C_1 and C_2 are arbitrary constants, $(C_1 + C_2)$ and $(C_1 - C_2)$ are also arbitrary constants. Therefore, Eq. (2.58) can be written as:

$$y(x) = e^{-bx/2}\left[D_1\sin\left(x\frac{\sqrt{4c - b^2}}{2}\right) + D_2\cos\left(x\frac{\sqrt{4c - b^2}}{2}\right)\right] \quad (2.60)$$

where $D_1 = i(C_1 - C_2)$ and $D_2 = C_1 + C_2$.

Additional details on methods for analytical solutions are available in many calculus books and books on differential equations. Many ODEs that arise in practical applications, however, cannot be solved analytically and instead require the use of numerical methods for their solution.

2.6 FUNCTIONS OF TWO OR MORE INDEPENDENT VARIABLES

A function has one dependent variable but can have one, two, or more independent variables. For example, the function $z = f(x, y) = \frac{x^2}{2^2} + \frac{y^2}{3^2}$ (equation of an elliptic paraboloid) has two independent variables x and y and one dependent variable z. The function associates a unique number z (**dependent variable**) with each combination of values of x and y (**independent variables**). This section reviews several topics that are related to differentiation of functions with two or more independent variables.

2.6.1 Definition of the Partial Derivative

For a function $z = f(x, y)$, the first partial derivative of f with respect to x is denoted $\frac{\partial f}{\partial x}$ or f_x and is defined by:

$$\frac{\partial f}{\partial x} = \lim_{\Delta x \to 0} \frac{f(x + \Delta x, y) - f(x, y)}{\Delta x} \quad (2.61)$$

provided, of course, that the limit exists. Similarly, the partial derivative of f with respect to y is denoted by $\frac{\partial f}{\partial y}$ or f_y and is defined by:

$$\frac{\partial f}{\partial y} = \lim_{\Delta y \to 0} \frac{f(x, y + \Delta y) - f(x, y)}{\Delta y} \quad (2.62)$$

again provided that the limit exists. In practice, the definitions in Eqs. (2.61) and (2.62) imply that if $z = f(x, y)$, then f_x is determined by differentiating the function with respect to x and treating y as a constant. In the same way, f_y is determined by differentiating the function with respect to y and treating x as a constant.

Partial derivatives of higher order

It is possible to take the second, third, and higher order partial derivatives of a function of several variables (providing that they exist). For example, the function $f(x, y)$ of two variables has two first partial derivatives f_x and f_y. Each of the first partial derivatives has two partial derivatives. f_x can be differentiated w.r.t x to give $f_{xx} = \dfrac{\partial^2 f}{\partial x^2}$, or w.r.t y to give $f_{xy} = \dfrac{\partial^2 f}{\partial y \partial x}$. In the same way f_y can be differentiated w.r.t y to give $f_{yy} = \dfrac{\partial^2 f}{\partial y^2}$, or w.r.t x to give $f_{yx} = \dfrac{\partial^2 f}{\partial x \partial y}$. The second partial derivatives f_{xy} and f_{yx} are called mixed partial derivatives of $f(x, y)$. If the function $f(x, y)$ and both its second mixed partial derivatives are continuous, then it can be shown that the order of differentiation does not matter, that is $\dfrac{\partial^2 f}{\partial x \partial y} = \dfrac{\partial^2 f}{\partial y \partial x}$.

2.6.2 Chain Rules

The **total differential** of a function of two variables, for example, $f(x, y)$, is given by:

$$df = \frac{\partial f}{\partial x}dx + \frac{\partial f}{\partial y}dy \tag{2.63}$$

Equation (2.63) holds whether or not x and y are independent of each other. All that is required is that the partial derivatives in Eq. (2.63) be continuous. Note that Eq. (2.63) can easily be generalized to a function of more than two variables.

There are several ways in which the function $f(x, y)$ can depend on its arguments x and y. First, both x and y can be dependent on a single independent variable such as t. In other words, $f(x, y) = f(x(t), y(t))$. Second, x and y themselves may be dependent on two other independent variables, say u and v, $f(x, y) = f(x(u, v), y(u, v))$. Third, x may be the independent variable and y may depend on x, $f(x, y) = f(x, y(x))$. Fourth, the function f depend on three variables x, y, and z, but z depends on both x and y, which are independent of each other. Differentiation of these cases is considered next.

(1) If x and y each depend on a single variable t, then $f(x, y)$ may be considered to be a function of the single independent variable t. In this case, the **total derivative** of f with respect to t can be determined simply by:

$$\frac{df}{dt} = \frac{\partial f}{\partial x}\frac{dx}{dt} + \frac{\partial f}{\partial y}\frac{dy}{dt} \qquad (2.64)$$

(2) If x and y each depend on two other independent variables u and v, then the partial derivative of f with respect to v but holding u constant, that is, $\left.\frac{\partial f}{\partial v}\right|_u$ is obtained by:

$$\left.\frac{\partial f}{\partial v}\right|_u = \left.\frac{\partial f}{\partial x}\frac{\partial x}{\partial v}\right|_u + \left.\frac{\partial f}{\partial y}\frac{\partial y}{\partial v}\right|_u \qquad (2.65)$$

Usually, the fact that u is held constant while differentiating partially with respect to v is implicitly understood, and the subscript u is dropped from Eq. (2.65):

$$\frac{\partial f}{\partial v} = \frac{\partial f}{\partial x}\frac{\partial x}{\partial v} + \frac{\partial f}{\partial y}\frac{\partial y}{\partial v} \qquad (2.66)$$

(3) If y depends on x, the function $f(x, y)$ is really a function of x, and the total derivative of f with respect to x can be defined:

$$\frac{df}{dx} = \frac{\partial f}{\partial x} + \frac{\partial f}{\partial y}\frac{\partial y}{\partial x} \qquad (2.67)$$

(4) If f is a function of x, y, and z, and if z in turn depends on x and y which are independent, then the partial derivative of f with respect to x is:

$$\left.\frac{\partial f}{dx}\right|_y = \frac{\partial f}{\partial x} + \left.\frac{\partial f}{\partial z}\frac{\partial z}{\partial x}\right|_y \qquad (2.68)$$

or simply:

$$\frac{\partial f}{dx} = \frac{\partial f}{\partial x} + \frac{\partial f}{\partial z}\frac{\partial z}{\partial x} \qquad (2.69)$$

where the term $\frac{\partial f}{\partial y}$ does not appear because it is multiplied by $\frac{\partial y}{\partial x}$, which is zero since x and y are independent variables.

In the same way, the partial derivative of f with respect to y is:

$$\frac{\partial f}{dy} = \frac{\partial f}{\partial y} + \frac{\partial f}{\partial z}\frac{\partial z}{\partial y} \qquad (2.70)$$

2.6.3 The Jacobian

The Jacobian is a quantity that arises when solving systems of nonlinear simultaneous equations. If $f_1(x, y) = a$ and $f_2(x, y) = b$ are two simultaneous equations that need to be solved for x and y, where a and b are constants, then the ***Jacobian matrix*** is defined as:

$$[J] = \begin{bmatrix} \dfrac{\partial f_1}{\partial x} & \dfrac{\partial f_1}{\partial y} \\ \dfrac{\partial f_2}{\partial x} & \dfrac{\partial f_2}{\partial y} \end{bmatrix} \qquad (2.71)$$

The ***Jacobian determinant*** or simply the ***Jacobian***, $J(f_1, f_2)$, is just the determinant of the Jacobian matrix:

$$J(f_1, f_2) = \det\left(\begin{bmatrix} \dfrac{\partial f_1}{\partial x} & \dfrac{\partial f_1}{\partial y} \\ \dfrac{\partial f_2}{\partial x} & \dfrac{\partial f_2}{\partial y} \end{bmatrix}\right) = \left(\dfrac{\partial f_1}{\partial x}\right)\left(\dfrac{\partial f_2}{\partial y}\right) - \left(\dfrac{\partial f_1}{\partial y}\right)\left(\dfrac{\partial f_2}{\partial x}\right) \tag{2.72}$$

This can be easily generalized to a system of n equations:

$$J(f_1, f_2, ..., f_n) = \det\left(\begin{bmatrix} \dfrac{\partial f_1}{\partial x_1} & \dfrac{\partial f_1}{\partial x_2} & \cdots & \dfrac{\partial f_1}{\partial x_n} \\ \dfrac{\partial f_2}{\partial x_1} & \dfrac{\partial f_2}{\partial x_2} & \cdots & \dfrac{\partial f_2}{\partial x_n} \\ \cdots & \cdots & \cdots & \cdots \\ \dfrac{\partial f_n}{\partial x_1} & \dfrac{\partial f_n}{\partial x_2} & \cdots & \dfrac{\partial f_n}{\partial x_n} \end{bmatrix}\right) \tag{2.73}$$

2.7 TAYLOR SERIES EXPANSION OF FUNCTIONS

Taylor series expansion of a function is a way to find the value of a function near a known point, that is, a point where the value of the function is known. The function is represented by a sum of terms of a convergent series. In some cases (if the function is a polynomial) the Taylor series can give the exact value of the function. In most cases, however, a sum of an infinite number of terms is required for the exact value. If only a few terms are used, the value of the function that is obtained from the Taylor series is an approximation. Taylor series expansion of functions is used extensively in numerical methods.

2.7.1 Taylor Series for a Function of One Variable

Given a function $f(x)$ that is differentiable $(n + 1)$ times in an interval containing a point $x = x_0$, Taylor's theorem states that for each x in the interval, there exists a value $x = \xi$ between x and $x = x_0$ such that:

$$f(x) = f(x_0) + (x - x_0)\dfrac{df}{dx}\bigg|_{x = x_0} + \dfrac{(x - x_0)^2}{2!}\dfrac{d^2 f}{dx^2}\bigg|_{x = x_0} + \dfrac{(x - x_0)^3}{3!}\dfrac{d^3 f}{dx^3}\bigg|_{x = x_0}$$
$$+ ... + \dfrac{(x - x_0)^n}{n!}\dfrac{d^n f}{dx^n}\bigg|_{x = x_0} + R_n(x) \tag{2.74}$$

where R_n, called the remainder, is given by:

$$R_n = \dfrac{(x - x_0)^{n + 1}}{(n + 1)!}\dfrac{d^{n + 1} f}{dx^{n + 1}}\bigg|_{x = \xi} \tag{2.75}$$

The proof of this theorem may be found in any textbook on calculus.

Note that for $n = 1$ Taylor's theorem reduces to:

$$f(x) = f(x_0) + (x - x_0)\frac{df}{dx}\bigg|_{x = \xi} \quad \text{or} \quad \frac{df}{dx}\bigg|_{x = \xi} = \frac{f(x) - f(x_0)}{(x - x_0)} \quad (2.76)$$

which is the mean value theorem for derivatives (Eq. (2.4)) given in Section 2.2.

The value of the remainder, R_n, cannot be actually calculated since the value of ξ is not known. When the Taylor series is used for approximating the value of the function at x, two or more terms are used. The accuracy of the approximation depends on how many terms of the Taylor series are used and on the closeness of point x to point x_0. The accuracy increases as x is closer to x_0 and as the number or terms increases. This is illustrated in Example 2-3 where the Taylor series is used for approximating the function $y = \sin(x)$.

Example 2-3: Approximation of a function with Taylor series expansion.

Approximate the function $y = \sin(x)$ by using Taylor series expansion about $x = 0$, using two, four, and six terms.

(a) In each case calculate the approximate value of the function at $x = \frac{\pi}{12}$, and at $x = \frac{\pi}{2}$.

(b) Using MATLAB, plot the function and the three approximations for $0 \le x \le \pi$.

SOLUTION

The first five derivatives of the function $y = \sin(x)$ are:

$y' = \cos(x)$, $y'' = -\sin(x)$, $y^{(3)} = -\cos(x)$, $y^{(4)} = \sin(x)$,   and $y^{(5)} = \cos(x)$

At $x = 0$ the values of these derivatives are:

$y' = 1$, $y'' = 0$, $y^{(3)} = -1$, $y^{(4)} = 0$,   and $y^{(5)} = 1$

Substituting this information, and $y(0) = \sin(0) = 0$ in Eq. (2.74) gives:

$$y(x) = 0 + x + 0 - \frac{x^3}{3!} + 0 + \frac{x^5}{5!} \quad (2.77)$$

(a) For $x = \frac{\pi}{12}$ the exact value of the function is $y = \sin\left(\frac{\pi}{12}\right) = \frac{1}{4}(\sqrt{6} + \sqrt{2}) = 0.2588190451$

The approximate values using two, four, and six terms of the Taylor series expansion are:

Using two terms in Eq. (2.77) gives: $y(x) = x = \frac{\pi}{12} = 0.2617993878$

Using four terms in Eq. (2.77) gives: $y(x) = x - \frac{x^3}{3!} = \frac{\pi}{12} - \frac{(\pi/12)^3}{3!} = 0.2588088133$

Using six terms in Eq. (2.77) gives: $y(x) = x - \frac{x^3}{3!} + \frac{x^5}{5!} = \frac{\pi}{12} - \frac{(\pi/12)^3}{3!} + \frac{(\pi/12)^5}{5!} = 0.2588190618$

For $x = \frac{\pi}{2}$ the exact value of the function is $y = \sin\left(\frac{\pi}{2}\right) = 1$

Using two terms in Eq. (2.77) gives: $y(x) = x = \frac{\pi}{2} = 1.570796327$

Using four terms in Eq. (2.77) gives: $y(x) = x - \dfrac{x^3}{3!} = \dfrac{\pi}{2} - \dfrac{(\pi/2)^3}{3!} = 0.9248322293$

Using six terms in Eq. (2.77) gives: $y(x) = x - \dfrac{x^3}{3!} + \dfrac{x^5}{5!} = \dfrac{\pi}{2} - \dfrac{(\pi/2)^3}{3!} + \dfrac{(\pi/2)^5}{5!} = 1.004524856$

(b) Using a MATLAB program, listed in the following script file, the function and the three approximations were calculated for the domain $0 \le x \le \pi$. The program also plots the results.

```
x = linspace(0,pi,40);
y = sin(x);
y2 = x;
y4 = x - x.^3/factorial(3);
y6 = x - x.^3/factorial(3) + x.^5/factorial(5);
plot(x,y,'r',x,y2,'k--',x,y4,'k-.',x,y6,'r--')
axis([0,4,-2,2])
legend('Exact','Two terms','Four terms','Six terms')
xlabel('x'); ylabel('y')
```

The plot produced by the program is shown on the right. The results from both parts show, as expected, that the approximation of the function with the Taylor series is more accurate when more terms are used and when the point at which the value of the function is desired is close to the point about which the function is expanded.

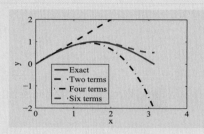

2.7.2 Taylor Series for a Function of Two Variables

Taylor's expansion for a function of two variables is done in the same way as for a function with one independent variable, except that the differentiation involves partial derivatives. Taylor's formula for the expansion of $f(x, y)$ about a point (x_0, y_0) is given by:

$$f(x, y) = f(x_0, y_0) + \frac{1}{1!}\left[(x - x_0)\frac{\partial f}{\partial x}\bigg|_{x_0, y_0} + (y - y_0)\frac{\partial f}{\partial y}\bigg|_{x_0, y_0}\right] +$$

$$\frac{1}{2!}\left[(x - x_0)^2\frac{\partial^2 f}{\partial x^2}\bigg|_{x_0, y_0} + 2(x - x_0)(y - y_0)\frac{\partial^2 f}{\partial x \partial y}\bigg|_{x_0, y_0} + (y - y_0)^2\frac{\partial^2 f}{\partial y^2}\bigg|_{x_0, y_0}\right] + \quad (2.78)$$

$$+ \dots + \frac{1}{n!}\left[\sum_{k=0}^{n}\frac{n!}{k!(n-k)!}(x - x_0)^k(y - y_0)^{n-k}\frac{\partial^n f}{\partial x^k \partial y^{n-k}}\bigg|_{x_0, y_0}\right]$$

2.8 PROBLEMS

Problems to be solved by hand
Solve the following problems by hand. When needed, use a calculator, or write a MATLAB script file to carry out the calculations.

2.1   Apply the intermediate value theorem to show that the polynomial $f(x) = -x^4 + 2x + 4$ has a root in the interval $[1, 2]$.

2.2   Apply the intermediate value theorem to show that the function $f(x) = \cos x - x^2$ has a root in the interval $[0, \pi/2]$.

2.3   Use the formal definition of the derivative (Eq. (2.2)) and associated terminology to show that the derivative of $\cos x$ is $-\sin x$.

2.4   Use the definition of the derivative (Eq. (2.2)) to show that:

$(a)\quad \dfrac{d}{dx}(u(x)v(x)) = u\dfrac{dv}{dx} + v\dfrac{du}{dx}.\qquad (b)\qquad \dfrac{d}{dx}\!\left(\dfrac{u(x)}{v(x)}\right) = \dfrac{v\dfrac{du}{dx} - u\dfrac{dv}{dx}}{v^2}.$

2.5   Use the chain rule (Eq. (2.3)) to find the second derivative of $f(x) = \sin(e^x)$. (Hint: define $u(x) = e^x$ and then apply the chain rule.

2.6   As a highway patrol officer, you are participating in a speed trap. A car passes your patrol car which you clock at 55 mph. One minute later, your partner in another patrol car situated one mile away from you, clocks the same car at 50 mph. Using the mean value theorem for derivatives (Eq. (2.4)), show that the car must have been exceeded the speed limit of 55 mph at some point during the one minute it traveled between the two patrol cars.

2.7   Coughing causes the windpipe in the throat to contract and forces the flowing air to pass with increasing velocity. Suppose the velocity, v, of the flowing air during the cough is given by:

$$v = C(R - r)r^2$$

where C is a constant, and R is the normal radius of the windpipe (i.e. when not coughing) which is also a constant, and r is the variable radius of the windpipe during the cough. Find the radius of the windpipe that produces the largest velocity of airflow during the cough.

2.8   Using the mean value theorem for integrals, find the average value of the function $f(x) = \sin x$ in the interval $[0, \pi]$. Show that the product of this average value times the width of the interval is exactly equal to the area under the curve.

2.9   Use the second fundamental theorem of calculus along with the chain rule to find $\dfrac{d}{dx}\!\left[\displaystyle\int_{1}^{\sqrt{x}} \sin t \, dt\right]$.

2.10 Given the following system of equations:

$$2x + 2y - 3z = 2$$
$$-1x + 3y + 2z = 0$$
$$3x + y - 3z = 1$$

Determine the unknowns x, y, and z using Cramer's rule.

2.11 The temperature distribution in a solid is given by $T(x, y) = e^{-y^2} \sin x$. The heat flux in the y direction is given by $q_y = -k \dfrac{\partial T}{\partial y}$. Using the definition of the partial derivative (Eq. (2.62)), find the heat flux q_y at the point $(1, 2)$.

2.12 Given the function $f(x) = (\sin x)(\cos y) \ln z$, find the total derivative with respect to x, $\dfrac{df}{dx}$ at the point $(1, 2, 3)$.

2.13 Find the determinant of the following matrix:

$$\begin{bmatrix} 2 & 0 & 0 & 3 \\ 1 & 1 & 1 & 0 \\ 5 & 1 & 1 & 9 \\ 1 & 1 & 0 & 0 \end{bmatrix}$$

2.14 Determine the order of the following ODEs and whether they are linear, nonlinear, homogeneous or non-homogeneous:

(a) $EI \dfrac{d^2 y}{dx^2} = M(x)$, where E and I are constants, and $M(x)$ is a known function of x.

(b) $\dfrac{d\phi}{dt} + \dfrac{\phi}{\tau} = \dfrac{CF(t)}{\tau}$, where C and τ are constants, and $F(t)$ is a known function of t.

(c) $\dfrac{dh}{dt} = \dfrac{-r^2 \sqrt{2gh}}{2hR - h^2}$, where r, g, and R are constants.

2.15 When transforming from Cartesian coordinates (x, y) to polar coordinates (r, θ), the following relations hold: $x(r, \theta) = r \cos \theta$ and $y(r, \theta) = r \sin \theta$. Find the Jacobian matrix $[J] = \begin{bmatrix} \dfrac{\partial x}{\partial r} & \dfrac{\partial x}{\partial \theta} \\ \dfrac{\partial y}{\partial r} & \dfrac{\partial y}{\partial \theta} \end{bmatrix}$. What is the Jacobian determinant?

2.16 Write the Taylor's series expansion of the function $f(x) = \ln(x + n)$ about $x = 0$, where $n \neq 0$ is a known constant.

2.17 Write the Taylor's series expansion of the function $f(x, y) = e^{-x^2} \sin y$ about the point $(1, 3)$.

Problems to be programmed in MATLAB
Solve the following problems using the MATLAB environment. Do not use MATLAB's built-in functions that execute the operations that are being asked in the problems.

2.18 Write a MATLAB program in a script file that evaluates the derivative of the function $f(x) = x^3 \sin x$ at the point $x = 3$ by using Eq. (2.2). The value of the derivative is calculated sixteen times by using $a = 3$ and sixteen values of x, $x = 2.6, 2.65, ..., 2.95, 3.05, ..., 3.35, 3.4$. The program should also plot the values of the derivative versus x.

2.19 Write a user-defined MATLAB function that evaluates the definite integral of a function by using the Riemann sum (see Eq. (2.7)). For function name and arguments use `I=RiemannSum('Fun-Name',a,b)`. `'FunName'` is a string with the name of a function file that calculates the value of the function to be integrated for a given value of x. `a` and `b` are the limits of integration, and `I` is the value of the integral. The Riemann sum is calculated by dividing the integration interval $[a, b]$ into ten subintervals. Use `RiemannSum` for evaluating the definite integral $\int_0^1 e^x dx$. Compare the result with the exact value of the integral of 1.71828.

2.20 Write a user-defined MATLAB function that carries out multiplication of two matrices $[c] = [a][b]$. For function name and arguments use `C = MatrixMult(A,B)`. The input arguments `A` and `B` are the matrices that are multiplied. The output argument `C` is the result. Do not use the matrix multiplication of MATLAB. The function `MatrixMult` should first check if the two matrices can be multiplied, and if not the output C should be the message "The matrices cannot be multiplied since the number of row in [b] is not equal to the number of columns in [a]". Use `MatrixMult` to carry out the multiplication that is illustrated in Fig. 2.14.

2.21 Write a user-defined MATLAB function that determines the cross product of two vectors $\vec{W} = \vec{V} \otimes \vec{U}$. For function name and arguments use `W = Cross(V,U)`. The input arguments `V` and `U` are the vectors that are multiplied. The output argument `W` is the result (three-element vector). Use `Cross` to determine the cross product of the vectors $v = 3i + 6.5j - 2k$ and $u = -5i + 4k + 10k$.

2.22 Write a user-defined MATLAB function to unfurl an $m \times n$ matrix into a vector of size $1 \times m \cdot n$. The vector consists of the matrix rows in order. For example, if the matrix is: $\begin{bmatrix} a_{11} & a_{12} & a_{13} \\ a_{21} & a_{22} & a_{23} \\ a_{31} & a_{32} & a_{33} \end{bmatrix}$, then the vector is:

$[a_{11} \; a_{12} \; a_{13} \; a_{21} \; a_{22} \; a_{23} \; a_{31} \; a_{32} \; a_{33}]$. For the function name and arguments use `v=unfurl(A)`, where the input argument `A` is a matrix of any size, and the output argument `v` is the vector.
     Use the function (in the Command Window) to unfurl the matrix:

$$\begin{bmatrix} 2 & -5 & 1 & 0 & 6 \\ -9 & 4 & 6 & 10 & -4 \\ 11 & 0 & -12 & 7 & 3 \end{bmatrix}$$

2.23 Write a user-defined MATLAB function that determines the transpose of any sized $m \times n$ matrix. Do not use the MATLAB built-in command for the transpose. For the function name and arguments use `At=transp(A)`, where the input argument `A` is a matrix of any size, and the output argument `At` is the transpose of `A`.

Use the function (in the Command Window) to determine the transpose of the matrix:

$$\begin{bmatrix} 2 & -5 & 1 & 0 & 6 \\ -9 & 4 & 6 & 10 & -4 \\ 11 & 0 & -12 & 7 & 3 \end{bmatrix}$$

2.24 Write a user-defined MATLAB function that calculates the determinant of a square ($n \times n$) matrix, where n can be 2, 3, or 4. For function name and arguments use `D = Determinant(A)`. The input argument `A` is the matrix whose determinant is calculated. The function `Determinant` should first check if the matrix is square. If it is not, the output `D` should be the message "The matrix has to be square".

Use `Determinant` to calculate the determinant of the following two matrices:

(a) $\begin{bmatrix} 2 & 2 & -3 \\ -1 & 3 & 2 \\ 3 & 1 & -1 \end{bmatrix}$ (b) $\begin{bmatrix} 2 & 1 & 4 & -2 \\ -3 & 4 & 2 & -1 \\ 3 & 5 & -2 & 1 \\ -2 & 3 & 2 & 4 \end{bmatrix}$

Problems in math, science, and engineering
Solve the following problems using the MATLAB environment.

2.25 One important application involving the total differential of a function of several variables is estimation of uncertainty.
(a) The electrical power P dissipated by a resistance R is related to the voltage V and resistance by

$P = \dfrac{V^2}{R}$. Write the total differential dP in terms of the differentials dV and dR, using Eq. (2.63).

(b) dP is interpreted as the uncertainty in the power, dV as the uncertainty in the voltage, and dR as the uncertainty in the resistance. Using the answer of part (a), determine the maximum percent uncertainty in the power P for $V = 200\,\text{V}$ with an uncertainty of 2%, and $R = 5000\,\Omega$ with an uncertainty of 3%.

2.26 An aircraft begins its descent at a distance $x = L$ ($x = 0$ is the spot at which the plane touches down) and an altitude of H. Suppose a cubic polynomial of the following form is used to describe the landing:

$$y = ax^3 + bx^2 + cx + d$$

where y is the altitude and x is the horizontal distance to the aircraft. The aircraft begins its descent from a level position, and lands at a level position.

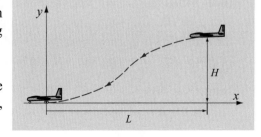

(a) Solve for the coefficients a, b, c, and d.

(b) If the aircraft maintains a constant forward speed ($\dfrac{dx}{dt} = u = constant$) and the magnitude of the vertical acceleration ($\dfrac{d^2y}{dt^2}$) is not to exceed a constant A, show that $\dfrac{6Hu^2}{L^2} \le A$.

(c) If A = 0.3 ft/s^2, H = 10000 ft, and u = 150 mph, how far from the airport should the pilot begin the descent?

2.27 An artery that branches from another more major artery has a resistance for blood flow that is given by:

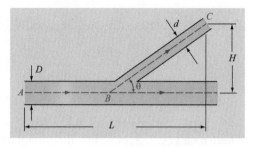

$$R_{flow} = K\left(\frac{L - H\cot\theta}{D^4} + \frac{H\csc\theta}{d^4}\right)$$

where R_{flow} is the resistance to blood flow from the major to the branching artery along path ABC (see diagram), d is the diameter of the smaller, branching artery, D is the diameter of the major artery, θ is the angle that the branching vessel makes with the horizontal, or axis, of the major artery, and L and H are the distances shown in the figure. Find the angle θ that minimizes the flow resistance in terms of d and D.

2.28 A rope with a length of 10 m is to be used to enclose a square area with side x and a circular area with radius r. How much rope should be used for the square and how much for the circle if the total area enclosed by the two shapes is to be a maximum.

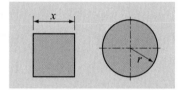

2.29 There are mechanical, electrical, and chemical systems that are described by the same mathematics as second-order, forced, damped harmonic motion. The resulting differential equation obtained after applying a force balance or conservation of momentum is of the form:

$$m\frac{d^2x}{dt^2} + \gamma\frac{dx}{dt} + kx = A_0\sin(\omega t)$$

where x is the displacement, t is time, m is the mass, γ is the damping coefficient, k is the restoring force (spring) constant, A_0 is the amplitude of the driving force, and ω is the frequency of the driving force.

(a) Determine the order of the ODE and whether it is linear, nonlinear, homogeneous or non-homogeneous.

(b) Find the homogeneous solution of the ODE by hand.

(c) Find the particular solution of the ODE by hand. Find $x(t)$ after a long time ($t \to \infty$). This is sometimes called the "steady state" response, even though it is actually time varying.

(d) Using MATLAB, plot the maximum amplitude of $x(t)$ from the steady state response as a function of the excitation frequency ω ($0 \le \omega \le 5$ rad/s) for A_0 = 1 N, k = 1 N/m, and m = 1 kg, for three values of γ γ = 0.5, γ = 1.0, and γ = 2.0 N–s/m (three plots on the same figure). Discuss the results. What happens at $\omega = \sqrt{\frac{k}{m}}$ when γ = 0?

Chapter 3

Solving Nonlinear Equations

3.1 BACKGROUND

Equations need to be solved in all areas of science and engineering. An equation of one variable can be written in the form:

$$f(x) = 0 \tag{3.1}$$

A solution to the equation (also called a **root** of the equation) is a numerical value of x that satisfies the equation. Graphically, as shown in Fig. 3-1, the solution is the point where the function $f(x)$ crosses or touches the x axis. An equation might have no solution or can have one or several (possibly many) roots.

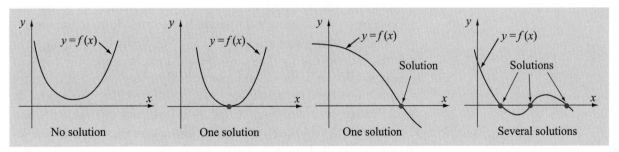

No solution   One solution   One solution   Several solutions

Figure 3-1: Illustration of equations with no, one, or several solutions.

When the equation is simple, the value of x can be determined analytically. This is the case when x can be written explicitly by applying mathematical operations, or when a known formula (such as the formula for solving a quadratic equation) can be used to determine the exact value of x. In many situations, however, it is impossible to determine the root of an equation analytically. For example, the area of a segment A_S of a circle with radius r (shaded area in Fig. 3-2) is given by:

$$A_S = \frac{1}{2}r^2(\theta - \sin\theta) \qquad (3.2)$$

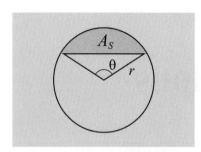

Figure 3-2: Segment of a circle.

To determine the angle θ if A_S and r are given, Eq. (3.2) has to be solved for θ. Obviously, θ cannot be written explicitly in terms of A_S and r, and the equation cannot be solved analytically.

A numerical solution of an equation $f(x) = 0$ is a value of x that satisfies the equation approximately. This means that when x is substituted in the equation, the value of $f(x)$ is close to zero, but not exactly zero. For example, to determine the angle θ for a circle with $r = 3$ m and $A_S = 8$ m^2, Eq. (3.2) can be written in the form:

$$f(\theta) = 8 - 4.5(\theta - \sin\theta) = 0 \qquad (3.3)$$

A plot of $f(\theta)$, (see Fig. 3-3), shows that the solution is between 2 and 3. Substituting $\theta = 2.4$ rad in Eq. (3.3) gives $f(\theta) = 0.2396$, and the solution $\theta = 2.43$ rad gives $f(\theta) = 0.003683$. Obviously, the latter is a more accurate, but not an exact, solution. It is possible to determine values of θ that give values of $f(\theta)$ that are closer to zero, but it is impossible to determine a numerical value of θ for which $f(\theta)$ is exactly zero. When solving an equation numerically, one has to select the desired accuracy of the solution.

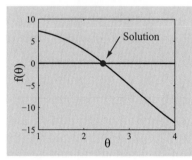

Figure 3-3: A plot of
$f(\theta) = 8 - 4.5(\theta - \sin\theta)$.

Overview of approaches in solving equations numerically

The process of solving an equation numerically is different from the procedure used to find an analytical solution. An analytical solution is obtained by deriving an expression that has an exact numerical value. A numerical solution is obtained in a process that starts by finding an approximate solution and is followed by a numerical procedure in which a better (more accurate) solution is determined. An initial numerical solution of an equation $f(x) = 0$ can be estimated by plotting $f(x)$ versus x and looking for the point where the graph crosses the x axis. It is also possible to write and execute a computer program that looks for a domain that contains a solution. Such a program looks for a solution by evaluating $f(x)$ at different values of x. It starts at one value of x and then changes the value of x in small increments. A change in the sign of $f(x)$ indicates that there is a root within the last increment. In most cases, when the equation that is solved is related to an application in science or engineering, the range of x that includes the solution can be estimated and used in the initial plot of $f(x)$, or for a numerical search of a small domain that contains a solution. When an equation has more than

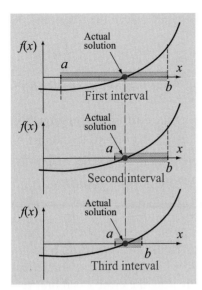

Figure 3-4: Illustration of a bracketing method.

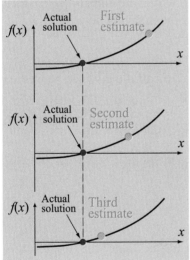

Figure 3-5: Illustration of an open method.

one root, a numerical solution is obtained one root at a time.

The methods used for solving equations numerically can be divided into two groups: ***bracketing methods*** and ***open methods***. In bracketing methods, illustrated in Fig. 3-4, an interval that includes the solution is identified. By definition, the endpoints of the interval are the upper bound and lower bound of the solution. Then, by using a numerical scheme, the size of the interval is successively reduced until the distance between the endpoints is less than the desired accuracy of the solution. In open methods, illustrated in Fig. 3-5, an initial estimate (one point) for the solution is assumed. The value of this initial guess for the solution should be close to the actual solution. Then, by using a numerical scheme, better (more accurate) values for the solution are calculated. Bracketing methods always converge to the solution. Open methods are usually more efficient but sometimes might not yield the solution.

As mentioned previously, since numerical solutions are generally not exact, there is a need for estimating the error. Several options are presented in Section 3.2. Sections 3.3 through 3.7 describe four numerical methods for finding a root of a single equation. Two bracketing methods: the bisection method and the regula falsi method, are presented in Sections 3.3 and 3.4, respectively. Three open methods: Newton's method, secant method, and fixed-point iteration, are introduced in the following three sections. Section 3.8 describes how to use MATLAB's built-in functions for obtaining numerical solutions, and Section 3.9 discusses how to deal with equations that have multiple roots. The last section in this chapter (3.10) deals with numerical methods for solving systems of nonlinear equations. The need to solve such systems arises in many problems in science and engineering and when numerical methods are used for solving ordinary differential equations (see Section 9.3).

3.2 ESTIMATION OF ERRORS IN NUMERICAL SOLUTIONS

Since numerical solutions are not exact, some criterion has to be applied in order to determine whether an estimated solution is accurate enough. Several measures can be used to estimate the accuracy of an approximate solution. The decision as to which measure to use depends on the application and has to be made by the person solving the equation.

Let x_{TS} be the true (exact) solution such that $f(x_{TS}) = 0$, and let x_{NS} be a numerically approximated solution such that $f(x_{NS}) = \varepsilon$ (where ε is a small number). Four measures that can be considered for estimating the error are:

True error: The true error is the difference between the true solution, x_{TS}, and a numerical solution, x_{NS}:

$$TrueError = x_{TS} - x_{NS} \tag{3.4}$$

Unfortunately, however, the true error cannot be calculated because the true solution is generally not known.

Tolerance in $f(x)$: Instead of considering the error in the solution, it is possible to consider the deviation of $f(x_{NS})$ from zero (the value of $f(x)$ at x_{TS} is obviously zero). The tolerance in $f(x)$ is defined as the absolute value of the difference between $f(x_{TS})$ and $f(x_{NS})$:

$$ToleranceInf = |f(x_{TS}) - f(x_{NS})| = |0 - \varepsilon| = |\varepsilon| \tag{3.5}$$

The tolerance in $f(x)$ then is the absolute value of the function at x_{NS}.

Tolerance in the solution: A tolerance is the maximum amount by which the true solution can deviate from an approximate numerical solution. A tolerance is useful for estimating the error when bracketing methods are used for determining the numerical solution. In this case, if it is known that the solution is within the domain $[a, b]$, then the numerical solution can be taken as the midpoint between a and b:

$$x_{NS} = \frac{a + b}{2} \tag{3.6}$$

plus or minus a tolerance that is equal to half the distance between a and b:

$$Tolerance = \left| \frac{b - a}{2} \right| \tag{3.7}$$

Relative error: If x_{NS} is an estimated numerical solution, then the ***True Relative Error*** is given by:

$$TrueRelativeError = \left| \frac{x_{TS} - x_{NS}}{x_{TS}} \right| \tag{3.8}$$

This True Relative Error cannot be calculated since the true solution x_{TS} is not known. Instead it is possible to calculate an ***Estimated Relative Error*** when two numerical estimates for the solution are known. This is the case when numerical solutions are calculated iteratively, where in each new iteration a more accurate solution is calculated. If $x_{NS}^{(n)}$ is the estimated numerical solution in the last iteration and $x_{NS}^{(n-1)}$ is the estimated numerical solution in the preceding iteration, then an Estimated Relative Error can be defined by:

$$EstimatedRelativeError = \left| \frac{x_{NS}^{(n)} - x_{NS}^{(n-1)}}{x_{NS}^{(n-1)}} \right| \tag{3.9}$$

When the estimated numerical solutions are close to the true solution, it is anticipated that the difference $x_{NS}^{(n)} - x_{NS}^{(n-1)}$ is small compared to the value of $x_{NS}^{(n)}$ and the Estimated Relative Error is approximately the same as the True Relative Error.

3.3 BISECTION METHOD

The bisection method is a bracketing method for finding a numerical solution of an equation of the form $f(x) = 0$ when it is known that within a given interval $[a, b]$, $f(x)$ is continuous and the equation has a solution. When this is the case, $f(x)$ will have opposite signs at the endpoints of the interval. As shown in Fig. 3-6, if $f(x)$ is continuous and

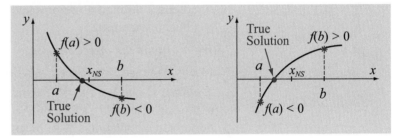

Figure 3-6: Solution of $f(x) = 0$ between $x = a$ and $x = b$.

has a solution between the points $x = a$ and $x = b$, then either $f(a) > 0$ and $f(b) < 0$ or $f(a) < 0$ and $f(b) > 0$. In other words, if there is a solution between $x = a$ and $x = b$, then $f(a)f(b) < 0$.

The process of finding a solution with the bisection method is illustrated in Fig. 3-7. It starts by finding points a and b that define an interval where a solution exists. Such an interval is found either by plotting $f(x)$ and observing a zero crossing, or by examining the function for sign change. The midpoint of the interval x_{NS1} is then taken as the first estimate for the numerical solution. The true solution is either in the section between points a and x_{NS1} or in the section between points x_{NS1} and b. If the numerical solution is not accurate enough, a new interval that contains the true solution is defined. The new interval is the half of the original interval that contains the true solution, and its midpoint is taken as the new (second) estimate of the numerical solution. The process continues until the numerical solution is accurate enough according to a criterion that is selected.

The procedure (or algorithm) for finding a numerical solution with the bisection method is summarized as follows.

Algorithm for the bisection method

1. Choose the first interval by finding points a and b such that a solution exists between them. This means that $f(a)$ and $f(b)$ have different signs such that $f(a)f(b) < 0$. The points can be determined by examining the plot of $f(x)$ versus x.

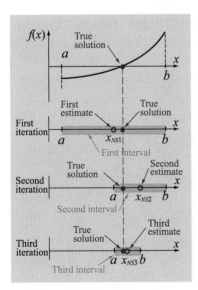

Figure 3-7: Bisection method.

2. Calculate the first estimate of the numerical solution x_{NS1} by:

$$x_{NS1} = \frac{(a+b)}{2}$$

3. Determine whether the true solution is between a and x_{NS1}, or between x_{NS1} and b. This is done by checking the sign of the product $f(a) \cdot f(x_{NS1})$:

    If $f(a) \cdot f(x_{NS1}) < 0$, the true solution is between a and x_{NS1}.

    If $f(a) \cdot f(x_{NS1}) > 0$, the true solution is between x_{NS1} and b.

4. Select the subinterval that contains the true solution (a to x_{NS1}, or x_{NS1} to b) as the new interval $[a, b]$, and go back to step 2.

Steps 2 through 4 are repeated until a specified tolerance or error bound is attained.

When should the bisection process be stopped?

Ideally, the bisection process should be stopped when the true solution is obtained. This means that the value of x_{NS} is such that $f(x_{NS}) = 0$. In reality, as discussed in Section 3.1, this true solution generally cannot be found computationally. In practice therefore, the process is stopped when the estimated error, according to one of the measures listed in Section 3.2, is smaller than some predetermined value. The choice of termination criteria may depend on the problem that is actually solved.

A MATLAB program written in a script file that determines a numerical solution by applying the bisection method is shown in the solution of the following example. (Rewriting this program in a form of a user-defined function is assigned as a homework problem.)

Example 3-1: Solution of a nonlinear equation using the bisection method.

Write a MATLAB program, in a script file, that determines the solution of the equation $8 - 4.5(x - \sin x) = 0$ by using the bisection method. The solution should have a tolerance of less than 0.001 rad. Create a table that displays the values of a, b, x_{NS}, $f(x_{NS})$, and the tolerance for each iteration of the bisection process.

SOLUTION

To find the approximate location of the solution, a plot of the function $f(x) = 8 - 4.5(x - \sin x)$ is made by using the fplot command of MATLAB. The plot (Fig. 3-8), shows that the solution is between $x = 2$ and $x = 3$. The initial interval is chosen as $a = 2$ and $b = 3$.

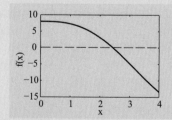

Figure 3-8: A plot of the function $f(x) = 8 - 4.5(x - \sin x)$.

A MATLAB program that solves the problem is as follows.

Program 3-1: Script file. Bisection method.

```
clear all
F = inline('8 - 4.5*(x - sin(x))');
```
Define $f(x)$ as an inline function.

```
a = 2; b = 3; imax = 20; tol = 0.001;
Fa = F(a); Fb = F(b);
if  Fa*Fb > 0
    disp('Error: The function has the same sign at points a and b.')
else
    disp('iteration   a      b   (xNS) Solution  f(xNS)  Tolerance')
    for i = 1:imax
        xNS = (a + b)/2;
        toli = (b - a)/2;
        FxNS = F(xNS);
        fprintf('%3i   %11.6f %11.6f %11.6f  %11.6f %11.6f\n', i, a, b, xNS, FxNS, toli)
        if FxNS == 0
            fprintf('An exact solution x =%11.6f was found',xNS)
            break
        end
        if toli < tol
            break
        end
        if i == imax
            fprintf('Solution was not obtained in %i iterations',imax)
            break
        end
        if F(a)*FxNS < 0
            b = xNS;
        else
            a = xNS;
        end
    end
end
```

Assign initial values to *a* and *b*, define max number of iterations and tolerance.

Stop the program if the function has the same sign at points *a* and *b*.

Calculate the numerical solution of the iteration, *xNS*.

Calculate the current tolerance.

Calculate the value of *f*(*xNS*) of the iteration.

Stop the program if the true solution, $f(x) = 0$, is found.

Stop the iterations if the tolerance of the iteration is smaller than the desired tolerance.

Stop the iterations if the solution was not obtained and the number of the iteration reaches `imax`.

Determine whether the true solution is between *a* and *xNS*, or between *xNS* and *b*, and select *a* and *b* for the next iteration.

When the program is executed, the display in the Command Window is:

iteration	a	b	(xNS) Solution	f(xNS)	Tolerance
1	2.000000	3.000000	2.500000	-0.556875	0.500000
2	2.000000	2.500000	2.250000	1.376329	0.250000
3	2.250000	2.500000	2.375000	0.434083	0.125000
4	2.375000	2.500000	2.437500	-0.055709	0.062500
5	2.375000	2.437500	2.406250	0.190661	0.031250
6	2.406250	2.437500	2.421875	0.067838	0.015625
7	2.421875	2.437500	2.429688	0.006154	0.007813
8	2.429688	2.437500	2.433594	-0.024755	0.003906
9	2.429688	2.433594	2.431641	-0.009295	0.001953
10	2.429688	2.431641	2.430664	-0.001569	0.000977

The numerical solution.

The value of the function at the numerical solution.

The last tolerance (satisfies the prescribed tolerance).

The output shows that the solution with the desired tolerance is obtained in the 10th iteration.

Additional notes on the bisection method

- The method always converges to an answer, provided a root was trapped in the interval $[a, b]$ to begin with.

- The method may fail when the function is tangent to the axis and does not cross the x axis at $f(x) = 0$.

- The method converges slowly relative to other methods.

3.4 REGULA FALSI METHOD

The regula falsi method (also called false position and linear interpolation methods) is a bracketing method for finding a numerical solution of an equation of the form $f(x) = 0$ when it is known that, within a given interval $[a, b]$, $f(x)$ is continuous and the equation has a solution. As illustrated in Fig. 3-9, the solution starts by finding an initial interval $[a_1, b_1]$ that brackets the solution. The values of the function at the endpoints are $f(a_1)$ and $f(b_1)$. The endpoints are then connected by a straight line, and the first estimate of the numerical solution, x_{NS1}, is the point where the straight line crosses the x axis. This is in contrast to the bisection method, where the midpoint of the interval was taken as the solution. For the second iteration a new interval $[a_2, b_2]$ is defined. The

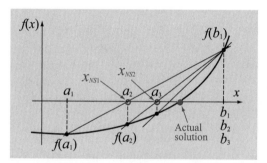

Figure 3-9: Regula Falsi method.

new interval is a subsection of the first interval that contains the solution. It is either $[a_1, x_{NS1}]$ (a_1 is assigned to a_2 and x_{NS1} to b_2) or $[x_{NS1}, b_1]$ (x_{NS1} is assigned to a_2 and b_1 to b_2). The endpoints of the second interval are next connected with a straight line, and the point where this new line crosses the x axis is the second estimate of the solution, x_{NS2}. For the third iteration a new subinterval $[a_3, b_3]$ is selected, and the iterations continue in the same way until the numerical solution is deemed accurate enough.

For a given interval $[a, b]$, the equation of a straight line that connects point $(b, f(b))$ to point $(a, f(a))$ is given by:

$$y = \frac{f(b) - f(a)}{b - a}(x - b) + f(b) \qquad (3.10)$$

The point x_{NS} where the line intersects the x axis is determined by substituting $y = 0$ in Eq. (3.10), and solving the equation for x:

$$x_{NS} = \frac{af(b) - bf(a)}{f(b) - f(a)} \tag{3.11}$$

The procedure (or algorithm) for finding a solution with the regula falsi method is almost the same as that for the bisection method.

Algorithm for the regula falsi method

1. Choose the first interval by finding points a and b such that a solution exists between them. This means that $f(a)$ and $f(b)$ have different signs such that $f(a)f(b) < 0$. The points can be determined by looking at a plot of $f(x)$ versus x.

2. Calculate the first estimate of the numerical solution x_{NS1} by using IQ. (3.11).

3. Determine whether the actual solution is between a and x_{NS1} or between x_{NS1} and b. This is done by checking the sign of the product $f(a) \cdot f(x_{NS1})$:

   If $f(a) \cdot f(x_{NS1}) < 0$, the solution is between a and x_{NS1}.

   If $f(a) \cdot f(x_{NS1}) > 0$, the solution is between x_{NS1} and b.

4. Select the subinterval that contains the solution (a to x_{NS1}, or x_{NS1} to b) as the new interval $[a, b]$, and go back to step 2.

Steps 2 through 4 are repeated until a specified tolerance or error bound is attained.

When should the iterations be stopped?

The iterations are stopped when the estimated error, according to one of the measures listed in Section 3.2, is smaller than some predetermined value.

Additional notes on the regula falsi method

- The method always converges to an answer, provided a root is initially trapped in the interval $[a, b]$.

- Frequently, as in the case shown in Fig. 3-9, the function in the interval $[a, b]$ is either concave up or concave down. In this case one of the endpoints of the interval stays the same in all the iterations, while the other endpoint advances toward the root. In other words, the numerical solution advances toward the root only from one side. The convergence toward the solution could be faster if the other endpoint would also "move" toward the root. Several modifications have been introduced to the regula falsi method that make the subinterval in successive iterations approach the root from both sides (see Problem 3.18).

3.5 NEWTON'S METHOD

Newton's method (also called the Newton–Raphson method) is a scheme for finding a numerical solution of an equation of the form $f(x) = 0$ where $f(x)$ is continuous and differentiable and the equation is known to have a solution near a given point. The method is illustrated in Fig. 3.10.

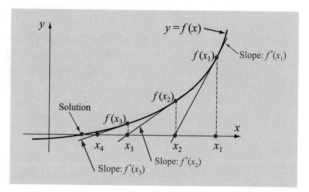

Figure 3-10: Newton's method.

The solution process starts by choosing point x_1 as the first estimate of the solution. The second estimate x_2 is obtained by taking the tangent line to $f(x)$ at the point $(x_1, f(x_1))$ and finding the intersection point of the tangent line with the x axis. The next estimate x_3 is the intersection of the tangent line to $f(x)$ at the point $(x_2, f(x_2))$ with the x axis, and so on. Mathematically, for the first iteration, the slope, $f'(x_1)$, of the tangent at point $(x_1, f(x_1))$ is given by:

$$f'(x_1) = \frac{f(x_1) - 0}{x_1 - x_2} \tag{3.12}$$

Solving Eq. (3.12) for x_2 gives:

$$x_2 = x_1 - \frac{f(x_1)}{f'(x_1)} \tag{3.13}$$

Equation 3.13 can be generalized for determining the "next" solution x_{i+1} from the present solution x_i:

$$x_{i+1} = x_i - \frac{f(x_i)}{f'(x_i)} \tag{3.14}$$

Equation (3.14) is the general iteration formula for Newton's method. It is called an iteration formula because the solution is found by repeated application of Eq. (3.14) for each successive value of i.

   Newton's method can also be derived by using Taylor series. Taylor series expansion of $f(x)$ about x_1 is given by:

$$f(x) = f(x_1) + (x - x_1)f'(x_1) + \frac{1}{2!}(x - x_1)^2 f''(x_1) + \dots \tag{3.15}$$

If x_2 is a solution of the equation $f(x) = 0$ and x_1 is a point near x_2, then:

$$f(x_2) = 0 = f(x_1) + (x_2 - x_1)f'(x_1) + \frac{1}{2!}(x_2 - x_1)^2 f''(x_1) + \dots \quad (3.16)$$

By considering only the first two terms of the series, an approximate solution can be determined by solving Eq. (3.16) for x_2:

$$x_2 = x_1 - \frac{f(x_1)}{f'(x_1)} \quad (3.17)$$

The result is the same as Eq. (3.13). In the next iteration the Taylor expansion is written about point x_2, and an approximate solution x_3 is calculated. The general formula is the same as that given in Eq. (3.14).

Algorithm for Newton's method

1. Choose a point x_1 as an initial guess of the solution.

2. For $i = 1, 2, \dots$, until the error is smaller than a specified value, calculate x_{i+1} by using Eq. (3.14).

When are the iterations stopped?

Ideally, the iterations should be stopped when an exact solution is obtained. This means that the value of x is such that $f(x) = 0$. Generally, as discussed in Section 3.1, this exact solution cannot be found computationally. In practice therefore, the iterations are stopped when an estimated error is smaller than some predetermined value. A tolerance in the solution, as in the bisection method, cannot be calculated since bounds are not known. Two error estimates that are typically used with Newton's method are:

Estimated relative error: The iterations are stopped when the estimated relative error (Eq. (3.9)) is smaller than a specified value ε:

$$\left| \frac{x_{i+1} - x_i}{x_i} \right| \leq \varepsilon \quad (3.18)$$

Tolerance in $f(x)$: The iterations are stopped when the absolute value of $f(x_i)$ is smaller than some number δ:

$$|f(x_i)| \leq \delta \quad (3.19)$$

The programming of Newton's method is very simple. A MATLAB user-defined function (called `NewtonRoot`) that finds the root of $f(x) = 0$ is listed in Fig. 3-11. The program consists of one loop in which the next solution `Xi` is calculated from the present solution `Xest` using Eq. (3.14). The looping stops if the error is small enough according to Eq. (3.18). To avoid the situation where the looping continues indefinitely (either because the solution does not converge or because of a programming error), the number of passes in the loop is limited to `imax`. The functions $f(x)$ and $f'(x)$ (that appear in Eq. (3.14)) have to be supplied as separate user-defined functions. Their names are typed in the arguments of `NewtonRoot` as strings.

Program 3-2: User-defined function. Newton's method.

function Xs = NewtonRoot(Fun,FunDer,Xest,Err,imax)
% NewtonRoot finds the root of Fun = 0 near the point Xest using Newton's method.
% Input variables:
% Fun   Name (string) of a function file that calculates Fun for a given x.
% FunDer   Name (string) of a function file that calculates the derivative of
%   Fun for a given x.
% Xest   Initial estimate of the solution.
% Err   Maximum error.
% imax   Maximum number of iterations
% Output variable:
% Xs   Solution

for i = 1:imax
    Xi = Xest - feval(Fun,Xest)/feval(FunDer,Xest);   Eq. (3.14).
    if abs((Xi - Xest)/Xest) < Err   Eq. (3.18).
        Xs = Xi;
        break
    end
    Xest = Xi;
end
if i == imax
    fprintf('Solution was not obtained in %i iterations.\n',imax)
    Xs = ('No answer');
end

Figure 3-11: MATLAB function file for solving equation using the Newton's method.

Example 3-2 shows how Eq. (3.14) is used, and how to use the user-defined function `NewtonRoot` to solve a specific problem.

Example 3-2: Solution of equation using Newton's method.

Find the solution of the equation $8 - 4.5(x - \sin x) = 0$ (the same equation as in Example 3-1) by using Newton's method in the following two ways:

(a) Using a nonprogrammable calculator, calculate the first two iterations on paper using six significant figures.

(b) Use MATLAB with the function `NewtonRoot` that is listed in Fig. 3-11. Use 0.0001 for the maximum relative error and 10 for the maximum number of iterations.

In both parts use $x = 2$ as the initial guess of the solution.

SOLUTION

In the present problem $f(x) = 8 - 4.5(x - \sin x)$ and $f'(x) = -4.5(1 - \cos x)$.

(*a*) To start the iterations, $f(x)$ and $f'(x)$ are substituted in Eq. (3.14):

$$x_{i+1} = x_i - \frac{8 - 4.5(x_i - \sin x_i)}{-4.5(1 - \cos x_i)} \tag{3.20}$$

In the first iteration $i = 1$ and $x_1 = 2$, and Eq. (3.20) gives:

$$x_2 = 2 - \frac{8 - 4.5(2 - \sin(2))}{-4.5(1 - \cos(2))} = 2.48517 \tag{3.21}$$

For the second iteration $i = 2$ and $x_2 = 2.48517$, and Eq. (3.20) gives:

$$x_3 = 2.48517 - \frac{8 - 4.5(2.48517 - \sin(2.48517))}{-4.5(1 - \cos(2.48517))} = 2.43099 \tag{3.22}$$

(*b*) To solve the equation with MATLAB using the function `NewtonRoot`, the user must create user-defined functions for $f(x)$ and $f'(x)$. The two functions, called `FunExample2` and `FunDerExample2`, are:

```
function y = FunExample2(x)
y = 8 - 4.5*(x - sin(x));
```

and

```
function y = FunDerExample2(x)
y = -4.5 + 4.5*cos(x);
```

Once the functions are created and saved, the `NewtonRoot` function can be used in the Command Window:

```
>> format long
>> xSolution = NewtonRoot('FunExample2','FunDerExample2',2,0.0001,10)
xSolution =
   2.43046574172363
```

A comparison of the results from parts *a* and *b* shows that the first four digits of the solution (2.430) are obtained in the second iteration. (In part *b* the solution process stops in the fourth iteration, see Problem 3.19.) This shows, as was mentioned before, that Newton's method usually converges fast. In Example 3-1 (bisection method) the first four digits are obtained only after 10 bisections.

Notes on Newton's method

- The method, when successful, works well and converges fast. When it does not converge, it is usually because the starting point is not close enough to the solution. Convergence problems typically occur when the value of $f'(x)$ is close to zero in the vicinity of the solution (where $f(x) = 0$). It is possible to show that Newton's method converges, if the function $f(x)$, and its first and second derivatives $f'(x)$ and $f''(x)$ are all continuous, if $f'(x)$ is not zero at the solution, and if the starting value x_1 is near the actual solution. Illustrations of two cases where Newton's method does not converge (i.e., diverges) are shown in Fig. 3-12.

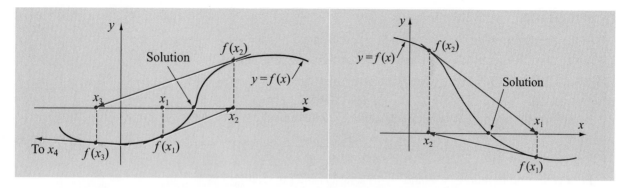

Figure 3-12: Cases where Newton's method diverges.

- A function $f'(x)$, which is the derivative of the function $f(x)$, has to be substituted in the iteration formula, Eq. (3.14). In many cases it is simple to write the derivative, but sometimes it can be difficult to determine. When an expression for the derivative is not available, it might be possible to determine the slope numerically or to find a solution by using the secant method (Section 3.6), which is somewhat similar to Newton's method but does not require an expression for the derivative.

Next, Example 3-3 illustrates the effect that the starting point can have on a numerical solution with Newton's method.

Example 3-3: Convergence of Newton's method.

Find the solution of the equation $\frac{1}{x} - 2 = 0$ by using Newton's method. For the starting point (initial estimate of the solution) use:

(*a*) $x = 1.4$,   (*b*) $x = 1$,   and   (*c*) $x = 0.4$

SOLUTION

The equation can easily be solved analytically, and the exact solution is $x = 0.5$.

For a numerical solution with Newton's method the function, $f(x) = \frac{1}{x} - 2$, and its derivative, $f'(x) = -\frac{1}{x^2}$, are substituted in Eq. (3.14):

$$x_{i+1} = x_i - \frac{f(x_i)}{f'(x_i)} = x_i - \frac{\frac{1}{x_i} - 2}{-\frac{1}{x_i^2}} = 2(x_i - x_i^2) \tag{3.23}$$

(*a*) When the starting point for the iterations is $x_1 = 1.4$, the next two iterations, using Eq. (3.23), are:

$$x_2 = 2(x_1 - x_1^2) = 2(1.4 - 1.4^2) = -1.12 \quad \text{and} \quad x_3 = 2(x_2 - x_2^2) = 2[(-1.12) - (-1.12)^2] = -4.7488$$

These results indicate that Newton's method diverges. This case is illustrated in Fig. 3-13*a*.

(b) When the starting point for the iterations is $x = 1$, the next two iterations, using Eq. (3.23), are:

$$x_2 = 2(x_1 - x_1^2) = 2(1 - 1^2) = 0 \quad \text{and} \quad x_3 = 2(x_2 - x_2^2) = 2(0 - 0^2) = 0$$

From these results it looks like the solution converges to $x = 0$, which is not a solution. At $x = 0$ the function is actually not defined (it is a singular point). A solution is obtained from Eq. (3.23) because the equation was simplified. This case is illustrated in Fig. 3-13b.

(c) When the starting point for the iterations is $x = 0.4$, the next two iterations, using Eq. (3.23), are:

$$x_2 = 2(x_1 - x_1^2) = 2(0.4 - 0.4^2) = 0.48 \quad \text{and} \quad x_3 = 2(x_2 - x_2^2) = 2(0.48 - 0.48^2) = 0.4992$$

In this case Newton's method converges to the correct solution. This case is illustrated in Fig. 3-13c. This example also shows that if the starting point is close enough to the true solution, Newton's method converges.

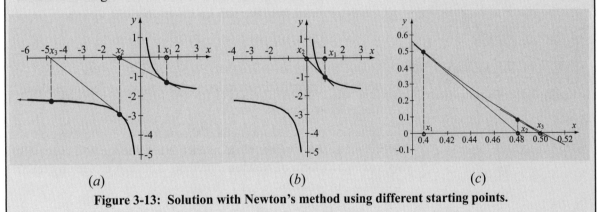

Figure 3-13: Solution with Newton's method using different starting points.

3.6 SECANT METHOD

The secant method is a scheme for finding a numerical solution of an equation of the form $f(x) = 0$. The method uses two points in the neighborhood of the solution to determine a new estimate for the solution (Fig. 3-14). The two points (marked as x_1 and x_2 in the figure) are used to define a straight line (secant line), and the point where the line intersects the x axis (marked as x_3 in the figure) is the new estimate for the solution. As shown, the two points can be on one side of the solution

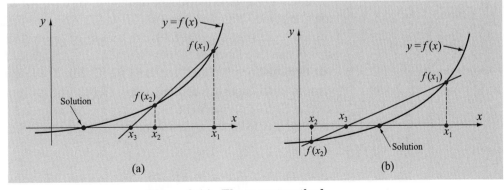

Figure 3-14: The secant method.

(Fig.3-14*a)* or the solution can be between the two points (Fig. 3-14*b*). The slope of the secant line is given by:

$$\frac{f(x_1) - f(x_2)}{x_1 - x_2} = \frac{f(x_2) - 0}{x_2 - x_3} \tag{3.24}$$

which can be solved for x_3:

$$x_3 = x_2 - \frac{f(x_2)(x_1 - x_2)}{f(x_1) - f(x_2)} \tag{3.25}$$

Once point x_3 is determined, it is used together with point x_2 to calculate the next estimate of the solution, x_4. Equation (3.25) can be generalized to an iteration formula in which a new estimate of the solution x_{i+1} is determined from the previous two solutions x_i and x_{i-1}.

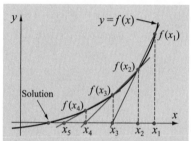

Figure 3-15: Secant method.

$$x_{i+1} = x_i - \frac{f(x_i)(x_{i-1} - x_i)}{f(x_{i-1}) - f(x_i)} \tag{3.26}$$

Figure 3-15 illustrates the iteration process with the secant method.

Relationship to Newton's method

Examination of the secant method shows that when the two points that define the secant line are close to each other, the method is actually an approximated form of Newton's method. This can be seen by rewriting Eq. (3.26) in the form:

$$x_{i+1} = x_i - \frac{f(x_i)}{\dfrac{f(x_{i-1}) - f(x_i)}{(x_{i-1} - x_i)}} \tag{3.27}$$

This equation is almost identical to Eq. (3.14) of Newton's method. In Eq. (3.27) the denominator of the second term on the right-hand side of the equation is an approximation of the value of the derivative of $f(x)$ at x_i. In Eq. (3.14) the denominator is actually the derivative $f'(x_i)$. In the secant method (unlike Newton's method), it is not necessary to know the analytical form of $f'(x)$.

Programming of the secant method is very similar to that of Newton's method. Figure 3-16 lists a MATLAB user-defined function (called `SecantRoot`) that finds the root of $f(x) = 0$. The program consists of one loop in which the next solution `Xi` is calculated from the previous two solutions, `Xb` and `Xa`, using Eq. (3.26). The looping stops if the error is small enough according to Eq. (3.18). The function $f(x)$ (that is used in Eq. (3.26)) has to be supplied as a separate user-defined function. Its name is typed in the argument of `SecantRoot` as a string.

Program 3-3: User-defined function. Secant method.

```
function Xs = SecantRoot(Fun,Xa,Xb,Err,imax)
% SecantRoot finds the root of Fun = 0 using the secant method.
% Input variables:
% Fun   Name (string) of a function file that calculates Fun for a given x.
% a, b  Two points in the neighborhood of the root (on either side or the
%       same side of the root).
% Err   Maximum error.
% imax  Maximum number of iterations
% Output variable:
% Xs    Solution

for i = 1:imax
    FunXb = feval(Fun,Xb);
    Xi = Xb - FunXb*(Xa - Xb)/(feval(Fun,Xa) - FunXb);      Eq. (3.26).
    if abs((Xi - Xb)/Xb) < Err                              Eq. (3.18).
        Xs = Xi;
        break
    end
    Xa = Xb;
    Xb = Xi;
end
if i == imax
    fprintf('Solution was not obtained in %i iterations.\n',imax)
    Xs = ('No answer');
end
```

Figure 3-16: MATLAB function file for solving equation using the secant method.

As an example, the function from Examples 3-1 and 3-2 is solved with the `SecantRoot` user-defined function. The two starting points are taken as $a = 2$ and $b = 3$.

```
>> format long
>> xSolution = SecantRoot('FunExample2',2,3,0.0001,10)
xSolution =
   2.43046572658875
```

The user-defined function `SecantRoot` is also used in the solution of Example 3-4.

3.7 FIXED-POINT ITERATION METHOD

Fixed-point iteration is a method for solving an equation of the form $f(x) = 0$. The method is carried out by rewriting the equation in the form:

$$x = g(x) \tag{3.28}$$

Obviously, when x is the solution of $f(x) = 0$, the left side and the right side of Eq. (3.28) are equal. This is illustrated graphically by plotting $y = x$ and $y = g(x)$, as shown in Fig. 3-17. The point of intersection of the two plots, called the ***fixed point***, is the solution. The numerical value of the solution is determined by an iterative process. It starts by taking a value of x near the fixed point as the first guess for the solution and substituting it in $g(x)$. The value of $g(x)$ that is obtained is the new (second) estimate for the solution. The second value is then substituted back in $g(x)$, which then gives the third estimate of the solution. The iteration formula is thus given by:

$$x_{i+1} = g(x_i) \tag{3.29}$$

The function $g(x)$ is called the ***iteration function***.

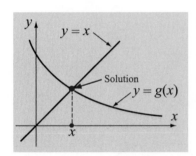

Figure 3-17: Fixed-point iteration method.

- When the method works, the values of x that are obtained are successive iterations that progressively converge toward the solution. Two such cases are illustrated graphically in Fig. 3-18. The solution process starts by choosing point x_1 on the x axis and drawing a vertical line that intersects the curve $y = g(x)$ at point $g(x_1)$. Since $x_2 = g(x_1)$, a horizontal line is drawn from point $(x_1, g(x_1))$ toward the line $y = x$. The intersection point gives the location of x_2. From x_2 a vertical line is drawn toward the curve $y = g(x)$. The intersection point is now $(x_2, g(x_2))$, and $g(x_2)$ is also the value of x_3. From point $(x_2, g(x_2))$ a horizontal line is drawn again toward $y = x$, and

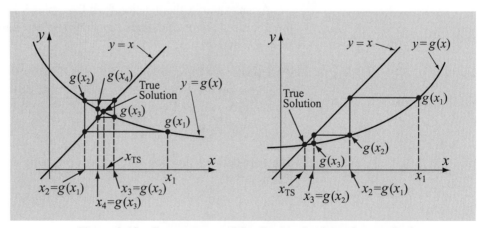

Figure 3-18: Convergence of the fixed-point iteration method.

the intersection point gives the location of x_3. As the process continues the intersection points converge toward the fixed point, or the true solution x_{TS}.

- It is possible, however, that the iterations will not converge toward the fixed point, but rather diverge away. This is shown in Fig. 3-19. The figure shows that even though the starting point is close to the solution, the subsequent points are moving further away from the solution.

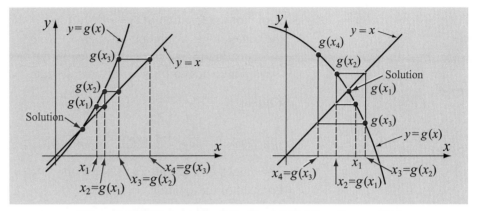

Figure 3-19: Divergence of the fixed-point iteration method.

- Sometimes, the form $f(x) = 0$ does not lend itself to deriving an iteration formula of the form $x = g(x)$. In such a case, one can always add and subtract x to $f(x)$ to obtain $x + f(x) - x = 0$. The last equation can be rewritten in the form that can be used in the fixed-point iteration method:

$$x = x + f(x) = g(x)$$

Choosing the appropriate iteration function g(x)

For a given equation $f(x) = 0$, the iteration function is not unique since it is possible to change the equation into the form $x = g(x)$ in different ways. This means that several iteration functions $g(x)$ can be written for the same equation. A $g(x)$ that should be used in Eq. (3.29) for the iteration process is one for which the iterations converge toward the solution. There might be more than one form that can be used, or it may be that none of the forms are appropriate so that the fixed-point iteration method cannot be used to solve the equation. In cases where there are multiple solutions, one iteration function may yield one root, while a different function yields other roots. Actually, it is possible to determine ahead of time if the iterations converge or diverge for a specific $g(x)$.

The fixed-point iteration method converges if in the neighborhood of the fixed point the derivative of $g(x)$ has an absolute value that is smaller than 1 (also called Lipschitz continuous):

$$|g'(x)| < 1 \tag{3.30}$$

As an example, consider the equation:

$$xe^{0.5x} + 1.2x - 5 = 0 \tag{3.31}$$

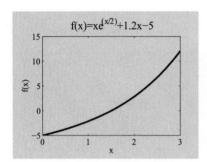

$f(x)=xe^{(x/2)}+1.2x-5$

Figure 3-20: A plot of
$f(x) = xe^{x/2} + 1.2x - 5$.

A plot of the function $f(x) = xe^{0.5x} + 1.2x - 5$ (see Fig. 3-20) shows that the equation has a solution between $x = 1$ and $x = 2$.

Equation (3.31) can be rewritten in the form $x = g(x)$ in different ways. Three possibilities are discussed next.

Case a:
$$x = \frac{5 - xe^{0.5x}}{1.2} \tag{3.32}$$

In this case $g(x) = \dfrac{5 - xe^{0.5x}}{1.2}$ and $g'(x) = -(e^{0.5x} + 0.5xe^{0.5x})/1.2$.

The values of $g'(x)$ at points $x = 1$ and $x = 2$, which are in the neighborhood of the solution, are:

$$g'(1) = -(e^{0.5 \cdot 1} + 0.5 \cdot 1e^{0.5 \cdot 1})/1.2 = -2.0609$$

$$g'(2) = -(e^{0.5 \cdot 2} + 0.5 \cdot 2e^{0.5 \cdot 2})/1.2 = -4.5305$$

Case b:
$$x = \frac{5}{e^{0.5x} + 1.2} \tag{3.33}$$

In this case $g(x) = \dfrac{5}{e^{0.5x} + 1.2}$ and $g'(x) = \dfrac{-5e^{0.5x}}{2(e^{0.5x} + 1.2)^2}$.

The values of $g'(x)$ at points $x = 1$ and $x = 2$, which are in the neighborhood of the solution, are:

$$g'(1) = \frac{-5e^{0.5 \cdot 1}}{2(e^{0.5 \cdot 1} + 1.2)^2} = -0.5079$$

$$g'(2) = \frac{-5e^{0.5 \cdot 2}}{2(e^{0.5 \cdot 2} + 1.2)^2} = -0.4426$$

Case c:
$$x = \frac{5 - 1.2x}{e^{0.5x}} \tag{3.34}$$

In this case $g(x) = \dfrac{5 - 1.2x}{e^{0.5x}}$ and $g'(x) = \dfrac{-3.7 + 0.6x}{e^{0.5x}}$.

The values of $g'(x)$ at points $x = 1$ and $x = 2$, which are in the neighborhood of the solution, are:

$$g'(1) = \frac{-3.7 + 0.6 \cdot 1}{e^{0.5 \cdot 1}} = -1.8802$$

$$g'(2) = \frac{-3.7 + 0.6 \cdot 2}{e^{0.5 \cdot 2}} = -0.9197$$

These results show that the iteration function $g(x)$ from Case b is the one that should be used since in this case $|g'(1)| < 1$ and $|g'(2)| < 1$. Substituting $g(x)$ from Case b in Eq. (3.29) gives:

$$x_{i+1} = \frac{5}{e^{0.5x_i} + 1.2} \tag{3.35}$$

Starting with $x_1 = 1$, the first few iterations are:

$$x_2 = \frac{5}{e^{0.5 \cdot 1} + 1.2} = 1.7552 \qquad\qquad x_3 = \frac{5}{e^{0.5 \cdot 1.7552} + 1.2} = 1.3869$$

$$x_4 = \frac{5}{e^{0.5 \cdot 1.3869} + 1.2} = 1.5622 \qquad\qquad x_5 = \frac{5}{e^{0.5 \cdot 1.5622} + 1.2} = 1.4776$$

$$x_6 = \frac{5}{e^{0.5 \cdot 1.4776} + 1.2} = 1.5182 \qquad\qquad x_7 = \frac{5}{e^{0.5 \cdot 1.5182} + 1.2} = 1.4986$$

As expected, the values calculated in the iterations are converging toward the actual solution, which is $x = 1.5050$.

On the contrary, if the function $g(x)$ from Case a is used in the iteration, the first few iterations are:

$$x_2 = \frac{5 - 1e^{0.5 \cdot 1}}{1.2} = 2.7927 \qquad\qquad x_3 = \frac{5 - 2.7927e^{0.5 \cdot 2.7927}}{1.2} = -5.2364$$

$$x_4 = \frac{5 - (-5.2364)e^{0.5 \cdot (-5.2364)}}{1.2} = 4.4849$$

$$x_5 = \frac{5 - 4.4849e^{0.5 \cdot 4.4849}}{1.2} = -31.0262$$

In this case the iterations give values that diverge from the solution.

When should the iterations be stopped?

The true error (the difference between the true solution and the estimated solution) cannot be calculated since the true solution in general is not known. As with Newton's method, the iterations can be stopped either when the relative error or the tolerance in $f(x)$ is smaller than some predetermined value (Eqs. (3.18) or (3.19)).

3.8 USE OF MATLAB BUILT-IN FUNCTIONS FOR SOLVING NONLINEAR EQUATIONS

MATLAB has two built-in functions for solving equations with one variable. The `fzero` command can be used to find a root of any equation, and the `roots` command can be used for finding the roots of a polynomial.

3.8.1 The *fzero* Command

The fzero command can be used to solve an equation (in the form $f(x) = 0$) with one variable. The user needs to know approximately where the solution is, or if there are multiple solutions, which one is desired. The form of the command is:

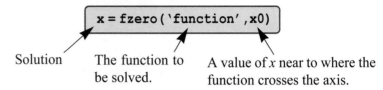

x = fzero('function',x0)

Solution   The function to   A value of x near to where the
                 be solved.   function crosses the axis.

- x is the solution, which is a scalar.

- 'function' is the function whose root is desired. It can be entered in three different ways:
  1. The simplest way is to enter the mathematical expression as a string.
  2. The function can also be supplied as a user-defined function, and the name of the function is typed as a string.
  3. The function can be defined as an inline function, and its name is then typed as a string.

- The function has to be written in a standard form. For example, if the function to be solved is $xe^{-x} = 0.2$, it has to be written as $f(x) = xe^{-x} - 0.2 = 0$. If this function is entered into the fzero command as a string, it is typed as: 'x*exp(-x)-0.2'.

- When a function is entered as a string, it cannot include predefined variables. For example, if the function to be entered is $f(x) = xe^{-x} - 0.2$, it is not possible to first define b=0.2 and then enter 'x*exp(-x)-b'.

- x0 can be a scalar or a two-element vector. If it is entered as a scalar, it has to be a value of x near the point where the function crosses the x axis. If x0 is entered as a vector, the two elements have to be points on opposite sides of the solution such that $f(x0(1))$ has a different sign than $f(x0(2))$. When a function has more than one solution, each solution can be determined separately by using the fzero function and entering values for x0 that are near each of the solutions.

Usage of the fzero command is illustrated next for solving the equation in Examples 3-1 and 3-2:

```
>> format long
>> Sol=fzero('8 - 4.5*(x - sin(x))',2)      Solve the equation from Example 3-1.
Sol =
   2.43046574172363
```

3.8.2 *The* `roots` *Command*

The `roots` command can be used to find the roots of a polynomial. The form of the command is:

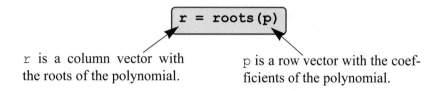

r is a column vector with the roots of the polynomial.

p is a row vector with the coefficients of the polynomial.

3.9 EQUATIONS WITH MULTIPLE SOLUTIONS

Many nonlinear equations of the form $f(x) = 0$ have multiple solutions or roots. As an example, consider the following equation:

$$f(x) = \cos(x)\cosh(x) + 1 \tag{3.36}$$

A plot of this function using MATLAB is shown in Fig. 3-21 over the interval $[0, 5]$. As can be seen, the function has zero crossings between $x = 1$ and $x = 2$, and between $x = 4$ and $x = 5$. Existence of multiple roots is typical of nonlinear equations. A general strategy for finding the roots in the interval $[0, 5]$ is:

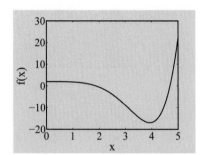

Figure 3-21: A plot of Eq. (3.36).

- Determine the approximate location of the roots by defining smaller intervals over which the roots exist. This can be done by plotting the function (as shown in Fig. 3-21) or by evaluating the function over a set of successive points and looking for sign changes of the function.

- Apply any of the methods described in Sections 3.3 through 3.7 over a restricted subinterval. For example, the first root that is contained within the interval $[1, 2]$ can be found by the bisection method or a similar method with $a = 1$ and $b = 2$. Alternatively, a starting value or initial guess can be used with Newton's method or fixed-point iteration method to determine the root. The `fzero` MATLAB built-in function can also be used to find the root. The process can then be repeated over the next interval $[4, 5]$ to find the next root.

The next example presents the solution of the function in Eq. (3.36) in a practical situation.

Example 3-4: Solution of equation with multiple roots.

The natural frequencies, ω_n, of free vibration of a cantilever beam are determined from the roots of the equation:

$$f(k_nL) = \cos(k_nL)\cosh(k_nL) + 1 = 0 \qquad (3.37)$$

where L is the length of the beam and the frequency ω_n is given by:

$$\omega_n = (k_nL)^2 \sqrt{\frac{EI}{\rho AL^4}}$$

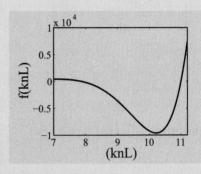

in which E is the elastic modulus, I is the moment of inertia, A is the cross-sectional area, and ρ is the density per unit length.

(a) Determine the value of the first root by defining smaller intervals over which the roots exist and using the secant method.

(b) Write a MATLAB program in a script file that determines the value of k_nL for the first four roots.

SOLUTION

Equation (3.37) is identical to Eq. (3.36), and a plot that shows the location of the first two roots is presented in Fig. 3-21. The location of the next two roots is shown in the figure on the right where the function is plotted over the interval $[7, 11.2]$. It shows that the third root is around 8 and the fourth root is near 11.

(a) The value of the first root is determined by using the SecantRoot user-defined function that is listed in Fig. 3-16. First, however, a user-defined function for the function $f(k_nL)$ in Eq. (3.37) is written (the function name is FunExample3):

```
function y = FunExample3(x)
y = cos(x)*cosh(x) + 1;
```

The first root is between 1 and 2. To find its solution numerically, the user-defined function SecantRoot (listed in Fig. 16) is used with $a = 1$, $b = 2$, $Err = 0.0001$, and $imax = 10$:

```
>> FirstSolution = SecantRoot('FunExample3',1,2,0.0001,10)
FirstSolution =
   1.87510406460241
```

format long is used in MATLAB.

(b) Next, a MATLAB program that automatically finds the four roots is written. The program evaluates the function over a set of successive intervals and looks for sign changes. It starts at $k_nL = 0$ and uses an increment of 0.2 up to a value of $k_nL = 11.2$. If a change in sign is detected, the root within that interval is determined by using MATLAB's built-in fzero function.

```
clear all
F = inline('cos(x)*cosh(x)+1');
Inc = 0.2;
```

```
i = 1;
KnLa = 0;                                    Define the left point of the first increment.
KnLb = KnLa + Inc;                           Define the right point of the first increment.
while KnLb <= 11.2
    if F(KnLa)*F(KnLb) < 0                   Check for a sign change in the value of the function.
        Roots(i) = fzero(F,[KnLa,KnLb]);     Determine the root within the interval if a sign change was detected.
        i = i + 1;
    end
    KnLa = KnLb;                             Define the left point of the next increment.
    KnLb = KnLa + Inc;                       Define the right point of the next increment.
end
Roots
```

When the program is executed, the display in the Command Window is:

Roots =

  1.87510406871196   4.69409113297418   7.85475743823761   10.99554073487547

These are the values of the first four roots.

3.10 SYSTEMS OF NONLINEAR EQUATIONS

A system of nonlinear equations consists of two, or more, nonlinear equations that have to be solved simultaneously. For example, Fig. 3-22 shows a catenary (hanging cable) curve given by the equation $y = \frac{1}{2}(e^{x/2} + e^{(-x)/2})$ and an ellipse specified by the equation $\frac{x^2}{5^2} + \frac{y^2}{3^2} = 1$. The point of intersection between the two curves is given by the solution of the following system of nonlinear equations:

$$f_1(x, y) = y - \frac{1}{2}(e^{x/2} + e^{(-x)/2}) = 0 \qquad (3.38)$$

$$f_2(x, y) = 9x^2 + 25y^2 - 225 = 0 \qquad (3.39)$$

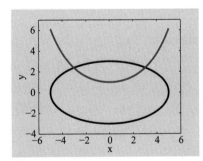

Figure 3-22: A plot of Eq. (3.38) and Eq. (3.39).

Analysis of many problems in science and engineering requires solution of systems of nonlinear equations. In addition, as shown in Chapter 9, one of the popular numerical methods for solving nonlinear ordinary differential equations (the finite difference method) requires the solution of a system of nonlinear algebraic equations.

In this section, two methods for solving systems of nonlinear equations are presented. The Newton method (also called the Newton–Raphson method), suitable for solving small systems, is described in Section 3.10.1. The fixed-point iteration method, which can also be used for solving large systems, is discussed in Section 3.10.2.

3.10.1 Newton's Method for Solving a System of Nonlinear Equations

Newton's method for solving a system of nonlinear equations is an extension of the method used for solving a single equation (Section 3.5). The method is first derived in detail for the solution of a system of two nonlinear equations. Subsequently, a general formulation is presented for the case of a system of n nonlinear equations.

Solving a system of two nonlinear equations

A system of two equations with two unknowns x and y can be written as:

$$f_1(x, y) = 0$$
$$f_2(x, y) = 0 \tag{3.40}$$

The solution process starts by choosing an estimated solution x_1 and y_1. If x_2 and y_2 are the true (unknown) solutions of the system and are sufficiently close to x_1 and y_1, then the value of f_1 and f_2 at x_2 and y_2 can be expressed using a Taylor series expansion of the functions $f_1(x, y)$ and $f_2(x, y)$ about (x_1, y_1) (see Section 2.7.2):

$$f_1(x_2, y_2) = f_1(x_1, y_1) + (x_2 - x_1)\frac{\partial f_1}{\partial x}\bigg|_{x_1, y_1} + (y_2 - y_1)\frac{\partial f_1}{\partial y}\bigg|_{x_1, y_1} + \dots \tag{3.41}$$

$$f_2(x_2, y_2) = f_2(x_1, y_1) + (x_2 - x_1)\frac{\partial f_2}{\partial x}\bigg|_{x_1, y_1} + (y_2 - y_1)\frac{\partial f_2}{\partial y}\bigg|_{x_1, y_1} + \dots \tag{3.42}$$

Since x_2 and y_2 are close to x_1 and y_1, approximate values for $f_1(x_2, y_2)$ and $f_2(x_2, y_2)$ can be calculated by neglecting the higher-order terms. Also, since $f_1(x_2, y_2) = 0$ and $f_2(x_2, y_2) = 0$, Eqs. (3.41) and (3.42) can be rewritten as:

$$\frac{\partial f_1}{\partial x}\bigg|_{x_1, y_1} \Delta x + \frac{\partial f_1}{\partial y}\bigg|_{x_1, y_1} \Delta y = -f_1(x_1, y_1) \tag{3.43}$$

$$\frac{\partial f_2}{\partial x}\bigg|_{x_1, y_1} \Delta x + \frac{\partial f_2}{\partial y}\bigg|_{x_1, y_1} \Delta y = -f_2(x_1, y_1) \tag{3.44}$$

where $\Delta x = x_2 - x_1$ and $\Delta y = y_2 - y_1$. Since all the terms in Eqs. (3.43) and (3.44) are known, except the unknowns Δx and Δy, these equations are a system of two linear equations. The system can be solved by using Cramer's rule (see Section 2.4.6):

$$\Delta x = \frac{-f_1(x_1, y_1)\dfrac{\partial f_2}{\partial y}\bigg|_{x_1, y_1} + f_2(x_1, y_1)\dfrac{\partial f_1}{\partial y}\bigg|_{x_1, y_1}}{J(f_1(x_1, y_1), f_2(x_1, y_1))} \tag{3.45}$$

$$\Delta y = \frac{-f_2(x_1, y_1)\frac{\partial f_1(x_1, y_1)}{\partial x}\bigg|_{x_1, y_1} + f_1(x_1, y_1)\frac{\partial f_2}{\partial x}\bigg|_{x_1, y_1}}{J(f_1(x_1, y_1), f_2(x_1, y_1))} \quad (3.46)$$

where

$$J(f_1, f_2) = det\begin{bmatrix} \dfrac{\partial f_1}{\partial x} & \dfrac{\partial f_1}{\partial y} \\ \dfrac{\partial f_2}{\partial x} & \dfrac{\partial f_2}{\partial y} \end{bmatrix} \quad (3.47)$$

is the Jacobian (see Section 2.6.3). Once Δx and Δy are known, the value of x_2 and y_2 is calculated by:

$$x_2 = x_1 + \Delta x$$
$$y_2 = y_1 + \Delta y \quad (3.48)$$

Obviously, the values of x_2 and y_2 that are obtained are not the true solution since the higher-order terms in Eqs. (3.41) and (3.42) were neglected. Nevertheless, these values are expected to be closer to the true solution than x_1 and y_1.

The solution process continues by using x_2 and y_2 as the new estimate for the solution and using Eqs. (3.43) and (3.44) to determine new Δx and Δy that give x_3 and y_3. The iterations continue until two successive answers differ by an amount smaller than a desired value.

An application of Newton's method is illustrated in Example 3-5 where the intersection point between the catenary curve and the ellipse in Fig. 3-22 is determined.

Example 3-5: Solution of a system of nonlinear equations using Newton's method.

The equations of the catenary curve and the ellipse, which are shown in the figure, are given by:

$$f_1(x, y) = y - \frac{1}{2}(e^{x/2} + e^{(-x)/2}) = 0 \quad (3.49)$$

$$f_2(x, y) = 9x^2 + 25y^2 - 225 = 0 \quad (3.50)$$

Use Newton's method to determine the point of intersection of the curves that resides in the first quadrant of the coordinate system.

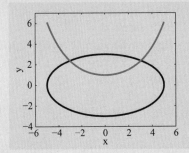

SOLUTION

Equations (3.49) and (3.50) are a system of two nonlinear equations. The points of intersection are given by the solution of the system. The solution with Newton's method is obtained by using Eqs. (3.43) and (3.44). In the present problem, the partial derivatives in the equations are given by:

$$\frac{\partial f_1}{\partial x} = -\frac{1}{4}(e^{x/2} - e^{(-x)/2}) \quad \text{and} \quad \frac{\partial f_1}{\partial y} = 1 \quad (3.51)$$

$$\frac{\partial f_2}{\partial x} = 18x \quad \text{and} \quad \frac{\partial f_2}{\partial y} = 50y \tag{3.52}$$

The Jacobian is given by:

$$J(f_1, f_2) = det\begin{bmatrix} \frac{\partial f_1}{\partial x} & \frac{\partial f_1}{\partial y} \\ \frac{\partial f_2}{\partial x} & \frac{\partial f_2}{\partial y} \end{bmatrix} = det\begin{bmatrix} -\frac{1}{4}(e^{x/2} - e^{(-x)/2}) & 1 \\ 18x & 50y \end{bmatrix} = -\frac{1}{4}(e^{x/2} - e^{(-x)/2})50y - 18x \tag{3.53}$$

Substituting Eqs. (3.51)–(3.53) in Eqs. (3.45) and (3.46) gives the solution for Δx and Δy.

The problem is solved in the MATLAB program that is listed below. The order of operations in the program is:

- The solution is initiated by the initial guess, $x_i = 2.5$, $y_i = 2.0$.
- The iterations start. Δy and Δx, are determined by substituting x_i and y_i in Eqs. (3.45) and (3.46).
- $x_{i+1} = x_i + \Delta x$, and $y_{i+1} = y_i + \Delta y$ are determined.
- If the estimated relative error (Eq. (3.9)) for both variables is smaller than 0.001, the iterations stop. Otherwise, the values of x_{i+1}, and y_{i+1} are assigned to x_i, and y_i, respectively, and the next iteration starts.

The program also displays the solution and the error at each iteration.

```
%  Solution of Chapter 3 Example 5
F1 = inline('y - 0.5*(exp(x/2) + exp(-x/2))');
F2 = inline('9*x^2 + 25*y^2 - 225');
F1x = inline('-(exp(x/2) - exp(-x/2))/4');
F2x = inline('18*x');
F2y = inline('50*y');
Jacob = inline('-(exp(x/2) - exp(-x/2))/4*50*y - 18*x');
xi = 2.5; yi = 2; Err = 0.001;            Assign the initial estimate of the solution.
for i = 1:5                                Start the iterations.
   Jac = Jacob(xi,yi);
   Delx = (-F1(xi,yi)*F2y(yi) + F2(xi,yi))/Jac;    Calculate Δx and Δy with
   Dely = (-F2(xi,yi)*F1x(xi) + F1(xi,yi)*F2x(xi))/Jac;   Eqs. (3.45) and (3.46).
   xip1 = xi + Delx;
   yip1 = yi + Dely;                       Calculate x_{i+1} and y_{i+1}.
   Errx = abs((xip1 - xi)/xi);
   Erry = abs((yip1 - yi)/yi);
   fprintf('i =%2.0f x = %-7.4f y = %-7.4f Error in x = %-7.4f Error in y = %-7.4f\n',i,xip1,yip1,Errx,Erry)
   if Errx < Err & Erry < Err
      break
   else
      xi = xip1; yi = yip1;     If the error is not small enough, assign x_{i+1} to x_i, and y_{i+1} to y_i.
   end
end
```

When the program is executed, the display in the Command Window is:

i = 1 x = 3.1388 y = 2.4001 Error in x = 0.25551 Error in y = 0.20003
i = 2 x = 3.0339 y = 2.3855 Error in x = 0.03340 Error in y = 0.00607
i = 3 x = 3.0312 y = 2.3859 Error in x = 0.00091 Error in y = 0.00016

These results show that the values converge quickly to the solution.

Solving a system of n nonlinear equations

Newton's method can easily be generalized to the case of a system of n nonlinear equations. With n unknowns, $x_1, x_2, ..., x_n$, a system of n simultaneous nonlinear equations has the form:

$$f_1(x_1, x_2, ..., x_n) = 0$$
$$f_2(x_1, x_2, ..., x_n) = 0$$
$$...$$
$$f_n(x_1, x_2, ..., x_n) = 0$$

(3.54)

The value of the functions at the next approximation of the solution, $x_{1, i+1}, x_{2, i+1}, ..., x_{n, i+1}$ is then obtained using a Taylor series expansion about the current value of the approximation of the solution, $x_{1, i}, x_{2, i}, ..., x_{n, i}$. Following the same procedure that led to Eqs. (3.43) and (3.44) results in the following system of n linear equations for the unknowns $\Delta x_1, \Delta x_2, ..., \Delta x_n$:

$$\begin{bmatrix} \dfrac{\partial f_1}{\partial x_1} & \dfrac{\partial f_1}{\partial x_2} & \cdots & \dfrac{\partial f_1}{\partial x_n} \\ \dfrac{\partial f_2}{\partial x_1} & \dfrac{\partial f_2}{\partial x_2} & \cdots & \dfrac{\partial f_2}{\partial x_n} \\ \cdots & \cdots & \cdots & \cdots \\ \dfrac{\partial f_n}{\partial x_1} & \dfrac{\partial f_n}{\partial x_2} & \cdots & \dfrac{\partial f_n}{\partial x_n} \end{bmatrix} \begin{bmatrix} \Delta x_1 \\ \Delta x_2 \\ \cdots \\ \Delta x_n \end{bmatrix} = \begin{bmatrix} -f_1 \\ -f_2 \\ \cdots \\ -f_n \end{bmatrix}$$

(3.55)

(The determinant of the matrix of the partial derivatives of the functions on the left-hand side of the equation is called the Jacobian, Section 2.6.3.) Once the system in Eq. (3.55) is solved, the new approximate solution is obtained from:

$$x_{1, i+1} = x_{1, i} + \Delta x_1$$
$$x_{2, i+1} = x_{2, i} + \Delta x_2$$
$$...$$
$$x_{n, i+1} = x_{n, i} + \Delta x_n$$

(3.56)

As with Newton's method for a single nonlinear equation, convergence is not guaranteed. Newton's iterative procedure for solving a system of nonlinear equations will likely converge provided the following three conditions are met:

(*i*) The functions $f_1, f_2, ..., f_n$ and their derivatives must be continuous and bounded near the solution (root).

(*ii*) The Jacobian must be nonzero, that is, $J(f_1, f_2, ..., f_n) \neq 0$, near the solution.

(*iii*) The initial estimate (guess) of the solution must be sufficiently close to the true solution.

Newton's method for solving a system of n nonlinear equations is summarized in the following algorithm.

Algorithm for Newton's method for solving a system of nonlinear equations

Given a system of n nonlinear equations,

1. Estimate (guess) an initial solution, $x_{1,i}, x_{2,i}, ..., x_{n,i}$.
2. Calculate the Jacobian and the value of the f s on the right-hand side of Eq. (3.55).
3. Solve Eq. (3.55) for $\Delta x_1, \Delta x_2, ..., \Delta x_n$.
4. Calculate a new estimate of the solution, $x_{1,i+1}, x_{2,i+1}, ..., x_{n,i+1}$, using Eq. (3.56).
5. Calculate the error. If the new solution is not accurate enough, assign the values of $x_{1,i+1}, x_{2,i+1}, ..., x_{n,i+1}$ to $x_{1,i}, x_{2,i}, ..., x_{n,i}$, and start a new iteration beginning with Step 2.

Additional comments on Newton's method for solving a system of nonlinear equations

- The method, when successful, converges fast. When it does not converge, it is usually because the initial guess is not close enough to the solution.

- The partial derivatives (the elements of the Jacobian matrix) have to be determined. This can be done analytically or numerically (numerical differentiation is covered in Chapter 6). However, for a large system of equations the determination of the Jacobian might be difficult.

- When the system of equations consists of more than three equations, the solution of Eq. (3.55) has to be done numerically. Methods for solving systems of linear equations are described in Chapter 4.

3.10.2 Fixed-Point Iteration Method for Solving a System of Nonlinear Equations

The fixed-point iteration method discussed in Section 3.7 for solving a single nonlinear equation can be extended to the case of a system of nonlinear equations. A system of n nonlinear equations with the unknowns, $x_1, x_2, ..., x_n$, has the form:

$$f_1(x_1, x_2, ..., x_n) = 0$$

$$f_2(x_1, x_2, ..., x_n) = 0$$

$$...$$

$$f_n(x_1, x_2, ..., x_n) = 0$$

(3.57)

The system can be rewritten in the form:

$$x_1 = g_1(x_1, x_2, ..., x_n)$$

$$x_2 = g_2(x_1, x_2, ..., x_n)$$

$$...$$

$$x_n = g_n(x_1, x_2, ..., x_n)$$

(3.58)

where the gs are the iteration functions. The solution process starts by guessing a solution, $x_{1,1}, x_{2,1}, ..., x_{n,1}$ which is substituted on the right-hand side of Eqs. (3.58). The values that are calculated by Eqs. (3.58) are the new (second) estimate of the solution, $x_{1,2}, x_{2,i+1}, ..., x_{n,i+1}$. The new estimate is substituted back on the right-hand side of Eqs. (3.58) to give a new solution, and so on. When the method works, the new estimates of the solution converge toward the true solution. In this case, the process is continued until the desired accuracy is achieved. For example, the estimated relative error is calculated for each of the variables, and the iterations are stopped when the largest relative error is smaller than a specified value.

Convergence of the method depends on the form of the iteration functions. For a given problem there are many possible forms of iteration functions since rewriting Eqs. (3.57) in the form of Eqs. (3.58) is not unique. In general, several forms might be appropriate for one solution, or in the case where several solutions exist, different iteration functions need to be used to find the multiple solutions. When using the fixed-point iteration method, one can try various forms of iteration functions, or it may be possible in some cases to determine ahead of time if the solution will converge for a specific choice of gs.

The fixed-point iteration method applied to a set of simultaneous nonlinear equations will converge under the following sufficient (but not necessary) conditions:

(i) $g_1, ..., g_n, \dfrac{\partial g_1}{\partial x_1}, ..., \dfrac{\partial g_1}{\partial x_n}, \dfrac{\partial g_2}{\partial x_1}, ..., \dfrac{\partial g_2}{\partial x_n}, ..., \dfrac{\partial g_n}{\partial x_n}$ are continuous in the

neighborhood of the solution.

$$\left|\frac{\partial g_1}{\partial x_1}\right| + \left|\frac{\partial g_1}{\partial x_2}\right| + \ldots + \left|\frac{\partial g_1}{\partial x_n}\right| \le 1$$

$$\left|\frac{\partial g_2}{\partial x_1}\right| + \left|\frac{\partial g_2}{\partial x_2}\right| + \ldots + \left|\frac{\partial g_2}{\partial x_n}\right| \le 1$$

(ii)

$$\ldots$$

$$\left|\frac{\partial g_n}{\partial x_1}\right| + \left|\frac{\partial g_n}{\partial x_2}\right| + \ldots + \left|\frac{\partial g_n}{\partial x_n}\right| \le 1$$

(iii) The initial guess, $x_{1,1}, x_{2,1}, \ldots, x_{n,1}$, is sufficiently close to the solution.

3.11 PROBLEMS

Problems to be solved by hand
Solve the following problems by hand. When needed, use a calculator, or write a MATLAB script file to carry out the calculations. If using MATLAB, do not use built-in functions for solving nonlinear equations.

3.1   The tolerance, *tol*, of the solution in the bisection method is given by $tol = \frac{1}{2}(b_n - a_n)$, where a_n and b_n are the endpoints of the interval after the *n*th iteration. The number of iterations *n* that are required for obtaining a solution with a tolerance that is equal to or smaller than a specified tolerance can be determined before the solution is calculated. Show that *n* is given by:

$$n \ge \frac{\log(b - a) - \log(tol)}{\log 2}$$

where *a* and *b* are the endpoints of the starting interval and *tol* is a user-specified tolerance.

3.2   Determine the root of $f(x) = x^2 - e^{-x}$ by:
(a) Using the bisection method. Start with $a = 0$ and $b = 1$, and carry out the first five iterations.
(b) Using the secant method. Start with the two points, $x_1 = 0$, and $x_2 = 1$, and carry out the first five iterations.
(c) Using Newton's method. Start at $x_1 = 0$ and carry out the first five iterations.

3.3   The location $\bar{x}$ of the centroid of a circular sector is given by:

$$\bar{x} = \frac{2r\sin\theta}{3\theta}$$

Determine the angle θ for which $\bar{x} = \frac{r}{2}$.

First, derive the equation that must be solved and then determine the root using the following methods:
(a) Use the bisection method. Start with $a = 1$ and $b = 2$, and carry out the first five iterations.
(b) Use the secant method. Start with the two points $x_1 = 1$ and $x_2 = 2$, and carry out the first five iterations.
(c) Use Newton's method. Start at $x_1 = 1$ and carry out the first five iterations.

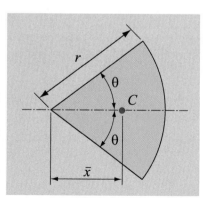

3.4   The lateral surface area, S, of a cone is given by:
$$S = \pi r \sqrt{r^2 + h^2}$$
where r is the radius of the base and h is the height. Determine the radius of a cone which has a surface area of 1200 m^2 and a height of 20 m, by calculating the first five iterations using the fixed-point iteration method. Use $r = S/(\pi \sqrt{r^2 + h^2})$ as the iteration function. Start with $r = 17$ m.

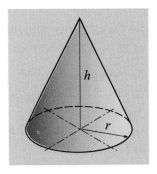

3.5   Determine the cube root of 155 by finding the numerical solution of the equation $x^3 - 155 = 0$. Use Newton's method. Start at $x = 155$ and carry out the first five iterations.

3.6   Determine the positive root of the polynomial $x^3 + 3.8x^2 - 8.6x - 24.4$.
(a) Plot the polynomial and choose a point near the root for the first estimate of the solution. Using Newton's method determine the approximate solution in the first five iterations. Start at $x = 2$.
(b) From the plot in part (a), choose two points near the root to start the solution process with the secant method. Determine the approximate solution in the first five iterations.

3.7   Find all the roots of the equation $x^3 + 12x^2 - 100x - 6 = 0$ using the bisection method (five iterations for each root).

3.8   Find all the roots of the equation $x^3 - 2.2x^2 - 2.15x + 5.1 = 0$ using the bisection method (five iterations for each root).

3.9   The equation $f(x) = -x^{1/3} + 0.5x^2 - 2 = 0$ has a root between $x = 2$ and $x = 3$. To find the root by using the fixed-point iteration method, the equation has to be written in the form $x = g(x)$. Derive two possible forms for $g(x)$ — one by solving for x from the first term of the equation and the next by solving for x from the second term of the equation.
(a) Determine which form should be used according to the condition in Eq. (3.30).
(b) Carry out the first five iterations using both forms of $g(x)$ to confirm your determination in part (a).

3.10   Find four different iteration functions for solving the equation $x^2 - 3x + e^x - 2 = 0$ by the fixed-point iteration method.

3.11   Find three different iteration functions for solving the equation $x - \tan x = 0$ by the fixed-point iteration method.

3.12   Solve the following system of nonlinear equations:
$$4x^2 - y^3 + 28 = 0$$
$$3x^3 + 4y^2 - 145 = 0$$
(a) Use Newton's method. Start at $x = 1$, $y = 1$, and carry out the first five iterations.
(b) Use the fixed-point iteration method. Use the iteration functions $y = (4x^2 + 28)^{1/3}$ and $x = \left[\dfrac{(145 - 4y^2)}{3} \right]^{1/3}$. Start at $x = 1$, $y = 1$, and carry out the first five iterations.

Problems to be programmed in MATLAB

Solve the following problems using the MATLAB environment. Do not use MATLAB's built-in functions for solving nonlinear equations.

3.13 In the program of Example 3-1 the iterations are executed in the for-end loop. In the loop, the inline function F is used twice (once in the command `FxNS = F(xNS)` and once in the command `if F(a)*FxNS < 0`). Rewrite the program such that the inline function F is used inside the loop only once. Execute the new program and show that the output is the same as in the example.

3.14 Write a MATLAB user-defined function that solves for a root of a nonlinear equation $f(x) = 0$ with the bisection method. Name the function `Xs = BisectionRoot(Fun, a, b, TolMax)`. The output argument `Xs` is the solution. The input argument `Fun` is the name (string) of a function file that calculates $f(x)$ for a given x, `a` and `b` are two points that bracket the root, and `TolMax` is the maximum tolerance. The program should include the following features:

- Check if points `a` and `b` are on opposite sides of the solution. If not, the program should stop and display an error message.

- The number of iterations should be determined (using the equation in Problem 3.1) before the iterations are carried out.

   Use `BisectionRoot` to solve the equation in Example 3-1.

3.15 Determining the square root of a number p, $\sqrt{p}$, is the same as finding a solution to the equation $f(x) = x^2 - p = 0$. Write a MATLAB user-defined function that determines the square root of a positive number by solving the equation using Newton's method. Name the function `[Xs] = SquareRoot(p)`. The output argument `Xs` is the answer, and the input argument `p` is the number whose square root is determined. The program should include the following features:

- It should check if the number is positive. If not, the program should stop and display an error message.

- The starting value of x for the iterations should be $x = p$.

- The iterations should stop when the estimated relative error (Eq. (3.9)) is smaller than 0.00001.

- The number of iterations should be limited to 20. If a solution is not obtained in 20 iterations, the program should stop and display an error message.

Use the function `SquareRoot` to determine the square root of (*a*) 729, (*b*) 1500, and (*c*) -72.

3.16 A new method for solving a nonlinear equation $f(x) = 0$, called trisection, is proposed. The method is similar to the bisection method. The solution starts by finding an interval $[a, b]$ that brackets the solution. The first estimate of the solution is the midpoint between $x = a$ and $x = b$. Then, as the name of the method suggests, the interval $[a, b]$ is divided into three equal sections. The section that contains the root is taken as the new interval for the next iteration.

   Write a MATLAB user-defined function that solves a nonlinear equation with the proposed new method. Name the function `Xs = TrisectionRoot(Fun, a, b, TolMax)`, where the output argument `Xs` is the solution. The input argument `Fun` is the name (string) of a function file that calculates $f(x)$ for a given x, `a` and `b` are two points that bracket the root, and `TolMax` is the maximum tolerance.

   Use the user-defined `TrisectionRoot` function to solve the equations in Problems 3.2 and 3.3. For the initial values of a and b take the values that are listed in part (*a*) of the problems, and for `TolMax` use 0.0001.

3.17 Write a MATLAB user-defined function that solves a nonlinear equation $f(x) = 0$ with the regula falsi method. Name the function Xs = RegulaRoot(Fun, a, b, ErrMax), where Xs is the solution, and the input arguments are: Fun: the name (string) of a function file that calculates $f(x)$ for a given x, a and b are two points that bracket the root, ErrMax: the maximum error according to Eq. (3.9).

The program should include the following features:

- Check if points a and b are on opposite sides of the solution. If not, the program should stop and display an error message.
- The number of iterations should be limited to 100 (to avoid an infinite loop). If a solution with the required accuracy is not obtained in 100 iterations the program should stop and display an error message.

Use the function RegulaRoot to solve the equation in Problems 3.3 (use $a = 1$, $b = 2$). For ErrMax use 0.0001.

3.18 Frequently, in the regula falsi method (see Section 3.4), one of the endpoints of the interval stays the same in all the iterations, while the other endpoint advances toward the root. In the modified regula falsi method, when this situation occurs, the straight line that connects the endpoints of the interval is replaced with a line that has a smaller slope. As shown in the figure, this is done by dividing the value of the function at the endpoint that stays the same by 2. Consequently, the line intersects the x axis closer to the root. Write a MATLAB user-defined

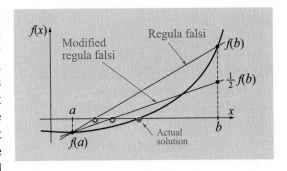

function that solves a nonlinear equation $f(x) = 0$ with the modified regula falsi method. The program should check if the endpoints stay the same; if an endpoint is unchanged in three iterations, the value of the function at that point should be divided by 2, when Eq. (3.11) is used. If the same point stays the same for three more iterations, the value of the function at that point should be divided again by 2 (the true value now is divided by 4), and so on. Name the function Xs = RegulaRootMod(Fun, a, b, ErrMax). The function arguments and features are the same as in Problem 3.17.

Use the function RegulaRootMod to solve the equation in Problems 3.3 (use $a = 1$, $b = 2$). For ErrMax use 0.0001.

3.19 Modify the function NewtonRoot that is listed in Fig. 3-11, such that the output will have three arguments. Name the function [Xs, FXs, iact] = NewtonRootMod(Fun, FunDer, Xest, Err, imax). The first output argument is the solution, the second is the value of the function at the solution, and the third is the actual number of iterations that are performed to obtain the solution. Use the function NewtonRootMod to solve the equation that is solved in Example 3-2.

3.20 Write a user-defined MATLAB function that solves for all the real roots in a specified domain of a non-linear function $f(x) = 0$ using the bisection method. Name the function R=BisecAll-Roots(fun, a, b, TolMax). The output argument R is a vector whose elements are the values of the roots. The input argument Fun is the name (string) of a function file that calculates $f(x)$ for a given x. The arguments a and b define the domain, and TolMax is the maximum tolerance that is used by the bisection method when the value of each root is calculated. Use the following algorithm:

1. Divide the domain $[a, b]$ into 10 equal subintervals of length h such that $h = (b-a)/10$.
2. Check for a sign change of $f(x)$ at the endpoints of each subinterval.
3. If a sign change is identified in a subinterval, use the bisection method for determining the root in that subinterval.
4. Divide the domain $[a, b]$ into 100 equal subintervals of length h such that $h = (b-a)/100$.
5. Repeat step 2. If a sign change is identified in a subinterval, check if it contains a root that was already obtained. If not, use the bisection method for determining the root in that subinterval.
6. If no new roots have been identified, stop the program.
7. If one or more new roots have been identified, repeat steps 4–6 where in each repetition the number of subintervals is multiplied by 10.

Use the function `BisecAllRoots`, with `TolMax` value of 0.0001, to find all the roots of the equation: $x^4 - 5.5x^3 - 7.2x^2 + 43x + 36 = 0$.

3.21 Examine the differences between the True Relative Error, Eq. (3.8), and the Estimated Relative Error, Eq. (3.9), by numerically solving the equation $f(x) = 0.5e^{(2+x)} - 40 = 0$. The exact solution of the equation is $x = \ln(80) - 2$. Write a MATLAB program in a script file that solves the equation by using Newton's method. Start the iterations at $x = 4$, and execute 11 iterations. In each iteration, calculate the True Relative Error (TRE) and the Estimated Relative Error (ERE). Display the results in a four column table (create a 2-dimensional array), with the number of iterations in the first column, the estimated numerical solution in the second, and TRE and ERE in the third and fourth columns, respectively.

Problems in math, science, and engineering
Solve the following problems using the MATLAB environment. As stated, use the MATLAB programs that are presented in the chapter, programs developed in previously solved problems, or MATLAB's built-in functions.

3.22 When calculating the payment of a mortgage, the relationship between the loan amount, *Loan*, the monthly payment, *MPay*, the duration of the loan in years, *Yrs*, and the annual interest rate, *Rate*, is given by the equation (annuity equation):

$$MPay = \frac{Loan \cdot Rate}{12\left(1 - \dfrac{1}{\left(1 + \dfrac{Rate}{12}\right)^{12 \cdot Yrs}}\right)}$$

Determine the *Rate* of a $170,000 loan for 20 years if the monthly payment is $1250.
(*a*) Use the user-defined function `NewtonRoot` given in Program 3-2. Use 0.001 for `Err`, and 1 for `Xest`.
(*b*) Use MATLAB's built-in function `fzero`.

3.23 The electrical resistance, $R(T)$, of a thermistor varies with temperature according to:

$$R(T) = 100(1 + AT - BT^2)$$

where R is in Ω, $A = 3.90802 \times 10^{-3}\,°\mathrm{C}^{-1}$, $B = 0.580195 \times 10^{-6}\,°\mathrm{C}^{-2}$, and T is the temperature in degrees C. Find the temperature corresponding to a resistance of 200 Ω.
(*a*) Use the user-defined function `BisectionRoot` given in Program 3-14. Use 0.001 for `TolMax`.
(*b*) Use MATLAB's built-in `fzero` function.

3.24 A quarterback is about to throw a pass to his wide receiver running a route. The quarterback is 6 ft tall and is supposed to hit the wide receiver straight down the field 60 ft away. The equation that describes the motion of the football is the familiar equation of projectile motion from physics:

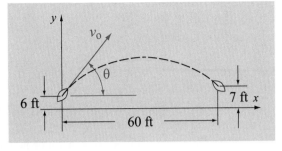

$$y = x\tan(\theta) - \frac{1}{2}\frac{x^2 g}{v_o^2}\frac{1}{\cos^2(\theta)} + h$$

where x and y are the horizontal and vertical distance, respectively, $g = 32.2$ ft/s^2 is the acceleration due to gravity, v_o is the initial velocity of the football as it leaves the quarterback's hand and θ is the angle the football makes with the horizontal just as it leaves the quarterback's throwing hand. For $v_o = 50$ ft/s, $x = 60$ ft, $h = 6$ ft, and $y = 7$ ft, find the angle θ at which the quarterback must launch the ball.

(*a*) Use the user-defined function `BisectionRoot` that was developed in Problem 3.14. For maximum tolerance use 10^{-3} rad.

(*b*) Use MATLAB built-in function `fzero`.

3.25 A pipe of length $L = 25$ m, and diameter $d = 10$ cm that carries steam, loses heat to the ambient air and surrounding surfaces by convection and radiation. If the total flow of heat per unit time, Q, emanating from the surface of the pipe is measured, then the surface temperature of the pipe, T_S, is determined by the following equation:

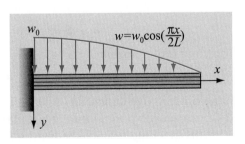

$$Q = \pi d L [h(T_S - T_{air}) + \varepsilon \sigma_{SB}(T_S^4 - T_{surr}^4)]$$

where $\varepsilon = 0.8$ is the radiative emissivity of the surface of the pipe, and $\sigma_{SB} = 5.67 \times 10^{-8}$ W/m^2/K^4 is the Stefan–Boltzmann constant. If $Q = 18405$ W, $h = 10$ W/m^2/K, and $T_{air} = T_{surr} = 298$ K, find the surface temperature of the pipe, T_S.

(*a*) Use the user-defined function `BisectionRoot` that was developed in Problem 3.14 with a starting interval of $[0, 1000]$. For maximum tolerance use 10^{-2}.

(*b*) Use MATLAB's built-in function `fzero`.

3.26 A cantilever I-beam is loaded with a distributed load, as shown. The deflection, y, of the centerline of the beam as a function of the position, x, is given by the equation:

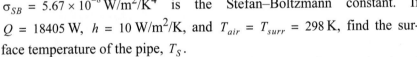

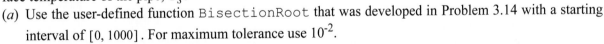

$$y = \frac{w_0 L}{3\pi^4 EI}\left(48L^3\cos\left(\frac{\pi}{2L}x\right) - 48L^3 + 3\pi^3 Lx^2 - \pi^3 x^3\right)$$

where $L = 3$ m is the length, $E = 70$ GPa is the elastic modulus, $I = 52.9 \times 10^{-6}$ m^4 is the moment of inertia, and $w_0 = 15$ kN/m.

Find the position x where the deflection of the beam is 9 mm.

(*a*) Use the user-defined function `NewtonRoot` given in Program 3-2. Use 0.0001 for `Err`, and 1.8 for `Xest`.

(*b*) Use the user-defined function `SecantRoot` given in Program 3-3. Use 0.0001 for `Err`, 1.5 for `Xa`. and 2.5 for `Xb`.

(*c*) Use MATLAB's built-in `fzero` function.

3.27 According to Archimedes' principle, the buoyancy force acting on an object that is partially immersed in a fluid is equal to the weight that is displaced by the portion of the object that is submerged.

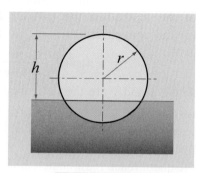

A spherical float with a mass of $m_f = 70$ kg and a diameter of 90 cm is placed in the ocean (density of sea water is approximately $\rho = 1030$ kg/m³. The height, h, of the portion of the float that is above water can be determined by solving an equation that equates the mass of the float to the mass of the water that is displaced by the portion of the float that is submerged:

$$\rho V_{cap} = m_f \qquad (3.59)$$

where for a sphere of radius r, the volume of a cap of depth d is given by:

$$V_{cap} = \frac{1}{3}\pi d^2 (3r - d)$$

Write Eq. (3.59) in terms of h and solve for h using:

(*a*) the user-defined function `NewtonRoot` given in Program 3-2. Use 0.0001 for `Err`, and 0.8 for `Xest`.

(*b*) MATLAB's built-in `fzero` function.

3.28 According to Archimedes' principle, the buoyancy force that is acting on an object that is partially immersed in a fluid is equal to the weight that is displaced by the portion of the body that is in the water.

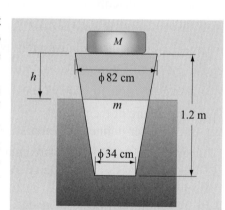

A buoy in the shape of a frustum and mass of $m = 50$ kg is placed in the ocean (density of sea water is approximately 1030 kg/m³). Equipment with a mass of M is attached on the top of the buoy as shown in the figure. The height, h, of the portion of the buoy that is above water can be determined from an equation that equates the total mass ($m + M$) to the mass of the water that is displaced by the volume V of the portion of the buoy that is submerged:

$$m + M = \rho V \qquad (3.60)$$

The volume of a frustum with base radii of a and b, and height H is given by:

$$V = \frac{1}{3}\pi H(a^2 + ab + b^2)$$

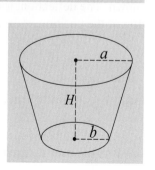

Write Eq. (3.60) in terms of h.

(*a*) Determine (by hand) the largest mass M that can be attached to the buoy, without letting M touch the water. (The mass for which $h = 0$.)

(*b*) Write a MATLAB program in a script file that determines (using `fzero`) the value of h for different values of M, starting with $M = 0$ kg with increments of 1 kg up to the largest mass M that can be attached, and then plot h versus M.

3.29 An *RLC* circuit consists of a resistor *R*, an inductor *L*, and a
capacitor *C* connected in series to an alternating voltage source *V.*
The current amplitude i_m is given by:

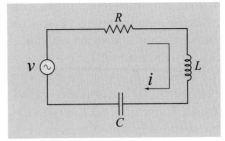

$$i_m = \frac{v_m}{\sqrt{R^2 + [\omega L - 1/(\omega C)]^2}}$$

where, ω, the angular frequency, is related to the frequency *f* by:
$\omega = 2\pi f$. Determine *f* for a circuit with $R = 140\ \Omega$, $L = 260$ mH,
$C = 25\ \mu F$, $v_m = 24$ V, and $i_m = 0.15$ A by using:

(*a*) Use the user-defined function `SecantRoot` given in Program 3-3. Use 0.0001 for `Err`, 30 for `Xa`.
and 50 for `Xb`.

(*b*) MATLAB's built-in `fzero` function.

3.30 The power output of a solar cell varies with the voltage it puts out. The voltage V_{mp} at which the output power is maximum is given by the equation:

$$e^{(qV_{mp}/k_BT)}\left(1 + \frac{qV_{mp}}{k_BT}\right) = e^{(qV_{OC}/k_BT)}$$

where V_{OC} is the open circuit voltage, *T* is the temperature in Kelvin, $q = 1.6022 \times 10^{-19}$ C is the charge on
an electron, and $k_B = 1.3806 \times 10^{-23}$ J/k is Boltzmann's constant. For $V_{OC} = 0.5$ V and room temperature
($T = 297$ K), determine the voltage V_{mp} at which the power output of the solar cell is a maximum.

(*a*) Write a program in a script file that uses the fixed-point iteration method to find the root. For starting
point use $V_{mp} = 0.5$ V. To terminate the iterations use Eq. (3.18) with $\varepsilon = 0.001$.

(b) Use MATLAB's `fzero` built-in function.

3.31 Viscous or frictional losses in pipe flow result in pressure drops that must be overcome. These are
expressed in terms of a friction factor, *f*. For turbulent flows in pipes, *f* is calculated from the Colebrook
equation:

$$\frac{1}{f^{0.5}} = -2\log_{10}\left(\frac{e/D}{3.7} + \frac{2.51}{(Re)f^{0.5}}\right)$$

where *e* is the roughness, *D* is the diameter of the pipe, and *Re* is the dimensionless Reynolds number. For
$e/D = 0.004$ and $Re = 2 \times 10^5$, solve this equation for *f.*

(*a*) Use the fixed-point iteration method with the following iteration function:

$$g(f) = \frac{1}{4\left[\log_{10}\left(\dfrac{e/D}{3.7} + \dfrac{2.51}{(Re)f^{0.5}}\right)\right]^2}$$

Stop the iterations when the maximum estimated relative error as given by Eq. (3.18) is smaller than
0.001.

(*b*) Use MATLAB's built-in `fzero` function.

3.32 A simplified model of the suspension of a car consists of a mass, m, a spring with stiffness k, and a dashpot with damping coefficient c, as shown in the figure. A bumpy road can be modeled by a sinusoidal up-and-down motion of the wheel $y = Y\sin(\omega t)$. From the solution of the equation of motion for this model, the steady-state up-and-down motion of the car (mass) is given by $x = X\sin(\omega t - \phi)$. The ratio between amplitude X and amplitude Y is given by:

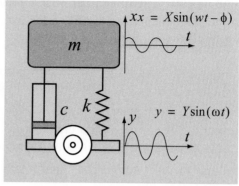

$$\frac{X}{Y} = \sqrt{\frac{mc\omega^3}{k(k - m\omega^2) + (\omega c)^2}}$$

Assuming $m = 2000\,\text{kg}$, $k = 500\,\text{kN/m}$, and $c = 38 \times 10^3\,\text{N-s/m}$, determine the frequency ω for which $X/Y = 0.2$. Rewrite the equation such that it is in the form of a polynomial in ω and solve by using

(a) Use the user-defined function `BisectionRoot` that was developed in Problem 3.14. For maximum tolerance use 10^{-2}.

(b) MATLAB's built-in `fzero` function.

3.33 A coating on the panel surface is cured by radiant energy from a heater. The temperature of the coating is determined by radiative and convective heat transfer processes. If the radiation is treated as diffuse and gray, the following nonlinear system of simultaneous equations determine the unknowns J_h, T_h, J_c, T_c:

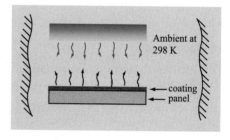

$$5.67 \times 10^{-8} T_c^4 + 17.41 T_c - J_c = 5188.18$$

$$J_c - 0.71 J_h + 7.46 T_c = 2352.71$$

$$5.67 \times 10^{-8} T_h^4 + 1.865 T_h - J_h = 2250$$

$$J_h - 0.71 J_c + 7.46 T_h = 11093$$

where J_h and J_c are the radiosities of the heater and coating surfaces, respectively, and T_h and T_c are the respective temperatures.

(a) Show that the following iteration functions can be used for solving the nonlinear system of equations with the fixed-point iteration method:

$$T_c = \left[\frac{J_c - 17.41 T_c + 5188.18}{5.67 \times 10^{-8}}\right]^{1/4} \qquad T_h = \left[\frac{2250 + J_h - 1.865 T_h}{5.67 \times 10^{-8}}\right]^{1/4}$$

$$J_c = 2352.71 + 0.71 J_h - 7.46 T_c \qquad J_h = 11093 + 0.71 J_c - 7.46 T_h$$

(b) Solve the nonlinear system of equations with the fixed-point iteration method using the iteration functions from part (a). Use the following initial values: $T_h = T_c = 298$ K, $J_c = 3000$ W/m^2, and $J_h = 5000$ W/m^2. Carry out 100 iterations, and plot the respective values to observe their convergence. The final answers should be: $T_c = 481$ K, $J_c = 6222$ W/m^2, $T_h = 671$ K, $J_h = 10504$ W/m^2.

Chapter 4

Solving a System of Linear Equations

4.1 BACKGROUND

Systems of linear equations that have to be solved simultaneously arise in problems that include several (possibly many) variables that are dependent on each other. Such problems occur not only in engineering and science, which are the focus of this book, but in virtually any discipline (business, statistics, economics, etc.). A system of two (or three) equations with two (or three) unknowns can be solved manually by substitution or other mathematical methods (e.g., Cramer's rule, Section 2.4.6). Solving a system in this way is practically impossible as the number of equations (and unknowns) increases beyond three.

An example of a problem in electrical engineering that requires a solution of a system of equations is shown in Fig. 4-1. Using Kirchhoff's law, the currents i_1, i_2, i_3, and i_4 can be determined by solving the following system of four equations:

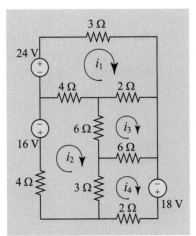

Figure 4-1: Electrical circuit.

$$9i_1 - 4i_2 - 2i_3 = 24$$
$$-4i_1 + 17i_2 - 6i_3 - 3i_4 = -16$$
$$-2i_1 - 6i_2 + 14i_3 - 6i_4 = 0 \qquad (4.1)$$
$$-3i_2 - 6i_3 + 11i_4 = 18$$

Obviously, more complicated circuits may require the solution of a system with a larger number of equations. Another example that requires a solution of a system of equations is calculating the force in members of a truss. The forces in the eight members of the truss shown in Fig. 4-2 are determined from the solution of the following system of eight equations (equilibrium equations of pins A, B, C, and D):

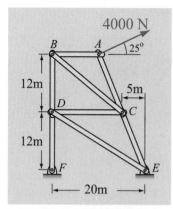

$$0.9231F_{AC} = 1690 \qquad\qquad -F_{AB} - 0.3846F_{AC} = 3625$$
$$F_{AB} - 0.7809F_{BC} = 0 \qquad\qquad 0.6247F_{BC} - F_{BD} = 0$$
$$F_{CD} + 0.8575F_{DE} = 0 \qquad\qquad F_{BD} - 0.5145F_{DE} - F_{DF} = 0 \quad (4.2)$$
$$0.3846F_{CE} - 0.3846F_{AC} - 0.7809F_{BC} - F_{CD} = 0$$
$$0.9231F_{AC} + 0.6247F_{BC} - 0.9231F_{CE} = 0$$

Figure 4-2: Eight-member

There are applications, for example, in finite element and finite difference analysis, where the system of equations that has to be solved contains thousands (or even millions) of simultaneous equations.

4.1.1 Overview of Numerical Methods for Solving a System of Linear Algebraic Equations

The general form of a system of n linear algebraic equations is:

$$a_{11}x_1 + a_{12}x_2 + \ldots + a_{1n}x_n = b_1$$
$$a_{21}x_1 + a_{22}x_2 + \ldots + a_{2n}x_n = b_2$$
$$\vdots \qquad \vdots \qquad \vdots \qquad \vdots \qquad (4.3)$$
$$a_{n1}x_1 + a_{n2}x_2 + \ldots + a_{nn}x_n = b_n$$

$$\begin{bmatrix} a_{11} & a_{12} & \ldots & a_{1n} \\ a_{21} & a_{22} & \ldots & a_{2n} \\ \ldots & \ldots & \ldots & \ldots \\ a_{n1} & a_{n2} & \ldots & a_{nn} \end{bmatrix} \begin{bmatrix} x_1 \\ x_2 \\ \ldots \\ x_n \end{bmatrix} = \begin{bmatrix} b_1 \\ b_2 \\ \ldots \\ b_n \end{bmatrix}$$

Figure 4-3: A system of n linear algebraic equations.

The matrix form of the equations is shown in Fig. 4-3. Two types of numerical methods, **direct** and **iterative**, are used for solving systems of linear algebraic equations. In direct methods, the solution is calculated by performing arithmetic operations with the equations. In iterative methods, an initial approximate solution is assumed and then used in an iterative process for obtaining successively more accurate solutions.

Direct methods

In direct methods, the system of equations that is initially given in the general form, Eqs. (4.3), is manipulated to an equivalent system of equations that can be easily solved. Three systems of equations that can be easily solved are the **upper triangular**, **lower triangular**, and **diagonal** forms.

The **upper triangular** form is shown in Eqs. (4.4), and is written in a matrix form for a system of four equations in Fig. 4-4.

$$\begin{bmatrix} a_{11} & a_{12} & a_{13} & a_{14} \\ 0 & a_{22} & a_{23} & a_{24} \\ 0 & 0 & a_{33} & a_{34} \\ 0 & 0 & 0 & a_{44} \end{bmatrix} \begin{bmatrix} x_1 \\ x_2 \\ x_3 \\ x_4 \end{bmatrix} = \begin{bmatrix} b_1 \\ b_2 \\ b_3 \\ b_4 \end{bmatrix}$$

$$a_{11}x_1 + a_{12}x_2 + a_{13}x_3 + \ldots + a_{1n}x_n = b_1$$
$$a_{22}x_2 + a_{23}x_3 + \ldots + a_{2n}x_n = b_2$$
$$a_{33}x_3 + \ldots + a_{3n}x_n = b_3$$
$$\vdots \qquad \vdots \qquad \vdots \qquad (4.4)$$
$$a_{n-1, n-1}x_{n-1} + a_{n-1, n}x_n = b_{n-1}$$
$$a_{nn}x_n = b_n$$

Figure 4-4: A system of four equations in upper triangular form.

The system in this form has all zero coefficients below the diagonal and is solved by a procedure called **back substitution**. It starts with the last equation, which is solved for x_n. The value of x_n is then substituted in the next-to-the-last equation, which is solved for x_{n-1}. The process continues in the same manner all the way up to the first equation. In the case of four equations, the solution is given by:

$$x_4 = \frac{b_4}{a_{44}}, \quad x_3 = \frac{b_3 - a_{34}b_4}{a_{33}}, \quad x_2 = \frac{b_2 - (a_{23}b_3 + a_{24}b_4)}{a_{22}} \quad \text{and}$$

$$x_4 = \frac{b_4 - (a_{12}b_2 + a_{13}b_3 + a_{14}b_4)}{a_{11}}$$

For a system of n equations in upper triangular form, a general formula for the solution using back substitution is:

$$x_n = \frac{b_n}{a_{nn}}$$

$$x_i = \frac{b_i - \sum\limits_{j=i+1}^{j=n} a_{ij}b_j}{a_{ii}} \qquad i = n-1, n-2, \ldots 1 \tag{4.5}$$

In Section 4.2 the upper triangular form and back substitution are used in the Gauss elimination method.

The **lower triangular** form is shown in Eqs. (4.6), and is written in matrix form for a system of four equations in Fig. 4-5.

$$\begin{bmatrix} a_{11} & 0 & 0 & 0 \\ a_{21} & a_{22} & 0 & 0 \\ a_{31} & a_{32} & a_{33} & 0 \\ a_{41} & a_{42} & a_{43} & a_{44} \end{bmatrix} \begin{bmatrix} x_1 \\ x_2 \\ x_3 \\ x_4 \end{bmatrix} = \begin{bmatrix} b_1 \\ b_2 \\ b_3 \\ b_4 \end{bmatrix}$$

Figure 4-5: A system of four equations in lower triangular form.

$$\begin{aligned} a_{11}x_1 &&&&&&= b_1 \\ a_{21}x_1 + a_{22}x_2 &&&&&&= b_2 \\ a_{31}x_1 + a_{32}x_2 + a_{33}x_3 &&&&&&= b_3 \\ &\vdots &\vdots &&&&\vdots \\ a_{n1}x_1 + a_{n2}x_2 + a_{n3}x_3 + \ldots + a_{nn}x_n &&&&&&= b_n \end{aligned} \tag{4.6}$$

The system in this form has zero coefficients above the diagonal. A system in lower triangular form is solved in the same way as the upper triangular form but in an opposite order. The procedure is called **forward substitution**. It starts with the first equation, which is solved for x_1. The value of x_1 is then substituted in the second equation, which is solved for x_2. The process continues in the same manner all the way down to the last equation. In the case of four equations, the solution is given by:

$$x_1 = \frac{b_1}{a_{11}}, \quad x_2 = \frac{b_2 - a_{21}b_1}{a_{22}}, \quad x_3 = \frac{b_3 - (a_{31}b_1 + a_{32}b_2)}{a_{33}} \quad \text{and}$$

$$x_4 = \frac{b_4 - (a_{41}b_1 + a_{42}b_2 + a_{43}b_3)}{a_{44}} \tag{4.7}$$

For a system of n equations in lower triangular form, a general formula for the solution using forward substitution is:

$$x_1 = \frac{b_1}{a_{11}}$$

$$x_i = \frac{b_i - \sum_{j=i}^{j=i-1} a_{ij}b_j}{a_{ii}} \qquad i = 2, 3, \dots n \tag{4.8}$$

In Section 4.5 the lower triangular form is used together with the upper triangular form in the *LU* decomposition method for solving a system of equations.

The ***diagonal*** form of a system of linear equations is shown in Eqs. (4.9) and is written in matrix form for a system of four equations in Fig. 4-6.

$$
\begin{aligned}
a_{11}x_1 &&&& = b_1 \\
& a_{12}x_2 &&& = b_2 \\
&& a_{13}x_3 && = b_3 \\
&& \vdots && \quad \vdots \\
&&&& a_n x_n = b_n
\end{aligned}
\tag{4.9}
$$

$$
\begin{bmatrix}
a_{11} & 0 & 0 & 0 \\
0 & a_{22} & 0 & 0 \\
0 & 0 & a_{33} & 0 \\
0 & 0 & 0 & a_{44}
\end{bmatrix}
\begin{bmatrix}
x_1 \\ x_2 \\ x_2 \\ x_4
\end{bmatrix}
=
\begin{bmatrix}
b_1 \\ b_2 \\ b_2 \\ b_4
\end{bmatrix}
$$

Figure 4-6: A system of four equations in diagonal form.

A system in diagonal form has nonzero coefficients along the diagonal and zeros everywhere else. Obviously, a system in this form can be easily solved. A similar form is used in the Gauss–Jordan method, which is presented in Section 4.4.

From the three forms of simultaneous linear equations (upper triangular, lower triangular, diagonal) it might appear that changing a given system of equations to the diagonal form is the best choice because the diagonal system is the easiest to solve. In reality, however, the total number of operations required for solving a system is smaller when other methods are used.

Three direct methods for solving systems of equations—Gauss elimination (Sections 4.2 and 4.3), Gauss–Jordan (Section 4.4), and *LU* decomposition (Section 4.5)—and two indirect (iterative) methods—Jacobi and Gauss–Seidel (Section 4.7)—are described in this chapter.

4.2 GAUSS ELIMINATION METHOD

The Gauss elimination method is a procedure for solving a system of linear equations. In this procedure, a system of equations that is given in a general form is manipulated to be in ***upper triangular*** form, which is then solved by using back substitution (see Section 4.1.1). For a set of four equations with four unknowns the general form is given by:

$$
\begin{bmatrix}
a_{11} & a_{12} & a_{13} & a_{14} \\
a_{21} & a_{22} & a_{23} & a_{24} \\
a_{31} & a_{32} & a_{33} & a_{34} \\
a_{41} & a_{42} & a_{43} & a_{44}
\end{bmatrix}
\begin{bmatrix}
x_1 \\ x_2 \\ x_3 \\ x_4
\end{bmatrix}
=
\begin{bmatrix}
b_1 \\ b_2 \\ b_3 \\ b_4
\end{bmatrix}
$$

Figure 4-7: Matrix form of a system of four equations.

$$
\begin{aligned}
a_{11}x_1 + a_{12}x_2 + a_{13}x_3 + a_{14}x_4 &= b_1 & (4.10a) \\
a_{21}x_1 + a_{22}x_2 + a_{23}x_3 + a_{24}x_4 &= b_2 & (4.10b) \\
a_{31}x_1 + a_{32}x_2 + a_{33}x_3 + a_{34}x_4 &= b_3 & (4.10c) \\
a_{41}x_1 + a_{42}x_2 + a_{43}x_3 + a_{44}x_4 &= b_4 & (4.10d)
\end{aligned}
\tag{4.10}
$$

The matrix form of the system is shown in Fig. 4-7. In the Gauss elimination method, the system of equations is manipulated into an equivalent system of equations that has the form:

$$a_{11}x_1 + a_{12}x_2 + a_{13}x_3 + a_{14}x_4 = b_1 \qquad (4.11a)$$
$$a'_{22}x_2 + a'_{23}x_3 + a'_{24}x_4 = b'_2 \qquad (4.11b)$$
$$a'_{33}x_3 + a'_{34}x_4 = b'_3 \qquad (4.11c)$$
$$a'_{44}x_4 = b'_4 \qquad (4.11d)$$

$$(4.11)$$

$$\begin{bmatrix} a_{11} & a_{12} & a_{13} & a_{14} \\ 0 & a'_{22} & a'_{23} & a'_{24} \\ 0 & 0 & a'_{33} & a'_{34} \\ 0 & 0 & 0 & a'_{44} \end{bmatrix} \begin{bmatrix} x_1 \\ x_2 \\ x_3 \\ x_4 \end{bmatrix} = \begin{bmatrix} b_1 \\ b'_2 \\ b'_3 \\ b'_4 \end{bmatrix}$$

Figure 4-8: Matrix form of the equivalent system.

The first equation in the equivalent system, (4.11a), is the same as (4.10a). In the second equation, (4.11b), the variable x_1 is eliminated. In the third equation, (4.11c), the variables x_1 and x_2 are eliminated. In the fourth equation, (4.11d), the variables x_1, x_2, and x_3 are eliminated. The matrix form of the equivalent system is shown in Fig. 4-8. The system of equations (4.11) is in upper triangular form, which can be easily solved by using back substitution.

In general, various mathematical manipulations can be used for converting a system of equations from the general form displayed in Eqs. (4.10) to the ***upper triangular*** form in Eqs. (4.11). One in particular, the Gauss elimination method, is described next. The procedure can be easily programmed in a computer code.

Gauss elimination procedure (forward elimination)

The Gauss elimination procedure is first illustrated for a system of four equations with four unknowns. The starting point is the set of equations that is given by Eqs. (4.10). Converting the system of equations to the form given in Eqs. (4.11) is done in steps.

Step 1: In the first step, the first equation is unchanged, and the terms that include the variable x_1 in all the other equations are eliminated. This is done one equation at a time by using the first equation that is called the ***pivot equation***. The coefficient a_{11} is called the ***pivot coefficient***, or the pivot element. To eliminate the term $a_{21}x_1$ in Eq. (4.10b), the pivot equation, Eq. (4.10a), is multiplied by $m_{21} = a_{21}/a_{11}$, and then the equation is subtracted from Eq. (4.10b):

$$\begin{array}{r} a_{21}x_1 + a_{22}x_2 + a_{23}x_3 + a_{24}x_4 = b_2 \\ - \quad m_{21}(a_{11}x_1 + a_{12}x_2 + a_{13}x_3 + a_{14}x_4) = m_{21}b_1 \\ \hline 0 + (a_{22} - m_{21}a_{12})x_2 + (a_{23} - m_{21}a_{13})x_3 + (a_{24} - m_{21}a_{14})x_4 = b_2 - m_{21}b_1 \end{array}$$

$$\underbrace{}_{a'_{22}} \qquad \underbrace{}_{a'_{23}} \qquad \underbrace{}_{a'_{24}} \qquad \underbrace{}_{b'_2}$$

$$\begin{bmatrix} a_{11} & a_{12} & a_{13} & a_{14} \\ 0 & a'_{22} & a'_{23} & a'_{24} \\ a_{31} & a_{32} & a_{33} & a_{34} \\ a_{41} & a_{42} & a_{43} & a_{44} \end{bmatrix} \begin{bmatrix} x_1 \\ x_2 \\ x_3 \\ x_4 \end{bmatrix} = \begin{bmatrix} b_1 \\ b'_2 \\ b_3 \\ b_4 \end{bmatrix}$$

Figure 4-9: Matrix form of the system after eliminating a_{21}.

It should be emphasized here that the pivot equation, Eq. (4.10a) itself, is not changed. The matrix form of the equations after this operation is shown in Fig. 4-9.

Next, the term $a_{31}x_1$ in Eq. (4.10c) is eliminated. The pivot equation, Eq. (4.10a), is multiplied by $m_{31} = a_{31}/a_{11}$ and then is subtracted

$$\begin{bmatrix} a_{11} & a_{12} & a_{13} & a_{14} \\ 0 & a'_{22} & a'_{23} & a'_{24} \\ 0 & a'_{32} & a'_{33} & a'_{34} \\ a_{41} & a_{42} & a_{43} & a_{44} \end{bmatrix} \begin{bmatrix} x_1 \\ x_2 \\ x_3 \\ x_4 \end{bmatrix} = \begin{bmatrix} b_1 \\ b'_2 \\ b'_3 \\ b_4 \end{bmatrix}$$

Figure 4-10: Matrix form of the system after eliminating a_{31}.

from Eq. (4.10c):

$$a_{31}x_1 + a_{32}x_2 + a_{33}x_3 + a_{34}x_4 = b_3$$

$$- \quad m_{31}(a_{11}x_1 + a_{12}x_2 + a_{13}x_3 + a_{14}x_4) = m_{31}b_1$$

$$0 + \underbrace{(a_{32} - m_{31}a_{12})}_{a'_{32}}x_2 + \underbrace{(a_{33} - m_{31}a_{13})}_{a'_{33}}x_3 + \underbrace{(a_{34} - m_{31}a_{14})}_{a'_{34}}x_4 = \underbrace{b_3 - m_{31}b_1}_{b'_3}$$

The matrix form of the equations after this operation is shown in Fig. 4-10.

Next, the term $a_{41}x_1$ in Eq. (4.10d) is eliminated. The pivot equation, Eq. (4.10a), is multiplied by $m_{41} = a_{41}/a_{11}$ and then is subtracted from Eq. (4.10d):

$$a_{41}x_1 + a_{42}x_2 + a_{43}x_3 + a_{44}x_4 = b_4$$

$$- \quad m_{41}(a_{11}x_1 + a_{12}x_2 + a_{13}x_3 + a_{14}x_4) = m_{41}b_1$$

$$0 + \underbrace{(a_{42} - m_{41}a_{12})}_{a'_{42}}x_2 + \underbrace{(a_{43} - m_{41}a_{13})}_{a'_{43}}x_3 + \underbrace{(a_{44} - m_{41}a_{14})}_{a'_{44}}x_4 = \underbrace{b_4 - m_{41}b_1}_{b'_4}$$

$$\begin{bmatrix} a_{11} & a_{12} & a_{13} & a_{14} \\ 0 & a'_{22} & a'_{23} & a'_{24} \\ 0 & a'_{32} & a'_{33} & a'_{34} \\ 0 & a'_{42} & a'_{43} & a'_{44} \end{bmatrix} \begin{bmatrix} x_1 \\ x_2 \\ x_3 \\ x_4 \end{bmatrix} = \begin{bmatrix} b_1 \\ b'_2 \\ b'_3 \\ b'_4 \end{bmatrix}$$

Figure 4-11: Matrix form of the system after eliminating a_{41}.

This is the end of ***Step 1***. The system of equations now has the following form:

$$\begin{array}{ll} a_{11}x_1 + a_{12}x_2 + a_{13}x_3 + a_{14}x_4 = b_1 & (4.12a) \\ 0 + a'_{22}x_2 + a'_{23}x_3 + a'_{24}x_4 = b'_2 & (4.12b) \\ 0 + a'_{32}x_2 + a'_{33}x_3 + a'_{34}x_4 = b'_3 & (4.12c) \\ 0 + a'_{42}x_2 + a'_{43}x_3 + a'_{44}x_4 = b'_4 & (4.12d) \end{array} \quad (4.12)$$

The matrix form of the equations after this operation is shown in Fig. 4-11. Note that the result of the elimination operation is to reduce the first column entries, except a_{11} (the pivot element), to zero.

Step 2: In this step Eqs. (4.12a) and (4.12b) are not changed, and the terms that include the variable x_2 in Eqs. (4.12c) and (4.12d) are eliminated. In this step Eq. (4.12b) is the pivot equation, and the coefficient a'_{22} is the pivot coefficient. To eliminate the term $a'_{32}x_2$ in Eq. (4.12c), the pivot equation, Eq. (4.12b), is multiplied by $m_{32} = a'_{32}/a'_{22}$ and then is subtracted from Eq. (4.12c):

$$a'_{32}x_2 + a'_{33}x_3 + a'_{34}x_4 = b'_3$$

$$- \quad m_{32}(a'_{22}x_2 + a'_{23}x_3 + a'_{24}x_4) = m_{32}b'_2$$

$$0 + \underbrace{(a'_{33} - m_{32}a'_{23})}_{a''_{33}}x_3 + \underbrace{(a'_{34} - m_{32}a'_{24})}_{a''_{34}}x_4 = \underbrace{b'_3 - m_{32}b'_2}_{b''_3}$$

$$\begin{bmatrix} a_{11} & a_{12} & a_{13} & a_{14} \\ 0 & a'_{22} & a'_{23} & a'_{24} \\ 0 & 0 & a''_{33} & a''_{34} \\ 0 & a'_{42} & a'_{43} & a'_{44} \end{bmatrix} \begin{bmatrix} x_1 \\ x_2 \\ x_3 \\ x_4 \end{bmatrix} = \begin{bmatrix} b_1 \\ b'_2 \\ b''_3 \\ b'_4 \end{bmatrix}$$

Figure 4-12: Matrix form of the system after eliminating a_{32}.

The matrix form of the equations after this operation is shown in Fig. 4-12.

Next, the term $a'_{42}x_2$ in Eq. (4.12d) is eliminated. The pivot equation, Eq. (4.12b), is multiplied by $m_{42} = a'_{42}/a'_{22}$ and then is subtracted from Eq. (4.12d):

$$a'_{42}x_2 + a'_{43}x_3 + a'_{44}x_4 = b'_4$$

$$m_{42}(a'_{22}x_2 + a'_{23}x_3 + a'_{24}x_4) = m_{42}b'_2$$

$$0 + (a'_{43} - m_{42}a'_{23})x_3 + (a'_{44} - m_{42}a'_{24})x_4 = b'_4 - m_{42}b'_2$$

$$\underbrace{\phantom{a'_{43} - m_{42}a'_{23}}}_{a''_{43}} \quad \underbrace{\phantom{a'_{44} - m_{42}a'_{24}}}_{a''_{44}} \quad \underbrace{\phantom{b'_4 - m_{42}b'_2}}_{b''_4}$$

This is the end of **Step 2**. The system of equations now has the following form:

$$
\begin{array}{ll}
a_{11}x_1 + a_{12}x_2 + a_{13}x_3 + a_{14}x_4 = b_1 & (4.13a) \\
0 + a'_{22}x_2 + a'_{23}x_3 + a'_{24}x_4 = b'_2 & (4.13b) \\
0 + 0 + a''_{33}x_3 + a''_{34}x_4 = b''_3 & (4.13c) \\
0 + 0 + a''_{43}x_3 + a''_{44}x_4 = b''_4 & (4.13d)
\end{array}
\quad (4.13)
$$

$$
\begin{bmatrix}
a_{11} & a_{12} & a_{13} & a_{14} \\
0 & a'_{22} & a'_{23} & a'_{24} \\
0 & 0 & a''_{33} & a''_{34} \\
0 & 0 & a''_{43} & a''_{44}
\end{bmatrix}
\begin{bmatrix}
x_1 \\ x_2 \\ x_3 \\ x_4
\end{bmatrix}
=
\begin{bmatrix}
b_1 \\ b'_2 \\ b''_3 \\ b''_4
\end{bmatrix}
$$

Figure 4-13: Matrix form of the system after eliminating a_{42}.

The matrix form of the equations at the end of **Step 2** is shown in Fig. 4-13.

Step 3: In this step Eqs. (4.13a), (4.13b), and (4.13c) are not changed, and the term that includes the variable x_3 in Eq. (4.13d) is eliminated. In this step Eq. (4.13c) is the pivot equation, and the coefficient a''_{33} is the pivot coefficient. To eliminate the term $a''_{43}x_3$ in Eq. (4.13d), the pivot equation is multiplied by $m_{43} = a''_{43}/a''_{33}$ and then is subtracted from Eq. (4.13d).

$$a''_{43}x_3 + a''_{44}x_4 = b''_4$$

$$m_{43}(a''_{33}x_3 + a''_{34}x_4) = m_{43}b''_3$$

$$(a''_{44} - m_{43}a''_{34})x_4 = b''_4 - m_{43}b''_3$$

$$\underbrace{\phantom{a''_{44} - m_{43}a''_{34}}}_{a'''_{44}} \quad \underbrace{\phantom{b''_4 - m_{43}b''_3}}_{b'''_4}$$

This is the end of **Step 3**. The system of equations is now in an upper triangular form:

$$
\begin{array}{ll}
a_{11}x_1 + a_{12}x_2 + a_{13}x_3 + a_{14}x_4 = b_1 & (4.14a) \\
0 + a'_{22}x_2 + a'_{23}x_3 + a'_{24}x_4 = b'_2 & (4.14b) \\
0 + 0 + a''_{33}x_3 + a''_{34}x_4 = b''_3 & (4.14c) \\
0 + 0 + 0 + a'''_{44}x_4 = b'''_4 & (4.14d)
\end{array}
\quad (4.14)
$$

$$
\begin{bmatrix}
a_{11} & a_{12} & a_{13} & a_{14} \\
0 & a'_{22} & a'_{23} & a'_{24} \\
0 & 0 & a''_{33} & a''_{34} \\
0 & 0 & 0 & a'''_{44}
\end{bmatrix}
\begin{bmatrix}
x_1 \\ x_2 \\ x_3 \\ x_4
\end{bmatrix}
=
\begin{bmatrix}
b_1 \\ b'_2 \\ b''_3 \\ b'''_4
\end{bmatrix}
$$

Figure 4-14: Matrix form of the system after eliminating a_{43}.

The matrix form of the equations is shown in Fig. 4-14. Once transformed to upper triangular form, the equations can be easily solved by using back substitution. The three steps of the Gauss elimination process are illustrated together in Fig. 4-15.

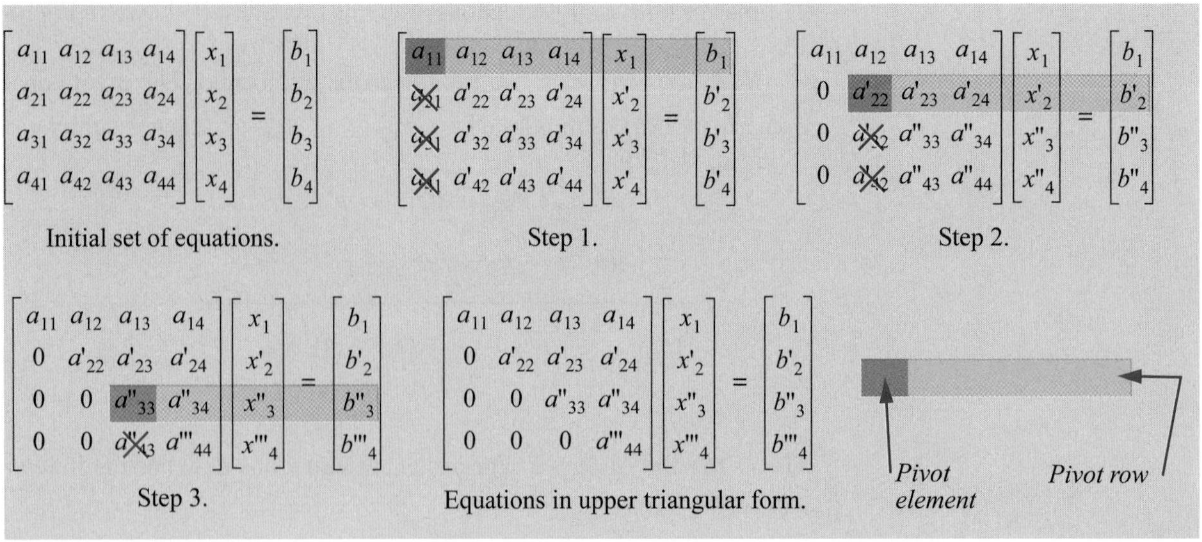

$$\begin{bmatrix} a_{11} & a_{12} & a_{13} & a_{14} \\ a_{21} & a_{22} & a_{23} & a_{24} \\ a_{31} & a_{32} & a_{33} & a_{34} \\ a_{41} & a_{42} & a_{43} & a_{44} \end{bmatrix} \begin{bmatrix} x_1 \\ x_2 \\ x_3 \\ x_4 \end{bmatrix} = \begin{bmatrix} b_1 \\ b_2 \\ b_3 \\ b_4 \end{bmatrix}$$

Initial set of equations.

$$\begin{bmatrix} a_{11} & a_{12} & a_{13} & a_{14} \\ \cancel{a_{21}} & a'_{22} & a'_{23} & a'_{24} \\ \cancel{a_{31}} & a'_{32} & a'_{33} & a'_{34} \\ \cancel{a_{41}} & a'_{42} & a'_{43} & a'_{44} \end{bmatrix} \begin{bmatrix} x_1 \\ x'_2 \\ x'_3 \\ x'_4 \end{bmatrix} = \begin{bmatrix} b_1 \\ b'_2 \\ b'_3 \\ b'_4 \end{bmatrix}$$

Step 1.

$$\begin{bmatrix} a_{11} & a_{12} & a_{13} & a_{14} \\ 0 & a'_{22} & a'_{23} & a'_{24} \\ 0 & \cancel{a'_{32}} & a''_{33} & a''_{34} \\ 0 & \cancel{a'_{42}} & a''_{43} & a''_{44} \end{bmatrix} \begin{bmatrix} x_1 \\ x'_2 \\ x''_3 \\ x''_4 \end{bmatrix} = \begin{bmatrix} b_1 \\ b'_2 \\ b''_3 \\ b''_4 \end{bmatrix}$$

Step 2.

$$\begin{bmatrix} a_{11} & a_{12} & a_{13} & a_{14} \\ 0 & a'_{22} & a'_{23} & a'_{24} \\ 0 & 0 & a''_{33} & a''_{34} \\ 0 & 0 & \cancel{a'''_{43}} & a'''_{44} \end{bmatrix} \begin{bmatrix} x_1 \\ x'_2 \\ x''_3 \\ x'''_4 \end{bmatrix} = \begin{bmatrix} b_1 \\ b'_2 \\ b''_3 \\ b'''_4 \end{bmatrix}$$

Step 3.

$$\begin{bmatrix} a_{11} & a_{12} & a_{13} & a_{14} \\ 0 & a'_{22} & a'_{23} & a'_{24} \\ 0 & 0 & a''_{33} & a''_{34} \\ 0 & 0 & 0 & a'''_{44} \end{bmatrix} \begin{bmatrix} x_1 \\ x'_2 \\ x''_3 \\ x'''_4 \end{bmatrix} = \begin{bmatrix} b_1 \\ b'_2 \\ b''_3 \\ b'''_4 \end{bmatrix}$$

Equations in upper triangular form.

Pivot element　　　　Pivot row

Figure 4-15: Gauss elimination procedure.

Example 4-1 shows a manual application of the Gauss elimination method for solving a system of four equations.

Example 4-1: Solving a set of four equations using Gauss elimination.

Solve the following system of four equations using the Gauss elimination method.

$$4x_1 - 2x_2 - 3x_3 + 6x_4 = 12$$
$$-6x_1 + 7x_2 + 6.5x_3 - 6x_4 = -6.5$$
$$x_1 + 7.5x_2 + 6.25x_3 + 5.5x_4 = 16$$
$$-12x_1 + 22x_2 + 15.5x_3 - x_4 = 17$$

SOLUTION

The solution follows the steps presented in the previous pages.

Step 1: The first equation is the pivot equation, and 4 is the pivot coefficient.

Multiply the pivot equation by $m_{21} = (-6)/4 = -1.5$ and subtract it from the second equation:

$$\begin{array}{r} -6x_1 + 7x_2 + 6.5x_3 - 6x_4 = -6.5 \\ \underline{(-1.5)(4x_1 - 2x_2 - 3x_3 + 6x_4) = (-6/4) \cdot 12} \\ 0x_1 + 4x_2 + 2x_3 + 3x_4 = 11.5 \end{array}$$

Multiply the pivot equation by $m_{31} = (1/4) = 0.25$ and subtract it from the third equation:

$$\begin{array}{r} x_1 + 7.5x_2 + 6.25x_3 + 5.5x_4 = 16 \\ \underline{(0.25)(4x_1 - 2x_2 - 3x_3 + 6x_4) = (1/4) \cdot 12} \\ 0x_1 + 8x_2 + 7x_3 + 4x_4 = 13 \end{array}$$

Multiply the pivot equation by $m_{41} = (-12)/4 = -3$ and subtract it from the fourth equation:

$$\begin{array}{r} -12x_1 + 22x_2 + 15.5x_3 - x_4 = 17 \\ \underline{(-3)(4x_1 - 2x_2 - 3x_3 + 6x_4) = -3 \cdot 12} \\ 0x_1 + 16x_2 + 6.5x_3 + 17x_4 = 53 \end{array}$$

At the end of *Step 1*, the four equations have the form:

$$4x_1 - 2x_2 - 3x_3 + 6x_4 = 12$$
$$4x_2 + 2x_3 + 3x_4 = 11.5$$
$$8x_2 + 7x_3 + 4x_4 = 13$$
$$16x_2 + 6.5x_3 + 17x_4 = 53$$

Step 2: The second equation is the pivot equation, and 4 is the pivot coefficient.
Multiply the pivot equation by $m_{32} = 8/4 = 2$ and subtract it from the third equation:

$$
\begin{array}{r}
8x_2 + 7x_3 + 4x_4 = 13 \\
- \quad 2(4x_2 + 2x_3 + 3x_4) = 2 \cdot 11.5 \\
\hline
0x_2 + 3x_3 - 2x_4 = -10
\end{array}
$$

Multiply the pivot equation by $m_{42} = 16/4 = 4$ and subtract it from the fourth equation:

$$
\begin{array}{r}
16x_2 + 6.5x_3 + 17x_4 = 53 \\
- \quad 4(4x_2 + 2x_3 + 3x_4) = 4 \cdot 11.5 \\
\hline
0x_2 - 1.5x_3 + 5x_4 = 7
\end{array}
$$

At the end of *Step 2*, the four equations have the form:

$$4x_1 - 2x_2 - 3x_3 + 6x_4 = 12$$
$$4x_2 + 2x_3 + 3x_4 = 11.5$$
$$3x_3 - 2x_4 = -10$$
$$-1.5x_3 + 5x_4 = 7$$

Step 3: The third equation is the pivot equation, and 3 is the pivot coefficient.
Multiply the pivot equation by $m_{43} = (-1.5)/3 = -0.5$ and subtract it from the fourth equation:

$$
\begin{array}{r}
-1.5x_3 + 5x_4 = 7 \\
- \quad -0.5(3x_3 - 2x_4) = -0.5 \cdot -10 \\
\hline
0x_3 + 4x_4 = 2
\end{array}
$$

At the end of *Step 3*, the four equations have the form:

$$4x_1 - 2x_2 - 3x_3 + 6x_4 = 12$$
$$4x_2 + 2x_3 + 3x_4 = 11.5$$
$$3x_3 - 2x_4 = -10$$
$$4x_4 = 2$$

Once the equations are in this form, the solution can be determined by back substitution. The value of x_4 is determined by solving the fourth equation:

$$x_4 = 2/4 = 0.5$$

Next, x_4 is substituted in the third equation, which is solved for x_3:

$$x_3 = \frac{-10 + 2x_4}{3} = \frac{-10 + 2 \cdot 0.5}{3} = -3$$

Next, x_4 and x_3 are substituted in the second equation, which is solved for x_2:

$$x_2 = \frac{11.5 - 2x_3 - 3x_4}{4} = \frac{11.5 - (2 \cdot -3) - (3 \cdot 0.5)}{4} = 4$$

Lastly, x_4, x_3 and x_2 are substituted in the first equation, which is solved for x_1:

$$x_1 = \frac{12 + 2x_2 + 3x_3 - 6x_4}{4} = \frac{12 + 2 \cdot 4 + 3 \cdot -3 - (6 \cdot 0.5)}{4} = 2$$

The extension of the Gauss elimination procedure to a system with n number of equations is straightforward. The elimination procedure starts with the first row as the pivot row and continues row after row down to one row before the last. At each step the pivot row is used to eliminate the terms that are below the pivot element in all the rows that are below. Once the original system of equations is changed to upper triangular form, back substitution is used for determining the solution.

When the Gauss elimination method is programmed, it is convenient and more efficient to create one matrix that includes the matrix of coefficients $[a]$ and the right-hand-side vector $[b]$. This is done by appending the vector $[b]$ to the matrix $[a]$, as shown in Example 4-2 where the Gauss elimination method is programmed in MATLAB.

Example 4-2: MATLAB user-defined function for solving a system of equations using Gauss elimination.

Write a user-defined MATLAB function for solving a system of linear equations, $[a][x] = [b]$, using the Gauss elimination method. For function name and arguments use x = Gauss (a,b), where a is the matrix of coefficients, b is the right-hand-side column vector of constants, and x is a column vector of the solution.

Use the user-defined function Gauss to:

(a) Solve the system of equations of Example 4-1.

(b) Solve the system of Eqs. (4.1).

SOLUTION

The following user-defined MATLAB function solves a system of linear equations. The program starts by appending the column vector $[b]$ to the matrix $[a]$. The new augmented matrix, named in the program ab, has the form:

$$\begin{bmatrix} a_{11} & a_{12} & a_{13} & \cdots & a_{1n} & b_1 \\ a_{21} & a_{22} & a_{23} & \cdots & a_{2n} & b_2 \\ a_{31} & a_{32} & a_{33} & \cdots & a_{3n} & b_3 \\ \cdots & \cdots & \cdots & \cdots & \cdots & \cdots \\ a_{n1} & a_{n2} & a_{n3} & \cdots & a_{nn} & b_n \end{bmatrix}$$

Next, the Gauss elimination procedure is applied (forward elimination). The matrix is changed such that all the elements below the diagonal of a are zero:

$$\begin{bmatrix} a_{11} & a_{12} & a_{13} & \cdots & a_{1n} & b_1 \\ 0 & a_{22} & a_{23} & \cdots & a_{2n} & b_2 \\ 0 & 0 & a_{33} & \cdots & \cdots & \cdots \\ \cdots & \cdots & \cdots & \cdots & \cdots & \cdots \\ 0 & 0 & 0 & \cdots & a_{nn} & b_n \end{bmatrix}$$

At the end of the program back substitution is used to solve for the unknowns, and the results are assigned to the column vector x.

Program 4-1: User-defined function. Gauss elimination.

```
function x = Gauss(a,b)
% The function solves a system of linear equations [a][x] = [b] using the Gauss
% elimination method.
% Input variables:
% a  The matrix of coefficients.
% b  Right-hand-side column vector of constants.
% Output variable:
% x  A column vector with the solution.
ab = [a,b];
[R, C] = size(ab);
for j = 1:R - 1
    for i = j + 1:R
        ab(i,j:C) = ab(i,j:C) - ab(i,j)/ab(j,j)*ab(j,j:C);
    end
end
x = zeros(R,1);
x(R) = ab(R,C)/ab(R,R);
for i = R - 1:-1:1
    x(i) = (ab(i,C) - ab(i,i + 1:R)*x(i + 1:R))/ab(i,i);
end
```

Append the column vector $[b]$ to the matrix $[a]$.

Pivot element.

Gauss elimination procedure (forward elimination).

The multiplier m_{ij}. Pivot equation.

Back substitution.

The user-defined function Gauss is next used in the Command Window, first to solve the system of equations of Example 4-1, and then to solve the system of Eqs. (4.1).

```
>> A=[4 -2 -3 6; -6 7 6.5 -6; 1 7.5 6.25 5.5; -12 22 15.5 -1];
>> B = [12; -6.5; 16; 17];
>> sola = Gauss(A,B)
sola =
    2.0000
    4.0000
   -3.0000
    0.5000
>> C = [9 -4 -2 0; -4 17 -6 -3; -2 -6 14 -6; 0 -3 -6 11];
>> D = [24; -16; 0; 18];
>> solb = Gauss(C,D)
solb =
    4.0343
    1.6545
    2.8452
    3.6395
```

Solution for part (a).

Solution for part (b).

4.2.1 Potential Difficulties When Applying the Gauss Elimination Method

The pivot element is zero

Since the pivot row is divided by the pivot element, a problem will arise during the execution of the Gauss elimination procedure if the value of the pivot element is equal to zero. As shown in the next section, this situation can be corrected by changing the order of the rows. In a procedure called pivoting, the pivot row that has the zero pivot element is exchanged with another row that has a nonzero pivot element.

The pivot element is small relative to the other terms in the pivot row

Significant errors due to rounding can occur when the pivot element is small relative to other elements in the pivot row. This is illustrated by the following example.

Consider the following system of simultaneous equations for the unknowns x_1 and x_2:

$$0.0003x_1 + 12.34x_2 = 12.343$$
$$0.4321x_1 + x_2 = 5.321$$

$$(4.15)$$

The exact solution of the system is $x_1 = 10$ and $x_2 = 1$.

The error due to rounding is illustrated by solving the system using Gaussian elimination on a machine with limited precision so that only four significant figures are retained with rounding. When the first equation of Eqs. (4.15) is entered, the constant on the right-hand side is rounded to 12.34.

The solution starts by using the first equation as the pivot equation and $a_{11} = 0.0003$ as the pivot coefficient. In the first step, the pivot equation is multiplied by $m_{21} = 0.4321/0.0003 = 1440$. With four significant figures and rounding this operation gives:

$$(1440)(0.0003x_1 + 12.34x_2) = 1440 \cdot 12.34$$

or:

$$0.4320x_1 + 17770x_2 = 17770$$

The result is next subtracted from the second equation in Eqs. (4.15):

$$
\begin{aligned}
0.4321x_1 + x_2 &= 5.321 \\
- \quad \underline{0.4320x_1 + 17770x_2} &= \underline{17770} \\
0.0001x_1 - 17770x_2 &= -17760
\end{aligned}
$$

After this operation the system is:

$$0.0003x_1 + 12.34x_2 = 12.34$$
$$0.0001x_1 - 17770x_2 = -17760$$

Note that the a_{21} element is not zero but a very small number. Next, the value of x_2 is calculated from the second equation:

$$x_2 = \frac{-17760}{-17770} = 0.9994$$

Then x_2 is substituted in the first equation, which is solved for x_1:

$$x_1 = \frac{12.34 - (12.34 \cdot 0.9994)}{0.0003} = \frac{12.34 - 12.33}{0.0003} = \frac{0.01}{0.0003} = 33.33$$

The solution that is obtained for x_1 is obviously incorrect. The incorrect value is obtained because the magnitude of a_{11} is small when compared to the magnitude of a_{12}. Consequently, a relatively small error (due to round-off arising from the finite precision of a computing machine) in the value of x_2 can lead to a large error in the value of x_1.

The problem can be easily remedied by exchanging the order of the two equations in Eq. (4.15):

$$0.4321x_1 + x_2 = 5.321$$
$$0.0003x_1 + 12.34x_2 = 12.343$$

(4.16)

Now as the first equation is used as the pivot equation, the pivot coefficient is $a_{11} = 0.4321$. In the first step, the pivot equation is multiplied by $m_{21} = 0.0003/0.4321 = 0.0006943$. With four significant figures and rounding this operation gives:

$$(0.0006943)(0.4321x_1 + x_2 = 5.321) = 0.0006943 \cdot 5.321$$

or:

$$0.0003x_1 + 0.0006943x_2 = 0.003694$$

The result is next subtracted from the second equation in Eqs. (4.16):

$$\begin{array}{r} 0.0003x_1 + 12.34x_2 = 12.34 \\ - \quad \underline{0.0003x_1 + 0.0006943x_2 = 0.003694} \\ 12.34x_2 = 12.34 \end{array}$$

After this operation the system is:

$$0.4321x_1 + x_2 = 5.321$$
$$0x_1 + 12.34x_2 = 12.34$$

Next, the value of x_2 is calculated from the second equation:

$$x_2 = \frac{12.34}{12.34} = 1$$

Then x_2 is substituted in the first equation that is solved for x_1:

$$x_1 = \frac{5.321 - 1}{0.4321} = 10$$

The solution that is obtained now is the exact solution.

In general, a more accurate solution is obtained when the equations are arranged (and rearranged every time a new pivot equation is used) such that the pivot equation has the largest possible pivot element. This is explained in more detail in the next section.

Round-off errors can also be significant when solving large systems of equations even when all the coefficients in the pivot row are of the same order of magnitude. This can be caused by a large number of operations (multiplication, division, addition, and subtraction) associated with large systems.

4.3 GAUSS ELIMINATION WITH PIVOTING

In the Gauss elimination procedure, the pivot equation is divided by the pivot coefficient. This, however, cannot be done if the pivot coefficient is zero. For example, for the following system of three equations:

$$0x_1 + 2x_2 + 3x_3 = 46$$
$$4x_1 - 3x_2 + 2x_3 = 16$$
$$2x_1 + 4x_2 - 3x_3 = 12$$

After the first step, the second equation has a pivot element that is equal to zero.

$$\begin{bmatrix} a_{11} & a_{12} & a_{13} & a_{14} \\ 0 & \boxed{0} & a'_{23} & a'_{24} \\ 0 & a'_{32} & a'_{33} & a'_{34} \\ 0 & a'_{42} & a'_{43} & a'_{44} \end{bmatrix} \begin{bmatrix} x_1 \\ x_2 \\ x_3 \\ x_4 \end{bmatrix} = \begin{bmatrix} b_1 \\ b'_2 \\ b'_3 \\ b'_4 \end{bmatrix}$$

Using pivoting, the second equation is exchanged with the fourth equation.

$$\begin{bmatrix} a_{11} & a_{12} & a_{13} & a_{14} \\ 0 & \boxed{a'_{32}} & a'_{33} & a'_{34} \\ 0 & 0 & a'_{23} & a'_{24} \\ 0 & a'_{42} & a'_{43} & a'_{44} \end{bmatrix} \begin{bmatrix} x_1 \\ x_2 \\ x_3 \\ x_4 \end{bmatrix} = \begin{bmatrix} b_1 \\ b'_3 \\ b'_2 \\ b'_4 \end{bmatrix}$$

Figure 4-16: Illustration of pivoting.

the procedure starts by taking the first equation as the pivot equation and the coefficient of x_1, which is 0, as the pivot coefficient. To eliminate the term $4x_1$ in the second equation, the pivot equation is supposed to be multiplied by $4/0$ and then subtracted from the second equation. Obviously, this is not possible when the pivot element is equal to zero. The division by zero can be avoided if the order in which the equations are written is changed such that in the first equation the first coefficient is not zero. For example, in the system above this can be done by exchanging the first two equations.

In the general Gauss elimination procedure, an equation (or a row) can be used as the pivot equation (pivot row) only if the pivot coefficient (pivot element) is not zero. If the pivot element is zero, the equation (i.e., the row) is exchanged with one of the equations (rows) that are below, which has a nonzero pivot coefficient. This exchange of rows, illustrated in Fig. 4-16, is called ***pivoting***.

Additional comments about pivoting

- If during the Gauss elimination procedure a pivot equation has a pivot element that is equal to zero, then if the system of equations that is being solved has a solution, an equation with a nonzero element in the pivot position can always be found.

- The numerical calculations are less prone to error and will have fewer round-off errors (see Section 4.2.1) if the pivot element has a larger numerical absolute value compared to the other elements in the same row. Consequently, among all the equations that can be exchanged to be the pivot equation, it is better to select the equation whose pivot element has the largest absolute numerical value. Moreover, it is good to employ pivoting for the purpose of having a pivot equation with the pivot element that has a largest absolute numerical value at all times (even when pivoting is not necessary).

The addition of pivoting to the programming of the Gauss elimination method is shown in the next example. The addition of pivoting every time a new pivot equation is used, such that the pivot row will have the largest absolute pivot element, is assigned as an exercise in Problem 4.23.

Example 4-3: MATLAB user-defined function for solving a system of equations using Gauss elimination with pivoting.

Write a user-defined MATLAB function for solving a system of linear equations $[a][x] = [b]$ using the Gauss elimination method with pivoting. Name the function x = GaussPivot(a,b), where a is the matrix of coefficients, b is the right-hand-side column vector of constants, and x is a column vector of the solution. Use the function to determine the forces in the loaded eight-member truss that is shown in the figure (same as in Fig. 4-2).

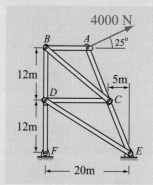

SOLUTION

The forces in the eight truss members are determined from the set of eight equations, Eq. (4.2). The equations are derived by drawing free body diagrams of pins A, B, C, and D and applying equations of equilibrium. The equations are rewritten here in a matrix form (intentionally, the equations are written in an order that requires pivoting):

$$
\begin{bmatrix}
0 & 0.9231 & 0 & 0 & 0 & 0 & 0 & 0 \\
-1 & -0.3846 & 0 & 0 & 0 & 0 & 0 & 0 \\
0 & 0 & 0 & 0 & 1 & 0 & 0.8575 & 0 \\
1 & 0 & -0.7809 & 0 & 0 & 0 & 0 & 0 \\
0 & -0.3846 & -0.7809 & 0 & -1 & 0.3846 & 0 & 0 \\
0 & 0.9231 & 0.6247 & 0 & 0 & -0.9231 & 0 & 0 \\
0 & 0 & 0.6247 & -1 & 0 & 0 & 0 & 0 \\
0 & 0 & 0 & 1 & 0 & 0 & -0.5145 & -1
\end{bmatrix}
\begin{bmatrix}
F_{AB} \\ F_{AC} \\ F_{BC} \\ F_{BD} \\ F_{CD} \\ F_{CE} \\ F_{DE} \\ F_{DF}
\end{bmatrix}
=
\begin{bmatrix}
1690 \\ 3625 \\ 0 \\ 0 \\ 0 \\ 0 \\ 0 \\ 0
\end{bmatrix}
\qquad (4.17)
$$

The function GaussPivot is created by modifying the function Gauss listed in the solution of Example 4-2.

> **Program 4-2: User-defined function. Gauss elimination with pivoting.**

```
function x = GaussPivot(a,b)
% The function solves a system of linear equations ax = b using the Gauss
% elimination method with pivoting.
% Input variables:
% a  The matrix of coefficients.
% b  Right-hand-side column vector of constants.
% Output variable:
% x  A column vector with the solution.

ab = [a,b];
[R, C] = size(ab);
for j = 1:R - 1
% Pivoting section starts
    if ab(j,j) = = 0
```

Check if the pivot element is zero.

```
        for k = j + 1:R
            if ab(k,j) ~ = 0
                abTemp = ab(j,:);
                ab(j,:) = ab(k,:);
                ab(k,:) = abTemp;
                break
            end
        end
    end
% Pivoting section ends
    for i = j + 1:R
        ab(i,j:C) = ab(i,j:C) - ab(i,j)/ab(j,j)*ab(j,j:C);
    end
end
x = zeros(R,1);
x(R) = ab(R,C)/ab(R,R);
for i = R - 1:-1:1
    x(i) = (ab(i,C) - ab(i,i + 1:R)*x(i + 1:R))/ab(i,i);
end
```

> If pivoting is required, search in the rows below for a row with nonzero pivot element.

> Switch the row that has a zero pivot element with the row that has a nonzero pivot element.

> Stop searching for a row with a nonzero pivot element.

The user-defined function `GaussPivot` is next used in a script file program to solve the system of equations Eq. (4.17).

```
% Example 4-3
a = [0 0.9231 0 0 0 0 0 0; -1 -0.3846 0 0 0 0 0 0; 0 0 0 0 1 0 0.8575 0; 1 0 -0.7809 0 0 0 0 0
    0 -0.3846 -0.7809 0 -1 0.3846 0 0; 0 0.9231 0.6247 0 0 -0.9231 0 0
    0 0 0.6247 -1 0 0 0 0; 0 0 0 1 0 0 -0.5145 -1];
b = [1690;3625;0;0;0;0;0;0];
Forces = GaussPivot(a,b)
```

When the script file is executed, the following solution is displayed in the Command Window.

$$
Forces =
\begin{bmatrix}
-4.3291e+003 \\
1.8308e+003 \\
-5.5438e+003 \\
-3.4632e+003 \\
2.8862e+003 \\
-1.9209e+003 \\
-3.3659e+003 \\
-1.7315e+003
\end{bmatrix}
\quad
\begin{bmatrix}
F_{AB} \\
F_{AC} \\
F_{BC} \\
F_{BD} \\
F_{CD} \\
F_{CE} \\
F_{DE} \\
F_{DF}
\end{bmatrix}
$$

```
>>
```

4.4 GAUSS–JORDAN ELIMINATION METHOD

The Gauss–Jordan elimination method is a procedure for solving a system of linear equations, $[a][x] = [b]$. In this procedure, a system of equations that is given in a general form is manipulated into an equivalent system of equations in *diagonal* form (see Section 4.1.1) with normalized elements along the diagonal. This means that when the diagonal form of the matrix of the coefficients, $[a]$, is reduced to the identity matrix, the new vector $[b']$ is the solution. The starting point of the procedure is a system of equations given in a general form (the illustration that follows is for a system of four equations):

$$
\begin{align}
a_{11}x_1 + a_{12}x_2 + a_{13}x_3 + a_{14}x_4 &= b_1 & (4.18a) \\
a_{21}x_1 + a_{22}x_2 + a_{23}x_3 + a_{24}x_4 &= b_2 & (4.18b) \\
a_{31}x_1 + a_{32}x_2 + a_{33}x_3 + a_{34}x_4 &= b_3 & (4.18c) \\
a_{41}x_1 + a_{42}x_2 + a_{43}x_3 + a_{44}x_4 &= b_4 & (4.18d)
\end{align}
\tag{4.18}
$$

$$
\begin{bmatrix}
a_{11} & a_{12} & a_{13} & a_{14} \\
a_{21} & a_{22} & a_{23} & a_{24} \\
a_{31} & a_{32} & a_{33} & a_{34} \\
a_{41} & a_{42} & a_{43} & a_{44}
\end{bmatrix}
\begin{bmatrix} x_1 \\ x_2 \\ x_3 \\ x_4 \end{bmatrix}
=
\begin{bmatrix} b_1 \\ b_2 \\ b_3 \\ b_4 \end{bmatrix}
$$

Figure 4-17: Matrix form of a system of four equations.

The matrix form of the system is shown in Fig. 4-17. In the Gauss–Jordan elimination method, the system of equations is manipulated to have the following diagonal form:

$$
\begin{align}
x_1 + 0 + 0 + 0 &= b'_1 & (4.19a) \\
0 + x_2 + 0 + 0 &= b'_2 & (4.19b) \\
0 + 0 + x_3 + 0 &= b'_3 & (4.19c) \\
0 + 0 + 0 + x_4 &= b'_4 & (4.19d)
\end{align}
\tag{4.19}
$$

$$
\begin{bmatrix}
1 & 0 & 0 & 0 \\
0 & 1 & 0 & 0 \\
0 & 0 & 1 & 0 \\
0 & 0 & 0 & 1
\end{bmatrix}
\begin{bmatrix} x_1 \\ x_2 \\ x_3 \\ x_4 \end{bmatrix}
=
\begin{bmatrix} b'_1 \\ b'_2 \\ b'_3 \\ b'_4 \end{bmatrix}
$$

Figure 4-18: Matrix form of the equivalent system after applying the Gauss–Jordan method.

The matrix form of the equivalent system is shown in Fig. 4-18. The terms on the right-hand side of the equations (column $[b']$) are the solution. In matrix form, the matrix of the coefficients is transformed into an identity matrix.

Gauss–Jordan elimination procedure

The Gauss–Jordan elimination procedure for transforming the system of equations from the form in Eqs. (4.18) to the form in Eqs. (4.19) is the same as the Gauss elimination procedure (see Section 4.2), except for the following two differences:

- The pivot equation is normalized by dividing all the terms in the equation by the pivot coefficient. This makes the pivot coefficient equal to 1.

- The pivot equation is used to eliminate the off-diagonal terms in **ALL** the other equations. This means that the elimination process is applied to the equations (rows) that are above and below the pivot equation. (In the Gaussian elimination method, only elements that are below the pivot element are eliminated.)

When the Gauss–Jordan procedure is programmed, it is convenient and more efficient to create a single matrix that includes the matrix of coefficients $[a]$ and the vector $[b]$. This is done by appending the vector $[b]$ to the matrix $[a]$. The augmented matrix at the starting point of the procedure is shown (for a system of four equations) in Fig. 4-19a. At the end of the procedure, shown in Fig. 4-19b, the elements of $[a]$ are replaced by an identity matrix, and the column $[b']$ is the solution.

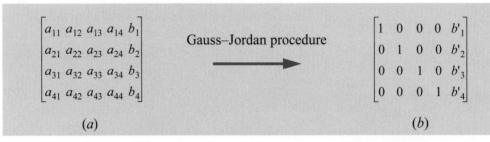

$$
\begin{bmatrix}
a_{11} & a_{12} & a_{13} & a_{14} & b_1 \\
a_{21} & a_{22} & a_{23} & a_{24} & b_2 \\
a_{31} & a_{32} & a_{33} & a_{34} & b_3 \\
a_{41} & a_{42} & a_{43} & a_{44} & b_4
\end{bmatrix}
\quad \xrightarrow{\text{Gauss–Jordan procedure}} \quad
\begin{bmatrix}
1 & 0 & 0 & 0 & b'_1 \\
0 & 1 & 0 & 0 & b'_2 \\
0 & 0 & 1 & 0 & b'_3 \\
0 & 0 & 0 & 1 & b'_4
\end{bmatrix}
$$

(a)   (b)

Figure 4-19: Schematic illustration of the Gauss–Jordan method.

The Gauss–Jordan method can also be used for solving several systems of equations $[a][x] = [b]$ that have the same coefficients $[a]$ but different right-hand-side vectors $[b]$. This is done by augmenting the matrix $[a]$ to include all of the vectors $[b]$. In Section 4.6.2 the method is used in this way for calculating the inverse of a matrix.

The Gauss–Jordan elimination method is demonstrated in Example 4-4 where it is used to solve the set of equations solved in Example 4-1.

Example 4-4: Solving a set of four equations using Gauss–Jordan elimination.

Solve the following set of four equations using the Gauss–Jordan elimination method.

$$4x_1 - 2x_2 - 3x_3 + 6x_4 = 12$$
$$-6x_1 + 7x_2 + 6.5x_3 - 6x_4 = -6.5$$
$$x_1 + 7.5x_2 + 6.25x_3 + 5.5x_4 = 16$$
$$-12x_1 + 22x_2 + 15.5x_3 - x_4 = 17$$

SOLUTION
The solution is carried out by using the matrix form of the equations. In matrix form, the system is:

$$
\begin{bmatrix}
4 & -2 & -3 & 6 \\
-6 & 7 & 6.5 & -6 \\
1 & 7.5 & 6.25 & 5.5 \\
-12 & 22 & 15.5 & -1
\end{bmatrix}
\begin{bmatrix}
x_1 \\ x_2 \\ x_3 \\ x_4
\end{bmatrix}
=
\begin{bmatrix}
12 \\ -6.5 \\ 16 \\ 17
\end{bmatrix}
$$

For the numerical procedure, a new matrix is created by augmenting the coefficient matrix to include the right-hand side of the equation:

$$
\begin{bmatrix}
4 & -2 & -3 & 6 & 12 \\
-6 & 7 & 6.5 & -6 & -6.5 \\
1 & 7.5 & 6.25 & 5.5 & 16 \\
-12 & 22 & 15.5 & -1 & 17
\end{bmatrix}
$$

The first pivoting row is the first row, and the first element in this row is the pivot element. The row is normalized by dividing it by the pivot element:

$$
\begin{bmatrix}
\dfrac{4}{4} & \dfrac{-2}{4} & \dfrac{-3}{4} & \dfrac{6}{4} & \dfrac{12}{4} \\
-6 & 7 & 6.5 & -6 & -6.5 \\
1 & 7.5 & 6.25 & 5.5 & 16 \\
-12 & 22 & 15.5 & -1 & 17
\end{bmatrix}
=
\begin{bmatrix}
1 & -0.5 & -0.75 & 1.5 & 3 \\
-6 & 7 & 6.5 & -6 & -6.5 \\
1 & 7.5 & 6.25 & 5.5 & 16 \\
-12 & 22 & 15.5 & -1 & 17
\end{bmatrix}
$$

Next, all the first elements in rows 2, 3, and 4 are eliminated:

$$
\begin{bmatrix}
1 & -0.5 & -0.75 & 1.5 & 3 \\
-6 & 7 & 6.5 & -6 & -6.5 \\
1 & 7.5 & 6.25 & 5.5 & 16 \\
-12 & 22 & 15.5 & -1 & 17
\end{bmatrix}
\begin{array}{l}
\\
\leftarrow -(-6)\begin{bmatrix}1 & -0.5 & -0.75 & 1.5 & 3\end{bmatrix} \\
\leftarrow -(1)\begin{bmatrix}1 & -0.5 & -0.75 & 1.5 & 3\end{bmatrix} \\
\leftarrow -(-12)\begin{bmatrix}1 & -0.5 & -0.75 & 1.5 & 3\end{bmatrix}
\end{array}
=
\begin{bmatrix}
1 & -0.5 & -0.75 & 1.5 & 3 \\
0 & 4 & 2 & 3 & 11.5 \\
0 & 8 & 7 & 4 & 13 \\
0 & 16 & 6.5 & 17 & 53
\end{bmatrix}
$$

The next pivot row is the second row, with the second element as the pivot element. The row is normalized by dividing it by the pivot element:

$$
\begin{bmatrix}
1 & -0.5 & -0.75 & 1.5 & 3 \\
0 & \dfrac{4}{4} & \dfrac{2}{4} & \dfrac{3}{4} & \dfrac{11.5}{4} \\
0 & 8 & 7 & 4 & 13 \\
0 & 16 & 6.5 & 17 & 53
\end{bmatrix}
=
\begin{bmatrix}
1 & -0.5 & -0.75 & 1.5 & 3 \\
0 & 1 & 0.5 & 0.75 & 2.875 \\
0 & 8 & 7 & 4 & 13 \\
0 & 16 & 6.5 & 17 & 53
\end{bmatrix}
$$

Next, all the second elements in rows 1, 3, and 4 are eliminated:

$$
\begin{bmatrix}
1 & -0.5 & -0.75 & 1.5 & 3 \\
0 & 1 & 0.5 & 0.75 & 2.875 \\
0 & 8 & 7 & 4 & 13 \\
0 & 16 & 6.5 & 17 & 53
\end{bmatrix}
\begin{array}{l}
\leftarrow -(-0.5)\begin{bmatrix}0 & 1 & 0.5 & 0.75 & 2.875\end{bmatrix} \\
\\
\leftarrow -(8)\begin{bmatrix}0 & 1 & 0.5 & 0.75 & 2.875\end{bmatrix} \\
\leftarrow -(16)\begin{bmatrix}0 & 1 & 0.5 & 0.75 & 2.875\end{bmatrix}
\end{array}
=
\begin{bmatrix}
1 & 0 & -0.5 & 1.875 & 4.4375 \\
0 & 1 & 0.5 & 0.75 & 2.875 \\
0 & 0 & 3 & -2 & -10 \\
0 & 0 & -1.5 & 5 & 7
\end{bmatrix}
$$

The next pivot row is the third row, with the third element as the pivot element. The row is normalized by dividing it by the pivot element:

$$
\begin{bmatrix}
1 & 0 & -0.5 & 1.875 & 4.4375 \\
0 & 1 & 0.5 & 0.75 & 2.875 \\
0 & 0 & \dfrac{3}{3} & \dfrac{-2}{3} & \dfrac{-10}{3} \\
0 & 0 & -1.5 & 5 & 7
\end{bmatrix}
=
\begin{bmatrix}
1 & 0 & -0.5 & 1.875 & 4.4375 \\
0 & 1 & 0.5 & 0.75 & 2.875 \\
0 & 0 & 1 & -0.667 & -3.333 \\
0 & 0 & -1.5 & 5 & 7
\end{bmatrix}
$$

Next, all the third elements in rows 1, 2, and 4 are eliminated:

$$
\begin{bmatrix}
1 & 0 & -0.5 & 1.875 & 4.4375 \\
0 & 1 & 0.5 & 0.75 & 2.875 \\
0 & 0 & 1 & -0.667 & -3.333 \\
0 & 0 & -1.5 & 5 & 7
\end{bmatrix}
\begin{array}{l}
\leftarrow -(-0.5)\begin{bmatrix}0 & 0 & 1 & -0.667 & -3.333\end{bmatrix} \\
\leftarrow -(0.5)\begin{bmatrix}0 & 0 & 1 & -0.667 & -3.333\end{bmatrix} \\
\\
\leftarrow -(-1.5)\begin{bmatrix}0 & 0 & 1 & -0.667 & -3.333\end{bmatrix}
\end{array}
=
\begin{bmatrix}
1 & 0 & 0 & 1.5417 & 2.7708 \\
0 & 1 & 0 & 1.0833 & 4.5417 \\
0 & 0 & 1 & -0.667 & -3.333 \\
0 & 0 & 0 & 4 & 2
\end{bmatrix}
$$

The next pivot row is the fourth row, with the fourth element as the pivot element. The row is normalized by dividing it by the pivot element:

$$\begin{bmatrix} 1 & 0 & 0 & 1.5417 & 2.7708 \\ 0 & 1 & 0 & 1.0833 & 4.5417 \\ 0 & 0 & 1 & -0.667 & -3.333 \\ 0 & 0 & 0 & \dfrac{4}{4} & \dfrac{2}{4} \end{bmatrix} = \begin{bmatrix} 1 & 0 & 0 & 1.5417 & 2.7708 \\ 0 & 1 & 0 & 1.0833 & 4.5417 \\ 0 & 0 & 1 & -0.667 & -3.333 \\ 0 & 0 & 0 & 1 & 0.5 \end{bmatrix}$$

Next, all the fourth elements in rows 1, 2, and 3 are eliminated

$$\begin{bmatrix} 1 & 0 & 0 & 1.5417 & 2.7708 \\ 0 & 1 & 0 & 1.0833 & 4.5417 \\ 0 & 0 & 1 & -0.667 & -3.333 \\ 0 & 0 & 0 & 1 & 0.5 \end{bmatrix} \begin{matrix} \longleftarrow -(1.5417)\begin{bmatrix}0 & 0 & 0 & 1 & 0.5\end{bmatrix} \\ \longleftarrow -(1.0833)\begin{bmatrix}0 & 0 & 0 & 1 & 0.5\end{bmatrix} \\ \longleftarrow -(-0.667)\begin{bmatrix}0 & 0 & 0 & 1 & 0.5\end{bmatrix} \\ {} \end{matrix} = \begin{bmatrix} 1 & 0 & 0 & 0 & 2 \\ 0 & 1 & 0 & 0 & 4 \\ 0 & 0 & 1 & 0 & -3 \\ 0 & 0 & 0 & 1 & 0.5 \end{bmatrix}$$

The solution is:

$$\begin{bmatrix} x_1 \\ x_2 \\ x_3 \\ x_4 \end{bmatrix} = \begin{bmatrix} 2 \\ 4 \\ -3 \\ 0.5 \end{bmatrix}$$

The Gauss–Jordan elimination method with pivoting

It is possible that the equations are written in such an order that during the elimination procedure a pivot equation has a pivot element that is equal to zero. Obviously, in this case it is impossible to normalize the pivot row (divide by the pivot element). As with the Gauss elimination method, the problem can be corrected by using pivoting. This is left as an exercise in Problem 4.24.

4.5 *LU* DECOMPOSITION METHOD

Background

The Gauss elimination method consists of two parts. The first part is the elimination procedure in which a system of linear equations that is given in a general form, $[a][x] = [b]$, is transformed into an equivalent system of equations $[a'][x] = [b']$ in which the matrix of coefficients $[a']$ is upper triangular. In the second part, the equivalent system is solved by using back substitution. The elimination procedure requires many mathematical operations and significantly more computing time than the back substitution calculations. During the elimination procedure, the matrix of coefficients $[a]$ and the vector $[b]$ are both changed. This means that if there is a need to solve systems of equations that have the same left-hand-side terms (same coefficient matrix $[a]$) but different right-hand-side constants (different vectors $[b]$), the elimination procedure has to be carried out for each $[b]$ again. Ideally, it would be better if the operations on the matrix of coefficients $[a]$ were dissociated from those on the vector of constants $[b]$. In this way, the elimination procedure with $[a]$ is done only once and then is used for solving systems of equations with different vectors $[b]$.

One option for solving various systems of equations $[a][x] = [b]$ that have the same coefficient matrices $[a]$ but different constant vectors $[b]$ is to first calculate the inverse of the matrix $[a]$. Once the inverse matrix $[a]^{-1}$ is known, the solution can be calculated by:

$$[x] = [a]^{-1}[b]$$

Calculating the inverse of a matrix, however, requires many mathematical operations, and is computationally inefficient. A more efficient method of solution for this case is the LU decomposition method.

In the LU decomposition method, the operations with the matrix $[a]$ are done without using, or changing, the vector $[b]$, which is only used in the substitution part of the solution. The LU decomposition method can be used for solving a single system of linear equations, but it is especially advantageous for solving systems that have the same coefficient matrices $[a]$ but different constant vectors $[b]$.

The LU decomposition method

The LU decomposition method is a method for solving a system of linear equations $[a][x] = [b]$. In this method the matrix of coefficients $[a]$ is decomposed (factored) into a product of two matrices $[L]$ and $[U]$:

$$[a] = [L][U] \tag{4.20}$$

where the matrix $[L]$ is a lower triangular matrix and $[U]$ is an upper triangular matrix. With this decomposition, the system of equations to be solved has the form:

$$[L][U][x] = [b] \tag{4.21}$$

To solve this equation, the product $[U][x]$ is defined as:

$$[U][x] = [y] \tag{4.22}$$

and is substituted in Eq. (4.21) to give:

$$[L][y] = [b] \tag{4.23}$$

Now, the solution $[x]$ is obtained in two steps. First, Eq. (4.23) is solved for $[y]$. Then, the solution $[y]$ is substituted in Eq. (4.22), and that equation is solved for $[x]$.

Since the matrix $[L]$ is a lower triangular matrix the solution $[y]$ in Eq. (4.23) is obtained by using the forward substitution method. Once $[y]$ is known and is substituted in Eq. (4.22), this equation is solved by using back substitution, since $[U]$ is an upper triangular matrix.

For a given matrix $[a]$ several methods can be used to determine the corresponding $[L]$ and $[U]$. Two of the methods, one related to the Gauss elimination method and another called Crout's method, are described next.

4.5.1 LU Decomposition Using the Gauss Elimination Procedure

When the Gauss elimination procedure is applied to a matrix $[a]$, the elements of the matrices $[L]$ and $[U]$ are actually calculated. The upper triangular matrix $[U]$ is the matrix of coefficients $[a]$ that is obtained at the end of the procedure, as shown in Figs. 4-8 and 4-14. The lower triangular matrix $[L]$ is not written explicitly during the procedure, but the elements that make up the matrix are actually calculated along the way. The elements of $[L]$ on the diagonal are all 1, and the elements below the diagonal are the multipliers m_{ij} that multiply the pivot equation when it is used to eliminate the elements below the pivot coefficient (see the **Gauss elimination procedure** in Section 4.2). For the case of a system of four equations, the matrix of coefficients $[a]$ is (4×4), and the decomposition has the form:

$$\begin{bmatrix} a_{11} & a_{12} & a_{13} & a_{14} \\ a_{21} & a_{22} & a_{23} & a_{24} \\ a_{31} & a_{32} & a_{33} & a_{34} \\ a_{41} & a_{42} & a_{43} & a_{44} \end{bmatrix} = \begin{bmatrix} 1 & 0 & 0 & 0 \\ m_{21} & 1 & 0 & 0 \\ m_{31} & m_{32} & 1 & 0 \\ m_{41} & m_{42} & m_{43} & 1 \end{bmatrix} \begin{bmatrix} a_{11} & a_{12} & a_{13} & a_{14} \\ 0 & a'_{22} & a'_{23} & a'_{24} \\ 0 & 0 & a''_{33} & a''_{34} \\ 0 & 0 & 0 & a'''_{44} \end{bmatrix} \quad (4.24)$$

A numerical example illustrating LU decomposition is given next. It uses the information in the solution of Example 4-1, where a system of four equations is solved by using the Gauss elimination method. The matrix $[a]$ can be written from the given set of equations in the problem statement, and the matrix $[U]$ can be written from the set of equations at the end of **step 3** (page 101). The matrix $[L]$ can be written by using the multipliers that are calculated in the solution. The decomposition has the form:

$$\begin{bmatrix} 4 & -2 & -3 & 6 \\ -6 & 7 & 6.5 & -6 \\ 1 & 7.5 & 6.25 & 5.5 \\ -12 & 22 & 15.5 & -1 \end{bmatrix} = \begin{bmatrix} 1 & 0 & 0 & 0 \\ -1.5 & 1 & 0 & 0 \\ 0.25 & 2 & 1 & 0 \\ -3 & 4 & -0.5 & 1 \end{bmatrix} \begin{bmatrix} 4 & -2 & -3 & 6 \\ 0 & 4 & 2 & 3 \\ 0 & 0 & 3 & -2 \\ 0 & 0 & 0 & 4 \end{bmatrix} \quad (4.25)$$

The decomposition in Eq. (4.25) can be verified by using MATLAB:

```
>> L = [1,0,0,0;-1.5,1,0,0;0.25,2,1,0;-3,4,-0.5,1]
L =
    1.0000        0        0        0
   -1.5000   1.0000        0        0
    0.2500   2.0000   1.0000        0
   -3.0000   4.0000  -0.5000   1.0000
>> U = [4,-2,-3,6;0,4,2,3;0,0,3,-2;0,0,0,4]
```

U =

4	-2	-3	6
0	4	2	3
0	0	3	-2
0	0	0	4

>> L*U

ans =

4.0000	-2.0000	-3.0000	6.0000
-6.0000	7.0000	6.5000	-6.0000
1.0000	7.5000	6.2500	5.5000
-12.0000	22.0000	15.5000	-1.0000

> Multiplication of the matrices L and U verifies that the answer is the matrix $[a]$.

4.5.2 LU Decomposition Using Crout's Method

In this method the matrix $[a]$ is decomposed into the product $[L][U]$, where the diagonal elements of the matrix $[U]$ are all 1s. It turns out that in this case, the elements of both matrices can be determined using formulas that can be easily programmed. This is illustrated for a system of four equations. In Crout's method the LU decomposition has the form:

$$\begin{bmatrix} a_{11} & a_{12} & a_{13} & a_{14} \\ a_{21} & a_{22} & a_{23} & a_{24} \\ a_{31} & a_{32} & a_{33} & a_{34} \\ a_{41} & a_{42} & a_{43} & a_{44} \end{bmatrix} = \begin{bmatrix} L_{11} & 0 & 0 & 0 \\ L_{21} & L_{22} & 0 & 0 \\ L_{31} & L_{32} & L_{33} & 0 \\ L_{41} & L_{42} & L_{43} & L_{44} \end{bmatrix} \begin{bmatrix} 1 & U_{12} & U_{13} & U_{14} \\ 0 & 1 & U_{23} & U_{24} \\ 0 & 0 & 1 & U_{34} \\ 0 & 0 & 0 & 1 \end{bmatrix} \qquad (4.26)$$

Executing the matrix multiplication on the right-hand side of the equation gives:

$$\begin{bmatrix} a_{11} & a_{12} & a_{13} & a_{14} \\ a_{21} & a_{22} & a_{23} & a_{24} \\ a_{31} & a_{32} & a_{33} & a_{34} \\ a_{41} & a_{42} & a_{43} & a_{44} \end{bmatrix} = \begin{bmatrix} L_{11} & (L_{11}U_{12}) & (L_{11}U_{13}) & (L_{11}U_{14}) \\ L_{21} & (L_{21}U_{12}+L_{22}) & (L_{21}U_{13}+L_{22}U_{23}) & (L_{21}U_{14}+L_{22}U_{24}) \\ L_{31} & (L_{31}U_{12}+L_{32}) & (L_{31}U_{13}+L_{32}U_{23}+L_{33}) & (L_{31}U_{14}+L_{32}U_{24}+L_{33}U_{34}) \\ L_{41} & (L_{41}U_{12}+L_{42}) & (L_{41}U_{13}+L_{42}U_{23}+L_{43}) & (L_{41}U_{14}+L_{42}U_{24}+L_{43}U_{34}+L_{44}) \end{bmatrix} \qquad (4.27)$$

The elements of the matrices $[L]$ and $[U]$ can be determined by solving Eq. (4.27). The solution is obtained by equating the corresponding elements of the matrices on both sides of the equation. Looking at Eq. (4.27) it can be observed that the elements of the matrices $[L]$ and $[U]$ can be easily determined row after row from the known elements of $[a]$ and the elements of $[L]$ and $[U]$ that are already calculated. Starting with the first row, the value of L_{11} is calculated from $L_{11} = a_{11}$. Once L_{11} is known, the values of U_{12}, U_{13}, and U_{14} are calculated by:

$$U_{12} = \frac{a_{12}}{L_{11}} \quad U_{13} = \frac{a_{13}}{L_{11}} \quad \text{and} \quad U_{14} = \frac{a_{14}}{L_{11}} \qquad (4.28)$$

Moving on to the second row, the value of L_{21} is calculated from $L_{21} = a_{21}$ and the value of L_{22} is calculated from:

$$L_{22} = a_{22} - L_{21}U_{12} \tag{4.29}$$

With the values of L_{21} and L_{22} known, the values of U_{23} and U_{24} are determined from:

$$U_{23} = \frac{a_{23} - L_{21}U_{13}}{L_{22}} \quad \text{and} \quad U_{24} = \frac{a_{24} - L_{21}U_{14}}{L_{22}} \tag{4.30}$$

In the third row:

$$L_{31} = a_{31}, \quad L_{32} = a_{32} - L_{31}U_{12}, \quad \text{and} \quad L_{33} = a_{33} - L_{31}U_{13} - L_{32}U_{23} \tag{4.31}$$

Once the values of L_{31}, L_{32}, and L_{33} are known, the value of U_{34} is calculated by:

$$U_{34} = \frac{a_{34} - L_{31}U_{14} - L_{32}U_{24}}{L_{33}} \tag{4.32}$$

In the fourth row the values of L_{41}, L_{42}, L_{43}, and L_{44} are calculated by:

$$L_{41} = a_{41}, \quad L_{42} = a_{42} - L_{41}U_{12}, \quad L_{43} = a_{43} - L_{41}U_{13} - L_{42}U_{23}, \quad \text{and}$$
$$L_{44} = a_{44} - L_{41}U_{14} - L_{42}U_{24} - L_{43}U_{34} \tag{4.33}$$

A procedure for determining the elements of the matrices $[L]$ and $[U]$ can be written by following the calculations in Eqs. (4.28) through (4.33). If $[a]$ is an $(n \times n)$ matrix, the elements of $[L]$ and $[U]$ are given by:

Step 1: Calculating the first column of $[L]$:

$$\text{for} \quad i = 1, 2, ..., n \qquad L_{i1} = a_{i1} \tag{4.34}$$

Step 2: Substituting 1s in the diagonal of $[U]$.

$$\text{for} \quad i = 1, 2, ..., n \qquad U_{ii} = 1 \tag{4.35}$$

Step 3: Calculating the elements in the first row of $[U]$ (except U_{11} which was already calculated):

$$\text{for} \quad j = 2, 3, ..., n \qquad U_{1j} = \frac{a_{1j}}{L_{11}} \tag{4.36}$$

Step 4: Calculating the rest of the elements row after row (i is the row number and j is the column number). The elements of $[L]$ are calculated first because they are used for calculating the elements of $[U]$:

for $i = 2, 3, ..., n$

$$\text{for} \quad j = 2, 3, ..., i \qquad L_{ij} = a_{ij} - \sum_{k=1}^{k=j-1} L_{ik}U_{kj} \tag{4.37}$$

$$\text{for} \quad j = (i+1), (i+2), ..., n \qquad U_{ij} = \frac{a_{ij} - \sum_{k=1}^{k=i-1} L_{ik}U_{kj}}{L_{ii}} \tag{4.38}$$

Examples 4-5 and 4-6 show how the *LU* decomposition with Crout's method is used for solving systems of equations. In Example 4-5 the calculations are done manually, and in Example 4-6 the decomposition is done with a user-defined MATLAB program.

Example 4-5: Solving a set of four equations using *LU* decomposition with Crout's method.

Solve the following set of four equations (the same as in Example 4-1) using *LU* decomposition with Crout's method.

$$4x_1 - 2x_2 - 3x_3 + 6x_4 = 12$$
$$-6x_1 + 7x_2 + 6.5x_3 - 6x_4 = -6.5$$
$$x_1 + 7.5x_2 + 6.25x_3 + 5.5x_4 = 16$$
$$-12x_1 + 22x_2 + 15.5x_3 - x_4 = 17$$

SOLUTION

First, the equations are written in matrix form:

$$\begin{bmatrix} 4 & -2 & -3 & 6 \\ -6 & 7 & 6.5 & -6 \\ 1 & 7.5 & 6.25 & 5.5 \\ -12 & 22 & 15.5 & -1 \end{bmatrix} \begin{bmatrix} x_1 \\ x_2 \\ x_3 \\ x_4 \end{bmatrix} = \begin{bmatrix} 12 \\ -6.5 \\ 16 \\ 17 \end{bmatrix} \tag{4.39}$$

Next, the matrix of coefficients $[a]$ is decomposed into the product $[L][U]$ as shown in Eq. (4.27). The decomposition is done by following the steps listed on the previous page:

Step 1: Calculating the first column of $[L]$:

for $i = 1, 2, 3, 4$ $L_{i1} = a_{i1}$: $L_{11} = 4$, $L_{21} = -6$, $L_{31} = 1$, $L_{41} = -12$

Step 2: Substituting 1s in the diagonal of $[U]$.

for $i = 1, 2, 3, 4$ $U_{ii} = 1$: $U_{11} = 1$, $U_{22} = 1$, $U_{33} = 1$, $U_{44} = 1$.

Step 3: Calculating the elements in the first row of $[U]$ (except U_{11} which was already calculated):

for $j = 2, 3, 4$ $U_{1j} = \dfrac{a_{1j}}{L_{11}}$: $U_{12} = \dfrac{a_{12}}{L_{11}} = \dfrac{-2}{4} = -0.5$, $U_{13} = \dfrac{a_{13}}{L_{11}} = \dfrac{-3}{4} = -0.75$,

$$U_{14} = \frac{a_{14}}{L_{11}} = \frac{6}{4} = 1.5$$

Step 4: Calculating the rest of the elements row after row, starting with the second row (i is the row number and j is the column number). In the present problem there are four rows, so i starts at 2 and ends with 4. For each value of i (each row), the elements of L are calculated first, and the elements of U are calculated subsequently. The general form of the equations is (Eqs. (4.37) and (4.38)):

for $i = 2, 3, 4$

$$\text{for } j = 2, 3, ..., i \quad L_{ij} = a_{ij} - \sum_{k=1}^{k=j-1} L_{ik} U_{kj} \tag{4.40}$$

$$\text{for } j = (i+1), (i+2), ..., n \quad U_{ij} = \frac{a_{ij} - \sum_{k=1}^{k=i-1} L_{ik} U_{kj}}{L_{ii}} \tag{4.41}$$

Starting with the second row $i = 2$:

for $j = 2$: $L_{22} = a_{22} - \sum_{k=1}^{k=1} L_{2k} U_{k2} = a_{22} - L_{21} U_{12} = 7 - (-6 \cdot -0.5) = 4$

for $j = 3, 4$: $U_{23} = \dfrac{a_{23} - \sum_{k=1}^{k=1} L_{2k} U_{k3}}{L_{22}} = \dfrac{a_{23} - (L_{21} U_{13})}{L_{22}} = \dfrac{6.5 - (-6 \cdot -0.75)}{4} = 0.5$

$U_{24} = \dfrac{a_{24} - \sum_{k=1}^{k=1} L_{2k} U_{k4}}{L_{22}} = \dfrac{a_{24} - (L_{21} U_{14})}{L_{22}} = \dfrac{-6 - (-6 \cdot 1.5)}{4} = 0.75$

Next, for the third row $i = 3$

for $j = 2, 3$: $L_{32} = a_{32} - \sum_{k=1}^{k=1} L_{3k} U_{k2} = a_{32} - L_{31} U_{12} = 7.5 - (1 \cdot -0.5) = 8$

$L_{33} = a_{33} - \sum_{k=1}^{k=2} L_{3k} U_{k3} = a_{33} - (L_{31} U_{13} + L_{32} U_{23}) = 6.25 - (1 \cdot -0.75 + 8 \cdot 0.5) = 3$

for $j = 4$: $U_{34} = \dfrac{a_{34} - \sum_{k=1}^{k=2} L_{3k} U_{k4}}{L_{33}} = \dfrac{a_{34} - (L_{31} U_{14} + L_{32} U_{24})}{L_{33}} = \dfrac{5.5 - (1 \cdot 1.5 + 8 \cdot 0.75)}{3} = -0.6667$

For the last row $i = 4$

for $j = 2, 3, 4$: $L_{42} = a_{42} - \sum_{k=1}^{k=1} L_{4k} U_{k2} = a_{42} - L_{41} U_{12} = 22 - (-12 \cdot -0.5) = 16$

$L_{43} = a_{43} - \sum_{k=1}^{k=2} L_{4k} U_{k3} = a_{43} - (L_{41} U_{13} + L_{42} U_{23}) = 15.5 - (-12 \cdot -0.75 + 16 \cdot 0.5) = -1.5$

$L_{44} = a_{44} - \sum_{k=1}^{k=4} L_{4k} U_{k4} = a_{44} - (L_{41} U_{14} + L_{42} U_{24} + L_{43} U_{34}) = -1 - (-12 \cdot 1.5 + 16 \cdot 0.75 + -1.5 \cdot -0.6667) = 4$

Writing the matrices $[L]$ and $[U]$ in a matrix form:

$$L = \begin{bmatrix} 4 & 0 & 0 & 0 \\ -6 & 4 & 0 & 0 \\ 1 & 8 & 3 & 0 \\ -12 & 16 & -1.5 & 4 \end{bmatrix} \quad \text{and} \quad U = \begin{bmatrix} 1 & -0.5 & -0.75 & 1.5 \\ 0 & 1 & 0.5 & 0.75 \\ 0 & 0 & 1 & -0.6667 \\ 0 & 0 & 0 & 1 \end{bmatrix}$$

To verify that the two matrices are correct, they are multiplied by using MATLAB:

```
>> L = [4 0 0 0; -6, 4 0 0; 1 8 3 0; -12 16 -1.5 4];
>> U = [1 -0.5 -0.75 1.5; 0 1 0.5 0.75; 0 0 1 -0.6667; 0 0 0 1];
>> L*U
ans =
    4.0000   -2.0000   -3.0000    6.0000
   -6.0000    7.0000    6.5000   -6.0000
    1.0000    7.5000    6.2500    5.4999
  -12.0000   22.0000   15.5000   -1.0000
```

This matrix is the same as the matrix of coefficients in Eq. (4.39) (except for round-off errors).

Once the decomposition is complete, a solution is obtained by using Eqs. (4.22) and (4.23). First, the matrix $[L]$ and the vector $[b]$ are used in Eq. (4.23), $[L][y] = [b]$, to solve for $[y]$:

$$\begin{bmatrix} 4 & 0 & 0 & 0 \\ -6 & 4 & 0 & 0 \\ 1 & 8 & 3 & 0 \\ -12 & 16 & -1.5 & 4 \end{bmatrix} \begin{bmatrix} y_1 \\ y_2 \\ y_3 \\ y_4 \end{bmatrix} = \begin{bmatrix} 12 \\ -6.5 \\ 16 \\ 17 \end{bmatrix} \qquad (4.42)$$

Using forward substitution the solution is:

$$y_1 = \frac{12}{4} = 3, \quad y_2 = \frac{-6.5 + 6y_1}{4} = 2.875, \quad y_3 = \frac{16 - y_1 - 8y_2}{3} = -3.333, \quad \text{and}$$

$$y_4 = \frac{17 + 12y_1 - 16y_2 + 1.5y_3}{4} = 0.5$$

Next, the matrix $[U]$ and the vector $[y]$ are used in Eq. (4.22), $[U][x] = [y]$, to solve for $[x]$:

$$\begin{bmatrix} 1 & -0.5 & -0.75 & 1.5 \\ 0 & 1 & 0.5 & 0.75 \\ 0 & 0 & 1 & -0.6667 \\ 0 & 0 & 0 & 1 \end{bmatrix} \begin{bmatrix} x_1 \\ x_2 \\ x_3 \\ x_4 \end{bmatrix} = \begin{bmatrix} 3 \\ 2.875 \\ -3.333 \\ 0.5 \end{bmatrix} \qquad (4.43)$$

Using back substitution the solution is:

$$x_4 = \frac{0.5}{1} = 0.5, \quad x_3 = -3.333 + 0.6667x_4 = -3, \quad x_2 = 2.875 - 0.5x_3 - 0.75x_4 = 4, \quad \text{and}$$

$$x_1 = 3 + 0.5x_2 + 0.75x_3 - 1.5x_4 = 2$$

Example 4-6: MATLAB user-defined function for solving a system of equations using *LU* decomposition with Crout's method.

Determine the currents i_1, i_2, i_3, and i_4 in the circuit shown in the figure (same as in Fig. 4-1). Write the system of equations that has to be solved in the form $[a][i] = [b]$. Solve the system by using the *LU* decomposition method, and use Crout's method for doing the decomposition.

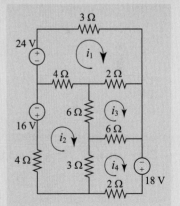

SOLUTION

The currents are determined from the set of four equations, Eq. (4.1). The equations are derived by using Kirchhoff's law. In matrix form, $[a][i] = [b]$, the equations are:

$$\begin{bmatrix} 9 & -4 & -2 & 0 \\ -4 & 17 & -6 & -3 \\ -2 & -6 & 14 & -6 \\ 0 & -3 & -6 & 11 \end{bmatrix} \begin{bmatrix} i_1 \\ i_2 \\ i_3 \\ i_4 \end{bmatrix} = \begin{bmatrix} 24 \\ -16 \\ 0 \\ 18 \end{bmatrix} \qquad (4.44)$$

To solve the system of equations, three user-defined functions are created. The functions are as follows:

`[L U]` = `LUdecompCrout(A)` This function decomposes the matrix `A` into lower triangular and upper triangular matrices `L` and `U`, respectively.

`y` = `ForwardSub(L,b)` This function solves a system of equations that is given in lower triangular form.

`x` = `BackwardSub(L,b)` This function solves a system of equations that is given in upper triangular form.

Listing of the user-defined function `LUdecompCrout`:

Program 4-3: User-defined function. *LU* decomposition using Crout's method.

```
function [L, U] = LUdecompCrout(A)
% The function decomposes the matrix A into a lower triangular matrix L
% and an upper triangular matrix U, using Crout's method, such that A = LU.
% Input variables:
% A  The matrix of coefficients.
% b  Right-hand-side column vector of constants.
% Output variable:
% L  Lower triangular matrix.
% U  Upper triangular matrix.

[R, C] = size(A);
for i = 1:R
    L(i,1) = A(i,1);          | Eq. (4.34). |          ┐
    U(i,i) = 1;               | Eq. (4.35). |          ├─ Steps 1 and 2 (page 116).
end                                                    ┘
for j = 2:R
    U(1,j) = A(1,j)/L(1,1);   | Eq. (4.36). |          ├─ Step 3 (page 116).
end
for i = 2:R
    for j = 2:i
        L(i,j) = A(i,j) - L(i,1:j - 1)*U(1:j - 1,j);   | Eq. (4.37). |
    end                                                                 ├─ Step 4 (page 116).
    for j = i + 1:R
        U(i,j) = (A(i,j) - L(i,1:i - 1)*U(1:i - 1,j))/L(i,i);  | Eq. (4.38). |
    end
end
```

Listing of the user-defined function `ForwardSub`:

Program 4-4: User-defined function. Forward substitution.

```
function y = ForwardSub(a,b)
% The function solves a system of linear equations ax = b
% where a is lower triangular matrix by using forward substitution.
% Input variables:
% a  The matrix of coefficients.
% b  A column vector of constants.
% Output variable:
% y  A column vector with the solution.
```

```
n = length(b);
y(1,1) = b(1)/a(1,1);
for i = 2:n
    y(i,1) = (b(i) - a(i,1:i - 1)*y(1:i - 1,1))./a(i,i);
end
```

Eq. (4.8).

Listing of the user-defined function `BackwardSub`:

Program 4-5: User-defined function. Back substitution.

```
function y = BackwardSub(a,b)
% The function solves a system of linear equations ax = b
% where a is an upper triangular matrix by using back substitution.
% Input variables:
% a  The matrix of coefficients.
% b  A column vector of constants.
% Output variable:
% y  A column vector with the solution.

n = length(b);
y(n,1) = b(n)/a(n,n);
for i = n - 1:-1:1
    y(i,1) = (b(i) - a(i,i + 1:n)*y(i + 1:n,1))./a(i,i);
end
```

Eq. (4.5).

The functions are then used in a MATLAB computer program (script file) that is used for solving the problem by following these steps:

- The matrix of coefficients [a] is decomposed into upper [U] and lower [L] triangular matrices (using the `LUdecompCrout` function).
- The matrix [L] and the vector [b] are used in Eq. (4.23), [L][y] = [b], to solve for [y], (using the `ForwardSub` function).
- The solution [y] and the matrix [U] are used in Eq. (4.22), [U][i] = [y], to solve for [i] (using the `BackwardSub` function).

Script file:

Program 4-6: Script file. Solving a system with Crout's *LU* decomposition.

```
% This script file solves a system of equations by using
% the Crout's LU decomposition method.
a = [9 -4 -2 0; -4 17 -6 -3; -2 -6 14 -6; 0 -3 -6 11];
b = [24; -16; 0; 18];
[L, U] = LUdecompCrout(a);
y = ForwardSub(L,b);
i = BackwardSub(U,y)
```

When the script file is executed, the following solution is displayed in the Command Window.

i =
    4.0343
    1.6545
    2.8452
    3.6395

The script file can be easily modified for solving the systems of equations $[a][i] = [b]$ for the same matrix $[a]$, but different values of $[b]$. The LU decomposition is done once, and only the last two steps have to be executed for each $[b]$.

4.5.3 LU Decomposition with Pivoting

Decomposition of a matrix $[a]$ into the matrices $[L]$ and $[U]$ means that $[a] = [L][U]$. In the presentation of Gauss and Crout's decomposition methods in the previous two subsections, it is assumed that it is possible to carry out all the calculations without pivoting. In reality, as was discussed in Section 4.3, pivoting may be required for a successful execution of the Gauss elimination procedure. Pivoting might also be needed with Crout's method. If pivoting is used, then the matrices $[L]$ and $[U]$ that are obtained are not the decomposition of the original matrix $[a]$. The product $[L][U]$ gives a matrix with rows that have the same elements as $[a]$, but due to the pivoting, the rows are in a different order. When pivoting is used in the decomposition procedure, the changes that are made have to be recorded and stored. This is done by creating a matrix $[P]$, called a permutation matrix, such that:

$$[P][a] = [L][U] \qquad (4.45)$$

If the matrices $[L]$ and $[U]$ are used for solving a system of equations $[a][x] = [b]$ (by using Eqs. (4.23) and (4.22)), then the order of the rows of $[b]$ have to be changed such that it is consistent with the pivoting. This is done by multiplying $[b]$ by the permutation matrix, $[P][b]$. Use of the permutation matrix is shown in Section 4.8.3, where the decomposition is done with MATLAB's built-in function.

4.6 INVERSE OF A MATRIX

The inverse of a square matrix $[a]$ is the matrix $[a]^{-1}$ such that the product of the two matrices gives the identity matrix $[I]$.

$$[a][a]^{-1} = [a]^{-1}[a] = [I] \qquad (4.46)$$

The process of calculating the inverse of a matrix is essentially the same as the process of solving a system of linear equations. This is illustrated for the case of a (4×4) matrix. If $[a]$ is a given matrix and $[x]$ is the unknown inverse of $[a]$, then:

$$
\begin{bmatrix} a_{11} & a_{12} & a_{13} & a_{14} \\ a_{21} & a_{22} & a_{23} & a_{24} \\ a_{31} & a_{32} & a_{33} & a_{34} \\ a_{41} & a_{42} & a_{43} & a_{44} \end{bmatrix} \begin{bmatrix} x_{11} & x_{12} & x_{13} & x_{14} \\ x_{21} & x_{22} & x_{23} & x_{24} \\ x_{31} & x_{32} & x_{33} & x_{34} \\ x_{41} & x_{42} & x_{43} & x_{44} \end{bmatrix} = \begin{bmatrix} 1 & 0 & 0 & 0 \\ 0 & 1 & 0 & 0 \\ 0 & 0 & 1 & 0 \\ 0 & 0 & 0 & 1 \end{bmatrix} \tag{4.47}
$$

Equation (4.47) can be rewritten as four separate systems of equations, where in each system one column of the matrix $[x]$ is the unknown:

$$
\begin{bmatrix} a_{11} & a_{12} & a_{13} & a_{14} \\ a_{21} & a_{22} & a_{23} & a_{24} \\ a_{31} & a_{32} & a_{33} & a_{34} \\ a_{41} & a_{42} & a_{43} & a_{44} \end{bmatrix} \begin{bmatrix} x_{11} \\ x_{21} \\ x_{31} \\ x_{41} \end{bmatrix} = \begin{bmatrix} 1 \\ 0 \\ 0 \\ 0 \end{bmatrix}, \quad
\begin{bmatrix} a_{11} & a_{12} & a_{13} & a_{14} \\ a_{21} & a_{22} & a_{23} & a_{24} \\ a_{31} & a_{32} & a_{33} & a_{34} \\ a_{41} & a_{42} & a_{43} & a_{44} \end{bmatrix} \begin{bmatrix} x_{12} \\ x_{22} \\ x_{32} \\ x_{42} \end{bmatrix} = \begin{bmatrix} 0 \\ 1 \\ 0 \\ 0 \end{bmatrix}
$$

$$
\begin{bmatrix} a_{11} & a_{12} & a_{13} & a_{14} \\ a_{21} & a_{22} & a_{23} & a_{24} \\ a_{31} & a_{32} & a_{33} & a_{34} \\ a_{41} & a_{42} & a_{43} & a_{44} \end{bmatrix} \begin{bmatrix} x_{13} \\ x_{23} \\ x_{33} \\ x_{43} \end{bmatrix} = \begin{bmatrix} 0 \\ 0 \\ 1 \\ 0 \end{bmatrix}, \quad
\begin{bmatrix} a_{11} & a_{12} & a_{13} & a_{14} \\ a_{21} & a_{22} & a_{23} & a_{24} \\ a_{31} & a_{32} & a_{33} & a_{34} \\ a_{41} & a_{42} & a_{43} & a_{44} \end{bmatrix} \begin{bmatrix} x_{14} \\ x_{24} \\ x_{34} \\ x_{44} \end{bmatrix} = \begin{bmatrix} 0 \\ 0 \\ 0 \\ 1 \end{bmatrix} \tag{4.48}
$$

Solving the four systems of equations in Eq. (4.48) gives the four columns of the inverse of $[a]$. The systems of equations can be solved by using any of the methods that have been introduced earlier in this chapter (or other methods). Two of the methods, the LU decomposition method and the Gauss–Jordan elimination method, are described in more detail next.

4.6.1 Calculating the Inverse with the LU Decomposition Method

The LU decomposition method is especially suitable for calculating the inverse of a matrix. As shown in Eqs. (4.48), the matrix of coefficients in all four systems of equations is the same. Consequently, the LU decomposition of the matrix $[A]$ is calculated only once. Then, each of the systems is solved by first using Eq. (4.23) (forward substitution) and then Eq. (4.22) (back substitution). This is illustrated, by using MATLAB, in Example 4-7.

Example 4-7: Determining the inverse of a matrix using the *LU* decomposition method.

Determine the inverse of the matrix $[a]$ by using the *LU* decomposition method.

$$[a] = \begin{bmatrix} 0.2 & -5 & 3 & 0.4 & 0 \\ -0.5 & 1 & 7 & -2 & 0.3 \\ 0.6 & 2 & -4 & 3 & 0.1 \\ 3 & 0.8 & 2 & -0.4 & 3 \\ 0.5 & 3 & 2 & 0.4 & 1 \end{bmatrix} \tag{4.49}$$

Do the calculations by writing a MATLAB user-defined function. Name the function `invA = InverseLU(A)`, where `A` is the matrix to be inverted, and `invA` is the inverse. In the function, use the functions `LUdecompCrout`, `ForwardSub`, and `BackwardSub` that were written in Example 4-6.

SOLUTION

If the inverse of $[a]$ is $[x]$ ($[x] = [a]^{-1}$), then $[a][x] = [I]$, which are the following five sets of five systems of equations that have to be solved. In each set of equations, one column of the inverse is calculated.

$$\begin{bmatrix} 0.2 & -5 & 3 & 0.4 & 0 \\ -0.5 & 1 & 7 & -2 & 0.3 \\ 0.6 & 2 & -4 & 3 & 0.1 \\ 3 & 0.8 & 2 & -0.4 & 3 \\ 0.5 & 3 & 2 & 0.4 & 1 \end{bmatrix}\begin{bmatrix} x_{11} \\ x_{21} \\ x_{31} \\ x_{41} \\ x_{51} \end{bmatrix} = \begin{bmatrix} 1 \\ 0 \\ 0 \\ 0 \\ 0 \end{bmatrix}, \quad \begin{bmatrix} 0.2 & -5 & 3 & 0.4 & 0 \\ -0.5 & 1 & 7 & -2 & 0.3 \\ 0.6 & 2 & -4 & 3 & 0.1 \\ 3 & 0.8 & 2 & -0.4 & 3 \\ 0.5 & 3 & 2 & 0.4 & 1 \end{bmatrix}\begin{bmatrix} x_{12} \\ x_{22} \\ x_{32} \\ x_{42} \\ x_{52} \end{bmatrix} = \begin{bmatrix} 0 \\ 1 \\ 0 \\ 0 \\ 0 \end{bmatrix}, \quad \begin{bmatrix} 0.2 & -5 & 3 & 0.4 & 0 \\ -0.5 & 1 & 7 & -2 & 0.3 \\ 0.6 & 2 & -4 & 3 & 0.1 \\ 3 & 0.8 & 2 & -0.4 & 3 \\ 0.5 & 3 & 2 & 0.4 & 1 \end{bmatrix}\begin{bmatrix} x_{13} \\ x_{23} \\ x_{33} \\ x_{43} \\ x_{53} \end{bmatrix} = \begin{bmatrix} 0 \\ 0 \\ 1 \\ 0 \\ 0 \end{bmatrix}$$

$$\begin{bmatrix} 0.2 & -5 & 3 & 0.4 & 0 \\ -0.5 & 1 & 7 & -2 & 0.3 \\ 0.6 & 2 & -4 & 3 & 0.1 \\ 3 & 0.8 & 2 & -0.4 & 3 \\ 0.5 & 3 & 2 & 0.4 & 1 \end{bmatrix}\begin{bmatrix} x_{14} \\ x_{24} \\ x_{34} \\ x_{44} \\ x_{54} \end{bmatrix} = \begin{bmatrix} 0 \\ 0 \\ 0 \\ 1 \\ 0 \end{bmatrix}, \quad \begin{bmatrix} 0.2 & -5 & 3 & 0.4 & 0 \\ -0.5 & 1 & 7 & -2 & 0.3 \\ 0.6 & 2 & -4 & 3 & 0.1 \\ 3 & 0.8 & 2 & -0.4 & 3 \\ 0.5 & 3 & 2 & 0.4 & 1 \end{bmatrix}\begin{bmatrix} x_{15} \\ x_{25} \\ x_{35} \\ x_{45} \\ x_{55} \end{bmatrix} = \begin{bmatrix} 0 \\ 0 \\ 0 \\ 0 \\ 1 \end{bmatrix} \tag{4.50}$$

The solution is obtained with the user-defined function `InverseLU` that is listed below. The function can be used for calculating the inverse of any sized square matrix.

The function executes the following operations:

- The matrix $[a]$ is decomposed into matrices $[L]$ and $[U]$ by applying Crout's method. This is done by using the function `LUdecompCrout` that was written in Example 4-6.
- Each system of equations in Eqs. (4.50) is solved by using Eqs. (4.23) and (4.22). This is done by first using the function `ForwardSub` and subsequently the function `BackwardSub` (see Example 4-6).

Program 4-7: User-defined function. Inverse of a matrix.

```
function invA = InverseLU(A)
% The function calculates the inverse of a matrix
% Input variables:
% A  The matrix to be inverted.
% Output variable:
% invA  The inverse of A.
```

```
[nR nC] = size(A);
I = eye(nR);
[L U] = LUdecompCrout(A);
for j = 1:nC
    y = ForwardSub(L,I(:,j));
    invA(:,j) = BackwardSub(U,y);
end
```

Create an identity matrix of the same size as $[A]$.

Decomposition of $[A]$ into $[L]$ and $[U]$.

In each pass of the loop, one set of the equations in Eqs. (4.50) is solved. Each solution is one column in the inverse of the matrix.

The function is then used in the Command Window for solving the problem.

```
>> F = [0.2 -5 3 0.4 0; -0.5 1 7 -2 0.3; 0.6 2 -4 3 0.1; 3 0.8 2 -0.4 3; 0.5 3 2 0.4 1];
>> invF = InverseLU(F)
invF =
   -0.7079    2.5314    2.4312    0.9666   -3.9023
   -0.1934    0.3101    0.2795    0.0577   -0.2941
    0.0217    0.3655    0.2861    0.0506   -0.2899
    0.2734   -0.1299    0.1316   -0.1410    0.4489
    0.7815   -2.8751   -2.6789   -0.7011    4.2338
>> invF*F
ans =
    1.0000   -0.0000    0.0000   -0.0000   -0.0000
    0.0000    1.0000    0.0000   -0.0000         0
         0   -0.0000    1.0000   -0.0000   -0.0000
   -0.0000    0.0000   -0.0000    1.0000   -0.0000
   -0.0000    0.0000   -0.0000   -0.0000    1.0000
```

The solution $[F]^{-1}$.

Check if $[F][F]^{-1} = [I]$.

4.6.2 Calculating the Inverse Using the Gauss–Jordan Method

The Gauss–Jordan method is easily adapted for calculating the inverse of a square $(n \times n)$ matrix $[a]$. This is done by first appending an identity matrix $[I]$ of the same size as the matrix $[a]$, to $[a]$ itself. This is shown schematically for a (4×4) matrix in Fig. 4-20a. Then, the Gauss–Jordan procedure is applied such that the elements of the matrix $[a]$ (the left half of the augmented matrix) are converted to 1s along the diagonal and 0s elsewhere. During this process, the terms of the identity matrix in Fig. 4-20a (the right half of the augmented matrix) are changed and become the elements $[a']$ in Fig. 4-20b, which constitute the inverse of $[a]$.

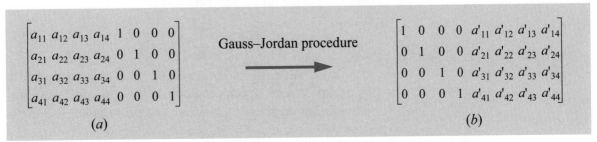

(a)   Gauss–Jordan procedure   (b)

Figure 4-20: Calculating the inverse with the Gauss–Jordan method.

4.7 ITERATIVE METHODS

A system of linear equations can also be solved by using an iterative approach. The process, in principle, is the same as in the fixed-point iteration method used for solving a single nonlinear equation (see Section 3.7). In an iterative process for solving a system of equations, the equations are written in an explicit form in which each unknown is written in terms of the other unknown. The explicit form for a system of four equations is illustrated in Fig. 4-21.

$$a_{11}x_1 + a_{12}x_2 + a_{13}x_3 + a_{14}x_4 = b_1$$
$$a_{21}x_1 + a_{22}x_2 + a_{23}x_3 + a_{24}x_4 = b_2$$
$$a_{31}x_1 + a_{32}x_2 + a_{33}x_3 + a_{34}x_4 = b_3$$
$$a_{41}x_1 + a_{42}x_2 + a_{43}x_3 + a_{44}x_4 = b_4$$

(a)

Writing the equations in an explicit form.
$\longrightarrow$

$$x_1 = [b_1 - (a_{12}x_2 + a_{13}x_3 + a_{14}x_4)]/a_{11}$$
$$x_2 = [b_2 - (a_{21}x_1 + a_{23}x_3 + a_{24}x_4)]/a_{22}$$
$$x_3 = [b_3 - (a_{31}x_1 + a_{32}x_2 + a_{34}x_4)]/a_{33}$$
$$x_4 = [b_4 - (a_{21}x_1 + a_{42}x_2 + a_{43}x_3)]/a_{44}$$

(b)

Figure 4-21: Standard (*a*) and explicit (*b*) form of a system of four equations.

The solution process starts by assuming initial values for the unknowns (first estimated solution). In the first iteration, the first assumed solution is substituted on the right-hand side of the equations, and the new values that are calculated for the unknowns are the second estimated solution. In the second iteration, the second solution is substituted back in the equations to give new values for the unknowns, which are the third estimated solution. The iterations continue in the same manner and when the method does work, the solutions that are obtained as successive iterations converge toward the actual solution. For a system with *n* equations, the explicit equations for the [x_i] unknowns are:

$$x_i = \frac{1}{a_{ii}}\left[b_i - \left(\sum_{\substack{j=1 \\ j \neq i}}^{j=n} a_{ij}x_j\right)\right] \qquad i = 1, 2, ..., n \qquad (4.51)$$

Condition for convergence

For a system of *n* equations $[a][x] = [b]$, a sufficient condition for convergence is that in each row of the matrix of coefficients $[a]$ the absolute value of the diagonal element is greater than the sum of the absolute values of the off-diagonal elements.

$$|a_{ii}| > \sum_{j=1, j \neq i}^{j=n} |a_{ij}| \qquad (4.52)$$

This condition is sufficient but not necessary for convergence when the iteration method is used. When condition (4.52) is satisfied, the matrix $[a]$ is classified as ***diagonally dominant***, and the iteration process converges toward the solution. The solution, however, might converge even when Eq. (4.52) is not satisfied.

Two specific iterative methods for executing the iterations, the Jacobi and Gauss–Seidel methods, are presented next. The difference between the two methods is in the way that the new calculated values of the unknowns are used. In the Jacobi method, the estimated values of the unknowns that are used on the right-hand side of Eq. (4.51) are updated all at once at the end of each iteration. In the Gauss–Seidel method, the value of each unknown is updated (and used in the calculation of the new estimate of the rest of the unknowns in the same iteration) when a new estimate for this unknown is calculated.

4.7.1 Jacobi Iterative Method

In the Jacobi method, an initial (first) value is assumed for each of the unknowns, $x_1^{(1)}, x_2^{(1)}, ..., x_n^{(1)}$. If no information is available regarding the approximate values of the unknown, the initial value of all the unknowns can be assumed to be zero. The second estimate of the solution $x_1^{(2)}, x_2^{(2)}, ..., x_n^{(2)}$ is calculated by substituting the first estimate in the right-hand side of Eqs. (4.51):

$$x_i^{(2)} = \frac{1}{a_{ii}}\left[b_i - \left(\sum_{\substack{j=1 \\ j \neq i}}^{j=n} a_{ij}x_j^{(1)} \right) \right] \qquad i = 1, 2, ..., n \qquad (4.53)$$

In general, the $(k+1)$th estimate of the solution is calculated from the (k)th estimate by:

$$x_i^{(k+1)} = \frac{1}{a_{ii}}\left[b_i - \left(\sum_{\substack{j=1 \\ j \neq i}}^{j=n} a_{ij}x_j^{(k)} \right) \right] \qquad i = 1, 2, ..., n \qquad (4.54)$$

The iterations continue until the differences between the values that are obtained in successive iterations are small. The iterations can be stopped when the absolute value of the estimated relative error (see Section 3.2) of all the unknowns is smaller than some predetermined value:

$$\left| \frac{x_i^{(k+1)} - x_i^{(k)}}{x_i^{(k)}} \right| < \varepsilon \qquad i = 1, 2, ..., n \qquad (4.55)$$

4.7.2 Gauss-Seidel Iterative Method

In the Gauss–Seidel method, initial (first) values are assumed for the unknowns $x_2, x_3, ..., x_n$ (all of the unknowns except x_1). If no information is available regarding the approximate value of the unknowns, the initial value of all the unknowns can be assumed to be zero. The first assumed values of the unknowns are substituted in Eq. (4.51) with $i = 1$ to calculate the value of x_1. Next, Eq. (4.51) with $i = 2$ is used for calculating a new value for x_2. This is followed by using Eq. (4.51) with $i = 3$ for calculating a new value for x_3. The process continues until $i = n$, which is the end of the first iteration. Then, the second iteration starts with $i = 1$ where a new value for x_1 is calculated, and so on. In the Gauss–Seidel method, the current values of the unknowns are used for calculating the new value of the next unknown. In other words, as a new value of an unknown is calculated, it is immediately used for the next application of Eq. (4.51). (In the Jacobi method, the values of the unknowns obtained in one iteration are used as a complete set for calculating the new values of the unknowns in the next iteration. The values of the unknowns are not updated in the middle of the iteration.)

Applying Eq. (4.51) to the Gauss–Seidel method gives the iteration formula:

$$x_1^{(k+1)} = \frac{1}{a_{11}}\left[b_1 - \sum_{j=2}^{j=n} a_{1j}x_j^{(k)}\right]$$

$$x_i^{(k+1)} = \frac{1}{a_{ii}}\left[b_i - \left(\sum_{j=1}^{j=i-1} a_{ij}x_j^{(k+1)} + \sum_{j=i+1}^{j=n} a_{ij}x_j^{(k)}\right)\right] \quad i = 2, 3, \ldots, n-1 \quad (4.56)$$

$$x_n^{(k+1)} = \frac{1}{a_{nn}}\left[b_n - \sum_{j=1}^{j=n-1} a_{nj}x_j^{(k+1)}\right]$$

Notice that the values of the unknowns in the $k + 1$ iteration, $x_i^{(k+1)}$, are calculated by using the values $x_j^{(k+1)}$ obtained in the $k + 1$ iteration for $j < i$ and using the values $x_j^{(k)}$ for $j > i$. The criterion for stopping the iterations is the same as in the Jacobi method, Eq. (4.55). The Gauss–Seidel method converges faster than the Jacobi method and requires less computer memory when programmed. The method is illustrated for a system of four equations in Example 4-8.

Example 4-8: Solving a set of four linear equations using Gauss–Seidel method.

Solve the following set of four linear equations using the Gauss–Seidel iteration method.

$$9x_1 - 2x_2 + 3x_3 + 2x_4 = 54.5$$
$$2x_1 + 8x_2 - 2x_3 + 3x_4 = -14$$
$$-3x_1 + 2x_2 + 11x_3 - 4x_4 = 12.5$$
$$-2x_1 + 3x_2 + 2x_3 + 10x_4 = -21$$

SOLUTION

First, the equations are written in an explicit form (see Fig. 4.21):

$$\begin{aligned}
x_1 &= [54.5 - (-2x_2 + 3x_3 + 2x_4)]/9 \\
x_2 &= [-14 - (2x_1 - 2x_3 + 3x_4)]/8 \\
x_3 &= [12.5 - (-3x_1 + 2x_2 - 4x_4)]/11 \\
x_4 &= [-21 - (-2x_1 + 3x_2 + 2x_3)]/10
\end{aligned} \qquad (4.57)$$

As a starting point, the initial value of all the unknowns, $x_1^{(1)}, x_2^{(1)}, x_3^{(1)}$, and $x_4^{(1)}$, is assumed to be zero. The first two iterations are calculated manually, and then a MATLAB program is used for calculating the values of the unknowns in seven iterations.

Manual calculation of the first two iterations:

The second estimate of the solution ($k = 2$) is calculated in the first iteration by using Eqs. (4.57). The values that are substituted for x_i in the right-hand side of the equations are the most recent known values. This means that when the first equation is used to calculate $x_1^{(2)}$ all the x_i values are zero. Then, when the second equation is used to calculate $x_2^{(2)}$, the new value $x_1^{(2)}$ is substituted for x_1, but the older values $x_3^{(1)}$ and $x_4^{(1)}$ are substituted for x_3 and x_4, and so on:

$$x_1^{(2)} = [54.5 - (-2 \cdot 0 + 3 \cdot 0 + 2 \cdot 0)]/9 = 6.056$$

$$x_2^{(2)} = [-14 - (2 \cdot 6.056 - (2 \cdot 0) + 3 \cdot 0)]/8 = -3.264$$

$$x_3^{(2)} = [12.5 - (-3 \cdot 6.056 + 2 \cdot -3.264 - (4 \cdot 0))]/11 = 3.381$$

$$x_4^{(2)} = [-21 - (-2 \cdot 6.056 + 3 \cdot -3.264 + 2 \cdot 3.381)]/10 = -0.5860$$

The third estimate of the solution ($k = 3$) is calculated in the second iteration:

$$x_1^{(3)} = [54.5 - (-2 \cdot -3.264 + 3 \cdot 3.381 + 2 \cdot -0.5860)]/9 = 4.333$$

$$x_2^{(3)} = [-14 - (2 \cdot 4.333 - (2 \cdot 3.381) + 3 \cdot -0.5860)]/8 = -1.768$$

$$x_3^{(3)} = [12.5 - (-3 \cdot 4.333 + 2 \cdot -1.768 - (4 \cdot -0.5860))]/11 = 2.427$$

$$x_4^{(3)} = [-21 - (-2 \cdot 4.333 + 3 \cdot -1.768 + 2 \cdot 2.427)]/10 = -1.188$$

MATLAB program that calculates the first seven iterations:

The following is a MATLAB program in a script file that calculates the first seven iterations of the solution by using Eqs. (4.57):

Program 4-8: Script file. Gauss–Seidel iteration.

```
k = 1; x1 = 0; x2 = 0; x3 = 0; x4 = 0;
disp('  k      x1      x2      x3      x4')
fprintf(' %2.0f     %-8.5f %-8.5f %-8.5f %-8.5f \n', k, x1, x2, x3, x4)
for k = 2 : 8
    x1 = (54.5 - (-2*x2 + 3*x3 + 2*x4))/9;
    x2 = (-14 - (2*x1 - 2*x3 + 3*x4))/8;
    x3 = (12.5 - (-3*x1 + 2*x2 - 4*x4))/11;
    x4 = (-21 - (-2*x1 + 3*x2 + 2*x3))/10;
    fprintf(' %2.0f     %-8.5f %-8.5f %-8.5f %-8.5f \n', k, x1, x2, x3, x4)
end
```

When the program is executed, the following results are displayed in the Command Window.

k	x1	x2	x3	x4
1	0.00000	0.00000	0.00000	0.00000
2	6.05556	-3.26389	3.38131	-0.58598
3	4.33336	-1.76827	2.42661	-1.18817

4	5.11778	-1.97723	2.45956	-0.97519
5	5.01303	-2.02267	2.51670	-0.99393
6	4.98805	-1.99511	2.49806	-1.00347
7	5.00250	-1.99981	2.49939	-0.99943
8	5.00012	-2.00040	2.50031	-0.99992

The results show that the solution converges toward the exact solution, which is:

$x_1 = 5$, $x_2 = -2$, $x_3 = 2.5$, and $x_4 = -1$.

4.8 USE OF MATLAB BUILT-IN FUNCTIONS FOR SOLVING A SYSTEM OF LINEAR EQUATIONS

MATLAB has mathematical operations and built-in functions that can be used for solving a system of linear equations and for carrying out other matrix operations that described in this chapter.

4.8.1 Solving a System of Equations Using MATLAB's Left and Right Division

Left division \ : Left division can be used to solve a system of n equations written in matrix form $[a][x] = [b]$, where $[a]$ is the $(n \times n)$ matrix of coefficients, $[x]$ is an $(n \times 1)$ column vector of the unknowns, and $[b]$ is an $(n \times 1)$ column vector of constants.

$$\boxed{x = a \backslash b}$$

For example, the solution of the system of equations in Examples 4-1 and 4-2 is calculated by (Command Window):

```
>> a = [4 -2 -3 6 ; -6 7 6.5 -6 ; 1 7.5 6.25 5.5 ; -12 22 15.5 -1];
>> b = [12; -6.5; 16; 17];
>> x = a\b
x =
    2.0000
    4.0000
   -3.0000
    0.5000
```

Right division / : Right division is used to solve a system of n equations written in matrix form $[x][a] = [b]$, where $[a]$ is the $(n \times n)$ matrix of coefficients, $[x]$ is a $(1 \times n)$ row vector of the unknowns, and $[b]$ is a $(1 \times n)$ row vector of constants.

$$\boxed{x = b / a}$$

For example, the solution of the system of equations in Examples 4-1 and 4-2 is calculated by (Command Window):

```
>> a = [4 -6 1 -12 ; -2 7 7.5 22 ; -3 6.5 6.25 15.5 ; 6 -6 5.5 -1];
>> b = [12 -6.5 16 17];
>> x = b/a
x =
    2.0000    4.0000   -3.0000    0.5000
```

Notice that the matrix $[a]$ used in the right division calculation is the transpose of the matrix used in the left division calculation.

4.8.2 Solving a System of Equations Using MATLAB's Inverse Operation

In matrix form, the system of equations $[a][x] = [b]$ can be solved for $[x]$. Multiplying both sides from the left by $[a]^{-1}$ (the inverse of $[a]$) gives:

$$[a]^{-1}[a][x] = [a]^{-1}[b] \tag{4.58}$$

Since $[a]^{-1}[a] = [I]$ (identity matrix), and $[I][x] = [x]$, Eq. (4.58) reduces to:

$$[x] = [a]^{-1}[b] \tag{4.59}$$

In MATLAB, the inverse of a matrix $[a]$ can be calculated either by raising the matrix to the power of -1 or by using the `inv(a)` function. Once the inverse is calculated, the solution is obtained by multiplying the vector $[b]$ by the inverse. This is demonstrated for the solution of the system in Examples 4-1 and 4-2.

```
>> a = [4 -2 -3 6 ; -6 7 6.5 -6 ; 1 7.5 6.25 5.5 ; -12 22 15.5 -1];
>> b = [12; -6.5; 16; 17];
>> x = a^-1*b            The same result is obtained by typing:  >> x = inv(a)*b.
x =
    2.0000
    4.0000
   -3.0000
    0.5000
```

4.8.3 MATLAB's Built-In Function for LU Decomposition

MATLAB has a built-in function, called lu, that decomposes a matrix $[a]$ into the product $[L][U]$, such that $[a] = [L][U]$ where $[L]$ is a lower triangular matrix and $[U]$ is an upper triangular matrix. One form of the function is:

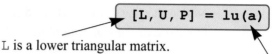

$$[L, U, P] = lu(a)$$

L is a lower triangular matrix.
U is an upper triangular matrix.
P is a permutation matrix.

a is the matrix to be decomposed.

MATLAB uses partial pivoting when determining the factorization. Consequently, the matrices $[L]$ and $[U]$ that are determined by MATLAB are the factorization of a matrix with rows that may be in a different order than in $[a]$. The permutation matrix $[P]$ (a matrix with 1s and 0s) contains the information about the pivoting. Multiplying $[a]$ by the matrix $[P]$ gives the matrix whose decomposition is given by $[L]$ and $[U]$ (see Section 4.5.3):

$$[L][U] = [P][a] \qquad (4.60)$$

The matrix $[P][a]$ has the same rows as $[a]$ but in a different order. If MATLAB does not use partial pivoting when the function lu is used, then the permutation matrix $[P]$ is the identity matrix.

If the matrices $[L]$ and $[U]$ that are determined by the lu function are subsequently used for solving a system of equations $[a][x] = [b]$ (by using Eq. (4.23) and (4.22)), then the vector $[b]$ has to be multiplied by the permutation matrix $[P]$. This pivots the rows in $[b]$ to be consistent with the pivoting in $[a]$. The following shows a MATLAB solution of the system of equations from Examples 4-1 and 4-2 using the function lu.

```
>> a = [4 -2 -3 6 ; -6 7 6.5 -6 ; 1 7.5 6.25 5.5 ; -12 22 15.5 -1];
>> b = [12; -6.5; 16; 17];
>> [L, U, P] = lu(a)          Decomposition of [a] using MATLAB's lu function.
L =
    1.0000         0         0         0
   -0.0833    1.0000         0         0
   -0.3333    0.5714    1.0000         0
    0.5000   -0.4286   -0.9250    1.0000
U =
  -12.0000   22.0000   15.5000   -1.0000
         0    9.3333    7.5417    5.4167
         0         0   -2.1429    2.5714
         0         0         0   -0.800
```

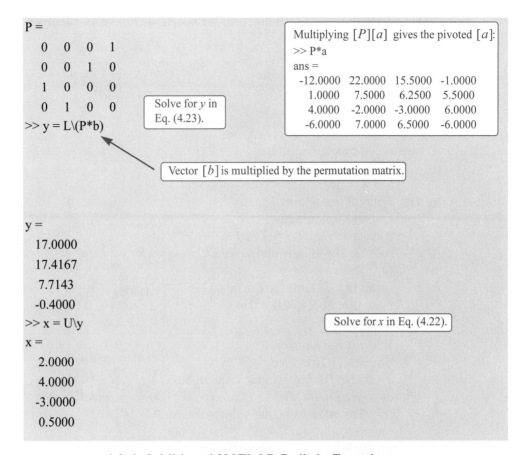

4.8.4 Additional MATLAB Built-In Functions

MATLAB has many built-in functions that can be useful in the analysis of systems of equations. Several of these functions are presented in Table 4-1. Note that the operations that are related to some of the functions in the table are discussed in Section 4.9.

Table 4-1: Built-in MATLAB functions for matrix operations and analysis.

Function	Description	Example

`inv(A)`	Inverse of a matrix. A is a square matrix. Returns the inverse of A.	>> A = [-3 1 0.6; 0.2 -4 3; 0.1 0.5 2];
		 >> Ain = inv(A)
		
		 Ain =
		
		  -0.3310 -0.0592 0.1882
		  -0.0035 -0.2111 0.3178
		   0.0174 0.0557 0.4111

Table 4-1: Built-in MATLAB functions for matrix operations and analysis. (Continued)

Function	Description	Example

`d=det(A)`	Determinant of a matrix `A` is a square matrix, `d` is the determinant of `A`.	`>> A = [-3 1 0.6; 0.2 -4 3; 0.1 0.5 2];`
		 `>> d = det(A)`
		 `d =`
		     `28.7000`

`n=norm(A)`
 	
 `n=norm(A,p)`
	Vector and matrix norm `A` is a vector or a matrix, `n` is its norm. ***When A is a vector:*** `norm(A,p)` returns: `sum(abs(A.^p)^(1/p)`. `p=inf` The infinity norm (see Eq. (4.70)). `norm(A)` Returns the Euclidean 2-norm (see Eq. (4.72)), same as `norm(A,2)`.	`>> A = [2 0 7 -9];`
		 `>> n=norm(A,1)`
		 `n =`
		     `18`
		 `>> n = norm(A,inf)`
		 `n =`
		     `9`
		 `>> n = norm(A,2)`
		 `n =`
		     `11.5758`

	When A is a matrix: `norm(A,p)` returns: `p=1` The 1-norm (largest column sum of `A` (see Eq. (4.74)). `p=2` The largest singular value, same as `norm(A)` (see Eq. (4.75)). This is not the Euclidean norm (see Eq. (4.76)). `p=inf` The infinity norm (see Eq. (4.73)).	`>> A = [1 3 -2; 0 -1 4; 5 2 3];`
		 `>> n = norm(A,1)`
		 `n =`
		     `9`
		 `>> n = norm(A,2)`
		 `n =`
		     `6.4818`
		 `>> n = norm(A,inf)`
		 `n =`
		     `10`

`c=cond(A)`
 	
 `c=cond(A,p)`
	Condition number (see Eq. (4.86)) `A` is a square matrix, `c` is the condition number of `A`. `cond(A)` The same as `p=2`. `p=1` The 1-norm condition number. `p=2` The 2-norm condition number. `p=inf` The infinity norm condition number.	`>> a = [9 -2 3 2; 2 8 -2 3; -3 2 11 -4; -2 3 2 10];`
		 `>> cond(a,inf)`
		 `ans =`
		     `3.8039` See the end of Example 4-10.

4.9 TRIDIAGONAL SYSTEMS OF EQUATIONS

Tridiagonal systems of linear equations have a matrix of coefficients with zero as their entries except along the diagonal, above-diagonal, and below-diagonal elements. A tridiagonal system of n equations in matrix form is shown in Eq. (4.61) and is illustrated for a system of five equations in Fig. 4-22.

$$\begin{bmatrix} A_{11} & A_{12} & 0 & 0 & 0 \\ A_{21} & A_{22} & A_{23} & 0 & 0 \\ 0 & A_{32} & A_{33} & A_{34} & 0 \\ 0 & 0 & A_{43} & A_{44} & A_{45} \\ 0 & 0 & 0 & A_{54} & A_{55} \end{bmatrix} \begin{bmatrix} x_1 \\ x_2 \\ x_3 \\ x_4 \\ x_5 \end{bmatrix} = \begin{bmatrix} B_1 \\ B_2 \\ B_3 \\ B_4 \\ B_5 \end{bmatrix}$$

Figure 4-22: Tridiagonal system of five equations.

$$\begin{bmatrix} A_{11} & A_{12} & 0 & 0 & 0 & 0 & 0 & 0 \\ A_{21} & A_{22} & A_{23} & 0 & 0 & 0 & 0 & 0 \\ 0 & A_{32} & A_{33} & A_{34} & 0 & 0 & 0 & 0 \\ & \cdots & \cdots & \cdots & & & & \\ & & \cdots & \cdots & \cdots & & & \\ 0 & 0 & 0 & 0 & A_{n-2,\,n-3} & A_{n-2,\,n-2} & A_{n-2,\,n-1} & 0 \\ 0 & 0 & 0 & 0 & 0 & A_{n-1,\,n-2} & A_{n-1,\,n-1} & A_{n-1,\,n} \\ 0 & 0 & 0 & 0 & 0 & 0 & A_{n,\,n-1} & A_{n,\,n} \end{bmatrix} \begin{bmatrix} x_1 \\ x_2 \\ x_3 \\ \cdots \\ \cdots \\ x_{n-2} \\ x_{n-1} \\ x_n \end{bmatrix} = \begin{bmatrix} B_1 \\ B_2 \\ B_3 \\ \cdots \\ \cdots \\ B_{n-2} \\ B_{n-1} \\ B_n \end{bmatrix} \quad (4.61)$$

The matrix of coefficients of tridiagonal systems has many elements that are zero (especially when the system contains a large number of equations). The system can be solved with the standard methods (Gauss, Gauss–Jordan, LU decomposition), but then a large number of zero elements are stored and a large number of needless operations (with zeros) are executed. To save computer memory and computing time, special numerical methods have been developed for solving tridiagonal systems of equations. One of these methods, the Thomas algorithm, is described in this section.

Many applications in engineering and science require the solution of tridiagonal systems of equations. Some numerical methods for solving differential equations also involve the solution of such systems.

Thomas algorithm for solving tridiagonal systems

The Thomas algorithm is a procedure for solving tridiagonal systems of equations. The method of solution in the Thomas algorithm is similar to the Gaussian elimination method in which the system is first changed to upper triangular form and then solved using back substitution. The Thomas algorithm, however, is much more efficient because only the non-zero elements of the matrix of coefficients are stored, and only the necessary operations are executed. (Unnecessary operations on the zero elements are eliminated.)

The Thomas algorithm starts by assigning the nonzero elements of the tridiagonal matrix of coefficients $[A]$ to three vectors. The diagonal elements A_{ii} are assigned to vector d (d stands for diagonal) such that $d_i = A_{ii}$. The above diagonal elements $A_{i,\,i+1}$ are assigned to vector a (a stands for above diagonal) such that $a_i = A_{i,\,i+1}$, and the below diagonal elements $A_{i-1,\,i}$ are assigned to vector b (b stands for below diagonal), such that $b_i = A_{i-1, i}$. With the nonzero elements in the matrix of coefficients stored as vectors, the system of equations has the form:

$$\begin{bmatrix} d_1 & a_1 & 0 & 0 & 0 & 0 & 0 & 0 \\ b_2 & d_2 & a_2 & 0 & 0 & 0 & 0 & 0 \\ 0 & b_3 & d_3 & a_3 & 0 & 0 & 0 & 0 \\ & \cdots & \cdots & \cdots & & & & \\ & & \cdots & \cdots & \cdots & & & \\ 0 & 0 & 0 & 0 & b_{n-2} & d_{n-2} & a_{n-2} & 0 \\ 0 & 0 & 0 & 0 & 0 & b_{n-1} & d_{n-1} & a_{n-1} \\ 0 & 0 & 0 & 0 & 0 & 0 & b_n & d_n \end{bmatrix} \begin{bmatrix} x_1 \\ x_2 \\ x_3 \\ \cdots \\ \cdots \\ x_{n-2} \\ x_{n-1} \\ x_n \end{bmatrix} = \begin{bmatrix} B_1 \\ B_2 \\ B_3 \\ \cdots \\ \cdots \\ B_{n-2} \\ B_{n-1} \\ B_n \end{bmatrix} \quad (4.62)$$

It should be emphasized here that in Eq. (4.62) the matrix of coefficients is displayed as a matrix (with the 0s), but in the Thomas algorithm only the vectors b, d, and a are stored.

Next, the first row is normalized by dividing the row by d_1. This makes the element d_1 (to be used as the pivot element) equal to 1:

$$\begin{bmatrix} 1 & a'_1 & 0 & 0 & 0 & 0 & 0 & 0 \\ b_2 & d_2 & a_2 & 0 & 0 & 0 & 0 & 0 \\ 0 & b_3 & d_3 & a_3 & 0 & 0 & 0 & 0 \\ & \cdots & \cdots & \cdots & & & & \\ & & \cdots & \cdots & \cdots & & & \\ 0 & 0 & 0 & 0 & b_{n-2} & d_{n-2} & a_{n-2} & 0 \\ 0 & 0 & 0 & 0 & 0 & b_{n-1} & d_{n-1} & a_{n-1} \\ 0 & 0 & 0 & 0 & 0 & 0 & b_n & d_n \end{bmatrix} \begin{bmatrix} x_1 \\ x_2 \\ x_3 \\ \cdots \\ \cdots \\ x_{n-2} \\ x_{n-1} \\ x_n \end{bmatrix} = \begin{bmatrix} B'_1 \\ B_2 \\ B_3 \\ \cdots \\ \cdots \\ B_{n-2} \\ B_{n-1} \\ B_n \end{bmatrix} \quad (4.63)$$

where $a'_1 = a_1/d_1$ and $B'_1 = B_1/d_1$.

Now the element b_2 is eliminated. The first row (the pivot row) is multiplied by b_2 and then is subtracted from the second row.

$$\begin{bmatrix} 1 & a'_1 & 0 & 0 & 0 & 0 & 0 & 0 \\ 0 & d'_2 & a_2 & 0 & 0 & 0 & 0 & 0 \\ 0 & b_3 & d_3 & a_3 & 0 & 0 & 0 & 0 \\ & \cdots & \cdots & \cdots & & & & \\ & & \cdots & \cdots & \cdots & & & \\ 0 & 0 & 0 & 0 & b_{n-2} & d_{n-2} & a_{n-2} & 0 \\ 0 & 0 & 0 & 0 & 0 & b_{n-1} & d_{n-1} & a_{n-1} \\ 0 & 0 & 0 & 0 & 0 & 0 & b_n & d_n \end{bmatrix} \begin{bmatrix} x_1 \\ x_2 \\ x_3 \\ \cdots \\ \cdots \\ x_{n-2} \\ x_{n-1} \\ x_n \end{bmatrix} = \begin{bmatrix} B'_1 \\ B'_2 \\ B_3 \\ \cdots \\ \cdots \\ B_{n-2} \\ B_{n-1} \\ B_n \end{bmatrix} \quad (4.64)$$

where $d'_2 = d_2 - b_2 a'_1$, and $B'_2 = B_2 - B_1 b_2$.

The operations performed with the first and second row are repeated with the second and third rows. The second row is normalized by dividing the row by d'_2. This makes the element d'_2 (to be used as the pivot element) equal to 1. The second row is then used to eliminate b_3 in the third row.

This process continues row after row until the system of equations is transformed to be upper triangular with 1s along the diagonal:

$$
\begin{bmatrix}
1 & a'_1 & 0 & 0 & 0 & 0 & 0 & 0 \\
0 & 1 & a'_2 & 0 & 0 & 0 & 0 & 0 \\
0 & 0 & 1 & a'_3 & 0 & 0 & 0 & 0 \\
& & \cdots & \cdots & \cdots & & & \\
& & & \cdots & \cdots & \cdots & & \\
0 & 0 & 0 & 0 & 0 & 1 & a'_{n-2} & 0 \\
0 & 0 & 0 & 0 & 0 & 0 & 1 & a'_{n-1} \\
0 & 0 & 0 & 0 & 0 & 0 & 0 & 1
\end{bmatrix}
\begin{bmatrix}
x_1 \\ x_2 \\ x_3 \\ \cdots \\ \cdots \\ x_{n-2} \\ x_{n-1} \\ x_n
\end{bmatrix}
=
\begin{bmatrix}
B'_1 \\ B''_2 \\ B''_3 \\ \cdots \\ \cdots \\ B''_{n-2} \\ B''_{n-1} \\ B''_n
\end{bmatrix}
\tag{4.65}
$$

Once the matrix of coefficients is in upper triangular form, the values of the unknowns are calculated by using back substitution.

In mathematical form, the Thomas algorithm can be summarized in the following steps:

Step 1: Define the vectors $b = [0, b_2, b_3, ..., b_n]$, $d = [d_1, d_2, ..., d_n]$, $a = [a_1, a_2, ..., a_{n-1}]$, and $B = [B_1, B_2, ..., B_n]$.

Step 2: Calculate: $a_1 = \dfrac{a_1}{d_1}$ and $B_1 = \dfrac{B_1}{d_1}$.

Step 3: For $i = 2, 3, ..., n-1$ calculate:

$$
a_i = \frac{a_i}{d_i - b_i a_{i-1}} \quad \text{and} \quad B_i = \frac{B_i - b_i B_{i-1}}{d_i - b_i a_{i-1}}
$$

Step 4: Calculate: $B_n = \dfrac{B_n - b_n B_{n-1}}{d_n - b_n a_{n-1}}$

Step 5: Calculate the solution using back substitution:

$x_n = B_n$ and for $i = n-1, n-2, n-3, ..., 2, 1,$ $x_i = B_i - a_i x_{i+1}$

A solution of a tridiagonal system of equations, using a user-defined MATLAB function, is shown in Example 4-9.

Example 4-9: Solving a tridiagonal system of equations using the Thomas algorithm.

Six springs with different spring constants k_i and unstretched lengths L_i are attached to each other in series. The endpoint B is then displaced such that the distance between points A and B is $L = 1.2$ m. Determine the positions $x_1, x_2, ..., x_5$ of the endpoints of the springs.

The spring constants and the unstretched lengths of the springs are:

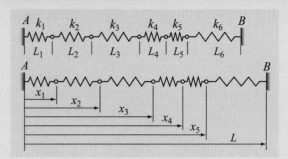

spring	1	2	3	4	5	6
k (kN/m)	8	9	15	12	10	18
L (m)	0.18	0.22	0.26	0.19	0.15	0.30

SOLUTION

The force, F, in a spring is given by:

$$F = k\delta$$

where k is the spring constant and δ is the extension of the spring beyond its unstretched length. Since the springs are connected in series, the force in all of the springs is the same. Consequently, it is possible to write five equations that equate the force in every two adjacent springs. For example, the condition that the force in the first spring is equal to the force in the second spring gives:

$$k_1(x_1 - L_1) = k_2[(x_2 - x_1) - L_2]$$

Similarly, four additional equations can be written:

$$k_2[(x_2 - x_1) - L_2] = k_3[(x_3 - x_2) - L_3]$$

$$k_3[(x_3 - x_2) - L_3] = k_4[(x_4 - x_3) - L_4]$$

$$k_4[(x_4 - x_3) - L_4] = k_5[(x_5 - x_4) - L_5]$$

$$k_5[(x_5 - x_4) - L_5] = k_6[(L - x_5) - L_6]$$

The five equations form a system that is tridiagonal. In matrix form the system is:

$$\begin{bmatrix} k_1 + k_2 & -k_2 & 0 & 0 & 0 \\ -k_2 & k_2 + k_2 & -k_3 & 0 & 0 \\ 0 & -k_3 & k_3 + k_4 & -k_4 & 0 \\ 0 & 0 & -k_4 & k_4 + k_5 & -k_5 \\ 0 & 0 & 0 & -k_5 & k_5 + k_6 \end{bmatrix} \begin{bmatrix} x_1 \\ x_2 \\ x_3 \\ x_4 \\ x_5 \end{bmatrix} = \begin{bmatrix} k_1 L_1 - k_2 L_2 \\ k_2 L_2 - k_3 L_3 \\ k_3 L_3 - k_4 L_4 \\ k_4 L_4 - k_5 L_5 \\ k_5 L_5 + k_6 L - k_6 L_6 \end{bmatrix} \quad (4.66)$$

The system of equations (4.66) is solved with a user-defined MATLAB function `Tridiagonal`, which is listed next.

Program 4-9: User-defined function. Solving a tridiagonal system of equations.

```
function x = Tridiagonal(A,B)
% The function solves a tridiagonal system of linear equations [a][x]=[b]
% using the Thomas algorithm.
% Input variables:
```

% A The matrix of coefficients.
% B Right-hand-side column vector of constants.
% Output variable:
% x A column vector with the solution.

```
[nR, nC] = size(A);
for i = 1:nR
    d(i) = A(i,i);
end
for i = 1:nR - 1
    ad(i) = A(i,i + 1);
end
for i = 2:nR
    bd(i) = A(i,i - 1);
end
ad(1) = ad(1)/d(1);
B(1) = B(1)/d(1);
for i = 2:nR - 1
    ad(i) = ad(i)/(d(i) - bd(i)*ad(i - 1));
    B(i) = (B(i) - bd(i)*B(i - 1))/(d(i) - bd(i)*ad(i - 1));
end
B(nR) = (B(nR) - bd(nR)*B(nR - 1))/(d(nR) - bd(nR)*ad(nR - 1));
x(nR,1) = B(nR);
for i = nR - 1:-1:1
    x(i,1) = B(i) - ad(i)*x(i + 1);
end
```

Define the vector d with the elements of the diagonal.

Define the vector ad with the above diagonal elements.

Define the vector bd with the below diagonal elements.

Step 1.

Step 2.

Step 3.

Step 4.

Step 5.

The user-defined function `Tridiagonal` is next used in a script file program to solve the system in Eq. (4.66).

```
% Example 4-9
k1 = 8000; k2 = 9000; k3 = 15000; k4 = 12000; k5 = 10000; k6 = 18000;
L = 1.5; L1 = 0.18; L2 = 0.22; L3 = 0.26; L4 = 0.19; L5 = 0.15; L6 = 0.30;
a = [k1 + k2, -k2, 0, 0, 0; -k2, k2+k3, -k3, 0, 0; 0, -k3, k3+k4, -k4, 0
    0, 0, -k4, k4+k5, -k5; 0, 0, 0, -k5, k5+k6];
b = [k1*L1 - k2*L2; k2*L2 - k3*L3; k3*L3 - k4*L4; k4*L4 - k5*L5; k5*L5 + k6*L - k6*L6];
Xs = Tridiagonal(a,b)
```

When the script file is executed, the following solution is displayed in the Command Window.

```
Xs =
    0.2262
    0.4872
    0.7718
    0.9926
    1.1795
>>
```

$$\begin{bmatrix} x_1 \\ x_2 \\ x_3 \\ x_4 \\ x_5 \end{bmatrix}$$

4.10 ERROR, RESIDUAL, NORMS, AND CONDITION NUMBER

A numerical solution of a system of equations is seldom an exact solution. Even though direct methods (Gauss, Gauss–Jordan, LU decomposition) can be exact, they are still susceptible to round-off errors when implemented on a computer. This is especially true with large systems and with ill-conditioned systems (see Section 4.11). Solutions that are obtained with iterative methods are approximate by nature. This section describes measures that can be used for quantifying the accuracy, or estimating the magnitude of the error, of a numerical solution.

4.10.1 Error and Residual

If $[x_{NS}]$ is a computed approximate numerical solution of a system of n equations $[a][x] = [b]$ and $[x_{TS}]$ is the true (exact) solution, then the true error is the vector:

$$[e] = [x_{TS}] - [x_{NS}] \qquad (4.67)$$

The true error, however, cannot in general be calculated because the true solution is not known.

An alternative measure of the accuracy of a solution is the residual $[r]$, which is defined by:

$$[r] = [a][x_{TS}] - [a][x_{NS}] = [b] - [a][x_{NS}] \qquad (4.68)$$

In words, $[r]$ measures how well the system of equations is satisfied when $[x_{NS}]$ is substituted for $[x]$. (This is equivalent to the tolerance in $f(x)$ when the solution of a single equation is considered. See Eq. (3.5) in Section 3.2.) The vector $[r]$ has n elements, and if the numerical solution is close to the true solution, then all the elements of $[r]$ are small. It should be remembered that $[r]$ does not really indicate how small the error is in the solution $[x]$. $[r]$ only shows how well the right-hand side of the equations is satisfied when $[x_{NS}]$ is substituted for $[x]$ in the original equations. This depends on the magnitude of the elements of the matrix $[a]$. As shown next in Example 4-10, it is possible to have an approximate numerical solution that has a large true error but gives a small residual.

A more accurate estimate of the error in a numerical solution can be obtained by using quantities that measure the size, or magnitude, of vectors and matrices. For numbers, it is easy to determine which one is large or small by comparing their absolute values. It is more difficult to measure the magnitude (size) of vectors and matrices. This is done by a quantity called **norm**, which is introduced next.

Example 4-10: Error and residual.

The true (exact) solution of the system of equations:

$$1.02x_1 + 0.98x_2 = 2$$
$$0.98x_1 + 1.02x_2 = 2$$

is $x_1 = x_2 = 1$.

Calculate the true error and the residual for the following two approximate solutions:

(a) $x_1 = 1.02$, $x_2 = 1.02$.

(b) $x_1 = 2$, $x_2 = 0$.

SOLUTION

In matrix form, the given system of equations is $[a][x] = [b]$, where $[a] = \begin{bmatrix} 1.02 & 0.98 \\ 0.98 & 1.02 \end{bmatrix}$ and $[b] = \begin{bmatrix} 2 \\ 2 \end{bmatrix}$.

The true solution is $[x_{TS}] = \begin{bmatrix} 1 \\ 1 \end{bmatrix}$.

The true error and the residual are given by Eq. (4.67) and (4.68), respectively. Applying these equations to the two approximate solutions gives:

(a) In this case $[x_{NS}] = \begin{bmatrix} 1.02 \\ 1.02 \end{bmatrix}$. Consequently, the error and residual are:

$$[e] = [x_{TS}] - [x_{NS}] = \begin{bmatrix} 1 \\ 1 \end{bmatrix} - \begin{bmatrix} 1.02 \\ 1.02 \end{bmatrix} = \begin{bmatrix} -0.02 \\ -0.02 \end{bmatrix} \quad \text{and}$$

$$[r] = [b] - [a][x_{NS}] = [b] - [a][x_{NS}] = \begin{bmatrix} 2 \\ 2 \end{bmatrix} - \begin{bmatrix} 1.02 & 0.98 \\ 0.98 & 1.02 \end{bmatrix}\begin{bmatrix} 1.02 \\ 1.02 \end{bmatrix} = \begin{bmatrix} -0.04 \\ -0.04 \end{bmatrix}$$

In this case both the error and the residual are small.

(b) In this case $[x_{NS}] = \begin{bmatrix} 2 \\ 0 \end{bmatrix}$. Consequently, the error and residual are:

$$[e] = [x_{TS}] - [x_{NS}] = \begin{bmatrix} 1 \\ 1 \end{bmatrix} - \begin{bmatrix} 2 \\ 0 \end{bmatrix} = \begin{bmatrix} -1 \\ 1 \end{bmatrix} \quad \text{and}$$

$$[r] = [b] - [a][x_{NS}] = [b] - [a][x_{NS}] = \begin{bmatrix} 2 \\ 2 \end{bmatrix} - \begin{bmatrix} 1.02 & 0.98 \\ 0.98 & 1.02 \end{bmatrix}\begin{bmatrix} 2 \\ 0 \end{bmatrix} = \begin{bmatrix} -0.04 \\ 0.04 \end{bmatrix}$$

In this case the error is large but the residual is small.

This example shows that a small residual does not necessarily guarantee a small error. Whether or not a small residual implies a small error depends on the "magnitude" of the matrix $[a]$.

4.10.2 Norms and Condition Number

A **norm** is a real number assigned to a matrix or vector that satisfies the following four properties:

(*i*) The norm of a vector or matrix denoted by $\|[a]\|$ is a positive quantity. It is equal to zero only if the object $[a]$ itself is zero. In other words, $\|[a]\| \geq 0$ and $\|[a]\| = 0$ only if $[a] = 0$. This statement means that all vectors or matrices except for the zero vector or zero matrix have a positive magnitude.

(*ii*) For all numbers α, $\|\alpha[a]\| = |\alpha|\|[a]\|$. This statement means that the two objects $[a]$ and $[-a]$ have the same "magnitude" and that the magnitude of $[10a]$ is 10 times the magnitude of $[a]$.

(*iii*) For matrices and vectors, $\|[a][x]\| \leq \|[a]\|\|[x]\|$, which means that the norm of a product of two matrices is equal to or smaller than the product of the norms of each matrix.

(*iv*) For any two vectors or matrices $[a]$ and $[b]$:

$$\|[a+b]\| \leq \|[a]\| + \|[b]\| \tag{4.69}$$

This statement is known as the **triangle inequality** because for vectors $[a]$ and $[b]$ it states that the sum of the lengths of two sides of a triangle can never be smaller than the length of the third side.

Any norm of a vector or a matrix must satisfy the four properties listed above in order to qualify as a legitimate measure of its "magnitude." Different ways of calculating norms for vectors and matrices are described next.

Vector norms

For a given vector $[v]$ of n elements, the **infinity norm** written as $\|v\|_\infty$ is defined by:

$$\|v\|_\infty = \max_{1 \leq i \leq n}|v_i| \tag{4.70}$$

In words, $\|v\|_\infty$ is a number equal to the element v_i with the largest absolute value.

The **1-norm** written as $\|v\|_1$ is defined by:

$$\|v\|_1 = \sum_{i=1}^{n}|v_i| \tag{4.71}$$

In words, $\|v\|_1$ is the sum of the absolute values of the elements of the vector.

The **Euclidean 2-norm** written as $\|v\|_2$ is defined by:

$$\|v\|_2 = \left(\sum_{i=1}^{n}v_i^2\right)^{1/2} \tag{4.72}$$

In words, $\|v\|_2$ is the square root of the sum of the square of the elements. It is also called the magnitude of the vector $[v]$.

Matrix norms

The matrix ***infinity norm*** is given by:

$$\|[a]\|_\infty = \max_{1 \le i \le n} \sum_{j=1}^{n} |a_{ij}| \tag{4.73}$$

In words, the absolute values of the elements in each row of the matrix are added. The value of the largest sum is assigned to $\|a\|_\infty$.

The matrix ***1-norm*** is calculated by:

$$\|[a]\|_1 = \max_{1 \le j \le n} \sum_{i=1}^{n} |a_{ij}| \tag{4.74}$$

It is similar to the infinity norm, except that the summation of the absolute values of the elements is done for each column, and the value of the largest sum is assigned to $\|a\|_1$.

The 2-norm of a matrix is evaluated as the spectral norm:

$$\|[a]\|_2 = max\left(\frac{\|[a][v]\|}{\|[v]\|}\right) \tag{4.75}$$

where $[v]$ is an eigenvector of the matrix $[a]$ corresponding to an eigenvalue λ. (Eigenvalues and eigenvectors are covered in Section 4.12.) The 2-norm of a matrix is calculated by MATLAB using a technique called singular value decomposition, where the matrix $[a]$ is factored into $[a] = [u][d][v]$, where $[u]$ and $[v]$ are orthogonal matrices (special matrices with the property $[u]^{-1} = [u]^T$), and $[d]$ is a diagonal matrix. The largest value of the diagonal elements of $[d]$ is used as the 2-norm of the matrix $[a]$.

The Euclidean norm for an $m \times n$ matrix $[a]$ (which is different from the 2-norm of a matrix) is given by:

$$\|[a]\|_{Euclidean} = \left(\sum_{i=1}^{m}\sum_{j=1}^{n} a_{ij}^2\right)^{1/2} \tag{4.76}$$

Using norms to determine bounds on the error of numerical solutions

From Eqs. (4.67) and (4.68), the residual can be written in terms of the error, $[e]$ as:

$$[r] = [a][x_{TS}] - [a][x_{NS}] = [a]([x_{TS}] - [x_{NS}]) = [a][e] \tag{4.77}$$

If the matrix $[a]$ is invertible (otherwise the system of equations does not have a solution), the error can be expressed as:

$$[e] = [a]^{-1}[r] \tag{4.78}$$

Applying property (*iii*) of the matrix norm to Eq. (4.78) gives:

$$\|[e]\| = \|[a]^{-1}[r]\| \le \|[a]^{-1}\|\|[r]\| \tag{4.79}$$

From Eq. (4.77), the residual $[r]$ is:

$$[r] = [a][e] \qquad (4.80)$$

Applying property (*iii*) of the matrix norm to Eq. (4.80) gives:

$$\|[r]\| = \|[a][e]\| \le \|[a]\|\|[e]\| \qquad (4.81)$$

The last equation can be rewritten as:

$$\frac{\|[r]\|}{\|[a]\|} \le \|[e]\| \qquad (4.82)$$

Equations (4.79) and (4.82) can be combined and written in the form:

$$\frac{\|[r]\|}{\|[a]\|} \le \|[e]\| = \left\| [a]^{-1}[r] \right\| \le \left\| [a]^{-1} \right\|\|[r]\| \qquad (4.83)$$

To use Eq. (4.83), two new quantities are defined. One is the **relative error** defined by $\|[e]\|/\|[x_{TS}]\|$, and the second is the **relative residual** defined by $\|[r]\|/\|[b]\|$. For an approximate numerical solution, the residual can be calculated from Eq. (4.68). With the residual known, Eq. (4.83) can be used for obtaining an upper bound and a lower bound on the relative error in terms of the relative residual. This is done by dividing Eq. (4.83) by $\|[x_{TS}]\|$, and rewriting the equation in the form:

$$\frac{1}{\|[a]\|}\frac{\|b\|}{\|[x_{TS}]\|}\frac{\|[r]\|}{\|[b]\|} \le \frac{\|[e]\|}{\|[x_{TS}]\|} \le \left\| [a]^{-1}\right\|\frac{\|[b]\|}{\|[x_{TS}]\|}\frac{\|[r]\|}{\|[b]\|} \qquad (4.84)$$

Since $[a][x_{TS}]=[b]$, property (*iii*) of matrix norms gives: $\|[b]\| \le \|[a]\|\|[x_{TS}]\|$ or $\frac{\|[b]\|}{\|[x_{TS}]\|} \le \|[a]\|$, and this means that $\|[a]\|$ can be substituted for $\frac{\|[b]\|}{\|[x_{TS}]\|}$ in the right-hand side of Eq. (4.84). Similarly, since $[x_{TS}] = [a]^{-1}[b]$, property (*iii*) of matrix norms gives $\|[x_{TS}]\| \le \|[a]^{-1}\|\|[b]\|$ or $\frac{1}{\|[a]^{-1}\|} \le \frac{\|[b]\|}{\|[x_{TS}]\|}$, and this means that $\frac{1}{\|[a]^{-1}\|}$ can be substituted for $\frac{\|[b]\|}{\|[x_{TS}]\|}$ in the left-hand side of Eq. (4.84). With these substitutions Eq. (4.84) becomes:

$$\frac{1}{\|[a]\|\|[a]^{-1}\|}\frac{\|[r]\|}{\|[b]\|} \le \frac{\|[e]\|}{\|[x_{TS}]\|} \le \|[a]^{-1}\|\|[a]\|\frac{\|[r]\|}{\|[b]\|} \qquad (4.85)$$

Equation (4.85) is the main result of this section. It provides a means for bounding the error in a numerical solution of a system of equations. Equation (4.85) states that the true relative error, $\frac{\|[e]\|}{\|[x_{TS}]\|}$ (which is not known), is bounded between $\dfrac{1}{\|[a]\|\|[a]^{-1}\|}$ times the relative residual,

$\dfrac{\|[r]\|}{\|[b]\|}$ (lower bound), and $\|[a]^{-1}\|\|[a]\|$ times the relative residual (upper bound). The relative residual can be calculated from the approximate numerical solution so that the true relative error can be bounded if the quantity $\|[a]\|\|[a]^{-1}\|$ (called condition number) can be calculated.

Condition number

The number $\|[a]\|\|[a]^{-1}\|$ is called the ***condition number*** of the matrix $[a]$. It is written as:

$$Cond[a] = \|[a]\|\|[a]^{-1}\| \tag{4.86}$$

- The condition number of the identity matrix is 1. The condition number of any other matrix is 1 or greater.

- If the condition number is approximately 1, then the true relative error is of the same order of magnitude as the relative residual.

- If the condition number is much larger than 1, then a small relative residual does not necessarily imply a small true relative error.

- For a given matrix, the value of the condition number depends on the matrix norm that is used.

- The inverse of a matrix has to be known in order to calculate the condition number of the matrix.

Example 4-11 illustrates the calculation of error, residual, norms, and condition number.

Example 4-11: Calculating error, residual, norm and condition number.

Consider the following set of four equations (the same that was solved in Example 4-8).

$$9x_1 - 2x_2 + 3x_3 + 2x_4 = 54.5$$
$$2x_1 + 8x_2 - 2x_3 + 3x_4 = -14$$
$$-3x_1 + 2x_2 + 11x_3 - 4x_4 = 12.5$$
$$-2x_1 + 3x_2 + 2x_3 + 10x_4 = -21$$

The true solution of this system is: $x_1 = 5$, $x_2 = -2$, $x_3 = 2.5$, and $x_4 = -1$. When this system was solved in Example 4-8 with the Gauss–Seidel iteration method, the numerical solution in the sixth iteration was: $x_1 = 4.98805$, $x_2 = -1.99511$, $x_3 = 2.49806$, and $x_4 = -1.00347$.

(a) Determine the true error, $[e]$, and the residual, $[r]$.

(b) Determine the infinity norms of the true solution, $[x_{TS}]$, the error, $[e]$, the residual, $[r]$, and the vector $[b]$.

(c) Determine the inverse of $[a]$, the infinity norm of $[a]$ and $[a]^{-1}$, and the condition number of the matrix $[a]$.

(d) Substitute the quantities from parts (b) and (c) in Eq. (4.85) and discuss the results.

SOLUTION

First, the equations are written in matrix form:

$$\begin{bmatrix} 9 & -2 & 3 & 2 \\ 2 & 8 & -2 & 3 \\ -3 & 2 & 11 & -4 \\ -2 & 3 & 2 & 10 \end{bmatrix} \begin{bmatrix} x_1 \\ x_2 \\ x_3 \\ x_4 \end{bmatrix} = \begin{bmatrix} 54.5 \\ -14 \\ 12.5 \\ -21 \end{bmatrix}$$

(*a*) The true solution is $x_{TS} = \begin{bmatrix} 5 \\ -2 \\ 2.5 \\ -1 \end{bmatrix}$, and the approximate numerical solution is $x_{NS} = \begin{bmatrix} 4.98805 \\ -1.99511 \\ 2.49806 \\ -1.00347 \end{bmatrix}$.

The error is then: $[e] = [x_{TS}] - [x_{NS}] = \begin{bmatrix} 5 \\ -2 \\ 2.5 \\ -1 \end{bmatrix} - \begin{bmatrix} 4.98805 \\ -1.99511 \\ 2.49806 \\ -1.00347 \end{bmatrix} = \begin{bmatrix} 0.0119 \\ -0.0049 \\ 0.0019 \\ 0.0035 \end{bmatrix}$.

The residual is given by Eq. (4.77) $[r] = [a][e]$. It is calculated with MATLAB (Command Window):

```
>> a = [9 -2 3 2; 2 8 -2 3; -3 2 11 -4; -2 3 2 10];
>> e = [0.0119; -0.0049; 0.0019; 0.0035];
>> r = a*e
r =
   0.13009000000000
  -0.00869000000000
  -0.03817000000000
   0.00001000000000
```

(*b*) The infinity norm of a vector is defined in Eq. (4.70): $\|v\|_\infty = \max\limits_{1 \le i \le n} |v_i|$. Using this equation to calculate infinity norm of the true solution, the residual, and the vector $[b]$ gives:

$$\|x_{TS}\|_\infty = \max\limits_{1 \le i \le 4} |x_{TS_i}| = max[|5|, |-2|, |2.5|, |-1|] = 5$$

$$\|e\|_\infty = \max\limits_{1 \le i \le 4} |e_i| = max[|0.0119|, |-0.0049|, |0.0019|, |0.0035|] = 0.0119$$

$$\|r\|_\infty = \max\limits_{1 \le i \le 4} |r_i| = max[|0.13009|, |-0.00869|, |-0.03817|, 0.00001] = 0.13009$$

$$\|b\|_\infty = \max\limits_{1 \le i \le 4} |b_i| = max[|54.5|, |-14|, |12.5|, |-21|] = 54.5$$

(*c*) The inverse of $[a]$ is calculated by using MATLAB's `inv` function (Command Window):

```
>> aINV = inv(a)
aINV =
   0.0910    0.0386   -0.0116   -0.0344
  -0.0206    0.1194    0.0308   -0.0194
   0.0349   -0.0200    0.0727    0.0281
   0.0174   -0.0241   -0.0261    0.0933
```

The infinity norms of $[a]$ and $[a]^{-1}$ are calculated by using Eq. (4.73), $\|[a]\|_\infty = \max\limits_{1 \le i \le n} \sum\limits_{j=1}^{n} |a_{ij}|$:

$$\|[a]\|_\infty = \max_{1 \le i \le 4} \sum_{j=1}^{n} |a_{ij}| = max[|9|+|-2|+|3|+|2|, \ |2|+|8|+|-2|+|3|, \ |-3|+|2|+|11|+|-4|, \ |9|+|-2|+|3|+|2|]$$

$$\|[a]\|_\infty = max[16, 15, 20, 16] = 20$$

$$\|[a]^{-1}\|_\infty = \max_{1 \le i \le 4} \sum_{j=1}^{n} |a_{ij}^{-1}| = max[|0.091| + |0.0386| + |-0.0116| + |-0.0344|, \ |-0.0206| + |0.1194| +$$

$$|0.0308| + |-0.0194|, \ |0.0349| + |-0.02| + |0.0727| + |0.0281|, \ |0.0174| + |-0.0241| + |-0.0261| + |0.0933|]$$

$$\|[a]^{-1}\|_\infty = max[0.1756, 0.1902, 0.1557, 0.1609] = 0.1902$$

The condition number of the matrix $[a]$ is calculated by using Eq. (4.86):

$$Cond[a] = \|[a]\|\|[a]^{-1}\| = 20 \cdot 0.1902 = 3.804$$

(d) Substituting all the variables calculated in parts (b) and (c) in Eq.(4.85) gives:

$$\frac{1}{\|[a]\|\|[a]^{-1}\|} \frac{\|[r]\|}{\|[b]\|} \le \frac{\|[e]\|}{\|[x_{TS}]\|} \le \|[a]^{-1}\|\|[a]\| \frac{\|[r]\|}{\|[b]\|}$$

$$\frac{1}{3.804} \frac{0.13009}{54.5} \le \frac{\|[e]\|}{\|[x_{TS}]\|} \le 3.804 \frac{0.13009}{54.5}$$

$$\frac{1}{3.804} 0.002387 \le \frac{\|[e]\|}{\|[x_{TS}]\|} \le 3.804 \cdot 0.002387, \quad \text{or} \quad 6.275 \times 10^{-4} \le \frac{\|[e]\|}{\|[x_{TS}]\|} \le 0.00908$$

These results indicate that the magnitude of the true relative error is between 6.275×10^{-4} and 0.00908. In this problem, the magnitude of the true relative error can be calculated because the true solution is known.

The magnitude of the true relative error is:

$$\frac{\|[e]\|}{\|[x_{TS}]\|} = \frac{0.0119}{5} = 0.00238, \text{ which is within the bounds calculated by Eq. (4.85).}$$

4.11 ILL-CONDITIONED SYSTEMS

An ill-conditioned system of equations is one in which small variations in the coefficients cause large changes in the solution. The matrix of coefficients of ill-conditioned systems generally has a condition number that is significantly greater than 1. As an example, consider the system:

$$\begin{align} 6x_1 - 2x_2 &= 10 \\ 11.5x_1 - 3.85x_2 &= 17 \end{align} \tag{4.87}$$

The solution of this system is:

$$x_1 = \frac{a_{12}b_2 - a_{22}b_1}{a_{12}a_{21} - a_{11}a_{22}} = \frac{-2 \cdot 17 - (-3.85 \cdot 10)}{-2 \cdot 11.5 - (6 \cdot -3.85)} = \frac{4.5}{0.1} = 45$$

$$x_2 = \frac{a_{21}b_1 - a_{11}b_2}{a_{12}a_{21} - a_{11}a_{22}} = \frac{11.5 \cdot 10 - (6 \cdot 17)}{-2 \cdot 11.5 - (6 \cdot -3.85)} = \frac{13}{0.1} = 130$$

If a small change is made in the system by changing a_{22} to 3.84:

$$\begin{align} 6x_1 - 2x_2 &= 10 \\ 11.5x_1 - 3.84x_2 &= 17 \end{align} \tag{4.88}$$

then the solution is:

$$x_1 = \frac{a_{12}b_2 - a_{22}b_1}{a_{12}a_{21} - a_{11}a_{22}} = \frac{-2 \cdot 17 - (-3.84 \cdot 10)}{-2 \cdot 11.5 - (6 \cdot -3.84)} = \frac{4.4}{0.04} = 110$$

$$x_2 = \frac{a_{21}b_1 - a_{11}b_2}{a_{12}a_{21} - a_{11}a_{22}} = \frac{11.5 \cdot 10 - (6 \cdot 17)}{-2 \cdot 11.5 - (6 \cdot -3.84)} = \frac{13}{0.04} = 325$$

It can be observed that there is a very large difference between the solutions of the two systems. A careful examination of the solutions of Eqs. (4.87) and (4.88) shows that the numerator of the equation for x_2 in both solutions is the same and that there is only a small difference in the numerator of the equation for x_1. At the same time, there is a large difference (a factor of 2.5) between the denominators of the two equations. The denominators of both equations are the determinants of the matrices of coefficients $[a]$.

The fact that the system in Eqs. (4.87) is ill-conditioned is evident from the value of the condition number. For this system:

$$[a] = \begin{bmatrix} 6 & -2 \\ 11.5 & -3.85 \end{bmatrix} \quad \text{and} \quad [a]^{-1} = \begin{bmatrix} 38.5 & -20 \\ 115 & -60 \end{bmatrix}$$

Using the infinity norm, Eq. (4.73), the condition number for the system is:

$$Cond[a] = \|[a]\| \|[a]^{-1}\| = 15.35 \cdot 175 = 2686.25$$

Using the 1-norm, Eq. (4.74), the condition number for the system is:

$$Cond[a] = \|[a]\| \|[a]^{-1}\| = 17.5 \cdot 153.5 = 2686.25$$

Using the 2-norm, the condition number for the system is (the norms were calculated with MATLAB built-in `norm(a,2)` function):

$$Cond[a] = \|[a]\| \|[a]^{-1}\| = 13.6774 \cdot 136.774 = 1870.7$$

These results show that with any norm used, the condition number of the matrix of coefficients of the system in Eqs. (4.87) is much larger than 1. This means that the system is likely ill-conditioned.

When an ill-conditioned system of equations is being solved numerically, there is a high probability that the solution obtained will have a large error or that a solution will not be obtained at all. In general, it is difficult to quantify the value of the condition number that can precisely identify an ill-conditioned system. This depends on the precision of the computer used and other factors. Thus, in practice one needs only to worry about whether or not the condition number is much larger than 1, and not about its exact value. Furthermore, it might not be possible to calculate the determinant and the condition number for an ill-conditioned system anyway because the mathematical operations done in these calculations are similar to the operations required in solving the system.

4.12 EIGENVALUES AND EIGENVECTORS

For a given $(n \times n)$ matrix $[a]$, the number λ is an eigenvalue[1] of the matrix if:

$$[a][u] = \lambda[u] \qquad (4.89)$$

The vector $[u]$ is a column vector with n elements called the eigenvector, associated with the eigenvalue λ.

Equation (4.89) can be viewed in a more general way. The multiplication $[a][u]$ is a mathematical operation and can be thought of as the matrix $[a]$ operating on the operand $[u]$. With this terminology, Eq. (4.89) can be read as: "$[a]$ operates on $[u]$ to yield λ times $[u]$", and Eq. (8.49) can be generalized to any mathematical operation as:

$$Lu = \lambda u \qquad (4.90)$$

where L is an operator that can represent multiplication by a matrix, differentiation, integration, and so on, u is a vector or function, and λ is a scalar constant. For example, if L represents second differentiation with respect to x, y is a function of x, and k is a constant, then Eq. (4.90) can have the form:

$$\frac{d^2 y}{dx^2} = k^2 y \qquad (4.91)$$

Equation (4.90) is a general statement of an ***eigenvalue problem***, where λ is called the ***eigenvalue*** associated with the operator L and u is the ***eigenvector*** or ***eigenfunction*** corresponding to the eigenvalue λ and the operator L.

Eigenvalues and eigenvectors arise in numerical methods and have special importance in science and engineering. For example, in the study of vibrations, the eigenvalues represent the natural frequencies of a system or component, and the eigenvectors represent the modes of these vibrations. It is important to identify these natural frequencies because when the system or component is subjected to periodic external loads (forces) at or near these frequencies, resonance can cause the response (motion) of the structure to be amplified, potentially leading to failure of the component. In mechanics of materials, the principal stresses are the eigenvalues of the stress matrix, and the principal directions are the directions of the associated eigenvectors. In quantum mechanics, eigenvalues are especially important. In Heisenberg's formulation of quantum mechanics, there exists an operator L corresponding to every observable quantity (i.e., any quantity that can be measured or inferred experimentally such as position, velocity, or energy). This

---

1. The word *eigenvalue* is derived from the German word *eigenwert*, which means "proper or characteristic value."

operator L operates on an operand Ψ called the wave function, and if the result is proportional to the wave function—that is, if $L\Psi = c\Psi$,—then the value of the observable, c, is the eigenvalue and is said to be *certain* (i.e., can be known very precisely). In other words, the eigenvalues c corresponding to the observable are those values of the observable that have a nonzero probability of occurring (and therefore being observed). Examples of such operators from quantum mechanics are: $\dfrac{ih}{2\pi}\dfrac{\partial\Psi}{\partial t} = E\Psi$, where $\dfrac{ih}{2\pi}\dfrac{\partial}{\partial t}(\ )$ is the energy operator and E is the energy; $-i\dfrac{h}{2\pi}\vec{\nabla}\Psi = \vec{p}\Psi$, where $-i\dfrac{h}{2\pi}\vec{\nabla}(\ )$ is the momentum operator and $\vec{p}$ is the linear momentum, where $i = \sqrt{-1}$, and h is Planck's constant. The eigenvectors, also known as eigenstates, represent one of many states in which an object or a system may exist corresponding to a particular eigenvalue.

There is a link between eigenvalue problems involving differential equations and eigenvalue problems involving matrices (4.89), which are the focus in this section. Numerical solution of eigenvalue problems involving ordinary differential equations (ODEs) results in systems of simultaneous equations of the form (4.89). In other words, numerical determination of the eigenvalues in a problem involving an ODE reduces to finding the eigenvalues of an associated matrix $[a]$, resulting in a problem of the form (4.89).

Beyond the physical importance of eigenvalues in science and engineering, the eigenvalues of a matrix can also provide useful information about its properties in numerical calculations involving that matrix. Section 4.7 showed that the Jacobi and Gauss–Seidel iterative methods can be written in the form of:

$$x_i^{(k+1)} = b'_i - [a]x_i^{(k)}$$

It turns out that whether or not these iterative methods converge to a solution depends on the eigenvalues of the matrix $[a]$. Moreover, how quickly the iterations converge depends on the magnitudes of the eigenvalues of $[a]$.

Determination of eigenvalues and eigenvectors

Determination of the eigenvalues of a matrix from Eq. (4.89) is accomplished by rewriting it in the form:

$$[a - \lambda I][u] = 0 \qquad (4.92)$$

where $[I]$ is the identity matrix with the same dimensions as $[a]$. Written in this homogeneous form, it can be seen that if the matrix $[a - \lambda I]$ is nonsingular (i.e., if it has an inverse), then multiplying both sides of (4.92) by $[a - \lambda I]^{-1}$ yields the trivial solution $[u] = 0$. On the other hand, if $[a - \lambda I]$ is singular, that is, if it does not have an inverse, then a nontrivial solution for $[u]$ is possible. Another way of stating this criterion is based on Cramer's rule (see Chapter 2): the matrix $[a - \lambda I]$ is singular if its determinant is zero:

$$\det[a - \lambda I] = 0 \qquad (4.93)$$

Equation (4.93) is called the ***characteristic equation***. For a given matrix $[a]$, it yields a polynomial equation for λ, whose roots are the eigenvalues. Once the eigenvalues are known, the eigenvectors can be determined. This is done by substituting the eigenvalues (one at a time) in Eq. (4.89) and solving the equation for $[u]$. For a small matrix $[a]$ $((2 \times 2)$, or $(3 \times 3))$ the eigenvalues can be determined directly by calculating the determinant and solving for the roots of the characteristic equation. This is shown in Example 4-12 where the eigenvalue problem approach is used for calculating the principal moments of inertia and the directions of the principal axes of an asymmetric cross-sectional area.

Example 4-12: Principal moments of inertia.

Determine the principal moments of inertia and the orientation of the principal axes of inertia for the cross-sectional area shown.
The moment of inertia I_x, I_y and the product of inertia I_{xy} are:

$I_x = 10228.5 \text{ mm}^4$, $I_y = 1307.34 \text{ mm}^4$,   and $I_{xy} = -2880 \text{ mm}^4$

SOLUTION

In matrix form, the two-dimensional moment of inertia matrix is given by:

$$I_{Iner} = \begin{bmatrix} I_x & -I_{xy} \\ -I_{xy} & I_y \end{bmatrix} = \begin{bmatrix} 10228.5 & 2880 \\ 2880 & 1307.34 \end{bmatrix} \qquad (4.94)$$

The principal moments of inertia and the orientation of the principal

axes of inertia can be calculated by solving the following eigenvalue problem:

$$[I_{Iner}][u] = \lambda[u] \qquad (4.95)$$

where the eigenvalues λ are the principal moments of inertia and the associated eigenvectors $[u]$ are unit vectors in the direction of the principal axes of inertia. The eigenvalues are determined by calculating the determinant in Eq. (4.93):

$$det[I_{Iner} - \lambda I] = 0 \qquad (4.96)$$

$$det \begin{bmatrix} (10228.5 - \lambda) & 2880 \\ 2880 & (1307.34 - \lambda) \end{bmatrix} = 0 \qquad (4.97)$$

The polynomial equation for λ is:

$$(10228.5 - \lambda)(1307.34 - \lambda) - 2880^2 = 0 \quad \text{or} \quad \lambda^2 - 11535.84\lambda + 5077727.19 = 0 \qquad (4.98)$$

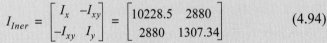

The solutions of the quadratic polynomial equation are the eigenvalues: $\lambda_1 = 11077.46$ mm^4 and $\lambda_2 = 458.38$ mm^4, which are the principal moments of inertia.

The eigenvectors that correspond to each eigenvalue are calculated by substituting the eigenvalues in Eq. (4.95). For the first eigenvector $u^{(1)}$:

$$\begin{bmatrix} 10228.5 & 2880 \\ 2880 & 1307.34 \end{bmatrix} \begin{bmatrix} u_1^{(1)} \\ u_2^{(1)} \end{bmatrix} = 11077.46 \begin{bmatrix} u_1^{(1)} \\ u_2^{(1)} \end{bmatrix} \quad \text{or} \quad \begin{bmatrix} -848.96 & 2880 \\ 2880 & -9770.12 \end{bmatrix} \begin{bmatrix} u_1^{(1)} \\ u_2^{(1)} \end{bmatrix} = \begin{bmatrix} 0 \\ 0 \end{bmatrix} \qquad (4.99)$$

The two equations in Eqs. (4.99) give $u_2^{(1)} = 0.29478 u_1^{(1)}$. By using the additional condition that the eigenvector in this problem is a unit vector, $(u_1^{(1)})^2 + (u_2^{(1)})^2 = 1$, the eigenvector associated with the first eigenvalue, $\lambda_1 = 11077$, is determined to be $u^{(1)} = 0.95919i + 0.28275j$.

For the second eigenvector $u^{(2)}$:

$$\begin{bmatrix} 10228.5 & 2880 \\ 2880 & 1307.34 \end{bmatrix} \begin{bmatrix} u_1^{(2)} \\ u_2^{(2)} \end{bmatrix} = 458.38 \begin{bmatrix} u_1^{(2)} \\ u_2^{(2)} \end{bmatrix} \quad \text{or} \quad \begin{bmatrix} 9770.12 & 2880 \\ 2880 & 848.96 \end{bmatrix} \begin{bmatrix} u_1^{(2)} \\ u_2^{(2)} \end{bmatrix} = \begin{bmatrix} 0 \\ 0 \end{bmatrix} \qquad (4.100)$$

The two equations in Eqs. (4.100) give $u_2^{(2)} = -3.3924 u_1^{(2)}$. By using the additional condition that the eigenvector is a unit vector, $(u_1^{(2)})^2 + (u_2^{(2)})^2 = 1$, the eigenvector associated with the second eigenvalue, $\lambda_2 = 458.38$, is determined to be $u^{(1)} = -0.28275i + 0.95919j$.

Determining the eigenvalues of larger matrices is more difficult. Various numerical methods for solving eigenvalue problems have been introduced. Two of them, the power method and the QR factorization method, are described next.

4.12.1 The Basic Power Method

The power method is an iterative procedure for determining the largest real eigenvalue and the corresponding eigenvector of a matrix. Consider an $(n \times n)$ matrix $[a]$ that has n distinct real eigenvalues $\lambda_1, \lambda_2, ..., \lambda_n$ and n associated eigenvectors $[u]_1, [u]_2, ..., [u]_n$. The eigenvalues are numbered from the largest to the smallest such that:

$$|\lambda_1| > |\lambda_2| > ... > |\lambda_n| \qquad (4.101)$$

Since the eigenvectors are linearly independent, they are a set of basis vectors. This means that any vector, belonging to the same space (i.e., group) as the eigenvectors, can be written as a linear combination of the basis vectors. Suppose that $[x]$ is a column vector in the same space as the eigenvectors. Then any vector $[x]$ in the same space as the eigenvectors $[u]_1, [u]_2, ..., [u]_n$ can be expressed as a linear combination of the eigenvectors:

$$[x] = c_1[u]_1 + c_2[u]_2 + ... + c_n[u]_n \qquad (4.102)$$

where the $c_i \neq 0$ are scalar constants. Let $[x]_1 = [x]$. Multiplying Eq.

(4.102) by $[a]$ yields:

$$[a][x]_1 = c_1[a][u]_1 + c_2[a][u]_2 + \ldots + c_n[a][u]_n = \lambda_1 c_1[x]_2 \quad (4.103)$$

where $[x]_2 = [u]_1 + \dfrac{c_2 \lambda_2}{c_1 \lambda_1}[u]_2 + \ldots + \dfrac{c_n \lambda_n}{c_1 \lambda_1}[u]_n$. since the $[u_i]$s are eigenvectors, $[a][u]_i = \lambda_i[u]_i$, so that:

$$[a][x]_2 = \lambda_1[u]_1 + \frac{c_2 \lambda_2^2}{c_1 \lambda_1}[u]_2 + \ldots + \frac{c_n \lambda_n^2}{c_1 \lambda_1}[u]_n$$

$$= \lambda_1 \left\{ [u]_1 + \frac{c_2 \lambda_2^2}{c_1 \lambda_1^2}[u]_2 + \ldots + \frac{c_n \lambda_n^2}{c_1 \lambda_1^2}[u]_n \right\} = \lambda_1[x]_3 \quad (4.104)$$

where $[x]_3 = [u]_1 + \dfrac{c_2 \lambda_2^2}{c_1 \lambda_1^2}[u]_2 + \ldots + \dfrac{c_n \lambda_n^2}{c_1 \lambda_1^2}[u]_n$. Multiplying Eq. (4.104) again by $[a]$ and using $[A][u]_i = \lambda_i[u]_i$ gives:

$$[a][x]_3 = \lambda_1[u]_1 + \frac{c_2 \lambda_2^3}{c_1 \lambda_1^2}[u]_2 + \ldots + \frac{c_n \lambda_n^3}{c_1 \lambda_1^2}[u]_n = \lambda_1[x]_4 \quad (4.105)$$

where $[x]_4 = [u]_1 + \dfrac{c_2 \lambda_2^3}{c_1 \lambda_1^3}[u]_2 + \ldots + \dfrac{c_n \lambda_n^3}{c_1 \lambda_1^3}[u]_n$.

It can be seen that each successive iteration yields:

$$[x]_{k+1} = [u]_1 + \frac{c_2 \lambda_2^k}{c_1 \lambda_1^k}[u]_2 + \ldots + \frac{c_n \lambda_n^k}{c_1 \lambda_1^k}[u]_n \quad (4.106)$$

Recall that λ_1 is the largest eigenvalue (see Eq. (4.101)), which means that $\dfrac{\lambda_i}{\lambda_1} < 1$ for all $i > 1$. Thus, when k is sufficiently large, all the terms on the right-hand side of Eq. (4.106) that contain $\left(\dfrac{\lambda_i}{\lambda_1} \right)^k$ can be neglected relative to the term $[u]_1$ so that:

$$[a][x]_{k+1} \to \lambda_1[u]_1 \quad \text{as} \quad k \to \infty \quad \text{and} \quad [x]_{k+1} \to [u]_1 \quad (4.107)$$

Equation (4.107) shows that the vector $[x]_k$ that is obtained from Eq. (4.106) is $[u]_1$ (the eigenvector). When the power method is implemented, the vector $[x]_k$ is normalized at each step by dividing the elements of the vector by the value of the largest element (see Eq. (4.102) through Eq. (4.106)). This makes the largest element of the vector equal to 1. It is because of this scaling at each step that the power method yields the eigenvalue and associated eigenvector simultaneously.

A numerical procedure for determining the largest eigenvalue of a $(n \times n)$ matrix $[a]$ with the power method is given in the following algorithm.

Algorithm for the power method

1. Start with a column eigenvector $[x]_i$ of length n. The vector can be any nonzero vector.

2. Multiply the vector $[x]_i$ by the matrix $[a]$. The result is a column vector $[x]_{i+1}$, $[x]_{i+1} = [a][x]_i$.

3. Normalize the resulting vector $[x]_{i+1}$. This is done by factoring out the largest element in the vector. The result of this operation is a multiplicative factor (scalar) times a normalized vector. The normalized vector has the value 1 for the element that used to be the largest, while the absolute value of the rest of the elements is less than 1.

4. Assign the normalized vector (without the multiplicative factor) to $[x]_i$ and go back to *1*.

The iterations continue in this manner until the difference between the vector $[x]_i$ and the normalized vector $[x]_{i+1}$ is less than some specified tolerance. The difference can be measured in different ways. One possibility is to use the infinity norm (see Section 4.10.2):

$$\left\| [x]_{i+1} - [x]_i \right\|_\infty \leq Tolerance \tag{4.108}$$

The last multiplicative factor is the largest eigenvalue, and the normalized vector is the associated eigenvector.

Example 4-13 illustrates how the power method works.

Example 4-13: Using the power method to determine the largest eigenvalue of a matrix.

Determine the largest eigenvalue of the following matrix:

$$\begin{bmatrix} 4 & 2 & -2 \\ -2 & 8 & 1 \\ 2 & 4 & -4 \end{bmatrix} \tag{4.109}$$

Use the power method, and start with the vector $x = [1, 1, 1]^T$.

SOLUTION

Starting with $i = 1$, $x_1 = [1, 1, 1]^T$. With the power method the vector $[x]_2$ is first calculated by $[x]_2 = [a][x]_1$ (Step 2) and then is normalized (Step 3):

$$[x]_2 = [a][x]_1 = \begin{bmatrix} 4 & 2 & -2 \\ -2 & 8 & 1 \\ 2 & 4 & -4 \end{bmatrix} \begin{bmatrix} 1 \\ 1 \\ 1 \end{bmatrix} = \begin{bmatrix} 4 \\ 7 \\ 2 \end{bmatrix} = 7 \begin{bmatrix} 0.5714 \\ 1 \\ 0.2857 \end{bmatrix} \tag{4.110}$$

For $i = 2$ the normalized vector $[x]_2$ (without the multiplicative factor) is multiplied by $[a]$. This results in $[x]_3$, which is then normalized:

$$[x]_3 = [a][x]_2 = \begin{bmatrix} 4 & 2 & -2 \\ -2 & 8 & 1 \\ 2 & 4 & -4 \end{bmatrix} \begin{bmatrix} 0.5714 \\ 1 \\ 0.2857 \end{bmatrix} = \begin{bmatrix} 3.7143 \\ 7.1429 \\ 4 \end{bmatrix} = 7.1429 \begin{bmatrix} 0.52 \\ 1 \\ 0.56 \end{bmatrix} \tag{4.111}$$

The next three iterations are:

$$i = 3: \qquad [x]_4 = [a][x]_3 = \begin{bmatrix} 4 & 2 & -2 \\ -2 & 8 & 1 \\ 2 & 4 & -4 \end{bmatrix} \begin{bmatrix} 0.52 \\ 1 \\ 0.56 \end{bmatrix} = \begin{bmatrix} 2.96 \\ 7.52 \\ 2.8 \end{bmatrix} = 7.52 \begin{bmatrix} 0.3936 \\ 1 \\ 0.3723 \end{bmatrix} \tag{4.112}$$

$$i = 4: \qquad [x]_5 = [a][x]_4 = \begin{bmatrix} 4 & 2 & -2 \\ -2 & 8 & 1 \\ 2 & 4 & -4 \end{bmatrix} \begin{bmatrix} 0.3936 \\ 1 \\ 0.3723 \end{bmatrix} = \begin{bmatrix} 2.8298 \\ 7.5851 \\ 3.2979 \end{bmatrix} = 7.5851 \begin{bmatrix} 0.3731 \\ 1 \\ 0.4348 \end{bmatrix} \tag{4.113}$$

$$i = 5: \qquad [x]_6 = [a][x]_5 = \begin{bmatrix} 4 & 2 & -2 \\ -2 & 8 & 1 \\ 2 & 4 & -4 \end{bmatrix} \begin{bmatrix} 0.3731 \\ 1 \\ 0.4348 \end{bmatrix} = \begin{bmatrix} 2.6227 \\ 7.6886 \\ 3.0070 \end{bmatrix} = 7.6886 \begin{bmatrix} 0.3411 \\ 1 \\ 0.3911 \end{bmatrix} \tag{4.114}$$

After three more iterations, the results are:

$$i = 8 \qquad [x]_9 = [a][x]_8 = \begin{bmatrix} 4 & 2 & -2 \\ -2 & 8 & 1 \\ 2 & 4 & -4 \end{bmatrix} \begin{bmatrix} 0.3272 \\ 1 \\ 0.3946 \end{bmatrix} = \begin{bmatrix} 2.5197 \\ 7.7401 \\ 3.0760 \end{bmatrix} = 7.7401 \begin{bmatrix} 0.3255 \\ 1 \\ 0.3974 \end{bmatrix} \tag{4.115}$$

The results show that the differences between the vector $[x_i]$ and the normalized vector $[x_{i+1}]$ are getting smaller. The value of the multiplicative factor (7.7401) is an estimate of the largest eignvalue. As shown in Section 4.12.3, a value of 7.7504 is obtained for the eigenvalue by MATLAB's built-in function `eig`.

Convergence of the power method

The power method generally converges very slowly, unless the starting vector $[x]$ is close to the eigenvector $[u]_1$. It can be seen from Eq. (4.106) that the ratio of the two largest eigenvalues determines how quickly the power method converges to an answer. A problem can arise when the starting vector $[x]$ is such that the value of c_1 in Eq. (4.102) is zero. This means that $[x]$ has no components in the direction of the corresponding eigenvector $[u]_1$. Theoretically, the power method in this case will fail. In practice, however, the method can still converge (very slowly) because the accumulation of round-off errors during the repeated multiplication with the matrix $[a]$ will produce components in the direction of $[u]_1$.

When can the power method be used?

The power method can be used under the following conditions:

- Only the largest eigenvalue is desired.

- The largest eigenvalue cannot be a repeated root of the characteristic equation. In other words, there cannot be other eigenvalues with the same magnitude as the largest eigenvalue.

- The largest eigenvalue must be real. This is also implied from the bullet above, because if the largest eigenvalue is complex, then the complex conjugate is also an eigenvalue, which means that there are two eigenvalues with the same magnitude.

Additional note on the power method

The matrix for which the eigenvalue is determined cannot be modified in any manner (i.e., the matrix cannot be changed to an upper triangular form, lower triangular form, etc.) before finding the largest eigenvalue. Modifying the matrix and then applying the power method will result in a different matrix with different eigenvalues than the original matrix.

4.12.2 The Inverse Power Method

The inverse power method can be used to determine the smallest eigenvalue. This is done by applying the power method to the inverse of the given matrix $[a]$ (i.e., $[a]^{-1}$). This works because the eigenvalues of the inverse matrix $[a]^{-1}$ are the reciprocals of the eigenvalues of $[a]$. Starting from $[a][x] = \lambda[x]$, multiplying both sides from the left by $[a]^{-1}$ gives:

$$[a]^{-1}[a][x] = [a]^{-1}\lambda[x] = \lambda[a]^{-1}[x] \tag{4.116}$$

Since $[a]^{-1}[a] = [I]$, Eq. (4.116) reduces to:

$$[x] = \lambda[a]^{-1}[x] \text{ or } [a]^{-1}[x] = \frac{[x]}{\lambda} \tag{4.117}$$

This shows that $1/\lambda$ is the eigenvalue of the inverse matrix $[a]^{-1}$. Thus, the power method can be applied for finding the largest eigenvalue of $[a]^{-1}$, and the result will be the largest value of $1/\lambda$, which corresponds to the smallest value of λ for the matrix $[a]$. Applying the power method to the inverse of $[a]$ is called the **inverse power method**.

The procedure of applying the inverse power method is in principle the same as the power method. A starting vector $[x]_i$ is multiplied by $[a]^{-1}$ to give $[x]_{i+1}$ which is then normalized and multiplied again:

$$[x]_{i+1} = [a]^{-1}[x]_i \tag{4.118}$$

Obviously, the inverse matrix $[a]^{-1}$ has to be calculated before iterations with Eq. (4.118) can be carried out. Numerical methods for calculating an inverse of a matrix are described in Section 4.6. In practice, however, calculating the inverse of a matrix is computationally inefficient and not desirable. To avoid the need for calculating the inverse of $[a]$, Eq. (4.118) can be rewritten as:

$$[a][x]_{i+1} = [x]_i \tag{4.119}$$

Now, for a given $[x]_i$ Eq. (4.119) is solved for $[x]_{i+1}$. This can best be done by using the LU decomposition method (Section 4.5).

Thus far, the power method has been used for finding the largest and smallest eigenvalues of a matrix. In some instances, it is necessary to find all the eigenvalues. The next two sections describe two numerical methods, the shifted power method and the QR factorization method, which can be used for finding all eigenvalues.

4.12.3 The Shifted Power Method

Once the largest or the smallest eigenvalue is known, the shifted power method can be used for finding the other eigenvalues. The shifted power method uses an important property of matrices and their eigenvalues. Given $[a][x] = \lambda[x]$, if λ_1 is the largest (or smallest) eigenvalue obtained by using the power method (or the inverse power method), then the eigenvalues of a new *shifted* matrix formed by $[a - \lambda_1 I]$ are $0, \lambda_2 - \lambda_1, \lambda_3 - \lambda_1, \lambda_4 - \lambda_1, \dots, \lambda_n - \lambda_1$. This can be seen easily because the eigenvalues of $[a - \lambda_1 I]$ are found from:

$$[a - \lambda_1 I][x] = \alpha[x] \tag{4.120}$$

where the α s are the eigenvalues of the shifted matrix $[a - \lambda_1 I]$. But $[a][x] = \lambda[x]$ so that Eq. (4.120) becomes:

$$(\lambda - \lambda_1)[x] = \alpha[x] \tag{4.121}$$

where $\lambda = \lambda_1, \lambda_2, \lambda_3, \dots, \lambda_n$. Therefore, the eigenvalues of the shifted matrix are $\alpha = 0, \lambda_2 - \lambda_1, \lambda_3 - \lambda_1, \lambda_4 - \lambda_1, \dots, \lambda_n - \lambda_1$. The eigenvectors of the shifted matrix $[a - \lambda_1 I]$ are the same as the eigenvectors of the original matrix $[a]$. Now, if the basic power method is applied to the shifted matrix $[a - \lambda_1 I]$ (after it is applied to $[a]$ to determine λ_1), then the largest eigenvalue of the shifted matrix, α_k, can be determined. Then, the eigenvalue λ_k can be determined since $\alpha_k = \lambda_k - \lambda_1$. All the other eigenvalues can be determined by repeating this process $k - 2$ times, where at each time the shifted matrix is $[a - \lambda_k I]$, where λ_k is the eigenvalue obtained from the previous shift.

The shifted power method is a tedious and inefficient process. A preferred method for finding all the eigenvalues of a matrix is the QR factorization method, which is described in the next section.

4.12.4 The QR Factorization and Iteration Method

The QR factorization and iteration method is a popular means for finding all the eigenvalues of a matrix. The method is based on the fact that similar matrices (see definition below) have the same eigenvalues and associated eigenvectors, and the fact that the eigenvalues of an upper triangular matrix are the elements along the diagonal. To find the eigenvalues (all real) of a matrix $[a]$, the strategy of the QR factorization method is to eventually transform the matrix into a similar matrix that is upper triangular. In actuality, this is not done in one step but, as described later in the section, in an iterative process.

The QR factorization method finds all the eigenvalues of a matrix but cannot find the corresponding eigenvectors. If the eigenvalues of the given matrix are all real, the QR factorization method eventually factors the given matrix into an orthogonal matrix and an upper triangular matrix. If the eigenvalues are complex (not covered in this book), the matrix is factored into an orthogonal matrix and a (2×2) block diagonal matrix (i.e., a matrix whose diagonal elements themselves are (2×2) block matrices).

Similar matrices

Two square matrices $[a]$ and $[b]$ are similar if:

$$[a] = [c]^{-1}[b][c] \tag{4.122}$$

where $[c]$ is an invertible matrix. The operation in Eq. (4.122) is called a ***similarity transformation***. Similar matrices have the same eigenvalues and associated eigenvectors.

The QR factorization and iteration procedure

The QR factorization procedure starts with the matrix $[a]_1$ whose eigenvalues are to be determined. The matrix is factored into two matrices $[Q]_1$ and $[R]_1$:

$$[a]_1 = [Q]_1[R]_1 \tag{4.123}$$

where $[Q]_1$ is an orthogonal matrix and $[R]_1$ is an upper triangular matrix. (An orthogonal matrix is a matrix whose inverse is the same as its transpose, $[Q]^{-1} = [Q]^T$, that is, $[Q]^T[Q] = [Q]^{-1}[Q] = [I]$).

The matrix $[R]_1$ is then multiplied from the right by $[Q]_1$ to give the matrix $[a]_2$

$$[a]_2 = [R]_1[Q]_1 \tag{4.124}$$

Since from Eq. (4.123) $[R]_1 = [Q]_1^T[a]_1$, Eq. (4.124) reduces to:

$$[a]_2 = [Q]_1^T[a]_1[Q]_1 \tag{4.125}$$

This means (see Eq. (4.122)) that the matrices $[a]_1$ and $[a]_2$ are similar, thus having the same eigenvalues. This completes the first iteration in the QR factorization and iteration procedure.

The second iteration starts by factoring the matrix $[a]_2$ into $[Q]_2$ (orthogonal) and $[R]_2$ (upper triangular) such that $[a]_2 = [Q]_2[R]_2$, and then calculating $[a]_3$ by $[a]_3 = [R]_2[Q]_2$. Again, since $[R]_2 = [Q]_2^T[a]_1$, the matrix $[a]_3$ is given by $[a]_3 = [Q]_2^T[R]_2[Q]_2$. The matrices $[a]_3$ and $[a]_2$ are similar, thus having the same eigenvalues (which are the same as the eigenvalues of $[a]_1$).

The iterations continue until the sequence of matrices generated, $[a]_1, [a]_2, [a]_3, \ldots$, results in an upper triangular matrix of the form:

$$\begin{bmatrix} \lambda_1 & X & X & X \\ 0 & \lambda_2 & X & X \\ 0 & 0 & \lambda_3 & X \\ 0 & 0 & 0 & \lambda_4 \end{bmatrix}$$

where the eigenvalues of the given matrix $[a]_1$ appear along the diagonal. The eigenvalues $\lambda_1, \lambda_2, \lambda_3, \ldots$, are not in any particular order.

In each iteration of the QR method, factoring a matrix $[a]$ into an orthogonal matrix $[Q]$ and an upper triangular matrix $[R]$, such that $[a] = [Q][R]$, is done in steps by using a special matrix $[H]$ called the **Householder matrix**.

The Householder matrix $[H]$

The $(n \times n)$ Householder matrix $[H]$ has the form:

$$[H] = [I] - \frac{2}{[v]^T[v]}[v][v]^T \tag{4.126}$$

where $[I]$ is the $(n \times n)$ identity matrix and $[v]$ is an n-element column vector given by:

$$[v] = [c] + \|c\|_2[e] \tag{4.127}$$

In Eq. (4.127) $[e]$ and $[c]$ are n-element column vectors, and $\|c\|_2$ is the Euclidean norm (length) of $[c]$:

$$\|c\|_2 = \sqrt{c_1^2 + c_2^2 + c_3^2 + \ldots + c_n^2} \tag{4.128}$$

Note that $[v]^T[v]$ is a scalar (number) and $[v][v]^T$ is an $(n \times n)$ matrix. The vectors $[c]$ and $[e]$ are described in detail in the next subsection.

The Householder matrix has special properties. First, it is symmetric. Second, it is orthogonal. Thus, $[H]^{-1} = [H]^T = [H]$. This means that $[H][a][H]$ yields a matrix that is similar to $[a]$.

Factoring a matrix $[a]$ *into an orthogonal matrix* $[Q]$ *and an upper triangular matrix* $[R]$

Factoring an $(n \times n)$ matrix $[a]$ into an orthogonal matrix $[Q]$ and an upper triangular matrix $[R]$, such that $[a] = [Q][R]$, is done in $(n-1)$ steps.

Step 1: The vector $[c]$, which has n elements, is defined as the first column of the matrix $[a]$:

$$[c] = \begin{bmatrix} a_{11} \\ a_{21} \\ \dots \\ a_{n1} \end{bmatrix} \tag{4.129}$$

The vector $[e]$ is defined as the following column vector of length n:

$$[e] = \begin{bmatrix} \pm 1 \\ 0 \\ 0 \\ \dots \\ 0 \end{bmatrix} \tag{4.130}$$

The first element of $[e]$ is $+1$ if the first element of $[c]$ (which is a_{11}) is positive and is -1 if the first element of $[c]$ is negative. The rest of the elements are zeros.

Once the vectors $[c]$ and $[e]$ are defined, the $(n \times n)$ Householder matrix $[H]^{(1)}$ can be constructed by using Eqs. (4.126)–(4.128). Using $[H]^{(1)}$ the matrix $[a]$ is factored into $[Q]^{(1)}[R]^{(1)}$ where:

$$[Q]^{(1)} = [H]^{(1)} \tag{4.131}$$

and

$$[R]^{(1)} = [H]^{(1)}[a] \tag{4.132}$$

The matrix $[Q]^{(1)}$ is orthogonal because it is a Householder matrix, and $[R]^{(1)}$ is a matrix with zeros in the elements of the first column that are below the $(1, 1)$ element. The matrix $[R]^{(1)}$ is illustrated in Fig. 4-23 for the case of a (5×5) matrix.

Step 2: The vector $[c]$, which has n elements, is defined as the second column of the $[R]^{(1)}$ matrix with its first entry set to 0:

$$[c] = \begin{bmatrix} 0 \\ R_{22}^{(1)} \\ R_{32}^{(1)} \\ \dots \\ R_{n2}^{(1)} \end{bmatrix} \tag{4.133}$$

The vector $[e]$ is defined as the following column vector of length n:

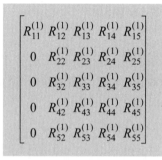

$$\begin{bmatrix} R_{11}^{(1)} & R_{12}^{(1)} & R_{13}^{(1)} & R_{14}^{(1)} & R_{15}^{(1)} \\ 0 & R_{22}^{(1)} & R_{23}^{(1)} & R_{24}^{(1)} & R_{25}^{(1)} \\ 0 & R_{32}^{(1)} & R_{33}^{(1)} & R_{34}^{(1)} & R_{35}^{(1)} \\ 0 & R_{42}^{(1)} & R_{43}^{(1)} & R_{44}^{(1)} & R_{45}^{(1)} \\ 0 & R_{52}^{(1)} & R_{53}^{(1)} & R_{54}^{(1)} & R_{55}^{(1)} \end{bmatrix}$$

Figure 4-23: The matrix $[R]^{(1)}$ after Step *1*.

$$[e] = \begin{bmatrix} 0 \\ \pm 1 \\ 0 \\ \dots \\ 0 \end{bmatrix} \qquad (4.134)$$

The second element of $[e]$ is $+1$ if the second element of $[c]$ in Eq. (4.133) (which is $R_{22}^{(1)}$) is positive and is -1 if the second element of $[c]$ is negative. The rest of the elements are zeros.

Once the vectors $[c]$ and $[e]$ are identified, the next $(n \times n)$ Householder matrix $[H]^{(2)}$ can be constructed by using Eqs. (4.126)–(4.128). Next, by using $[H]^{(2)}$, the matrix $[a]$ is factored into $[Q]^{(2)}[R]^{(2)}$ where:

$$[Q]^{(2)} = [Q]^{(1)}[H]^{(2)} \qquad (4.135)$$

and

$$[R]^{(2)} = [H]^{(2)}[R]^{(1)} \qquad (4.136)$$

The matrix $[Q]^{(2)}$ is orthogonal, and $[R]^{(2)}$ is a matrix with zeros as the elements of the first and second columns that are below the diagonal elements in these columns. The matrix $[R]^{(2)}$ is illustrated in Fig. 4-24 for the case of a (5×5) matrix.

$$\begin{bmatrix} R_{11}^{(2)} & R_{12}^{(2)} & R_{13}^{(2)} & R_{14}^{(2)} & R_{15}^{(2)} \\ 0 & R_{22}^{(2)} & R_{23}^{(2)} & R_{24}^{(2)} & R_{25}^{(2)} \\ 0 & 0 & R_{33}^{(2)} & R_{34}^{(2)} & R_{35}^{(2)} \\ 0 & 0 & R_{43}^{(2)} & R_{44}^{(2)} & R_{45}^{(2)} \\ 0 & 0 & R_{53}^{(2)} & R_{54}^{(2)} & R_{55}^{(2)} \end{bmatrix}$$

Figure 4-24: The matrix $[R]^{(2)}$ after *Step 2*.

Step 3: Moving to the third column of $[a]$, the vector $[c]$, which has n elements, is defined as:

$$[c] = \begin{bmatrix} 0 \\ 0 \\ R_{33}^{(2)} \\ R_{34}^{(2)} \\ \dots \\ R_{n3}^{(2)} \end{bmatrix} \qquad (4.137)$$

The vector $[e]$ is defined as the following column vector of length n:

$$[e] = \begin{bmatrix} 0 \\ 0 \\ \pm 1 \\ 0 \\ \dots \\ 0 \end{bmatrix} \qquad (4.138)$$

The third element of $[e]$ is $+1$ if the third element of $[c]$ (which is $R_{33}^{(2)}$)

is positive and is -1 if the third element of $[c]$ is negative. The rest of the elements are zeros.

Once the vectors $[c]$ and $[e]$ are identified, the next $(n \times n)$ Householder matrix $[H]^{(3)}$ can be constructed by using Eqs. (4.126)–(4.128). Next, by using $[H]^{(3)}$, the matrix $[a]$ is factored into $[Q]^{(3)}[R]^{(3)}$ where:

$$[Q]^{(3)} = [Q]^{(2)}[H]^{(3)} \tag{4.139}$$

and

$$[R]^{(3)} = [H]^{(3)}[R]^{(2)} \tag{4.140}$$

The matrix $[Q]^{(3)}$ is orthogonal, and $[R]^{(3)}$ is a matrix with zeros as the elements of the first, second, and third columns that are below the diagonal elements in these columns. The matrix $[R]^{(3)}$ is illustrated in Fig. 4-25 for the case of a (5×5) matrix.

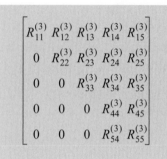

$$\begin{bmatrix} R_{11}^{(3)} & R_{12}^{(3)} & R_{13}^{(3)} & R_{14}^{(3)} & R_{15}^{(3)} \\ 0 & R_{22}^{(3)} & R_{23}^{(3)} & R_{24}^{(3)} & R_{25}^{(3)} \\ 0 & 0 & R_{33}^{(3)} & R_{34}^{(3)} & R_{35}^{(3)} \\ 0 & 0 & 0 & R_{44}^{(3)} & R_{45}^{(3)} \\ 0 & 0 & 0 & R_{54}^{(3)} & R_{55}^{(3)} \end{bmatrix}$$

Figure 4-25: The matrix $[R]^{(3)}$ after Step 3.

Steps 4 through (n–1): The factoring of matrix $[a]$ into orthogonal and upper triangular matrices continues in the same way as in Steps 1 through 3. In a general Step i, the vector $[c]$ has zeros in elements 1 through $i - 1$, and in elements i through n, the values of the elements $R_{ii}^{(i-1)}$ through $R_{ni}^{(i-1)}$ of the matrix $[R]^{(i-1)}$ that was calculated in the previous step. The vector $[e]$ has $+1$ for the ith element if the ith element of $[c]$ is positive and is -1 if the ith element of $[c]$ is negative. The rest of its elements are zeros.

Once the vectors $[c]$ and $[e]$ are identified, the next $(n \times n)$ Householder matrix $[H]^{(i)}$ can be constructed by using Eqs. (4.126)–(4.128). Next, by using $[H]^{(i)}$, the matrix $[a]$ is factored into $[Q]^{(i)}[R]^{(i)}$ where:

$$[Q]^{(i)} = [Q]^{(i-1)}[H]^{(i)} \tag{4.141}$$

and

$$[R]^{(i)} = [H]^{(i)}[R]^{(i-1)} \tag{4.142}$$

The matrix $[Q]^{(i)}$ is orthogonal, and $[R]^{(i)}$ is a matrix with zeros in the elements of the first through the ith columns that are below the diagonal elements in these columns.

After the last step (Step $n - 1$), the matrix $[R]^{(n-1)}$ is upper triangular. The matrices $[Q]^{(n-1)}$ and $[R]^{(n-1)}$ obtained in the last step are the orthogonal and upper triangular matrices that the matrix $[A]$ is factored into in the iterative process.

$$[A] = [Q]^{(n-1)}[R]^{(n-1)} \tag{4.143}$$

Example 4-14 shows a hand calculation of QR factorization of a matrix.

Example 4-14: QR factorization of a matrix.

Factor the following matrix $[a]$ into an orthogonal matrix $[Q]$ and an upper triangular matrix $[R]$:

$$[a] = \begin{bmatrix} 6 & -7 & 2 \\ 4 & -5 & 2 \\ 1 & -1 & 1 \end{bmatrix} \tag{4.144}$$

SOLUTION

The solution follows the steps listed in pages 160–162. Since the matrix $[a]$ is (3×3), the factorization requires only two steps.

Step 1: The vector $[c]$ is defined as the first column of the matrix $[a]$:

$$[c] = \begin{bmatrix} 6 \\ 4 \\ 1 \end{bmatrix}$$

The vector $[e]$ is defined as the following three element column vector:

$$[e] = \begin{bmatrix} 1 \\ 0 \\ 0 \end{bmatrix}$$

Using Eq. (4.128), the Euclidean norm, $\|c\|_2$, of $[c]$ is:

$$\|c\|_2 = \sqrt{c_1^2 + c_2^2 + c_3^2} = \sqrt{6^2 + 4^2 + 1^2} = 7.2801$$

Using Eq. (4.127), the vector $[v]$ is:

$$[v] = [c] + \|c\|_2[e] = \begin{bmatrix} 6 \\ 4 \\ 1 \end{bmatrix} + 7.2801 \begin{bmatrix} 1 \\ 0 \\ 0 \end{bmatrix} = \begin{bmatrix} 13.2801 \\ 4 \\ 1 \end{bmatrix}$$

Next, the products $[v]^T[v]$ and $[v][v]^T$ are calculated:

$$[v]^T[v] = \begin{bmatrix} 13.2801 & 4 & 1 \end{bmatrix} \begin{bmatrix} 13.2801 \\ 4 \\ 1 \end{bmatrix} = 193.3611$$

$$[v][v]^T = \begin{bmatrix} 13.2801 \\ 4 \\ 1 \end{bmatrix} \begin{bmatrix} 13.2801 & 4 & 1 \end{bmatrix} = \begin{bmatrix} 176.3611 & 53.1204 & 13.2801 \\ 53.1204 & 16 & 4 \\ 13.2801 & 4 & 1 \end{bmatrix}$$

The Householder matrix $[H]^{(1)}$ is then:

$$[H]^{(1)} = [I] - \frac{2}{[v]^T[v]}[v][v]^T = \begin{bmatrix} 1 & 0 & 0 \\ 0 & 1 & 0 \\ 0 & 0 & 1 \end{bmatrix} - \frac{2}{193.3611}\begin{bmatrix} 176.3611 & 53.1204 & 13.2801 \\ 53.1204 & 16 & 4 \\ 13.2801 & 4 & 1 \end{bmatrix} = \begin{bmatrix} -0.8242 & -0.5494 & -0.1374 \\ -0.5494 & 0.8345 & -0.0414 \\ -0.1374 & -0.0414 & 0.9897 \end{bmatrix}$$

Once the Housholder matrix $[H]^{(1)}$ is constructed, $[a]$ can be factored into $[Q]^{(1)}[R]^{(1)}$ where:

$$[Q]^{(1)} = [H]^{(1)} = \begin{bmatrix} -0.8242 & -0.5494 & -0.1374 \\ -0.5494 & 0.8345 & -0.0414 \\ -0.1374 & -0.0414 & 0.9897 \end{bmatrix}$$

and

$$[R]^{(1)} = [H]^{(1)}[a] = \begin{bmatrix} -0.8242 & -0.5494 & -0.1374 \\ -0.5494 & 0.8345 & -0.0414 \\ -0.1374 & -0.0414 & 0.9897 \end{bmatrix}\begin{bmatrix} 6 & -7 & 2 \\ 4 & -5 & 2 \\ 1 & -1 & 1 \end{bmatrix} = \begin{bmatrix} -7.2801 & 8.6537 & -2.8846 \\ 0 & -0.2851 & 0.5288 \\ 0 & 0.1787 & 0.6322 \end{bmatrix}$$

This completes the first step.

Step 2: The vector $[c]$, which has three elements, is defined as:

$$[c] = \begin{bmatrix} 0 \\ R_{22}^{(1)} \\ R_{32}^{(1)} \end{bmatrix} = \begin{bmatrix} 0 \\ -0.2851 \\ 0.1787 \end{bmatrix}$$

The vector $[e]$ is defined as the following three element column vector:

$$[e] = \begin{bmatrix} 0 \\ -1 \\ 0 \end{bmatrix}$$

Using Eq. (4.128), the Euclidean norm, $\|c\|_2$, of $[c]$ is:

$$\|c\|_2 = \sqrt{c_1^2 + c_2^2 + c_3^2} = \sqrt{0^2 + (-0.2851)^2 + 0.1787^2} = 0.3365$$

Using Eq. (4.127), the vector $[v]$ is:

$$[v] = [c] + \|c\|_2[e] = \begin{bmatrix} 0 \\ -0.2851 \\ 0.1787 \end{bmatrix} + 0.3365\begin{bmatrix} 0 \\ -1 \\ 0 \end{bmatrix} = \begin{bmatrix} 0 \\ -0.6215 \\ 0.1787 \end{bmatrix}$$

Next, the products $[v]^T[v]$ and $[v][v]^T$ are calculated:

$$[v]^T[v] = \begin{bmatrix} 0 & -0.6215 & 0.1787 \end{bmatrix}\begin{bmatrix} 0 \\ -0.6215 \\ 0.1787 \end{bmatrix} = 0.4183$$

$$[v][v]^T = \begin{bmatrix} 0 \\ -0.6215 \\ 0.1787 \end{bmatrix}\begin{bmatrix} 0 & -0.6215 & 0.1787 \end{bmatrix} = \begin{bmatrix} 0 & 0 & 0 \\ 0 & 0.3864 & -0.1111 \\ 0 & 0.1111 & 0.0319 \end{bmatrix}$$

The Householder matrix $[H]^{(2)}$ is then:

$$[H]^{(2)} = [I] - \frac{2}{[v]^T[v]}[v][v]^T = \begin{bmatrix} 1 & 0 & 0 \\ 0 & 1 & 0 \\ 0 & 0 & 1 \end{bmatrix} - \frac{2}{0.4183}\begin{bmatrix} 0 & 0 & 0 \\ 0 & 0.3864 & -0.1111 \\ 0 & 0.1111 & 0.0319 \end{bmatrix} = \begin{bmatrix} 1 & 0 & 0 \\ 0 & -0.8474 & 0.5311 \\ 0 & 0.5311 & 0.8473 \end{bmatrix}$$

Once the Housholder matrix $[H]^{(2)}$ is constructed, $[a]$ can be factored into $[Q]^{(2)}[R]^{(2)}$ where:

$$[Q]^{(2)} = [Q]^{(1)}[H]^{(2)} = \begin{bmatrix} -0.8242 & -0.5494 & -0.1374 \\ -0.5494 & 0.8345 & -0.0414 \\ -0.1374 & -0.0414 & 0.9897 \end{bmatrix}\begin{bmatrix} 1 & 0 & 0 \\ 0 & -0.8474 & 0.5311 \\ 0 & 0.5311 & 0.8473 \end{bmatrix} = \begin{bmatrix} -0.8242 & 0.3927 & -0.4082 \\ -0.5494 & -0.7291 & 0.4082 \\ -0.1374 & 0.5607 & 0.8166 \end{bmatrix}$$

and

$$[R]^{(2)} = [H]^{(2)}[R]^{(1)} = \begin{bmatrix} 1 & 0 & 0 \\ 0 & -0.8474 & 0.5311 \\ 0 & 0.5311 & 0.8473 \end{bmatrix}\begin{bmatrix} -7.2801 & 8.6537 & -2.8846 \\ 0 & -0.2851 & 0.5288 \\ 0 & 0.1787 & 0.6322 \end{bmatrix} = \begin{bmatrix} -7.2801 & 8.6537 & -2.8846 \\ 0 & 0.3365 & -0.1123 \\ 0 & 0 & 0.8165 \end{bmatrix}$$

This completes the factorization, which means that:

$$[a] = [Q]^{(2)}[R]^{(2)} \quad \text{or} \quad \begin{bmatrix} 6 & -7 & 2 \\ 4 & -5 & 2 \\ 1 & -1 & 1 \end{bmatrix} = \begin{bmatrix} -0.8242 & 0.3927 & -0.4082 \\ -0.5494 & -0.7291 & 0.4082 \\ -0.1374 & 0.5607 & 0.8166 \end{bmatrix}\begin{bmatrix} -7.2801 & 8.6537 & -2.8846 \\ 0 & 0.3365 & -0.1123 \\ 0 & 0 & 0.8165 \end{bmatrix}$$

The results can be verified by using MATLAB. First, it is verified that the matrix $[Q]^{(2)}$ is orthogonal. This is done by calculating the inverse of $[Q]^{(2)}$ with MATLAB's built-in function, inv, and verifying that it is equal to the transpose of $[Q]^{(2)}$. Then the multiplication $[Q]^{(2)}[R]^{(2)}$ is done with MATLAB, and the result is compared with $[a]$.

```
>> Q2 = [-0.8242 0.3927 -0.4082; -0.5494 -0.7291 0.4084; -0.1374 0.5607 0.8166];
>> R2 = [-7.2801 8.6537 -2.8846; 0 0.3365 -0.1123; 0 0 0.8165];
>> invQ2 = inv(Q2)
invQ2 =
  -0.8242  -0.5494  -0.1372
   0.3924  -0.7290   0.5607
  -0.4081   0.4081   0.8165
>> a = Q2*R2
a =
   6.0003  -7.0002   2.0001
   3.9997  -4.9997   2.0001
   1.0003  -1.0003   1.0001
```

The results (other than errors due to rounding) verify the factorization.

The QR factorization and iteration method for finding the eigenvalues of a matrix is summarized in the following algorithm:

Algorithm for finding the eigenvalues with the QR factorization and iteration method

Given a $(n \times n)$ matrix $[a]_1$ whose eigenvalues are to be determined.

1. Factor $[a]_1$ into an orthogonal matrix $[Q]_1$ and an upper triangular matrix $[R]_1$, such that $[a]_1 = [Q]_1[R]_1$. This is done in $n - 1$ steps as described in pages 160–162 (*Step 1*).

2. Calculate $[a]_2$ by: $[a]_2 = [R]_1[Q]_1$.

3. Repeat the first two steps to obtain a sequence of matrices $[a]_2$, $[a]_3$, $[a]_4$, ..., until the last matrix in the sequence is upper triangular. The elements along the diagonal are then the eigenvalues.

Example 4-15 shows implementation of the QR factorization method in MATLAB.

Example 4-15: Calculating eigenvalues using the QR factorization and iteration method.

The three-dimensional state of stress at a point inside a loaded structure is given by:

$$\sigma_{ij} = \begin{bmatrix} 45 & 30 & -25 \\ 30 & -24 & 68 \\ -25 & 68 & 80 \end{bmatrix} \text{ MPa}$$

Determine the principal stresses at the point by determining the eigenvalues of the stress matrix, using the QR factorization method.

SOLUTION

The problem is solved with MATLAB. First, a user defined function named `QRFactorization` is written. Then, the function is used in a MATLAB program written in a script file for determining the eigenvalues using the QR factorization and iteration method.

The user-defined MATLAB function `QRFactorization`, which is listed below, uses the Householder matrix in the procedure that is described in pages 160–162 to calculate the QR factorization of a square matrix.

Program 4-10: User-defined function, QR factorization of a matrix.

```
function [Q R] = QRFactorization(R)
% The function factors a matrix [A] into an orthogonal matrix [Q] and an upper-triangular matrix [R].
% Input variables:
% A  The (square) matrix to be factored.
% Output variables:
% Q Orthogonal matrix.
% R Upper-triangular matrix.

nmatrix = size(R);
n = nmatrix(1);
I = eye(n);
Q = I;
for j = 1:n - 1
   c = R(:,j);
   c(1:j - 1) = 0;              ⎤— Define the vector [c].
   e(1:n,1) = 0;
   if c(j) > 0
```

```
      e(j) = 1;
   else
      e(j) = -1;
   end
   clength = sqrt(c'*c);
   v = c + clength*e;
   H = I - 2/(v'*v)*v*v';
   Q = Q*H;
   R = H*R;
end
```

Define the vector [e].

Eq. (4.128).

Generate the vector [v], Eq. (4.127).

Construct the Householder matrix [H], Eq. (4.126).

The determination of the eigenvalues follows the procedure in the algorithm.

```
A = [45 30 -25; 30 -24 68; -25 68 80]
for i = 1:40
   [q R] = QRFactorization(A);
   A = R*q;
end
A
e = diag(A)
```

The program repeats the QR factorization 40 times and then displays (in the Command Window) the last matrix [A] that is obtained. The diagonal elements of the matrix are the eigenvalues of the original matrix [A].

```
A =
   45   30  -25
   30  -24   68
  -25   68   80
A =
 114.9545   0.0000   0.0000
   0.0000 -70.1526  -1.5563
   0.0000  -1.5563  56.1981
e =
 114.9545
 -70.1526
  56.1981
```

The results show that after 40 iterations the matrix [A] is nearly upper triangular. Actually, in this case, QR factorization results in a diagonal matrix because the original matrix [σ] is symmetric.

4.12.5 Use of MATLAB Built-In Functions for Determining Eigenvalues and Eigenvectors

MATLAB has built-in functions that determine the eigenvalues and eigenvectors of a matrix, and a built-in function that performs QR factorization.

The eigenvalues and eigenvectors of a matrix can be determined with the built-in function `eig`. If only the eigenvalues are desired, the function has the form:

$$\boxed{\texttt{d = eig(A)}}$$

`d` is a vector with the eigen-values of `A`.   `A` is the matrix whose eigen-values are to be determined.

For determining the eigenvalues and the eigenvectors, the built-in function has the following form:

$$\boxed{\texttt{[V,D] = eig(A)}}$$

`V` is a matrix whose columns are the eigenvectors of `A`. `D` is a diagonal matrix whose diagonal elements are the eigenvalues.   `A` is the matrix whose eigen-values and eigenvectors are to be determined.

With this notation, `A*V=V*D`. For example, if $A = \begin{bmatrix} 6 & -7 & 2 \\ 4 & -5 & 2 \\ 1 & -1 & 1 \end{bmatrix}$ the eigen-values together with the eigenvectors can be determined by typing:

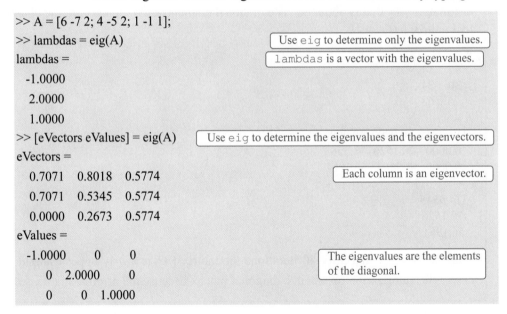

```
>> A = [6 -7 2; 4 -5 2; 1 -1 1];
>> lambdas = eig(A)          Use eig to determine only the eigenvalues.
lambdas =                    lambdas is a vector with the eigenvalues.
  -1.0000
   2.0000
   1.0000
>> [eVectors eValues] = eig(A)   Use eig to determine the eigenvalues and the eigenvectors.
eVectors =
   0.7071   0.8018   0.5774    Each column is an eigenvector.
   0.7071   0.5345   0.5774
   0.0000   0.2673   0.5774
eValues =
  -1.0000        0        0     The eigenvalues are the elements
        0   2.0000        0     of the diagonal.
        0        0   1.0000
```

MATLAB also has a built-in function to perform QR factorization of matrices. The function is called `qr`, and its simplest format is:

$$[Q,R] = qr(A)$$

Q is an orthogonal matrix, and R is   A is the matrix that is factored.
an upper-triangular matrix such
that A=Q*R.

As an example, the matrix that was factored in Example 4-14 is factored
below by using the function `qr`.

```
>> A = [6 -7 2; 4 -5 2; 1 -1 1];
>> [Q R] = qr(A)                          Use MATLAB's built-in function qr to factor the matrix A.
Q =
   -0.8242    0.3925   -0.4082            Q is an orthogonal matrix.
   -0.5494   -0.7290    0.4082
   -0.1374    0.5608    0.8165
R =                                       R is an upper triangular matrix.
   -7.2801    8.6537   -2.8846
        0    0.3365   -0.1122
        0         0    0.8165
```

4.13 PROBLEMS

Problems to be solved by hand
Solve the following problems by hand. When needed, use a calculator, or write a MATLAB script file to carry out the calculations. If using MATLAB, do not use built-in functions for operations with matrices.

4.1 Solve the following system of equations using the Gauss elimination method
$$\begin{aligned} x_1 + 2x_2 - 2x_3 &= 9 \\ 2x_1 + 3x_2 + x_3 &= 23 \\ 3x_1 + 2x_2 - 4x_3 &= 11 \end{aligned}.$$

4.2   Given the system of equations $[a][x] = [b]$, where $a = \begin{bmatrix} 2 & -4 & 1 \\ 6 & 2 & -1 \\ -2 & 6 & -2 \end{bmatrix}$, $x = \begin{bmatrix} x_1 \\ x_2 \\ x_3 \end{bmatrix}$, and $b = \begin{bmatrix} 4 \\ 10 \\ -6 \end{bmatrix}$, deter-

mine the solution using the Gauss elimination method.

4.3   Consider the following system of two linear equations:
$$\begin{aligned} 0.0003x_1 + 1.566x_2 &= 1.569 \\ 0.3454x_1 - 2.436x_2 &= 1.018 \end{aligned}.$$

(*a*) Solve the system with the Gauss elimination method using rounding with four significant figures.
(*b*) Switch the order of the equations, and solve the system with the Gauss elimination method using rounding with four significant figures.
    Check the answers by substituting the solution back in the equations.

4.4   Solve the following system of equations using the Gauss elimination method:

$$4x_1 + 3x_2 + 2x_3 + x_4 = 1$$
$$3x_1 + 4x_2 + 3x_3 + 2x_4 = 1$$
$$2x_1 + 3x_2 + 4x_3 + 3x_4 = -1$$
$$x_1 + 2x_2 + 3x_3 + 4x_4 = -1$$

4.5   Solve the following system of equations with the Gauss elimination method.

$$2x_1 + x_2 + 4x_3 - 2x_4 = 19$$
$$-3x_1 + 4x_2 + 2x_3 - x_4 = 1$$
$$3x_1 + 5x_2 - 2x_3 + x_4 = 8$$
$$-2x_1 + 3x_2 + 2x_3 + 4x_4 = 13$$

4.6   Solve the following system of equations using the Gauss–Jordan method:

$$x_1 + 2x_2 - 2x_3 = 9$$
$$2x_1 + 3x_2 + x_3 = 23$$
$$3x_1 + 2x_2 - 4x_3 = 11$$

4.7   Given the system of equations $[a][x] = [b]$, where $a = \begin{bmatrix} 2 & -4 & 1 \\ 6 & 2 & -1 \\ -2 & 6 & -2 \end{bmatrix}$, $x = \begin{bmatrix} x_1 \\ x_2 \\ x_3 \end{bmatrix}$, and $b = \begin{bmatrix} 4 \\ 10 \\ -6 \end{bmatrix}$, deter-

mine the solution using the Gauss–Jordan method.

4.8   Solve the following system of equations with the Gauss–Jordan elimination method.

$$2x_1 + x_2 + 4x_3 - 2x_4 = 19$$
$$-3x_1 + 4x_2 + 2x_3 - x_4 = 1$$
$$3x_1 + 5x_2 - 2x_3 + x_4 = 8$$
$$-2x_1 + 3x_2 + 2x_3 + 4x_4 = 13$$

4.9   Determine the LU decomposition of the matrix $a = \begin{bmatrix} 1 & 2 & 3 \\ 4 & 5 & 6 \\ 3 & 2 & 2 \end{bmatrix}$ using the Gauss elimination procedure.

4.10   Determine the LU decomposition of the matrix $a = \begin{bmatrix} 1 & 2 & 3 \\ 4 & 5 & 6 \\ 3 & 2 & 2 \end{bmatrix}$ using Crout's method.

4.11   Solve the following system with LU decomposition using Crout's method.

$$\begin{bmatrix} 5 & -1 & 0 \\ -1 & 5 & -1 \\ 0 & -1 & 5 \end{bmatrix} \begin{bmatrix} x \\ y \\ z \end{bmatrix} = \begin{bmatrix} 9 \\ 4 \\ -6 \end{bmatrix}$$

4.12 Find the inverse of the matrix $\begin{bmatrix} -4/5 & -3/5 & -2/5 \\ -3/5 & -6/5 & -4/5 \\ -2/5 & -4/5 & -6/5 \end{bmatrix}$ using the Gauss–Jordan method.

4.13 Given the matrix $a = \begin{bmatrix} 6 & 3 & 11 \\ 3 & 2 & 7 \\ 3 & 2 & 6 \end{bmatrix}$, determine the inverse of $[a]$ using the Gauss–Jordan method.

4.14 Carry out the first three iterations of the solution of the following system of equations using the Gauss–Seidel iterative method. For the first guess of the solution, take the value of all the unknowns to be zero.

$$8x_1 + 2x_2 + 3x_3 = 51$$
$$2x_1 + 5x_2 + x_3 = 23$$
$$-3x_1 + x_2 + 6x_3 = 20$$

4.15 Find the condition number of the matrix in Problem 4.13 using the infinity norm.

4.16 Find the condition number of the matrix in Problem 4.13 using the 1-norm.

4.17 Show that the eigenvalues of the $n \times n$ identity matrix is the number 1 repeated n times.

4.18 Show that the eigenvalues of the following matrix are $10, \sqrt{2}, -\sqrt{2}$.

$$\begin{bmatrix} 10 & 0 & 0 \\ 1 & -3 & -7 \\ 0 & 1 & 3 \end{bmatrix}$$

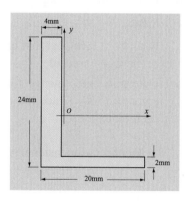

4.19 The moment of inertia I_x, I_y, and the product of inertia I_{xy} of the cross-sectional area shown in the figure are:

$I_x = 7523$ mm^4, $I_y = 3210$ mm^4, and $I_{xy} = -2640$ mm^4

The principal moments of inertia are the eigenvalues of the matrix $\begin{bmatrix} 7523 & -2640 \\ -2640 & 3210 \end{bmatrix}$, and the principal axes are in the direction of the eigen-vectors. Determine the principal moments of inertia by solving the characteristic equation. Determine the orientation of the principal axes of inertia (unit vectors in the directions of the eigenvectors).

4.20 Determine the principal moments of inertia of the cross-sectional area in Problem 4.19 by using the QR factorization and iteration method. Carry out the first four iterations.

4.21 The structure of a CO_2 molecule may be idealized as three masses connected by two springs, where the masses are the carbon and oxygen atoms and the springs represent the chemical bond between the carbon and oxygen atoms. The equation of motion for each atom (mass) may be written as:

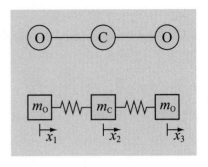

$$m_O \frac{d^2 x_1}{dt^2} = -kx_1 + kx_2$$

$$m_C \frac{d^2 x_2}{dt^2} = -2kx_2 + kx_1 + kx_3$$

$$m_O \frac{d^2 x_3}{dt^2} = kx_2 - kx_3$$

where k is the restoring force spring constant representing the C–O bonds. Since the molecule is free to vibrate, normal mode (i.e., along the axis) vibrations can be examined by substituting $x_j = A_j e^{i\omega t}$, where A_j is the amplitude of the jth mass, $i = \sqrt{-1}$, ω is the frequency, and t is time. This results in the following system of equations:

$$-\omega^2 A_1 = -\frac{k}{m_O} A_1 + \frac{k}{m_O} A_2$$

$$-\omega^2 A_2 = -\frac{2k}{m_C} A_2 + \frac{k}{m_C} A_1 + \frac{k}{m_C} A_3 \qquad (4.145)$$

$$-\omega^2 A_3 = \frac{k}{m_O} A_2 - \frac{k}{m_O} A_3$$

(a) Rewrite the system of equations in Eq. (4.145) as an eigenvalue problem, and show that the quantity ω^2 is the eigenvalue.

(b) Write the characteristic equation and solve (analytically) for the different frequencies.

(c) If $k = 14.2 \times 10^2$ kg/s^2, $m_O = 16$amu and $m_C = 12$amu (amu $= 1.6605 \times 10^{-27}$ kg), find the wavelengths $\lambda = \frac{2\pi c}{\omega}$ (where $c = 3 \times 10^8$ m/s is the speed of light) that correspond to the frequencies from part (b). It turns out that one of the wavelengths ($\lambda = 4.3$ μm) is observed in the absorption spectrum of the CO_2 molecule.

(d) Determine the eigenvectors corresponding to the eigenvalues found in parts (b) and (c). From the eigenvectors, deduce the relative motion of the atoms (i.e., are they moving towards, or away from each other.)

4.22 The three-dimensional state of stress at a point is given by the stress tensor:

$$\sigma_{ij} = \begin{bmatrix} 30 & 15 & 20 \\ 15 & 22 & 26 \\ 20 & 26 & 40 \end{bmatrix}$$

The principal stresses and the principal directions at the point are given by the eigenvalues and the eigenvectors. Use the power method for determining the value of the largest principal stress. Start with a column vector of 1s, and carry out the first three iterations.

Problems to be programmed in MATLAB
Solve the following problems using the MATLAB environment. Do not use MATLAB's built-in functions for operations with matrices.

4.23 Modify the user-defined function GaussPivot in Program 4-2 (Example 4-3) such that in each step of the elimination the pivot row is switched with the row that has a pivot element with the largest absolute numerical value. For the function name and arguments use x = GaussPivotLarge(a,b), where a is the matrix of coefficients, b is the right-hand-side column of constants, and x is the solution.
(*a*) Use the GaussPivotLarge function to solve the system of linear equations in Eq. (4.17).
(*b*) Use the GaussPivotLarge function to solve the system:

$$\begin{bmatrix} 0 & 3 & 8 & -5 & -1 & 6 \\ 3 & 12 & -4 & 8 & 5 & -2 \\ 8 & 0 & 0 & 10 & -3 & 7 \\ 3 & 1 & 0 & 0 & 0 & 4 \\ 0 & 0 & 4 & -6 & 0 & 2 \\ 3 & 0 & 5 & 0 & 0 & -6 \end{bmatrix} \begin{bmatrix} x_1 \\ x_2 \\ x_3 \\ x_4 \\ x_5 \\ x_6 \end{bmatrix} = \begin{bmatrix} 34 \\ 20 \\ 45 \\ 36 \\ 60 \\ 28 \end{bmatrix}$$

4.24 Write a user-defined MATLAB function that solves a system of n linear equations, $[a][x] = [b]$, with the Gauss–Jordan method. The program should include pivoting in which the pivot row is switched with the row that has a pivot element with the largest absolute numerical value. For the function name and arguments use x = GaussJordan(a,b), where a is the matrix of coefficients, b is the right-hand-side column of constants, and x is the solution.
(*a*) Use the GaussJordan function to solve the system:

$$4x_1 + 3x_2 + 2x_3 + x_4 = 1$$
$$x_1 + 4x_2 + 3x_3 + 2x_4 = 1$$
$$2x_1 + 3x_2 + 4x_3 + 3x_4 = -1$$
$$x_1 + 2x_2 + 3x_3 + 4x_4 = -1$$

(*b*) Use the GaussJordan function to solve the system:

$$\begin{bmatrix} 0 & 3 & 8 & -5 & -1 & 6 \\ 3 & 12 & -4 & 8 & 5 & -2 \\ 8 & 0 & 0 & 10 & -3 & 7 \\ 3 & 1 & 0 & 0 & 0 & 4 \\ 0 & 0 & 4 & -6 & 0 & 2 \\ 3 & 0 & 5 & 0 & 0 & -6 \end{bmatrix} \begin{bmatrix} x_1 \\ x_2 \\ x_3 \\ x_4 \\ x_5 \\ x_6 \end{bmatrix} = \begin{bmatrix} 34 \\ 20 \\ 45 \\ 36 \\ 60 \\ 28 \end{bmatrix}$$

4.25 Write a user-defined MATLAB function that determines the inverse of a matrix using the Gauss–Jordan method. For the function name and arguments use Ainv = Inverse(A), where A is the matrix to be inverted, and Ainv is the inverse of the matrix. Use the Inverse function to calculate the inverse of the matrix:

$$\begin{bmatrix} -0.04 & 0.04 & 0.12 \\ 0.56 & -1.56 & 0.32 \\ -0.24 & 1.24 & -0.28 \end{bmatrix}$$

4.26 Write a user-defined MATLAB function that calculates the 1-norm of any matrix. For the function name and arguments use N = OneNorm(A), where A is the matrix and N is the value of the norm. Use the function for calculating the 1-norm of:

(a) The matrix $A = \begin{bmatrix} 6 & 3 & 11 \\ 3 & 2 & 7 \\ 3 & 2 & 6 \end{bmatrix}$ (b)   The matrix $B = \begin{bmatrix} 6 & 3 & 11 & -1 & 2 \\ 3 & -2 & 7 & 0 & 4 \\ 3 & 2 & -6 & 5 & -3 \\ -5 & 7 & 1 & -4 & 0 \end{bmatrix}$.

4.27 Write a user-defined MATLAB function that calculates the infinity norm of any matrix. For the function name and arguments use N = InfinityNorm(A), where A is the matrix, and N is the value of the norm. Use the function for calculating the infinity norm of:

(a)   The matrix $A = \begin{bmatrix} 6 & 3 & 11 \\ 3 & 2 & 7 \\ 3 & 2 & 6 \end{bmatrix}$.   (b)   The matrix $B = \begin{bmatrix} 6 & 3 & 11 & -1 & 2 \\ 3 & -2 & 7 & 0 & 4 \\ 3 & 2 & -6 & 5 & -3 \\ -5 & 7 & 1 & -4 & 0 \end{bmatrix}$.

4.28 Write a user-defined MATLAB function that calculates the condition number of an $(n \times n)$ matrix by using the 1-norm. For the function name and arguments use c = CondNumb(A), where A is the matrix and c is the value of the condition number. Within the function, use the user-defined functions Inverse from Problem 4.25 and OneNorm from Problem 4.26. Use the function CondNumb for calculating the condition number of the matrix of the coefficients in Problem 4.25.

4.29 Write a user-defined MATLAB function that determines the largest eigenvalue of an $(n \times n)$ matrix by using the power method. For the function name and argument use e = MaxEig(A), where A is the matrix and e is the value of the largest eigenvalue. Use the function MaxEig for calculating the largest eigenvalue of the matrix of Problem 4.22. Check the answer by using MATLAB's built-in function for finding the eigenvalues of a matrix.

4.30 Write a user-defined MATLAB function that determines the smallest eigenvalue of an $(n \times n)$ matrix by using the inverse power method. For the function name and argument use e = MinEig(A), where A is the matrix and e is the value of the smallest eigenvalue. Inside MinEig use the user-defined function Inverse, that was written in Problem 4.25, for calculating the inverse of the matrix A. Use the function MinEig for calculating the smallest eigenvalue of the matrix of Problem 4.22. Check the answer by using MATLAB's built-in function for finding the eigenvalues of a matrix.

4.31 Write a user-defined MATLAB function that determines all the eigenvalues of an $(n \times n)$ matrix by using the QR factorization and iteration method. For the function name and argument use e=AllEig(A), where A is the matrix and e is a vector whose elements are the eigenvalues. Use the function AllEig for calculating the eigenvalues of the matrix of Problem 4.22. Check the answer by using MATLAB's built-in function for finding the eigenvalues of a matrix.

Problems in math, science, and engineering
Solve the following problems using the MATLAB environment. As stated, use the MATLAB programs that
are presented in the chapter, programs developed in previously solved problems, or MATLAB's built-in
functions.

4.32 The axial force F_i in each of the 13 member pin-connected truss, shown in the figure, can be calculated by solving the following system of 13 equations:

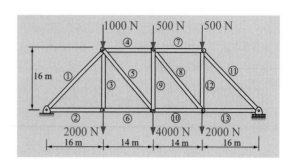

$F_2 + 0.7071F_1 = 0$, $-F_2 + F_6 = 0$, $F_3 - 2000 = 0$

$F_4 + 0.6585F_5 - 0.7071F_1 = 0$

$0.7071F_1 + F_3 + 0.7526F_5 + 1000 = 0$

$F_7 + 0.6585F_8 - F_4 = 0$, $0.7526F_8 + F_9 + 500 = 0$

$F_{10} - 0.6585F_5 - F_6 = 0$, $F_9 + 0.7526F_5 - 4000 = 0$

$0.7071F_{11} - F_7 = 0$, $0.7071F_{11} + F_{12} + 500 = 0$

$F_{12} + 0.7526F_8 - 2000 = 0$, $F_{13} + 0.7071F_{11} = 0$

(*a*) Solve the system of equations using the user-defined function `GaussPivotLarge` developed in Problem 4.23.
(*b*) Solve this system using Gauss–Seidel iteration. Does the solution converge for a starting (guess) vector whose elements are all zero?
(*c*) Solve the system of equations using MATLAB left division operation.

4.33 Mass spectrometry of a sample gives a series of peaks that represent various masses of ions of constituents within the sample. For each peak, the height of the peak I_i is influenced by the amounts of the various constituents:

$$I_j = \sum_{i=1}^{N} C_{ij} n_j$$

where C_{ij} is the contribution of ions of species i to the height of peak j and n_j is the amount of ions or concentration of species j. The coefficients C_{ij} for each peak are given by:

Peak identity	Species				
	CH_4	C_2H_4	C_2H_6	C_3H_6	C_3H_8
1	0.2	0.2	0.3	0.2	0.2
2	28	1	0	0	0.1
3		18	12	2.4	16
4			10	0	1
5				10	2
6					18

If a sample produces a mass spectrum with peak heights, $I_1 = 3.4$, $I_2 = 20.5$, $I_3 = 170$, $I_4 = 49$, $I_5 = 39.8$, $I_6 = 96.3$. Determine the concentrations of the different species in the sample.

4.34 The axial force F_i in each of the 21 member pin connected truss, shown in the figure, can be calculated by solving
the following system of 21 equations:

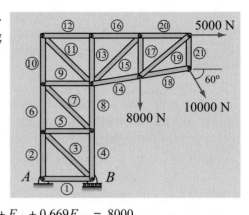

$-F_1 - 0.7071F_3 = 0$, $-F_2 + F_6 - 0.7071F_3 = 0$

$F_5 + 0.7071F_3 = 0$, $-F_5 - 0.7071F_7 = 0$

$-F_4 + F_8 + 0.7071F_7 = 0$, $-F_6 + F_{10} - 0.7071F_7 = 0$

$F_9 + 0.7071F_7 = 0$, $0.9806F_{14} + 0.7071F_{15} - 0.7071F_{11} - F_9 = 0$

$0.1961F_{14} + 0.7071F_{15} + 0.7071F_{11} + F_{13} - F_8 = 0$

$-F_{10} - 0.7071F_{11} = 0$, $F_{12} + 0.7071F_{11} = 0$

$-F_{12} + F_{16} = 0$, $F_{13} = 0$,

$0.9806F_{18} - 0.9806F_{14} + 0.7433F_{19} = 0$, $0.1961F_{18} - 0.1961F_{14} + F_{17} + 0.669F_{19} = 8000$

$F_{20} - F_{16} - 0.7071F_{15} = 0$, $-F_{17} - 0.7071F_{15} = 0$, $-F_{20} - 0.7433F_{19} = -5000$, $-F_{21} - 0.669F_{19} = 0$

$0.9806F_{11} = 10000\cos 60°$, $F_{21} - 0.1961F_{18} = 10000\sin 60°$

(*a*) Solve the system of equations using the user-defined function `GaussJordan` developed in Problem
      4.24.
(*b*) Solve the system of equations using MATLAB's left division operation.

4.35 A rigid bar ABC is supported by 3 pin-connected bars
as shown. A force $P = 40\,\text{kN}$ is applied to the rigid bar at a
distance d from A. The forces in the bars, F_{AD}, F_{BE}, and F_{CG}
can be determined from the solution of the following system
of three equations:

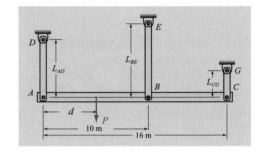

$$F_{AD} + F_{BE} + F_{CG} = P,   10F_{BE} + 16F_{CG} = d \cdot P$$

$$\frac{6L_{AD}}{E_{AD}A_{AD}}F_{AD} - \frac{16L_{BE}}{E_{BE}A_{BE}}F_{BE} + \frac{10L_{CG}}{E_{CG}A_{CG}}F_{CG} = 0$$

where L, E, and A, denote the length, elastic modulus, and
cross-sectional area, respectively, of the bars. Once the force in each of the bars is known, its elongation δ
can be determined with the formula $\delta = \dfrac{FL}{EA}$.

Write a MATLAB program in a script file that determines the forces in the three bars and their elongation
for $0 \le d \le 16$ m. The program displays the three forces as a function of d in one plot, and the elongation of
the bars as a function of d in a second plot (two plots on the same page). Also given: $L_{AD} = 4$ m,
$L_{BE} = 5$ m, $L_{CG} = 2$ m, $E_{AD} = 70$ GPa, $E_{BE} = 200$ GPa, $E_{CG} = 115$ GPa,   and
$A_{AD} = A_{BE} = A_{CG} = 5 \cdot 10^{-5}\,\text{m}^2$.

4.36 In dairy science and technology, ice cream is made such that it contains a desired amount of milk fat,
milk solids–nonfat (msnf), water, sugar, stabilizers, and emulsifiers. Suppose 10 kg of an ice cream mix is
to be made containing 18% fat, 9.5% msnf, 15% sucrose, 0.4% stabilizer, and 1% egg yolk, using the following ingredients: cream (containing 35% fat), milk (containing 3.5% fat), skim milk powder (containing
97% msnf), sucrose, stabilizer, and egg yolk. Determine how much (in kg) of each ingredient is used.
Assume that 9% of the skim portion of the milk and cream contain msnf. (Hint: Set up a system of three
equations where the unknowns are the amount of skim milk powder (*x*), the amount of milk (*y*), and the

amount of cream (z) using (1) the fact that all the components have to add up to 10 kg, (2) writing an equation for the total content of msnf, and (3) writing a balance equation for the fat.)

4.37 The currents, $i_1, i_2, i_3, i_4, i_5, i_6$, in the circuit that is shown can be determined from the solution of the following system of equations. (Obtained by applying Kirchhoff's law.)

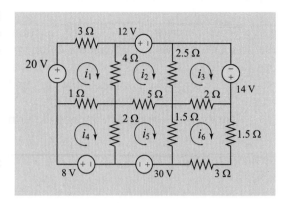

$$8i_1 - 4i_2 - i_4 = 20, \quad -4i_1 + 11.5i_2 - 2.5i_3 - 5i_5 = -12$$
$$-2.5i_2 + 4.5i_3 - 2i_6 = 14, \quad -i_1 + 3i_4 - 2i_5 = 8$$
$$-5i_2 - 2i_4 + 8.5i_5 - 1.5i_6 = -30, \quad -2i_3 - 1.5i_5 + 8i_6 = 0$$

Solve the system using the following methods.
(a) Use the user-defined function `GaussJordan` that was developed in Problem 4.24.
(b) Use MATLAB's built-in functions.

4.38 When balancing the following chemical reaction by conserving the number of atoms of each element between reactants and products:

$$P_2I_4 + aP_4 + bH_2O \leftrightharpoons cPH_4I + dH_3PO_4$$

the unknown stoichiometric coefficients a, b, c, and d are given by the solution of the following system of equations:

$$\begin{bmatrix} -4 & 0 & 1 & 1 \\ 0 & 0 & 1 & 0 \\ 0 & -2 & 4 & 3 \\ 0 & -1 & 0 & 4 \end{bmatrix} \begin{bmatrix} a \\ b \\ c \\ d \end{bmatrix} = \begin{bmatrix} 2 \\ 4 \\ 0 \\ 0 \end{bmatrix}$$

Solve for the unknown stoichiometric coefficients using
(a) The user-defined function `GaussJordan` that was developed in Problem 4.24.
(b) MATLAB's left division operation.

4.39 When balancing the following chemical reaction by conserving the number of atoms of each element between reactants and products:

$$(Cr(N_2H_4CO)_6)_4(Cr(CN)_6)_3 + aKMnO_4 + bH_2SO_4 \rightarrow$$
$$cK_2Cr_2O_7 + dMnSO_4 + eCO_2 + fKNO_3 + gK_2SO_4 + hHO_2$$

the unknown stoichiometric coefficients a through h are given by the solution of the following system of equations:

$$\begin{bmatrix} 0 & 0 & 2 & 0 & 0 & 0 & 0 & 0 \\ 0 & 0 & 0 & 0 & 0 & 1 & 0 & 0 \\ 0 & -2 & 0 & 0 & 0 & 0 & 0 & 2 \\ 0 & 0 & 0 & 0 & 1 & 0 & 0 & 0 \\ -4 & -4 & 7 & 4 & 2 & 3 & 4 & 1 \\ -1 & 0 & 2 & 0 & 0 & 1 & 2 & 0 \\ -1 & 0 & 0 & 1 & 0 & 0 & 0 & 0 \\ 0 & -1 & 0 & 1 & 0 & 0 & 1 & 0 \end{bmatrix} \begin{bmatrix} a \\ b \\ c \\ d \\ e \\ f \\ g \\ h \end{bmatrix} = \begin{bmatrix} 7 \\ 66 \\ 96 \\ 42 \\ 24 \\ 0 \\ 0 \\ 0 \end{bmatrix}$$

Solve for the unknown stoichiometric coefficients using
(*a*) The user-defined function `GaussJordan` that was developed in Problem 4.24.
(*b*) MATLAB's left division operation.

4.40 Write a user-defined MATLAB function that determines the principal stresses and the directions of the principal stresses for a given three-dimensional state of stress. For the function name and arguments, use `[Ps Pd] = PrinplStre(S)`, where S is a (3×3) matrix with the values of the stress tensor, Ps is a column vector with the values of the principal stresses, and Pd is a (3×3) matrix in which each row lists a unit vector in a principal direction. Use MATLAB built-in functions.

   Use the function for determining the principal stresses and principal directions for the state of stress given in Problem 4.22: $\quad \sigma_{ij} = \begin{bmatrix} 30 & 15 & 20 \\ 15 & 22 & 26 \\ 20 & 26 & 40 \end{bmatrix}$.

4.41 The structure of the C_2H_2 (acetylene) molecule may be idealized as four masses connected by two springs (see discussion in Problem 4.21). By applying the equation of motion, the following system of equations can be written for the amplitudes of vibration of each atom:

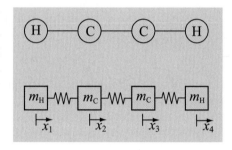

$$\begin{bmatrix} \dfrac{k_{CH}}{m_H} - \omega^2 & -\dfrac{k_{CH}}{m_H} & 0 & 0 \\[2ex] -\dfrac{k_{CH}}{m_C} & \dfrac{(k_{CH}+k_{CC})}{m_C} - \omega^2 & -\dfrac{k_{CC}}{m_C} & 0 \\[2ex] 0 & -\dfrac{k_{CC}}{m_C} & \dfrac{(k_{CH}+k_{CC})}{m_C} - \omega^2 & -\dfrac{k_{CH}}{m_C} \\[2ex] 0 & 0 & -\dfrac{k_{CH}}{m_H} & \dfrac{k_{CH}}{m_H} - \omega^2 \end{bmatrix} \begin{bmatrix} A_1 \\ A_2 \\ A_3 \\ A_4 \end{bmatrix} = \begin{bmatrix} 0 \\ 0 \\ 0 \\ 0 \end{bmatrix}$$

where ω is the frequency, $k_{CH} = 5.92 \times 10^2$ kg/s^2 and $k_{CC} = 15.8 \times 10^2$ kg/s^2 are the restoring force spring constants representing the C–H and C–C bonds, respectively, $m_H = 1\,\text{amu}$, and $m_C = 12\,\text{amu}$ ($\text{amu} = 1.6605 \times 10^{-27}$ kg) are the masses of the atoms.
(*a*) Determine the eigenvalues (frequencies), and the corresponding wavelengths ($\lambda = \dfrac{2\pi c}{\omega}$ (where $c = 3 \times 10^8$ m/s is the speed of light).
(*b*) Determine the eigenvectors corresponding to the eigenvalues found in part (*a*). From the eigenvectors, deduce the relative motion of the atoms (i.e., are they moving towards, or away from each other?)

Chapter 5

Curve Fitting and Interpolation

Core Topics

Curve fitting with a linear equation (5.2).

Curve fitting with nonlinear equation by writing the equation in linear form (5.3).

Curve fitting with quadratic and higher order polynomials (5.4).

Interpolation using a single polynomial (5.5).

Lagrange polynomials (5.5.1).

Newton's polynomials (5.5.2).

Piecewise (spline) interpolation (5.6).

Use of MATLAB built-in functions for curve fitting and interpolation (5.7).

Complementary Topics

Curve fitting with linear combination of nonlinear functions (5.8).

5.1 BACKGROUND

Many scientific and engineering observations are made by conducting experiments in which physical quantities are measured and recorded. The experimental records are typically referred to as data points. For example, the strength of many metals depends on the size of the grains. Testing specimens with different grain sizes yields a discrete set of numbers (d – average grain diameter, σ_y – yield strength) as shown in Table 5-1.

Table 5-1: Strength-grain size data.

d (mm)	0.005	0.009	0.016	0.025	0.040	0.062	0.085	0.110
σ_y (MPa)	205	150	135	97	89	80	70	67

Sometimes measurements are made and recorded continuously with analog devices, but in most cases, especially in recent years with the wide use of computers, the measured quantities are digitized and stored as a set of discrete points.

Once the data is known, scientists and engineers can use it in different ways. Often the data is used for developing, or evaluating, mathematical formulas (equations) that represent the data. This is done by curve fitting in which a specific form of an equation is assumed, or provided by a guiding theory, and then the parameters of the equation are determined such that the equation best fits the data points. Sometimes the data points are used for estimating the expected values between the known points, a procedure called ***interpolation***, or for predicting how the data might extend beyond the range over which it was measured, a procedure called ***extrapolation***.

Curve fitting

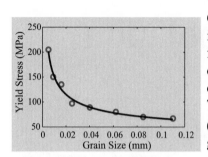

Figure 5-1: Curve fitting.

Curve fitting is a procedure in which a mathematical formula (equation) is used to best fit a given set of data points. The objective is to find a function that fits the data points overall. This means that the function does not have to give the exact value at any single point, but fits the data well overall. For example, Fig. 5-1 shows the data points from Table 5-1 and a curve that shows the best fit of a power function ($\sigma = Cd^m$) to the data points. It can be observed that the curve fits the general trend of the data but does not match any of the data points exactly. Curve fitting is typically used when the values of the data points have some error, or scatter. Generally, all experimental measurements have built-in errors or uncertainties, and requiring a curve fit to go through every data point is not beneficial. The procedure is also used for determining the validity of proposed equations used to represent the data and for determining the values of parameters (coefficients) in the equations. Curve fitting can be carried out with many types of functions and with polynomials of various orders.

Interpolation

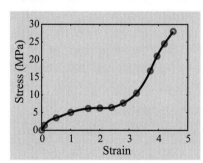

Figure 5-2: Interpolation.

Interpolation is a procedure for estimating a value ***between*** known values of data points. It is done by first determining a polynomial that gives the exact value at the data points, and then using the polynomial for calculating values between the points. When a small number of points is involved, a single polynomial might be sufficient for interpolation over the whole domain of the data points. Often, however, when a large number of points are involved, different polynomials are used in the intervals between the points in a process that is called spline interpolation. For example, Fig. 5-2 shows a plot of the stress–strain relationship for rubber. The red markers show experimental points that were measured very accurately, and the solid curve was obtained by using spline interpolation. It can be observed that the curve passes through the points precisely and gives a good estimate of values between the points.

The next three sections cover curve fitting. Section 5.2 describes how to curve-fit a set of data points with a linear function using least-squares regression analysis. In Section 5.3 data points are curve fit with nonlinear functions by rewriting the functions in a linear form. In Section 5.4 curve fitting is carried out with second and higher-order polynomials. Interpolation is covered in the next two sections. Section 5.5 shows how to find the equation of a single polynomial that passes through a given set of data points (Lagrange and Newton's polynomials), and Section 5.6 covers piecewise (spline) interpolation in which different polynomials are used for interpolation in the intervals between the data points. Section 5.7 describes the tools that MATLAB has for curve fitting and interpolation. In Section 5.8 curve fitting is done in a more general way by using a linear combination of nonlinear functions.

5.2 CURVE FITTING WITH A LINEAR EQUATION

Curve fitting using a linear equation (first degree polynomial) is the process by which an equation of the form:

$$y = a_1 x + a_0 \qquad (5.1)$$

is used to best fit given data points. This is done by determining the constants a_1 and a_0 that give the smallest error when the data points are substituted in Eq. (5.1). If the data comprise only two points, the constants can be determined such that Eq. (5.1) gives the exact values at the points. Graphically, as shown in Fig. 5-3, it means that the straight line that corresponds to Eq. (5.1) passes through the two points.

When the data consists of more than two points, obviously, a straight line cannot pass through all of the points. In this case, the constants a_1 and a_0 are determined such that the line has the best fit overall, as illustrated in Fig. 5-4.

The process of obtaining the constants that give the best fit first requires a definition of best fit (Section 5.2.1) and a mathematical procedure for deriving the value of the constants (Section 5.2.2).

5.2.1 Measuring How Good Is a Fit

A criterion that measures how good a fit is between given data points and an approximating linear function is a formula that calculates a number that quantifies the overall agreement between the points and the function. Such a criterion is needed for two reasons. First, it can be used to compare two different functions that are used for fitting the same data points. Second, and even more important, the criterion itself is used for determining the coefficients of the function that give the best fit. This is shown in Section 5.2.2.

The fit between given data points and an approximating linear function is determined by first calculating the error, also called the residual, which is the difference between a data point and the value of the approximating function, at each point. Subsequently, the residuals are used for calculating a total error for all the points. Figure 5-5 shows a general case of a linear function (straight line) that is used for curve fitting n points. The residual r_i at a point, (x_i, y_i), is the difference between the value y_i of the data point and the value of the function $f(x_i)$ used to approximate the data points:

$$r_i = y_i - f(x_i) \qquad (5.2)$$

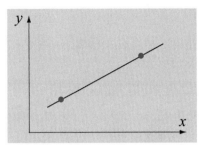

Figure 5-3: Two data points.

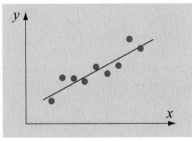

Figure 5-4: Many data points.

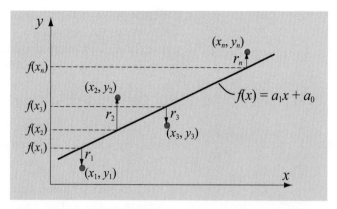

Figure 5-5: Curve-fitting points with a linear equation.

A criterion that measures how well the approximating function fits the given data can be obtained by calculating a total error E in terms of the residuals. The overall error can be calculated in different ways. One simple way is to add the residuals of all the points:

$$E = \sum_{i=1}^{n} r_i = \sum_{i=1}^{n} [y_i - (a_1 x_i + a_0)] \tag{5.3}$$

The error that is calculated in this way does not provide a good measure of the overall fit. This is because a bad fit with positive residuals and negative residuals (both can be large) can sum up to give a zero (or very close to zero) error, implying a good fit. A situation like this is shown in Fig. 5-6, where E according to Eq. (5.3) is zero since $r_1 = -r_4$ and $r_2 = -r_3$.

Another possibility is to make the overall error E equal to the sum of the absolute values of the residuals:

$$E = \sum_{i=1}^{n} |r_i| = \sum_{i=1}^{n} |y_i - (a_1 x_i + a_0)| \tag{5.4}$$

With this definition, the total error is always a positive number since the residuals cannot cancel each other. A smaller E in Eq. (5.4) indicates a better fit. This measure can be used to evaluate or compare proposed fits, but it cannot be used for determining the constants of the function that give the best fit. This is because the measure is not unique, which means that for the same set of points there can be several functions that give the same total error. This is shown in Fig. 5-7 where total error E according to Eq. (5.4) is the same for the two approximating lines.

A definition for the overall error E that gives a good measure of the total error and can also be used for determining a unique linear function that has the best fit (i.e., smallest total error) is obtained by making E equal to the sum of the squares of the residuals:

$$E = \sum_{i=1}^{n} r_i^2 = \sum_{i=1}^{n} [y_i - (a_1 x_i + a_0)]^2 \tag{5.5}$$

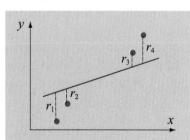

Figure 5-6: Fit with no error according to Eq. (5.3).

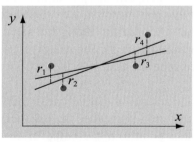

Figure 5-7: Two fits with the same error according to Eq. (5.4).

With this definition, the overall error is always a positive number (positive and negative residuals do not cancel each other). In addition, larger residuals have a relatively larger effect (weight) on the total error. As already mentioned, Eq. (5.5) can be used to calculate the coefficients a_1 and a_0 in the linear function $y = a_1 x + a_0$ that give the smallest total error. This is done by using a procedure called linear least-squares regression, which is presented in the next section.

5.2.2 Linear Least-Squares Regression

Linear least-squares regression is a procedure in which the coefficients a_1 and a_0 of a linear function $y = a_1 x + a_0$ are determined such that the function has the best fit to a given set of data points. The best fit is defined as the smallest possible total error that is calculated by adding the squares of the residuals according to Eq. (5.5).

For a given set of n data points (x_i, y_i), the overall error calculated by Eq. (5.5) is:

$$E = \sum_{i=1}^{n} [y_i - (a_1 x_i + a_0)]^2 \tag{5.6}$$

Since all the values x_i and y_i are known, E in Eq. (5.6) is a nonlinear function of the two variables a_1 and a_0. The function E has a minimum at the values of a_1 and a_0 where the partial derivatives of E with respect to each variable is equal to zero. Taking the partial derivatives and setting them equal to zero gives:

$$\frac{\partial E}{\partial a_0} = -2 \sum_{i=1}^{n} (y_i - a_1 x_i - a_0) = 0 \tag{5.7}$$

$$\frac{\partial E}{\partial a_1} = -2 \sum_{i=1}^{n} (y_i - a_1 x_i - a_0) x_i = 0 \tag{5.8}$$

Equations (5.7) and (5.8) are a system of two linear equations for the unknowns a_1 and a_0, and can be rewritten in the form:

$$n a_0 + \left(\sum_{i=1}^{n} x_i \right) a_1 = \sum_{i=1}^{n} y_i \tag{5.9}$$

$$\left(\sum_{i=1}^{n} x_i \right) a_0 + \left(\sum_{i=1}^{n} x_i^2 \right) a_1 = \sum_{i=1}^{n} x_i y_i \tag{5.10}$$

The solution of the system is:

$$a_1 = \frac{n \sum_{i=1}^{n} x_i y_i - \left(\sum_{i=1}^{n} x_i \right) \left(\sum_{i=1}^{n} y_i \right)}{n \sum_{i=1}^{n} x_i^2 - \left(\sum_{i=1}^{n} x_i \right)^2} \tag{5.11}$$

$$a_0 = \frac{\left(\sum_{i=1}^{n} x_i^2\right)\left(\sum_{i=1}^{n} y_i\right) - \left(\sum_{i=1}^{n} x_i y_i\right)\left(\sum_{i=1}^{n} x_i\right)}{n \sum_{i=1}^{n} x_i^2 - \left(\sum_{i=1}^{n} x_i\right)^2} \tag{5.12}$$

Since Eqs. (5.11) and (5.12) contain summations that are the same, it is convenient to calculate the summations first and then to substitute them in the equations. To do this the summations are defined by:

$$S_x = \sum_{i=1}^{n} x_i, \quad S_y = \sum_{i=1}^{n} y_i, \quad S_{xy} = \sum_{i=1}^{n} x_i y_i, \quad S_{xx} = \sum_{i=1}^{n} x_i^2 \tag{5.13}$$

With these definitions, the equations for the coefficients a_1 and a_0 are:

$$a_1 = \frac{n S_{xy} - S_x S_y}{n S_{xx} - (S_x)^2} \qquad a_0 = \frac{S_{xx} S_y - S_{xy} S_x}{n S_{xx} - (S_x)^2} \tag{5.14}$$

Equations (5.14) give the values of a_1 and a_0 in the equation $y = a_1 x + a_0$ that has the best fit to n data points (x_i, y_i). Example 5-1 shows how to use Eqs. (5.11) and (5.12) for fitting a linear equation to a set of data points.

Example 5-1: Determination of absolute zero temperature.

According to Charles's law for an ideal gas, at constant volume, a linear relationship exists between the pressure, p, and temperature, T. In the experiment shown in the figure, a fixed volume of gas in a sealed container is submerged in ice water ($T = 0°C$). The temperature of the gas is then increased in ten increments up to $T = 100°C$ by heating the water, and the pressure of the gas is measured at each temperature. The data from the experiment is:

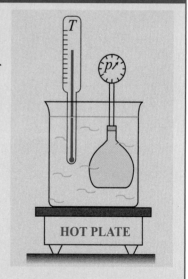

T (°C)	0	10	20	30	40	50	60
p (atm.)	0.94	0.96	1.0	1.05	1.07	1.09	1.14

T (°C)	70	80	90	100
p (atm.)	1.17	1.21	1.24	1.28

Extrapolate the data to determine the absolute zero temperature, T_0. This can be done using the following steps:

(a) Make a plot of the data (p versus T).

(b) Use linear least-squares regression to determine a linear function in the form $p = a_1 T + a_0$ that best fits the data points. First calculate the coefficients by hand using only the four data points: 0, 30, 70, and 100°C. Then write a user-defined MATLAB function that calculates the coefficients of the linear function for any number of data points and use it with all the data points to determine the coefficients.

(c) Plot the function, and extend the line (extrapolate) until it crosses the horizontal (T) axis. This point is an estimate of the absolute zero temperature, T_0. Determine the value of T_0 from the function.

SOLUTION

(a) A plot (p versus T) of the data is created by MATLAB (Command Window):

```
>> T = 0:10:100;
p = [0.94 0.96 1.0 1.05 1.07 1.09 1.14 1.17 1.21 1.24 1.28];
>> plot(T,p,'*r')
```

The plot that is obtained is shown on the right (axes titles were added using the Plot Editor). The plot shows, as expected, a nearly linear relationship between the pressure and the temperature.

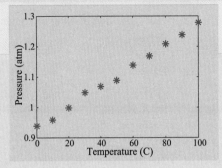

(b) Hand calculation of least-squares regression of the four data points:

$(0, 0.94)$, $(30, 1.05)$, $(70, 1.17)$, $(100, 1.28)$

The coefficients a_1 and a_0 of the equation $p = a_1 T + a_0$ that best fits the data points are determined by using Eqs. (5.14). The summations, Eqs. (5.13), are calculated first.

$$S_x = \sum_{i=1}^{4} x_i = 0 + 30 + 70 + 100 = 200 \qquad S_y = \sum_{i=1}^{4} y_i = 0.94 + 1.05 + 1.17 + 1.28 = 4.44$$

$$S_{xx} = \sum_{i=1}^{4} x_i^2 = 0^2 + 30^2 + 70^2 + 100^2 = 15800$$

$$S_{xy} = \sum_{i=1}^{4} x_i y_i = 0 \cdot 0.94 + 30 \cdot 1.05 + 70 \cdot 1.17 + 100 \cdot 1.28 = 241.4$$

Substituting the Ss in Eqs. (5.14) gives:

$$a_1 = \frac{n S_{xy} - S_x S_y}{n S_{xx} - (S_x)^2} = \frac{4 \cdot 241.4 - (200 \cdot 4.44)}{4 \cdot 15800 - 200^2} = 0.003345$$

$$a_0 = \frac{S_{xx} S_y - S_{xy} S_x}{n S_{xx} - (S_x)^2} = \frac{15800 \cdot 4.44 - (241.4 \cdot 200)}{4 \cdot 15800 - 200^2} = 0.9428$$

From this calculation, the equation that best fits the data is: $p = 0.003345 T + 0.9428$.

Next, the problem is solved by writing a MATLAB user-defined function that calculates the coefficients of the linear function for any number of data points. The inputs to the function are two vectors with the coordinates of the data points. The outputs are the coefficients a_1 and a_0 of the linear equation, which are calculated with Eqs. (5.14).

Program 5-1: User-defined function. Linear least-squares regression.

```
function [a1,a0] = LinearRegression(x, y)
% LinearRegression calculates the coefficients a1 and a0 of the linear
% equation y = a1*x + a0 that best fits n data points.
% Input variables:
% x   A vector with the coordinates x of the data points.
% y   A vector with the coordinates y of the data points.
% Output variables:
```

```
% a1   The coefficient a1.
% a0   The coefficient a0.


nx = length(x);
ny = length(y);
if nx ~ = ny                          ┌─────────────────────────────────────────────────────────────┐
                                      │ Check if the vectors x and y have the same number of elements. │
                                      └─────────────────────────────────────────────────────────────┘
   disp('ERROR: The number of elements in x must be the same as in y.')
   a1 = 'Error';                              ┌────────────────────────────────────────┐
   a0 = 'Error';                              │ If yes, MATLAB displays an error message │
                                              │ and the constants are not calculated.    │
else                                          └────────────────────────────────────────┘
   Sx = sum(x);
   Sy = sum(y);
   Sxy = sum(x.*y);                   ┌─────────────────────────────────────────────┐
   Sxx = sum(x.^2);                   │ Calculate the summation terms in Eqs. (5.13). │
   a1 = (nx*Sxy - Sx*Sy)/(nx*Sxx - Sx^2);   └─────────────────────────────────────────────┘
   a0 = (Sxx*Sy - Sxy*Sx)/(nx*Sxx - Sx^2);  ┌─────────────────────────────────────────────────┐
end                                          │ Calculate the coefficients $a_1$ and $a_0$ in Eqs. (5.14). │
                                             └─────────────────────────────────────────────────┘
```

The user-defined function `LinearRegression` is next used in Command Window for determining the best fit line to the given points in the problem.

```
>> T = 0:10:100;
>> p = [0.94 0.96 1.0 1.05 1.07 1.09 1.14 1.17 1.21 1.24 1.28];
>> [a1, a0]=LinearRegression(T,p)
a1 =
   0.0034                              ┌──────────────────────────────────┐
a0 =                                   │ The equation that best fit the data is: │
   0.9336                              │ $p = 0.0034T + 0.9336$           │
                                       └──────────────────────────────────┘
```

(c) The solution is done in the following script file that plots the function, the points, and calculates the value of T_0 from the function.

```
T = 0:10:100;
p = [0.94 0.96 1.0 1.05 1.07 1.09 1.14 1.17 1.21 1.24 1.28];
Ti = [-300 100];
pi = 0.0034*Ti+0.9336;
plot(T,p,'*r','markersize',12)
hold on
plot(Ti,pi,'k')
xlabel('Temperature (C)','fontsize',20)
ylabel('Pressure (atm)','fontsize',20)
T0 = -0.9336/0.0034
```

When this script file is executed, the figure shown on the right is displayed, and the value of the calculated absolute zero temperate is displayed in the Command Window, as shown below.

```
T0 =
   -274.5882
```

This result is close to the handbook value of -273.15 °C.

5.3 CURVE FITTING WITH NONLINEAR EQUATION BY WRITING THE EQUATION IN A LINEAR FORM

Many situations in science and engineering show that the relationship between the quantities that are being considered is not linear. For example, Fig. 5-8 shows a plot of data points that were measured in an experiment with an *RC* circuit. In this experiment, the voltage across the resistor is measured as a function of time, starting when the switch is closed.

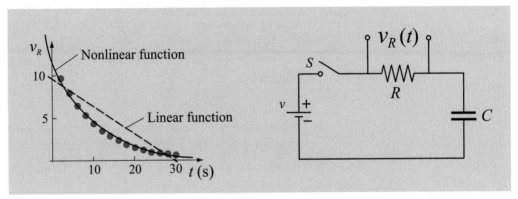

Figure 5-8: Curve-fitting points with linear equation.

The data points from the experiment are listed in Example 5-2. It is obvious from the plot that curve fitting the data points with a nonlinear function gives a much better fit than curve fitting with a linear function.

There are many kinds of nonlinear functions. This section shows curve fitting with nonlinear functions that can be written in a form for which the linear least-squares regression method can be used for determining the coefficients that give the best fit. Examples of nonlinear functions used for curve fitting in the present section are:

$$y = bx^m \qquad \text{(power function)}$$

$$y = be^{mx} \text{ or } y = b10^{mx} \qquad \text{(exponential function)}$$

$$y = \frac{1}{mx + b} \qquad \text{(reciprocal function)}$$

Polynomials of second, or higher, degree are also nonlinear functions. Curve fitting with such polynomials is covered separately in Section 5.4.

Writing a nonlinear equation in linear form

In order to be able to use linear regression, the form of a nonlinear equation of two variables is changed such that the new form is linear with terms that contain the original variables. For example, the power function $y = bx^m$ can be put into linear form by taking the natural logarithm (ln) of both sides:

$$\ln(y) = \ln(bx^m) = m\ln(x) + \ln(b) \qquad (5.15)$$

This equation is linear for $\ln(y)$ in terms $\ln(x)$. The equation is in the form $Y = a_1 X + a_0$ where $Y = \ln(y)$, $a_1 = m$, $X = \ln(x)$, and $a_0 = \ln(b)$:

$$\underbrace{\ln(y)}_{Y} = \underbrace{m}_{a_1}\underbrace{\ln(x)}_{X} + \underbrace{\ln(b)}_{a_0}$$

This means that linear least-squares regression can be used for curve fitting an equation of the form $y = bx^m$ to a set of data points x_i, y_i. This is done by calculating a_1 and a_0 using Eqs. (5.11) and (5.12) [or (5.13) and (5.14)] while substituting $\ln(y_i)$ for y_i and $\ln(x_i)$ for x_i. Once a_1 and a_0 are known, the constants b and m in the exponential equation are calculated by:

$$m = a_1 \text{ and } b = e^{(a_0)} \qquad (5.16)$$

Many other nonlinear equations can be transformed into linear form in a similar way. Table 5-2 lists several such equations.

Table 5-2: Transforming nonlinear equations to linear form.

Nonlinear equation	Linear form	Relationship to $Y = a_1 X + a_0$	Values for linear least-squares regression	Plot where data points appear to fit a straight line
$y = bx^m$	$\ln(y) = m\ln(x) + \ln(b)$	$Y = \ln(y), \quad X = \ln(x)$ $a_1 = m, \quad a_0 = \ln(b)$	$\ln(x_i)$ and $\ln(y_i)$	y vs. x plot on logarithmic y and x axes. $\ln(y)$ vs. $\ln(x)$ plot on linear x and y axes.
$y = be^{mx}$	$\ln(y) = mx + \ln(b)$	$Y = \ln(y), \quad X = x$ $a_1 = m, \quad a_0 = \ln(b)$	x_i and $\ln(y_i)$	y vs. x plot on logarithmic y and linear x axes. $\ln(y)$ vs. x plot on linear x and y axes.
$y = b10^{mx}$	$\log(y) = mx + \log(b)$	$Y = \log(y), \quad X = x$ $a_1 = m, \quad a_0 = \log(b)$	x_i and $\ln(y_i)$	y vs. x plot on logarithmic y and linear x axes. $\ln(y)$ vs. x plot on linear x and y axes.
$y = \dfrac{1}{mx + b}$	$\dfrac{1}{y} = mx + b$	$Y = \dfrac{1}{y}, \quad X = x$ $a_1 = m, \quad a_0 = b$	x_i and $1/y_i$	$1/y$ vs. x plot on linear x and y axes.
$y = \dfrac{mx}{b + x}$	$\dfrac{1}{y} = \dfrac{b}{m}\dfrac{1}{x} + \dfrac{1}{m}$	$Y = \dfrac{1}{y}, \quad X = \dfrac{1}{x}$ $a_1 = \dfrac{b}{m}, \quad a_0 = \dfrac{1}{m}$	$1/x_i$ and $1/y_i$	$1/y$ vs. $1/x$ plot on linear x and y axes.

How to choose an appropriate nonlinear function for curve fitting

A plot of the given data points can give an indication as to the relationship between the quantities. Whether the relationship is linear or nonlinear can be determined by plotting the points in a figure with linear axes. If in such a plot the points appear to line up along a straight line, then the relationship between the plotted quantities is linear.

A plot with linear axes in which the data points appear to line up along a curve indicates a nonlinear relationship between the plotted quantities. The question then is which nonlinear function to use for the curve fitting. Many times in engineering and science there is knowledge from a guiding theory of the physical phenomena and the form of the mathematical equation associated with the data points. For example, the process of charging a capacitor shown in Fig. 5-8 is modeled with an exponential function. If there is no knowledge of a possible form of the equation, choosing the most appropriate nonlinear function to curve-fit given data may be more difficult.

For given data points it is possible to foresee, to some extent, if a proposed nonlinear function has a potential for providing a good fit. This is done by plotting the data points in a specific way and examining whether the points appear to fit a straight line. For the functions listed in Table 5-2 this is shown in the fifth (last) column of the table. For power and exponential functions, this can be done by plotting the data using different combinations of linear and logarithmic axes. For all functions it can be done by plotting the transformed values of the data points in plots with linear axes.

For example, as was mentioned before, the data points from the experiment that are shown in Fig. 5-8 are expected to fit an exponential function. This means that a plot of the voltage v_R versus time t on a plot with a logarithmic vertical axis (for v_R) and linear horizontal axis (for t) should reveal that the data points will be fit by a straight line. Another option is to make a plot of $\ln(v_R)$ vs. t on linear vertical and horizontal axes, which is also expected to show that the points line up along a straight line. Both of these plots are shown in Fig. 5-9. The figures con-

The script file for making the plots is:

```
tx=2:2:30;
vexp=[9.7 8.1 6.6 5.1 4.4  3.7 2.8
2.4 2.0 1.6 1.4 1.1 0.85 0.69 0.6];
vexpLOG=log(vexp)
subplot(1,2,1)
semilogy(tx,vexp,'or')
subplot(1,2,2)
plot(tx,vexpLOG,'or')
```

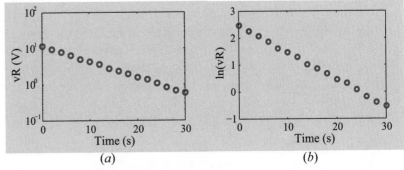

Figure 5-9: (*a*) **A plot of** v_R **vs.** t **in a plot with a logarithmic vertical axis and linear horizontal axis.** (*b*) **A plot of** $\ln(v_R)$ **vs.** t **in a plot with linear vertical and horizontal axes.**

firm that the data from the capacitor charging experiment can be curve fit with an exponential function. The actual curve fitting is shown in Example 5-2.

Other considerations when choosing a nonlinear function for curve fitting are as follows:

- Exponential functions cannot pass through the origin.
- Exponential functions can only fit data with all positive ys, or all negative ys.
- Logarithmic functions cannot include $x = 0$ or negative values of x.
- For power function $y = 0$ when $x = 0$.
- The reciprocal equation cannot include $y = 0$.

Example 5-2: Curve fitting with a nonlinear function by writing the equation in a linear form.

An experiment with an RC circuit is used for determining the capacitance of an unknown capacitor. In the circuit, shown on the right and in Fig. 5-8, a 5-MΩ resistor is connected in series to the unknown capacitor C and a battery. The experiment starts by closing the switch and measuring the voltages, v_R, across the resistor every 2 seconds for 30 seconds. The data measured in the experiment is:

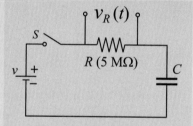

t (s)	2	4	6	8	10	12	14	16	18
v_R (V)	9.7	8.1	6.6	5.1	4.4	3.7	2.8	2.4	2.0

t (s)	20	22	24	26	28	30
v_R (V)	1.6	1.4	1.1	0.85	0.69	0.6

Theoretically, the voltage across the resistor as a function of time is given by the exponential function:

$$v_R = ve^{(-t/(RC))} \tag{5.17}$$

Determine the capacitance of the capacitor by curve fitting the exponential function to the data.

SOLUTION

It was shown in Fig. 5-9 that, as expected, an exponential function can fit the data well. The problem is solved by first determining the constants b and m in the exponential function $v = be^{mt}$ that give the best fit of the function to the data. This is done by changing the equation to have a linear form and then using linear least-squares regression.

The linear least-squares regression is applied by using the user-defined function `LinearRegression` that was developed in the solution of Example 5-1. The inputs to the function are the values t_i and $\ln((v_r)_i)$. Once b and m are known, the value of C is determined by equating the coefficients in the exponent of e:

$$\frac{-1}{RC} = m \text{ solving for } C \text{ gives: } C = \frac{-1}{Rm} \tag{5.18}$$

The calculations are done by executing the following MATLAB program (script file):

Program 5-2: Script file. Curve fitting with a nonlinear function.

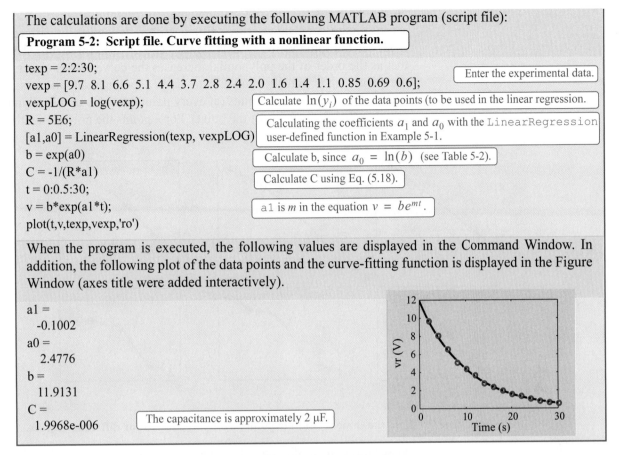

```
texp = 2:2:30;
vexp = [9.7 8.1 6.6 5.1 4.4 3.7 2.8 2.4 2.0 1.6 1.4 1.1 0.85 0.69 0.6];
vexpLOG = log(vexp);
R = 5E6;
[a1,a0] = LinearRegression(texp, vexpLOG)
b = exp(a0)
C = -1/(R*a1)
t = 0:0.5:30;
v = b*exp(a1*t);
plot(t,v,texp,vexp,'ro')
```

Enter the experimental data.

Calculate $\ln(y_i)$ of the data points (to be used in the linear regression.

Calculating the coefficients a_1 and a_0 with the `LinearRegression` user-defined function in Example 5-1.

Calculate b, since $a_0 = \ln(b)$ (see Table 5-2).

Calculate C using Eq. (5.18).

`a1` is m in the equation $v = be^{mt}$.

When the program is executed, the following values are displayed in the Command Window. In addition, the following plot of the data points and the curve-fitting function is displayed in the Figure Window (axes title were added interactively).

```
a1 =
  -0.1002
a0 =
   2.4776
b =
   11.9131
C =
   1.9968e-006
```

The capacitance is approximately 2 μF.

5.4 CURVE FITTING WITH QUADRATIC AND HIGHER-ORDER POLYNOMIALS

Background

Polynomials are functions that have the form:

$$f(x) = a_n x^n + a_{n-1} x^{n-1} + \dots + a_1 x + a_0 \qquad (5.19)$$

The coefficients $a_n, a_{n-1}, ..., a_1, a_0$ are real numbers, and n, which is a nonnegative integer, is the degree, or order, of the polynomial. A plot of the polynomial is a curve. A first-order polynomial is a linear function, and its plot is a straight line. Higher-order polynomials are nonlinear functions, and their plots are curves. A quadratic (second-order) polynomial is a curve that is either concave up or down (parabola). A third-order polynomial has an inflection point such that the curve can be concave up (or down) in one region, and concave down (or up) in another. In general, as the order of a polynomial increases, its curve can have more "bends."

A given set of n data points can be curve-fit with polynomials of different order up to an order of $(n-1)$. As shown later in this section, the coefficients of a polynomial can be determined such that the polynomial best fits the data by minimizing the error in a least squares sense. Figure 5-10 shows curve fitting with polynomials of different order for the same set of 11 data points. The plots in the figure show that as the order of the polynomial increases the curve passes closer to the points. It is actually possible to have a polynomial that passes exactly through all of the points (at every point the value of the polynomial is equal to the value of the point). For n points the polynomial that

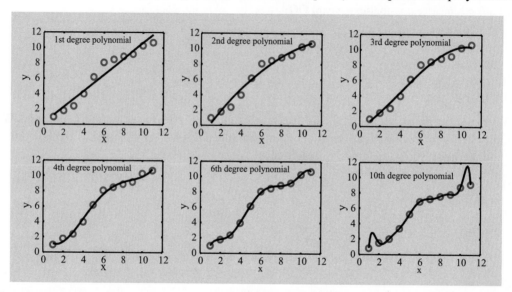

Figure 5-10: Curve fitting of the same set of data points with polynomials for different degrees.

passes through all of the points is one of order $(n-1)$. In Fig. 5-10 it is the tenth degree polynomial (since there are 11 points).

Figure 5-10 shows that the same set of data points can be curve fit with polynomials of different order. The question as to which of the polynomials gives the best fit does not have a simple answer. It depends on the type and source of data, the engineering or science application associated with the data, and the purpose of the curve fitting. For example, if the data points themselves are not accurate (there is possibly a large error when the quantity is measured), it does not make a lot of sense to use a higher-order polynomial that follows the points closely. On the other hand, if the values of the data points are very accurate and the curve fitting is used for representing the data, curve fitting with a higher-order polynomial might be more appropriate. However, as explained in the ***important note*** that follows, use of higher-order polynomials for curve fitting is not recommended.

Important note

As already mentioned, for any number of data points, n, it is possible to derive a polynomial (order of $(n-1)$) that passes exactly through all the points. However, when many points are involved, this polynomial is of a high degree. Although the high-order polynomial gives the exact values at all of the data points, often the polynomial deviates significantly **between** some of the points. This can be seen in the plot with the tenth order polynomial in Fig. 5-10, where between the first two points and between the last two points the curve of the polynomial wanders away and does not follow the general trend of the data points. This means that even though the high-order polynomial gives the exact values at all the data points, it cannot be used reliably for interpolation or extrapolation. Appropriate methods for interpolation are described in Sections 5.5 and 5.6.

Polynomial regression

Polynomial regression is a procedure for determining the coefficients of a polynomial of a second degree, or higher, such that the polynomial best fits a given set of data points. As in linear regression, the derivation of the equations that are used for determining the coefficients is based on minimizing the total error according to Eq. (5.5).

If the polynomial, of order m, that is used for the curve fitting is:

$$f(x) = a_m x^m + a_{m-1} x^{m-1} + \dots + a_1 x + a_0 \tag{5.20}$$

then, for a given set of n data points (x_i, y_i) (m is smaller than $n - 1$), the total error calculated by Eq. (5.5) is:

$$E = \sum_{i=1}^{n} [y_i - (a_m x_i^m + a_{m-1} x_i^{m-1} + \dots + a_1 x_i + a_0)]^2 \tag{5.21}$$

Since all the values x_i and y_i of the data points are known, E in Eq. (5.21) is a nonlinear function of the $m + 1$ variables (the coefficients a_0 through a_m). The function E has a minimum at the values of a_0 through a_m where the partial derivatives of E with respect to each of the variables is equal to zero. Taking the partial derivatives of E in Eq. (5.21) and setting them to zero gives a set of $m + 1$ linear equations for the coefficients. To simplify the presentation here, the derivation for the case of $m = 2$ (quadratic polynomial) is shown in detail. In this case Eq. (5.21) is:

$$E = \sum_{i=1}^{n} [y_i - (a_2 x_i^2 + a_1 x_i + a_0)]^2 \tag{5.22}$$

Taking the partial derivatives with respect to a_0, a_1, and a_2, and setting them equal to zero gives:

$$\frac{\partial E}{\partial a_0} = -2 \sum_{i=1}^{n} (y_i - a_2 x_i^2 - a_1 x_i - a_0) = 0 \tag{5.23}$$

$$\frac{\partial E}{\partial a_1} = -2 \sum_{i=1}^{n} (y_i - a_2 x_i^2 - a_1 x_i - a_0) x_i = 0 \tag{5.24}$$

$$\frac{\partial E}{\partial a_2} = -2 \sum_{i=1}^{n} (y_i - a_2 x_i^2 - a_1 x_i - a_0) x_i^2 = 0 \tag{5.25}$$

Equations (5.23) through (5.25) are a system of three linear equations for the unknowns a_0, a_1, and a_2, which can be rewritten in the form:

$$n a_0 + \left(\sum_{i=1}^{n} x_i \right) a_1 + \left(\sum_{i=1}^{n} x_i^2 \right) a_2 = \sum_{i=1}^{n} y_i \tag{5.26}$$

$$\left(\sum_{i=1}^{n} x_i \right) a_0 + \left(\sum_{i=1}^{n} x_i^2 \right) a_1 + \left(\sum_{i=1}^{n} x_i^3 \right) a_2 = \sum_{i=1}^{n} x_i y_i \tag{5.27}$$

$$\left(\sum_{i=1}^{n} x_i^2 \right) a_0 + \left(\sum_{i=1}^{n} x_i^3 \right) a_1 + \left(\sum_{i=1}^{n} x_i^4 \right) a_2 = \sum_{i=1}^{n} x_i^2 y_i \tag{5.28}$$

The solution of the system of equations (5.26)–(5.28) gives the values of the coefficients a_0, a_1, and a_2 of the polynomial $y = a_2 x_i^2 + a_1 x_i + a_0$ that best fits the n data points (x_i, y_i).

The coefficients for higher-order polynomials are derived in the same way. For an mth order polynomial, Eqs. (5.26)–(5.28) are extended to a set of $m + 1$ linear equations for the $m + 1$ coefficients. The equations for a fourth order polynomial are shown in Example 5-3.

Example 5-3: Using polynomial regression for curve fitting of stress–strain curve.

A tension test is conducted for determining the stress–strain behavior of rubber. The data points from the test are shown in the figure, and their values are given below. Determine the fourth order polynomial that best fits the data points. Make a plot of the data points and the curve that corresponds to the polynomial.

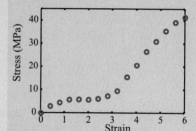

Strain ε	0	0.4	0.8	1.2	1.6	2.0	2.4
Stress σ (MPa)	0	3.0	4.5	5.8	5.9	5.8	6.2

Strain ε	2.8	3.2	3.6	4.0	4.4	4.8	5.2	5.6	6.0
Stress σ (MPa)	7.4	9.6	15.6	20.7	26.7	31.1	35.6	39.3	41.5

SOLUTION

A polynomial of the fourth order can be written as:

$$f(x) = a_4 x^4 + a_3 x^3 + a_2 x^2 + a_1 x + a_0 \tag{5.29}$$

Curve fitting of 16 data points with this polynomial is done by polynomial regression. The values of the five coefficients a_0, a_1, a_2, a_3, and a_4 are obtained by solving a system of five linear equations. The five equations can be written by extending Eqs. (5.26)–(5.28).

$$n a_0 + \left(\sum_{i=1}^{n} x_i \right) a_1 + \left(\sum_{i=1}^{n} x_i^2 \right) a_2 + \left(\sum_{i=1}^{n} x_i^3 \right) a_3 + \left(\sum_{i=1}^{n} x_i^4 \right) a_4 = \sum_{i=1}^{n} y_i \tag{5.30}$$

$$\left(\sum_{i=1}^{n} x_i\right)a_0 + \left(\sum_{i=1}^{n} x_i^2\right)a_1 + \left(\sum_{i=1}^{n} x_i^3\right)a_2 + \left(\sum_{i=1}^{n} x_i^4\right)a_3 + \left(\sum_{i=1}^{n} x_i^5\right)a_4 = \sum_{i=1}^{n} x_i y_i \tag{5.31}$$

$$\left(\sum_{i=1}^{n} x_i^2\right)a_0 + \left(\sum_{i=1}^{n} x_i^3\right)a_1 + \left(\sum_{i=1}^{n} x_i^4\right)a_2 + \left(\sum_{i=1}^{n} x_i^5\right)a_3 + \left(\sum_{i=1}^{n} x_i^6\right)a_4 = \sum_{i=1}^{n} x_i^2 y_i \tag{5.32}$$

$$\left(\sum_{i=1}^{n} x_i^3\right)a_0 + \left(\sum_{i=1}^{n} x_i^4\right)a_1 + \left(\sum_{i=1}^{n} x_i^5\right)a_2 + \left(\sum_{i=1}^{n} x_i^6\right)a_3 + \left(\sum_{i=1}^{n} x_i^7\right)a_4 = \sum_{i=1}^{n} x_i^3 y_i \tag{5.33}$$

$$\left(\sum_{i=1}^{n} x_i^4\right)a_0 + \left(\sum_{i=1}^{n} x_i^5\right)a_1 + \left(\sum_{i=1}^{n} x_i^6\right)a_2 + \left(\sum_{i=1}^{n} x_i^7\right)a_3 + \left(\sum_{i=1}^{n} x_i^8\right)a_4 = \sum_{i=1}^{n} x_i^4 y_i \tag{5.34}$$

The calculations and the plot are done with MATLAB in a script file that is listed below. The computer program follows these steps:

Step 1: Create vectors x and y with the data points.

Step 2: Create a vector xsum in which the elements are the summation terms of the powers of x_i.

For example, the fourth element is: $xsum(4) = \sum_{i=1}^{n} x_i^4$

Step 3: Set up the system of five linear equations (Eqs. (5.30)–(5.34)) in the form $[a][p] = [b]$, where $[a]$ is the matrix with the summation terms of the powers of x_i, $[p]$ is the vector of the unknowns (the coefficients of the polynomial), and $[b]$ is a vector of the summation terms on the right-hand side of Eqs. (5.30)–(5.34).

Step 4: Solve the system of five linear equations $[a][p] = [b]$ (Eqs. (5.30)–(5.34)) for p, by using MATLAB's left division. The solution is a vector with the coefficients of the fourth order polynomial that best fits the data.

Step 5: Plot the data points and the curve-fitting polynomial.

Program 5-3: Script file. Curve fitting using polynomial regression.

```
clear all
x = 0:0.4:6;                                              Assign the experimental data
y = [0 3 4.5 5.8 5.9 5.8 6.2 7.4 9.6 15.6 20.7 26.7 31.1 35.6 39.3 41.5];    points to vectors x and y.
n = length(x);                                           n is the number of data points.
m = 4;                                                   m is the order of the polynomial.
for i = 1:2*m
    xsum(i) = sum(x.^(i));                   Define a vector with the summation terms of the powers of x_i.
end
% Beginning of Step 3
a(1,1) = n;
b(1,1) = sum(y);                             Assign the first row of the matrix [a] and the first
for j = 2:m + 1                              element of the column vector [b].
    a(1,j) = xsum(j - 1);
end
```

```
for i = 2:m + 1
    for j = 1:m + 1
        a(i,j) = xsum(j + i - 2);
    end
    b(i,1) = sum(x.^(i - 1).*y);
end
```

Create rows 2 through 5 of the matrix [a] and elements 2 through 5 of the column vector [b].

```
% Step 4
p = (a\b)'
```

Solve the system $[a][p] = [b]$ for $[p]$. Transpose the solution such that $[p]$ is a row vector.

```
for i = 1:m + 1
    Pcoef(i) = p(m + 2 - i);
end
```

Create a new vector for the coefficients of the polynomial, to be used in MATLAB's `polyval` built-in function (see note at the end of the example).

```
epsilon = 0:0.1:6;
```

Define a vector of strain to be used for plotting the polynomial.

```
stressfit = polyval(Pcoef,epsilon);
```

Stress calculated by the polynomial.

```
plot(x,y,'ro',epsilon,stressfit,'k','linewidth',2)
xlabel('Strain','fontsize',20)
ylabel('Stress (MPa)','fontsize',20)
```

Plot the data points and the curve-fitting polynomial.

When the program is executed, the solution [p] is displayed in the Command Window. In addition, the plot of the data points and the curve-fitting polynomial is displayed in the Figure Window.

```
p =
  -0.2746   12.8780   -10.1927   3.1185   -0.2644
```

The curve-fitting polynomial is:

$$f(x) = (-0.2644)x^4 + 3.1185x^3 - 10.1927x^2 + 12.878x - 0.2746$$

Note: In MATLAB a polynomial is represented by a vector whose elements are the polynomial's coefficients. The first element in the vector is the coefficient of the highest order term in the polynomial, and the last element in the vector is the coefficient a_0.

5.5 INTERPOLATION USING A SINGLE POLYNOMIAL

Interpolation is a procedure in which a mathematical formula is used to represent a given set of data points, such that the formula gives the exact value at all the data points and an estimated value **between** the points. This section shows how this is done by using a single polynomial, regardless of the number of points. As was mentioned in the previous section, for any number of points n there is a polynomial of order $n - 1$ that passes through all of the points. For two points the polynomial is of first order (a straight line connecting the points). For three points the polynomial is of second order (a parabola that connects the points), and so on. This is illustrated in Fig. 5-11 which shows how first, second, third, and fourth-order polynomials connect two, three, four, and five points, respectively.

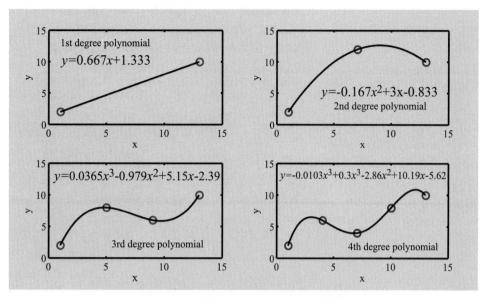

Figure 5-11: Various order polynomials.

Once the polynomial is determined, it can be used for estimating the y values between the known points simply by substituting for the x coordinate in the polynomial. Interpolation with a single polynomial gives good results for a small number of points. For a large number of points the order of the polynomial is high, and although the polynomial passes through all the points, it might deviate significantly between the points. This was shown in Fig. 5-10 for a polynomial of tenth degree and is shown later in Fig. 5-17, where a 15th-order polynomial is used for interpolation of a set of 16 data points. Consequently, interpolation with a single polynomial might not be appropriate for a large number of points. For a large number of points, better interpolation can be done by using piecewise (spline) interpolation (covered in Section 5.6) in which different lower-order polynomials are used for interpolation between different points of the same set of data points.

For a given set of n points, only one (unique) polynomial of order m ($m = n - 1$) passes exactly through all of the points. The polynomial, however, can be written in different mathematical forms. This section shows how to derive three forms of polynomials (standard, Lagrange, and Newton's). The different forms are suitable for use in different circumstances.

The standard form of an mth-order polynomial is:

$$f(x) = a_m x^m + a_{m-1} x^{m-1} + \dots + a_1 x + a_0 \tag{5.35}$$

The coefficients in this form are determined by solving a system of $m + 1$ linear equations. The equations are obtained by writing the polynomial explicitly for each point (substituting each point in the polynomial). For example, the five points ($n = 5$) in the fourth degree ($m = 4$)

polynomial plot in Fig. 5-11 are: $(1, 2)$, $(4, 6)$, $(7, 4)$, $(10, 8)$, and $(13, 10)$. Writing Eq. (5.35) for each of the points gives the following system of five equations for the unknowns a_0, a_1, a_2, a_3, and a_4:

$$
\begin{aligned}
a_4 1^4 + a_3 1^3 + a_2 1^2 + a_1 1 + a_0 &= 2 \\
a_4 4^4 + a_3 4^3 + a_2 4^2 + a_1 4 + a_0 &= 6 \\
a_4 7^4 + a_3 7^3 + a_2 7^2 + a_1 7 + a_0 &= 4 \\
a_4 10^4 + a_3 10^3 + a_2 10^2 + a_1 10 + a_0 &= 8 \\
a_4 13^4 + a_3 13^3 + a_2 13^2 + a_1 13 + a_0 &= 10
\end{aligned}
\tag{5.36}
$$

The solution of this system of equations gives the values of the coefficients. A MATLAB solution of Eqs. (5.36) is:

```
>> a = [1 1 1 1 1; 4^4 4^3 4^2 4 1; 7^4 7^3 7^2 7 1; 10^4 10^3 10^2 10 1; 13^4 13^3 13^2 13 1]
a =
          1         1         1         1         1
        256        64        16         4         1
       2401       343        49         7         1
      10000      1000       100        10         1
      28561      2197       169        13         1
>> b = [2; 6; 4; 8; 10]
>> A = a\b
A =
   -0.0103
    0.3004
   -2.8580
   10.1893
   -5.6214
```

> The polynomial that corresponds to these coefficients is:
> $$y = -0.0103x^4 + 0.3x^3 - 2.86x^2 + 10.19x - 5.62$$
> (see Fig. 5-11).

In practice, solving the system of equations, especially for higher-order polynomials, is not efficient, and frequently the matrix of the coefficients is ill conditioned (see Section 4.11).

It is possible to write the polynomial in other forms that may be easier to use. Two such forms, the Lagrange and Newton forms, are described in the next two subsections.

5.5.1 Lagrange Interpolating Polynomials

Lagrange interpolating polynomials are a particular form of polynomials that can be written to fit a given set of data points by using the values at the points. The polynomials can be written right away and do not require any preliminary calculations for determining coefficients.

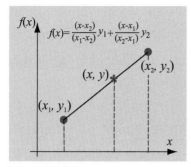

Figure 5-12: First-order Lagrange polynomial.

For two points, (x_1, y_1), and (x_2, y_2), the first-order Lagrange polynomial that passes through the points (Fig. 5-12) has the form:

$$f(x) = y = a_1(x - x_2) + a_2(x - x_1) \tag{5.37}$$

Substituting the two points in Eq. (5.37) gives:

$$y_1 = a_1(x_1 - x_2) + a_2(x_1 - x_1) \quad \text{or} \quad a_1 = \frac{y_1}{(x_1 - x_2)} \tag{5.38}$$

and

$$y_2 = a_1(x_2 - x_2) + a_2(x_2 - x_1) \quad \text{or} \quad a_2 = \frac{y_2}{(x_2 - x_1)} \tag{5.39}$$

Substituting the coefficients a_1 and a_2 back in Eq. (5.37) gives:

$$f(x) = \frac{(x - x_2)}{(x_1 - x_2)} y_1 + \frac{(x - x_1)}{(x_2 - x_1)} y_2 \tag{5.40}$$

Equation (5.40) is a linear function of x (an equation of a straight line that connects the two points). It is easy to see that if $x = x_1$ is substituted in Eq. (5.40), the value of the polynomial is y_1, and if $x = x_2$ is substituted, the value of the polynomial is y_2. Substituting a value of x between the points gives an interpolated value of y. Equation (5.40) can also be rewritten in the standard form $f(x) = a_1x + a_0$:

$$f(x) = \frac{(y_2 - y_1)}{(x_2 - x_1)} x + \frac{x_2 y_1 - x_1 y_2}{(x_2 - x_1)} \tag{5.41}$$

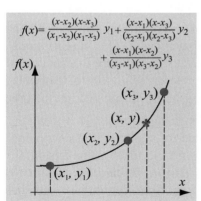

Figure 5-13: Second-order Lagrange polynomial.

For three points, (x_1, y_1), (x_2, y_2), and (x_3, y_3), the second-order Lagrange polynomial that passes through the points (Fig. 5-13) has the form:

$$f(x) = y = a_1(x - x_2)(x - x_3) + a_2(x - x_1)(x - x_3) + a_3(x - x_1)(x - x_2) \tag{5.42}$$

Once the coefficients are determined such that the polynomial passes through the three points, the polynomial is:

$$f(x) = \frac{(x - x_2)(x - x_3)}{(x_1 - x_2)(x_1 - x_3)} y_1 + \frac{(x - x_1)(x - x_3)}{(x_2 - x_1)(x_2 - x_3)} y_2 + \frac{(x - x_1)(x - x_2)}{(x_3 - x_1)(x_3 - x_2)} y_3 \tag{5.43}$$

Equation (5.43) is a quadratic function of x. When the coordinate x_1, x_2, or x_3 of one of the three given points is substituted in Eq. (5.43), the value of the polynomial is equal to y_1, y_2, or y_3, respectively. This is because the coefficient in front of the corresponding y_i is equal to 1 and the coefficient of the other two terms is equal to zero.

Following the format of the polynomials in Eqs. (5.41) and (5.43), the general formula of an $n - 1$ order Lagrange polynomial that passes through n points $(x_1, y_1), (x_2, y_2), ..., (x_n, y_n)$ is:

$$f(x) = \frac{(x-x_2)(x-x_3)\ldots(x-x_n)}{(x_1-x_2)(x_1-x_3)\ldots(x_1-x_n)}y_1 + \frac{(x-x_1)(x-x_3)\ldots(x-x_n)}{(x_2-x_1)(x_2-x_3)\ldots(x_2-x_n)}y_2 +$$

$$\ldots + \frac{(x-x_1)(x-x_2)\ldots(x-x_{i-1})(x-x_{i+1})\ldots(x-x_n)}{(x_i-x_1)(x_i-x_2)\ldots(x_i-x_{i-1})(x_i-x_{i+1})\ldots(x_i-x_n)}y_i + \ldots + \qquad (5.44)$$

$$\frac{(x-x_1)(x-x_2)\ldots(x-x_{n-1})}{(x_n-x_1)(x_n-x_2)\ldots(x_n-x_{n-1})}y_n$$

On the right-hand side of Eq. (5.44) the numerator of the ith term does not contain $(x-x_i)$, and the denominator does not contain (x_i-x_i). Consequently, when the coordinate x_i of one of the n points is substituted in Eq. (5.44), the value of the polynomial is equal to y_i. Equation (5.44) can be written in a compact form using summation and product notation as:

$$f(x) = \sum_{i=1}^{n} y_i L_i(x) = \sum_{i=1}^{n} y_i \prod_{\substack{j=1 \\ j \neq i}}^{n} \frac{(x-x_j)}{(x_i-x_j)} \qquad (5.45)$$

where $L_i(x) = \prod_{\substack{j=1 \\ j \neq i}}^{n} \frac{(x-x_j)}{(x_i-x_j)}$ are called the Lagrange functions. This

form can easily be implemented in a computer program, as shown in Example 5-4.

Additional notes about Lagrange polynomials

- The spacing between the data points does not have to be equal.

- For a given set of points, the whole expression of the interpolation polynomial has to be calculated for every value of x. In other words, the interpolation calculations for each value of x are independent of others. This is different from other forms (e.g., Eq. (5.35)) where once the coefficients of the polynomial are determined, they can be used for calculating different values of x.

- If an interpolated value is calculated for a given set of data points, and then the data set is enlarged to include additional points, all the terms of the Lagrange polynomial have to be calculated again. As shown in Section 5.5.2, this is different from Newton's polynomials where only the new terms have to be calculated if more data points are added.

Application of a Lagrange polynomial is shown in Example 5-4.

Example 5-4: Lagrange interpolating polynomial.

The set of the following five data points is given:

x	1	2	4	5	7
y	52	5	-5	-40	10

(a) Determine the fourth-order polynomial in the Lagrange form that passes through the points.
(b) Use the polynomial obtained in part (a) to determine the interpolated value for $x = 3$.
(c) Develop a MATLAB user-defined function that interpolates using a Lagrange polynomial. The input to the function are the coordinates of the given data points and the x coordinate at the point at which the interpolated value of y is to be calculated. The output from the function is the interpolated value of y at $x = 3$.

SOLUTION

(a) Following the form of Eq. (5.44), the Lagrange polynomial for the five given points is:

$$f(x) = \frac{(x-2)(x-4)(x-5)(x-7)}{(1-2)(1-4)(1-5)(1-7)}52 + \frac{(x-1)(x-4)(x-5)(x-7)}{(2-1)(2-4)(2-5)(2-7)}5 + \frac{(x-1)(x-2)(x-5)(x-7)}{(4-1)(4-2)(4-5)(4-7)}(-5) +$$

$$\frac{(x-1)(x-2)(x-4)(x-7)}{(5-1)(5-2)(5-4)(5-7)}(-40) + \frac{(x-1)(x-2)(x-4)(x-5)}{(7-1)(7-2)(7-4)(7-5)}10$$

(b) The interpolated value for $x = 3$ is obtained by substituting the x in the polynomial:

$$f(3) = \frac{(3-2)(3-4)(3-5)(3-7)}{(1-2)(1-4)(1-5)(1-7)}52 + \frac{(3-1)(3-4)(3-5)(3-7)}{(2-1)(2-4)(2-5)(2-7)}5 + \frac{(3-1)(3-2)(3-5)(3-7)}{(4-1)(4-2)(4-5)(4-7)}(-5) +$$

$$\frac{(3-1)(3-2)(3-4)(3-7)}{(5-1)(5-2)(5-4)(5-7)}(-40) + \frac{(3-1)(3-2)(3-4)(3-5)}{(7-1)(7-2)(7-4)(7-5)}10$$

$$f(3) = -5.778 + 2.667 - 4.444 + 13.333 + 0.222 = 6$$

(c) The MATLAB user-defined function for interpolation using Lagrange polynomials is named `Yint=LagrangeINT(x,y,Xint)`. x and y are vectors with the coordinates of the given data points, and `Xint` is the coordinate of the point at which y is to be interpolated.

- The program first calculates the product terms in the Lagrange functions in Eq. (5.45). The terms are assigned to a variable (vector) named L.

$$L_i = \prod_{\substack{j=1 \\ j \neq i}}^{n} \frac{(x - x_j)}{(x_i - x_j)} \quad \text{where} \quad x = \texttt{Xint}$$

- The program next calculates the value of the polynomial at $x = \texttt{Xint}$.

$$f(x) = \sum_{i=1}^{n} y_i L_i$$

Program 5-4: User-defined function. Interpolation using a Lagrange polynomial.

```
function Yint = LagrangeINT(x,y,Xint)
% LagrangeINT fits a Lagrange polynomial to a set of given points and
% uses the polynomial to determine the interpolated value of a point.
```

```
% Input variables:
% x  A vector with the x coordinates of the given points.
% y  A vector with the y coordinates of the given points.
% Xint  The x coordinate of the point at which y is to be interpolated.
% Output variable:
% Yint  The interpolated value of Xint.

n = length(x);                          The length of the vector x gives the number of terms in the polynomial.
for i = 1:n
   L(i) = 1;
   for j = 1:n
      if j ~= i                         Calculate the product terms L_i.
         L(i) = L(i)*(Xint - x(j))/(x(i) - x(j));
      end
   end
end                                     Calculate the value of the polynomial f(x) = sum_{i=1}^{n} y_i L_i.
Yint = sum(y .*L);
```

The `Lagrange(x,y,Xint)` function is then used in the Command Window for calculating the interpolated value of $x = 3$.

```
>> x = [1 2 4 5 7];
>> y = [52 5 -5 -40 10];
>> Yinterpolated = LagrangeINT(x,y,3)
Yinterpolated =
   6.0000
```

5.5.2 Newton's Interpolating Polynomials

Newton's interpolating polynomials are a popular means of exactly fitting a given set of data points. The general form of an $n-1$ order Newton's polynomial that passes through n points is:

$$f(x) = a_1 + a_2(x-x_1) + a_3(x-x_1)(x-x_2) + \ldots + a_n(x-x_1)(x-x_2)\ldots(x-x_{n-1}) \quad (5.46)$$

The special feature of this form of the polynomial is that the coefficients a_1 through a_n can be determined using a simple mathematical procedure. (Determination of the coefficients does not require a solution of a system of n equations.) Once the coefficients are known, the polynomial can be used for calculating an interpolated value at any x.

Newton's interpolating polynomials have additional desirable features that make them a popular choice. The data points do not have to be in descending or ascending order, or in any order. Moreover, after the n coefficients of an $n-1$ order Newton's interpolating polynomial are determined for n given points, more points can be added to the data set and only the new additional coefficients have to be determined.

First-order Newton's polynomial

For two given points, (x_1, y_1) and (x_2, y_2), the first-order Newton's polynomial has the form:

$$f(x) = a_1 + a_2(x - x_1) \qquad (5.47)$$

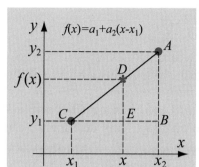

Figure 5-14: First-order Newton's polynomial.

As shown in Fig. 5-14, it is an equation of a straight line that passes through the points. The coefficients a_1 and a_2 can be calculated by considering the similar triangles in Fig. 5-14.

$$\frac{DE}{CE} = \frac{AB}{CB}, \quad \text{or} \quad \frac{f(x) - y_1}{x - x_1} = \frac{y_2 - y_1}{x_2 - x_1} \qquad (5.48)$$

Solving Eq. (5.48) for $f(x)$ gives:

$$f(x) = y_1 + \frac{y_2 - y_1}{x_2 - x_1}(x - x_1) \qquad (5.49)$$

Comparing Eq. (5.49) with Eq. (5.47) gives the values of the coefficients a_1 and a_2 in terms of the coordinates of the points:

$$a_1 = y_1, \quad \text{and} \quad a_2 = \frac{y_2 - y_1}{x_2 - x_1} \qquad (5.50)$$

Notice that the coefficient a_2 is the slope of the line that connects the two points. As shown in Chapter 6, a_2 is the two-point forward difference approximation for the first derivative at (x_1, y_1).

Second-order Newton's polynomial

For three given points, (x_1, y_1), (x_2, y_2), and (x_3, y_3), the second-order Newton's polynomial has the form:

$$f(x) = a_1 + a_2(x - x_1) + a_3(x - x_1)(x - x_2) \qquad (5.51)$$

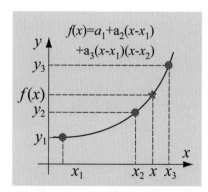

Figure 5-15: Second-order Newton's polynomial.

As shown in Fig. 5-15, it is an equation of a parabola that passes through the three points. The coefficients a_1, a_2, and a_3 can be determined by substituting the three points in Eq. (5.51). Substituting $x = x_1$ and $f(x_1) = y_1$ gives: $a_1 = y_1$. Substituting the second point, $x = x_2$ and $f(x_2) = y_2$, (and $a_1 = y_1$) in Eq. (5.51) gives:

$$y_2 = y_1 + a_2(x_2 - x_1) \quad \text{or} \quad a_2 = \frac{y_2 - y_1}{x_2 - x_1} \qquad (5.52)$$

Substituting the third point, $x = x_3$ and $f(x_3) = y_3$ (as well as $a_1 = y_1$ and $a_2 = \frac{y_2 - y_1}{x_2 - x_1}$) in Eq. (5.51) gives:

$$y_3 = y_1 + \frac{y_2 - y_1}{x_2 - x_1}(x_3 - x_1) + a_3(x_3 - x_1)(x_3 - x_2) \qquad (5.53)$$

Equation (5.53) can be solved for a_3 and rearranged to give (after some algebra):

$$a_3 = \frac{\dfrac{y_3-y_2}{x_3-x_2} - \dfrac{y_2-y_1}{x_2-x_1}}{(x_3-x_1)} \tag{5.54}$$

The coefficients a_1, and a_2 are the same in the first-order and second-order polynomials. This means that if two points are given and a first-order Newton's polynomial is fit to pass through those points, and then a third point is added, the polynomial can be changed to be of second-order and pass through the three points by only determining the value of one additional coefficient.

Third-order Newton's polynomial

For four given points, (x_1,y_1), (x_2,y_2), (x_3,y_3) and (x_4,y_4), the third-order Newton's polynomial that passes through the four points has the form:

$$f(x) = y = a_1 + a_2(x-x_1) + a_3(x-x_1)(x-x_2) + a_4(x-x_1)(x-x_2)(x-x_3) \tag{5.55}$$

The formulas for the coefficients a_1, a_2, and a_3 are the same as for the second order polynomial. The formula for the coefficient a_4 can be obtained by substituting (x_4,y_4), in Eq. (5.55) and solving for a_4, which gives:

$$a_4 = \frac{\dfrac{\left(\dfrac{y_4-y_3}{x_4-x_3} - \dfrac{y_3-y_2}{x_3-x_2}\right)}{(x_4-x_2)} - \dfrac{\left(\dfrac{y_3-y_2}{x_3-x_2} - \dfrac{y_2-y_1}{x_2-x_1}\right)}{(x_3-x_1)}}{(x_4-x_1)} \tag{5.56}$$

A general form of Newton's polynomial and its coefficients

A careful examination of the equations for the coefficients a_2 (Eq. (5.52)), a_3, (Eq. (5.54)) and a_4, (Eq. (5.56)) shows that the expressions follow a certain pattern. The pattern can be clarified by defining so-called ***divided differences***.

For two points, (x_1,y_1), and (x_2,y_2), the first divided difference, written as $f[x_2,x_1]$, is defined as the slope of the line connecting the two points:

$$f[x_2,x_1] = \frac{y_2-y_1}{x_2-x_1} = a_2 \tag{5.57}$$

The first divided difference is equal to the coefficient a_2.

For three points (x_1,y_1), (x_2,y_2), and (x_3,y_3) the second divided difference, written as $f[x_3,x_2,x_1]$, is defined as the difference between the first divided differences of points (x_3,y_3), and (x_2,y_2), and points (x_2,y_2), and (x_1,y_1) divided by (x_3-x_1):

$$f[x_3, x_2, x_1] = \frac{f[x_3, x_2] - f[x_2, x_1]}{x_3 - x_1} = \frac{\dfrac{y_3 - y_2}{x_3 - x_2} - \dfrac{y_2 - y_1}{x_2 - x_1}}{(x_3 - x_1)} = a_3 \quad (5.58)$$

The second divided difference is thus equal to the coefficient a_3.

For four points (x_1, y_1), (x_2, y_2), (x_3, y_3), and (x_4, y_4) the third divided difference, written as $f[x_4, x_3, x_2, x_1]$, is defined as the difference between the second divided differences of points (x_2, y_2), (x_3, y_3) and (x_4, y_4), and points (x_1, y_1), (x_2, y_2), and (x_3, y_3) divided by $(x_4 - x_1)$:

$$
\begin{aligned}
f[x_4, x_3, x_2, x_1] &= \frac{f[x_4, x_3, x_2] - f[x_3, x_2, x_1]}{x_4 - x_1} \\[2mm]
&= \frac{\dfrac{f[x_4, x_3] - f[x_3, x_2]}{x_4 - x_2} - \dfrac{f[x_3, x_2] - f[x_2, x_1]}{x_3 - x_1}}{(x_4 - x_1)} \\[2mm]
&= \frac{\dfrac{\left(\dfrac{y_4 - y_3}{x_4 - x_3} - \dfrac{y_3 - y_2}{x_3 - x_2}\right)}{(x_4 - x_2)} - \dfrac{\dfrac{y_3 - y_2}{x_3 - x_2} - \dfrac{y_2 - y_1}{x_2 - x_1}}{(x_3 - x_1)}}{(x_4 - x_1)} = a_4
\end{aligned}
\quad (5.59)
$$

The third divided difference is thus equal to the coefficient a_4.

The next (fourth) divided difference (when five data points are given) is:

$$f[x_5, x_4, x_3, x_2, x_1] = \frac{f[x_5, x_4, x_3, x_2] - f[x_4, x_3, x_2, x_1]}{x_5 - x_1} = a_5 \quad (5.60)$$

If more data points are given, the procedure for calculating higher differences continues in the same manner. In general, when n data points are given, the procedure starts by calculating $(n-1)$ first divided differences. Then, $(n-2)$ second divided differences are calculated from the first divided differences. This is followed by calculating $(n-3)$ third divided differences from the second divided differences. The process ends when one nth divided difference is calculated from two $(n-1)$ divided differences to give the coefficient a_n.

The procedure for finding the coefficients by using divided differences can be followed in a divided difference table. Such a table for the case of five data points is shown in Fig. 5-16.

In general terms, for n given data points, (x_1, y_1), (x_2, y_2), ..., (x_n, y_n), the first divided differences between two points (x_i, y_i), and (x_j, y_j) are given by:

$$f[x_j, x_i] = \frac{y_j - y_i}{x_j - x_i} \quad (5.61)$$

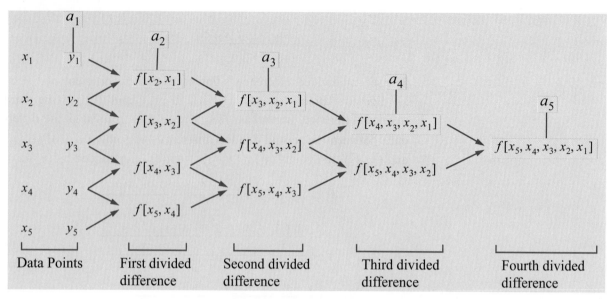

Figure 5-16: Table of divided differences for five data points.

The *k*th divided difference for second and higher divided differences up to the $(n-1)$ difference is given by:

$$f[x_k, x_{k-1}, ..., x_2, x_1] = \frac{f[x_k, x_{k-1}, ..., x_3, x_2] - f[x_{k-1}, x_{k-2}, ..., x_2, x_1]}{x_k - x_1} \qquad (5.62)$$

With these definitions, the $(n-1)$ order Newton's polynomial, Eq. (5.46) is given by:

$$f(x) = y = y_1 + f[x_2, x_1](x-x_1) + f[x_3, x_2, x_1](x-x_1)(x-x_2) + ... + f[x_n, x_{n-1}, ..., x_2, x_1](x-x_1)(x-x_2)...(x-x_{n-1})$$

$$\underbrace{\qquad}_{a_1} \underbrace{\qquad}_{a_2} \underbrace{\qquad\qquad}_{a_3} \underbrace{\qquad\qquad\qquad}_{a_n} \qquad (5.63)$$

Notes about Newton's polynomials

- The spacings between the data points do not have to be the same.

- For a given set of *n* points, once the coefficients a_1 through a_n are determined, they can be used for interpolation at any point between the data points.

After the coefficients a_1 through a_n are determined (for a given set of *n* points), additional data points can be added (they do not have to be in order), and only the additional coefficients have to be determined.

Example 5-5 shows application of Newton's interpolating polynomials.

Example 5-5: Newton's interpolating polynomial.

The set of the following five data points is given:

x	1	2	4	5	7
y	52	5	-5	-40	10

(a) Determine the fourth-order polynomial in Newton's form that passes through the points. Calculate the coefficients by using a divided difference table.

(b) Use the polynomial obtained in part (a) to determine the interpolated value for $x = 3$.

(c) Write a MATLAB user-defined function that interpolates using Newton's polynomial. The input to the function should be the coordinates of the given data points and the x coordinate of the point at which y is to be interpolated. The output from the function is the y value of the interpolated point.

SOLUTION

(a) Newton's polynomial for the given points has the form:

$$f(x) = y = a_1 + a_2(x-1) + a_3(x-1)(x-2) + a_4(x-1)(x-2)(x-4) + a_5(x-1)(x-2)(x-4)(x-5)$$

The coefficients can be determined by the following divided difference table:

With the coefficients determined, the polynomial is:

$$f(x) = y = 52 - 47(x-1) + 14(x-1)(x-2) - 6((x-1)(x-2)(x-4)) + 2(x-1)(x-2)(x-4)(x-5)$$

(b) The interpolated value for $x = 3$ is obtained by substituting for x in the polynomial:

$$f(3) = y = 52 - 47(3-1) + 14(3-1)(3-2) - 6((3-1)(3-2)(3-4)) + 2(3-1)(3-2)(3-4)(3-5) = 6$$

(c) The MATLAB user-defined function for Newton's interpolation is named `Yint=NewtonsINT(x,y,Xint)`. `x` and `y` are vectors with the coordinates of the given data points, and `Xint` is the coordinate of the point at which y is to be interpolated.

- The program starts by calculating the first divided differences, which are then used for calculating the higher divided differences. The values are assigned to a table named `divDIF`.
- The coefficients of the polynomial (first row of the table) are then assigned to a vector named `a`.

- The known polynomial is used for interpolation.

Program 5-5: User-defined function. Interpolation using Newton's polynomial.

```
function Yint = NewtonsINT(x,y,Xint)
% NewtonsINT fits a Newtons polynomial to a set of given points and
% uses the polynomial to determines the interpolated value of a point.
% Input variables:
% x  A vector with the x coordinates of the given points.
% y  A vector with the y coordinates of the given points.
% Xint  The x coordinate of the point to be interpolated.
% Output variable:
% Yint  The interpolated value of Xint.
```

```
n = length(x);
```
> The length of the vector x gives the number of coefficients (and terms) of the polynomial.

```
a(1) = y(1);
```
> The first coefficient a_1.

```
for i = 1:n - 1
    divDIF(i,1) = (y(i + 1) - y(i))/(x(i + 1) - x(i));
end
```
> Calculate the finite divided differences. They are assigned to the first column of `divDIF`.

```
for j = 2:n - 1
    for i = 1:n - j
        divDIF(i,j) = (divDIF(i + 1,j - 1) - divDIF(i,j - 1))/(x(j + i) - x(i));
    end
end
```
> Calculate the second and higher divided differences (up to an order of $(n-1)$). The values are assigned in columns to `divDIF`.

```
for j = 2:n
    a(j) = divDIF(1,j - 1);
end
```
> Assign the coefficients a_2 through a_n. to vector a.

```
Yint = a(1);
xn = 1;
for k = 2:n
    xn = xn*(Xint - x(k - 1));
    Yint = Yint + a(k)*xn;
end
```
> Calculate the interpolated value of `Xint`. The first term in the polynomial is a_1. The following terms are added by using a loop.

The `NewtonsINT(x,y,Xint)` Function is then used in the Command Window for calculating the interpolated value of $x = 3$.

```
>> x = [1 2 4 5 7];
>> y = [52 5 -5 -40 10];
>> Yinterpolated = NewtonsINT(x,y,3)
Yinterpolated =
    6
```

5.6 PIECEWISE (SPLINE) INTERPOLATION

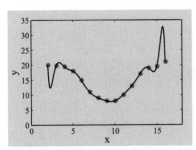

Figure 5-17: Fitting 16 data points with a 15th order polynomial.

When a set of n data points is given and a single polynomial is used for interpolation between the points, the polynomial gives the exact values at the points (passes through the points) and yields estimated (interpolated) values between the points. When the number of points is small such that the order of the polynomial is low, typically the interpolated values are reasonably accurate. However, as already mentioned in Section 5.4, large errors might occur when a high-order polynomial is used for interpolation involving a large number of points. This is shown in Fig. 5-17 where a polynomial of 15th order is used for interpolation with a set of 16 data points. It is clear from the figure that near the ends the polynomial deviates significantly from the trend of the data, and thus cannot be reliably used for interpolation.

When a large number of points is involved, a better interpolation can be obtained by using many low-order polynomials instead of a single high-order polynomial. Each low-order polynomial is valid in one interval between two or several points. Typically, all of the polynomials are of the same order, but the coefficients are different in each interval. When first-order polynomials are used, the data points are connected with straight lines. For second-order (quadratic), and third-order (cubic) polynomials, the points are connected by curves. Interpolation in this way is called *piecewise*, or *spline*, interpolation. The data points where the polynomials from two adjacent intervals meet are called *knots*. The name "spline" comes from a draftsman's spline, which is a thin flexible rod used to physically interpolate over discrete points marked by pegs.

The three types of spline interpolation are linear, quadratic, and cubic.

5.6.1 Linear Splines

With linear splines, interpolation is carried out by using a first-order polynomial (linear function) between the points (the points are connected with straight lines), as shown in Fig. 5-18. Using the Lagrange form, the equation of the straight line that connects the first two points is given by:

$$f_1(x) = \frac{(x - x_2)}{(x_1 - x_2)} y_1 + \frac{(x - x_1)}{(x_2 - x_1)} y_2 \qquad (5.64)$$

For n given points, there are $n - 1$ intervals. The interpolation in interval i, which is between points x_i and x_{i+1} ($x_i \leq x \leq x_{i+1}$), is done by using the equation of the straight line that connects point (x_i, y_i) with point (x_{i+1}, y_{i+1}):

$$f_i(x) = \frac{(x - x_{i+1})}{(x_i - x_{i+1})} y_i + \frac{(x - x_i)}{(x_{i+1} - x_i)} y_{i+1} \quad \text{for} \quad i = 1, 2, ..., n-1 \quad (5.65)$$

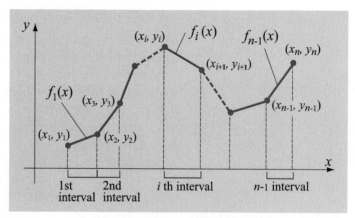

Figure 5-18: Linear splines.

It is obvious that linear splines give continuous interpolation since the two adjacent polynomials have the same value at a common knot. There is, however, a discontinuity in the slope of the linear splines at the knots.

Interpolation with linear splines is easy to calculate and program, and gives good results when the data points are closely spaced. Example 5-6 shows a numerical application of linear splines by hand and by using a user-defined MATLB function.

Example 5-6: Linear splines.

The set of the following four data points is given:

x	8	11	15	18
y	5	9	10	8

(*a*) Determine the linear splines that fit the data.

(*b*) Determine the interpolated value for $x = 12.7$.

(*c*) Write a MATLAB user-defined function for interpolation with linear splines. The inputs to the function are the coordinates of the given data points and the x coordinate of the point at which y is to be interpolated. The output from the function is the interpolated y value at the given point. Use the function for determining the interpolated value of y for $x = 12.7$.

SOLUTION

(*a*)   There are four points and thus three splines. Using Eq. (5.65) the equations of the splines are:

$$f_1(x) = \frac{(x - x_2)}{(x_1 - x_2)}y_1 + \frac{(x - x_1)}{(x_2 - x_1)}y_2 = \frac{(x - 11)}{(8 - 11)}5 + \frac{(x - 8)}{(11 - 8)}9 = \frac{5}{-3}(x - 11) + \frac{9}{2}(x - 8) \quad \text{for} \quad 8 \leq x \leq 11$$

$$f_2(x) = \frac{(x - x_3)}{(x_2 - x_3)}y_2 + \frac{(x - x_2)}{(x_3 - x_2)}y_3 = \frac{(x - 15)}{(11 - 15)}9 + \frac{(x - 11)}{(15 - 11)}10 = \frac{9}{-4}(x - 15) + \frac{10}{4}(x - 11) \quad \text{for} \quad 11 \leq x \leq 15$$

$$f_3(x) = \frac{(x - x_4)}{(x_3 - x_4)}y_3 + \frac{(x - x_3)}{(x_4 - x_3)}y_4 = \frac{(x - 18)}{(15 - 18)}10 + \frac{(x - 15)}{(18 - 15)}8 = \frac{10}{-3}(x - 18) + \frac{8}{3}(x - 15) \quad \text{for} \quad 15 \leq x \leq 18$$

(*b*) The interpolated value of y for $x = 12.7$ is obtained by substituting the value of x in the equation for $f_2(x)$ above:

$$f_2(x) = \frac{9}{-4}(12.7 - 15) + \frac{10}{4}(12.7 - 11) = 9.425$$

(c) The MATLAB user-defined function for linear spline interpolation is named `Yint=LinearSpline(x,y,Xint)`. `x` and `y` are vectors with the coordinates of the given data points, and `Xint` is the coordinate of the point at which y is to be interpolated.

> **Program 5-6: User-defined function. Linear splines.**

```
function Yint = LinearSpline(x, y, Xint)
% LinearSpline calculates interpolation using linear splines.
% Input variables:
% x   A vector with the coordinates x of the data points.
% y   A vector with the coordinates y of the data points.
% Xint   The x coordinate of the interpolated point.
% Output variable:
% Yint   The y value of the interpolated point.

n = length(x);                                    ⎤ The length of the vector x gives the number of terms in the data.
for i = 2:n
   if Xint < x(i)
      break                                        ⎬ Find the interval that includes Xint.
   end
end                                               ⎦
Yint = (Xint - x(i))*y(i - 1)/(x(i -1) - x(i)) + (Xint - x(i - 1))*y(i)/(x( i ) - x(i - 1));   Calculate Yint with Eq. (5.65).
```

The `LinearSpline(x,y,Xint)` function is then used in the Command Window for calculating the interpolated value of $x = 12.7$.

```
>> x = [8 11 15 18];
>> y = [5 9 10 8];
>> Yint = LinearSpline(x,y,12.7)
Yint =
   9.4250
```

5.6.2 Quadratic Splines

In quadratic splines, second-order polynomials are used for interpolation in the intervals between the points (Fig. 5-19). For n given points there are $n - 1$ intervals, and using the standard form, the equation of the polynomial in the ith interval, between points x_i and x_{i+1}, is given by:

$$f_i(x) = a_i x^2 + b_i x + c_i \quad \text{for} \quad i = 1, 2, ..., n - 1 \tag{5.66}$$

Overall, there are $n - 1$ equations, and since each equation has three coefficients, a total of $3(n - 1) = 3n - 3$ coefficients have to be determined. The coefficients are determined by applying the following conditions:

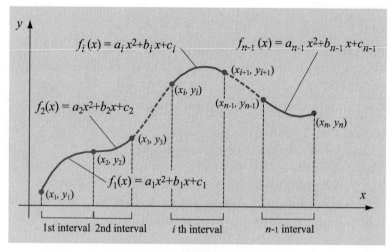

Figure 5-19: Quadratic splines.

1. Each polynomial $f_i(x)$ must pass through the endpoints of the interval, (x_i, y_i) and (x_{i+1}, y_{i+1}), which means that $f_i(x_i) = y_i$ and $f_i(x_{i+1}) = y_{i+1}$:

$$a_i x_i^2 + b_i x_i + c_i = y_i \quad \text{for} \quad i = 1, 2, ..., n-1 \tag{5.67}$$

$$a_i x_{i+1}^2 + b_i x_{i+1} + c_i = y_{i+1} \quad \text{for} \quad i = 1, 2, ..., n-1 \tag{5.68}$$

Since there are $n-1$ intervals, these conditions give $2(n-1) = 2n-2$ equations.

2. At the interior knots, the slopes (first derivative) of the polynomials from adjacent intervals are equal. This means that as the curve that passes through an interior knot switches from one polynomial to another, the slope is continuous. In general, the first derivative of the ith polynomial is:

$$f'(x) = \frac{df}{dx} = 2a_i x + b_i \tag{5.69}$$

For n points, the first interior point is $i = 2$, and the last is $i = n-1$. Equating the successive first derivatives at all of the interior points gives:

$$2a_{i-1}x_i + b_{i-1} = 2a_i x_i + b_i \quad \text{for} \quad i = 2, 3, ..., n-1 \tag{5.70}$$

Since there are $n-2$ interior points, this condition gives $n-2$ equations.

Together, the two conditions give $3n-4$ equations. However, the $n-1$ polynomials have $3n-3$ coefficients so that one more equation (condition) is needed in order to solve for all of the coefficients. An additional condition that is commonly applied is that the second derivative at either the first point or the last point is zero. Consider the first choice:

3. The second derivative at the first point, (x_1, y_1), is zero. The polynomial in the first interval (between the first and the second points) is:

$$f_1(x) = a_1 x^2 + b_1 x + c_1 \qquad (5.71)$$

The second derivative of the polynomial is $f_1''(x) = 2a_1$, and equating it to zero means that $a_1 = 0$. This condition actually means that a straight line connects the first two points (the slope is constant).

A note on quadratic and cubic splines

Quadratic splines have a continuous first derivative at the interior points (knots), and for n given points they require the solution of a linear system of $3n - 4$ equations for the coefficients of the polynomials. As is shown in the next section, cubic splines have continuous first and second derivatives at the interior points, and can be written in a form that requires the solution of a linear system of only $n - 2$ equations for the coefficients.

Example 5-7 shows an application of quadratic splines for interpolation of a given set of five points.

Example 5-7: Quadratic splines.

The set of the following five data points is given:

x	8	11	15	18	22
y	5	9	10	8	7

(a) Determine the quadratic splines that fit the data.
(b) Determine the interpolated value of y for $x = 12.7$.
(c) Make a plot of the data points and the interpolating polynomials.

SOLUTION

(a) There are five points ($n = 5$) and thus four splines ($i = 1, ..., 4$). The quadratic equation for the ith spline is:

$$f_i(x) = a_i x^2 + b_i x + c_i$$

There are four polynomials, and since each polynomial has three coefficients, a total of 12 coefficients have to be determined. The coefficients are $a_1, b_1, c_1, a_2, b_2, c_2, a_3, b_3, c_3, a_4, b_4$, and c_4. The coefficient a_1 is set equal to zero (see condition 3). The other 11 coefficients are determined from a linear system of 11 equations.

Eight equations are obtained from the condition that the polynomial in each interval passes through the endpoints, Eqs. (5.67) and (5.68):

$i = 1$ $f_1(x) = a_1 x_1^2 + b_1 x_1 + c_1 = b_1 8 + c_1 = 5$

 $f_1(x) = a_1 x_2^2 + b_1 x_2 + c_1 = b_1 11 + c_1 = 9$

$i = 2$ $f_2(x) = a_2 x_2^2 + b_2 x_2 + c_2 = a_2 11^2 + b_2 11 + c_2 = 9$

 $f_2(x) = a_2 x_3^2 + b_2 x_3 + c_2 = a_2 15^2 + b_2 15 + c_2 = 10$

$i = 3$ $\quad f_3(x) = a_3 x_3^2 + b_3 x_3 + c_3 = a_3\,15^2 + b_3\,15 + c_3 = 10$

$\qquad\quad f_3(x) = a_3 x_4^2 + b_3 x_4 + c_3 = a_3\,18^2 + b_3\,18 + c_3 = 8$

$i = 4$ $\quad f_4(x) = a_4 x_4^2 + b_4 x_4 + c_4 = a_4\,18^2 + b_4\,18 + c_4 = 8$

$\qquad\quad f_4(x) = a_4 x_5^2 + b_4 x_5 + c_4 = a_4\,22^2 + b_4\,22 + c_4 = 7$

Three equations are obtained from the condition that at the interior knots the slopes (first derivative) of the polynomials from adjacent intervals are equal, Eq. (5.70).

$i = 2$ $\quad 2a_1 x_2 + b_1 = 2a_2 x_2 + b_2 \longrightarrow b_1 = 2a_2 11 + b_2$ $\quad$ or: $\quad b_1 - 2a_2 11 - b_2 = 0$

$i = 3$ $\quad 2a_2 x_3 + b_2 = 2a_3 x_3 + b_3 \longrightarrow 2a_2 15 + b_2 = 2a_3 15 + b_3$ $\quad$ or: $\quad 2a_2 15 + b_2 - 2a_3 15 - b_3 = 0$

$i = 4$ $\quad 2a_3 x_4 + b_3 = 2a_4 x_3 + b_4 \longrightarrow 2a_3 18 + b_3 = 2a_4 18 + b_4$ $\quad$ or: $\quad 2a_3 18 + b_3 - 2a_4 18 - b_4 = 0$

The system of 11 linear equations can be written in a matrix form:

$$
\begin{bmatrix}
8 & 1 & 0 & 0 & 0 & 0 & 0 & 0 & 0 & 0 & 0 \\
11 & 1 & 0 & 0 & 0 & 0 & 0 & 0 & 0 & 0 & 0 \\
0 & 0 & 11^2 & 11 & 1 & 0 & 0 & 0 & 0 & 0 & 0 \\
0 & 0 & 15^2 & 15 & 1 & 0 & 0 & 0 & 0 & 0 & 0 \\
0 & 0 & 0 & 0 & 0 & 15^2 & 15 & 1 & 0 & 0 & 0 \\
0 & 0 & 0 & 0 & 0 & 18^2 & 18 & 1 & 0 & 0 & 0 \\
0 & 0 & 0 & 0 & 0 & 0 & 0 & 18^2 & 18 & 1 \\
0 & 0 & 0 & 0 & 0 & 0 & 0 & 22^2 & 22 & 1 \\
1 & 0 & -22 & -1 & 0 & 0 & 0 & 0 & 0 & 0 & 0 \\
0 & 0 & 30 & 1 & 0 & -30 & -1 & 0 & 0 & 0 & 0 \\
0 & 0 & 0 & 0 & 0 & 36 & 1 & 0 & -36 & -1 & 0
\end{bmatrix}
\begin{bmatrix}
b_1 \\ c_1 \\ a_2 \\ b_2 \\ c_2 \\ a_3 \\ b_3 \\ c_3 \\ a_4 \\ b_4 \\ c_4
\end{bmatrix}
=
\begin{bmatrix}
5 \\ 9 \\ 9 \\ 10 \\ 10 \\ 8 \\ 8 \\ 7 \\ 0 \\ 0 \\ 0
\end{bmatrix}
\qquad (5.72)
$$

The system in Eq. (5.72) is solved with MATLAB:

```
>> A = [8 1 0 0 0 0 0 0 0 0 0; 11 1 0 0 0 0 0 0 0 0 0; 0 0 11^2 11 1 0 0 0 0 0 0
0 0 15^2 15 1 0 0 0 0 0; 0 0 0 0 0 15^2 15 1 0 0 0; 0 0 0 0 0 18^2 18 1 0 0 0
0 0 0 0 0 0 0 18^2 18 1; 0 0 0 0 0 0 0 22^2 22 1; 1 0 -22 -1 0 0 0 0 0 0 0
0 0 30 1 0 -30 -1 0 0 0 0; 0 0 0 0 0 36 1 0 -36 -1 0];
>> B = [5; 9; 9; 10; 10; 8; 8; 7; 0; 0; 0];
>> coefficients = (A\B)'
coefficients =
    1.3333  -5.6667  -0.2708   7.2917  -38.4375   0.0556  -2.5000  35.0000   0.0625  -2.7500  37.2500
```

$\qquad b_1 \qquad c_1 \qquad a_2 \qquad b_2 \qquad c_2 \qquad a_3 \qquad b_3 \qquad c_3 \qquad a_4 \qquad b_4 \qquad c_4$

With the coefficients known, the polynomials are:

$f_1(x) = 1.333x - 5.6667$ for $8 \le x \le 11$, $\qquad f_2(x) = (-0.2708)x^2 + 7.2917x - 38.4375$ for $11 \le x \le 15$

$f_3(x) = 0.0556x^2 - 2.5x + 35$ for $15 \le x \le 18$, $\qquad f_4(x) = 0.0625x^2 - 2.75x + 37.25$ for $18 \le x \le 22$

(b) The interpolated value of y for $x = 12.7$ is calculated by substituting the value of x in $f_2(x)$:

$f_2(12.7) = (-0.2708)12.7^2 + 7.2917 \cdot 12.7 - 38.4375 = 10.4898$

(*c*) The plot on the right shows the data points and the polynomials. The plot clearly shows that the first spline is a straight line (constant slope).

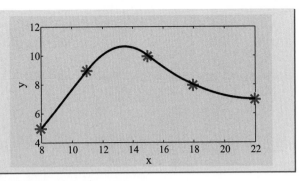

5.6.3 Cubic Splines

In cubic splines, third-order polynomials are used for interpolation in the intervals between the points. For n given points there are $n-1$ intervals, and since each third-order polynomial has four coefficients the determination of all of the coefficients may require a large number of calculations. As was explained earlier in this chapter, polynomials can be written in different forms (standard, Lagrange, Newton) and theoretically any of these forms can be used for cubic splines. Practically, however, the amount of calculations that have to be executed for determining all the coefficients varies greatly with the form of the polynomial that is used. The presentation that follows shows two derivations of cubic splines. The first uses the standard form of the polynomials, and the second uses a variation of the Lagrange form. The derivation with the standard form is easier to follow, understand, and use (it is similar to the derivation of the quadratic splines), but it requires the solution of a system of $4n-4$ linear equations. The derivation that is based on the Lagrange form is more sophisticated, but requires the solution of a system of only $n-2$ linear equations.

Cubic splines with standard form polynomials

For n given points, as shown in Fig. 5-20, there are $n-1$ intervals, and using the standard form, the equation of the polynomial in the ith interval, between points x_i and x_{i+1} is given by:

$$f_i(x) = a_i x^3 + b_i x^2 + c_i x + d_i \qquad (5.73)$$

Overall, there are $n-1$ equations, and since each equation has four coefficients, a total of $4(n-1) = 4n-4$ coefficients have to be determined. The coefficients are found by applying the following conditions:

1. Each polynomial $f_i(x)$ must pass through the endpoints of the interval, (x_i, y_i) and (x_{i+1}, y_{i+1}), which means that $f_i(x_i) = y_i$ and $f_i(x_{i+1}) = y_{i+1}$:

$$a_i x_i^3 + b_i x_i^2 + c_i x_i + d_i = y_i \quad \text{for} \quad i = 1, 2, ..., n-1 \qquad (5.74)$$

$$a_i x_{i+1}^3 + b_i x_{i+1}^2 + c_i x_{i+1} + d_i = y_{i+1} \quad \text{for} \quad i = 1, 2, ..., n-1 \qquad (5.75)$$

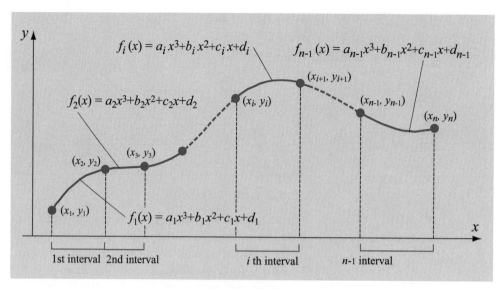

Figure 5-20: Cubic splines.

Since there are $n-1$ intervals, this condition gives $2(n-1) = 2n-2$ equations.

2. At the interior knots, the slopes (first derivatives) of the polynomials from the adjacent intervals are equal. This means that as the curve that passes through an interior knot switches from one polynomial to another, the slope must be continuous. The first derivative of the ith polynomial is:

$$f_i{}'(x) = \frac{df_i}{dx} = 3a_ix^2 + 2b_ix + c_i \qquad (5.76)$$

For n points the first interior point is $i = 2$, and the last is $i = n-1$. Equating the first derivatives at each interior point gives:

$$3a_{i-1}x_i^2 + 2b_{i-1}x_i + c_{i-1} = 3a_ix_i^2 + 2b_ix_i + c_i \text{ for } i = 2, 3, ..., n-1 \;(5.77)$$

Since there are $n-2$ interior points, this condition gives $n-2$ additional equations.

3. At the interior knots, the second derivatives of the polynomials from adjacent intervals must be equal. This means that as the curve that passes through an interior knot switches from one polynomial to another, the rate of change of the slope (curvature) must be continuous. The second derivative of the polynomial in the ith interval is:

$$f_i{}''(x) = \frac{d^2f_i}{dx^2} = 6a_ix + 2b_i \qquad (5.78)$$

For n points, the first interior point is $i = 2$, and the last is $i = n-1$. Equating the second derivatives at each interior point gives:

$$6a_{i-1}x_i + 2b_{i-1} = 6a_ix_i + 2b_i \text{ for } i = 2, 3, ..., n-1 \qquad (5.79)$$

Since there are $n-2$ interior points, this condition gives $n-2$ additional equations.

Together, the three conditions give $4n-6$ equations. However, the $n-1$ polynomials have $4n-4$ coefficients, and two more equations (conditions) are needed in order to solve for the coefficients. The additional conditions are usually taken to be that the second derivative is zero at the first point and at the last point. This gives two additional equations:

$$6a_1x_1 + 2b_1 = 0 \quad \text{and} \quad 6a_{n-1}x_n + 2b_{n-1} = 0 \qquad (5.80)$$

Cubic splines with the second derivatives at the endpoints set equal to zero are called **natural cubic splines**. Applying all the conditions gives a system of $4n-4$ linear equations for the $4n-4$ coefficients. The system can be solved using one of the methods from Chapter 4.

Cubic splines based on Lagrange form polynomials

The derivation of cubic splines using the Lagrange form starts with the second derivative of the polynomial. Figure 5-21 shows spline interpolation with cubic polynomials in (*a*), the first derivatives of the polynomials in (*b*), and their second derivatives in (*c*). The figure shows an *i*th interval with the adjacent $i-1$ and $i+1$ intervals. The second derivative of a third-order polynomial is a linear function. This means that within each spline the second derivative is a linear function of x (see Fig. 5-21*c*). For the *i*th interval, this linear function can be written in the Lagrange form:

$$f_i''(x) = \frac{x - x_{i+1}}{x_i - x_{i+1}} f_i''(x_i) + \frac{x - x_i}{x_{i+1} - x_i} f_i''(x_{i+1}) \qquad (5.81)$$

where the values of the second derivative of the third-ordered polynomial at the endpoints (knots) of the *i*th interval are $f_i''(x_i)$ and $f_i''(x_{i+1})$. The third order polynomial in interval i can be determined by integrating Eq. (5.81) twice. The resulting expression contains two constants of integration. These two constants can be determined from the condition that the values of the polynomial at the knots are known:

$$f_i(x_i) = y_i \quad \text{and} \quad f_i(x_{i+1}) = y_{i+1}$$

Once the constants of integration are determined, the equation of the third-order polynomial in interval i is given by:

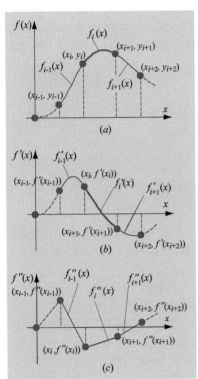

Figure 5-21: Third-order polynomial (*a*) and its first (*b*) and second (*c*) derivatives.

$$f_i(x) = \frac{f_i''(x_i)}{6(x_{i+1} - x_i)}(x_{i+1} - x)^3 + \frac{f_i''(x_{i+1})}{6(x_{i+1} - x_i)}(x - x_i)^3$$

$$+ \left[\frac{y_i}{x_{i+1} - x_i} - \frac{f_i''(x_i)(x_{i+1} - x_i)}{6} \right](x_{i+1} - x) \qquad (5.82)$$

$$+ \left[\frac{y_{i+1}}{x_{i+1} - x_i} - \frac{f_i''(x_{i+1})(x_{i+1} - x_i)}{6} \right](x - x_i)$$

$$\text{for } x_i \leq x \leq x_{i+1} \qquad \text{and } i = 1, 2, ..., n-1$$

For each interval Eq. (5.82) contains two unknowns, $f_i''(x_i)$ and $f_i''(x_{i+1})$. These are the values of the second derivative at the endpoints of the interval. Equations that relate the values of the second derivatives at the $n-2$ interior points can be derived by requiring continuity of the first derivatives of polynomials from adjacent intervals at the interior knots:

$$f_i'(x_{i+1}) = f_{i+1}'(x_{i+1}) \quad \text{for} \quad i = 1, 2, ..., n-2 \qquad (5.83)$$

This condition is applied by using Eq. (5.82) to write the expressions for $f_i(x)$ and $f_{i+1}(x)$, differentiating the expressions, and substituting the derivatives in Eq. (5.83). These operations give (after some algebra) the following equations:

$$(x_{i+1} - x_i)f''(x_i) + 2(x_{i+2} - x_i)f''(x_{i+1}) + (x_{i+2} - x_{i+1})f''(x_{i+2})$$

$$= 6 \left[\frac{y_{i+2} - y_{i+1}}{x_{i+2} - x_{i+1}} - \frac{y_{i+1} - y_i}{x_{i+1} - x_i} \right] \qquad (5.84)$$

$$\text{for} \quad i = 1, 2, ..., n-2$$

This is a system of $n-2$ linear equations that contains n unknowns.

How is the polynomial in each interval determined?

- For n given data points there are $n-1$ intervals. The cubic polynomial in each interval is given by Eq. (5.82), (total of $n-1$ polynomials).

- The $n-1$ polynomials contain n coefficients $f''(x_1)$, through $f''(x_n)$. These are the values of the second derivatives of the polynomials at the data points. The second derivative at the interior knots is taken to be continuous. This means that at the interior knots, the second derivatives of the polynomials from adjacent intervals are equal. Consequently, for n data points there are n values (the value of the second derivative at each point) that have to be determined.

- Equations (5.84) give a system of $n-2$ linear equations for the n unknown coefficients $f''(x_1)$ through $f''(x_n)$. To solve for the coefficients, two additional relations are required. Most commonly, the second derivative at the endpoints of the data (the first and last points) is set to zero (natural cubic splines):

$$f''(x_1) = 0 \quad \text{and} \quad f''(x_n) = 0 \tag{5.85}$$

With these conditions, the linear system of Eqs. (5.84) can be solved, and the coefficients can be substituted in the equations for the polynomials (Eqs. (5.82)). Cubic splines with the second derivatives at the endpoints set equal to zero are called ***natural cubic splines***.

Simplified form of the equations

The form of Eqs. (5.82) and (5.84) can be simplified by defining h_i as the length of the ith interval (the intervals do not have to be of the same length):

$$h_i = x_{i+1} - x_i \tag{5.86}$$

and defining a_i as the second derivative of the polynomial at point x_i:

$$a_i = f''(x_i) \tag{5.87}$$

With these definitions, the equation of the polynomial in the ith interval is:

$$
\begin{aligned}
f_i(x) &= \frac{a_i}{6h_i}(x_{i+1} - x)^3 + \frac{a_{i+1}}{6h_i}(x - x_i)^3 \\
&+ \left[\frac{y_i}{h_i} - \frac{a_i h_i}{6}\right](x_{i+1} - x) + \left[\frac{y_{i+1}}{h_i} - \frac{a_{i+1} h_i}{6}\right](x - x_i) \\
&\text{for } x_i \le x \le x_{i+1} \quad \text{and} \quad i = 1, 2, ..., n-1
\end{aligned}
\tag{5.88}
$$

and the system of linear equations that has to be solved for the a_is is given by:

$$
h_i a_i + 2(h_i + h_{i+1})a_{i+1} + h_{i+1}a_{i+2} = 6\left[\frac{y_{i+2} - y_{i+1}}{h_{i+1}} - \frac{y_{i+1} - y_i}{h_i}\right] \tag{5.89}
$$
$$\text{for} \quad i = 1, 2, ..., n-2$$

To carry out cubic spline interpolation, Eq. (5.89) is used for writing a system of $n-2$ equations with $n-2$ unknowns, a_2 through a_{n-1}. (Remember that with natural cubic splines a_1 and a_n are equal to zero.) Equations (5.89) result in a tridiagonal system of equations that can be solved efficiently using the Thomas algorithm (see Chapter 4). Once the system of equations is solved, the cubic polynomials for each interval can be written using Eq. (5.88). Example 5-8 shows a solution of the problem that is solved in Example 5-7 with cubic splines.

Note on using cubic splines in MATLAB

Cubic splines are available as a built-in function within MATLAB. However, the option labeled `cubic` (also called `pchip`) is ***not*** the method of cubic splines. Rather, the option labeled `spline` is the appropriate option to use for cubic splines. Even when using the option `spline`, the user is cautioned that it is not the natural splines described in this chapter. The cubic splines available in MATLAB under the option `'spline'` use the not-a-knot conditions at the end-points, that is, at the first and last data points. The not-a-knot condition refers to the fact that the third derivatives are continuous at the second point and at the second to last point.

Example 5-8: Cubic splines.

The set of the following five data points is given:

x	8	11	15	18	22
y	5	9	10	8	7

(a) Determine the natural cubic splines that fit the data.
(b) Determine the interpolated value of y for $x = 12.7$.
(c) Plot of the data points and the interpolating polynomials.

SOLUTION

(a)   There are five points ($n = 5$), and thus four splines ($i = 1, ..., 4$). The cubic equation in the ith spline is:

$$f_i(x) = \frac{a_i}{6h_i}(x_{i+1}-x)^3 + \frac{a_{i+1}}{6h_i}(x-x_i)^3 + \left[\frac{y_i}{h_i} - \frac{a_i h_i}{6}\right](x_{i+1}-x) + \left[\frac{y_{i+1}}{h_i} - \frac{a_{i+1} h_i}{6}\right](x-x_i) \quad \text{for} \quad i=1,...4$$

where $h_i = x_{i+1} - x_i$. The four equations contain five unknown coefficients a_1, a_2, a_3, a_4, and a_5. For natural cubic splines the coefficients a_1 and a_5 are set to be equal to zero. The other three coefficients are determined from a linear system of three equations given by Eq. (5.89).

The values of the h_is are: $h_1 = x_2 - x_1 = 11 - 8 = 3$, $h_2 = x_3 - x_2 = 15 - 11 = 4$,

$$h_3 = x_4 - x_3 = 18 - 15 = 3, \quad h_4 = x_5 - x_4 = 22 - 18 = 4$$

$i = 1$ $h_1 a_1 + 2(h_1 + h_2)a_2 + h_2 a_3 = 6\left[\frac{y_3 - y_2}{h_2} - \frac{y_2 - y_1}{h_1}\right]$

$$3 \cdot 0 + 2(3 + 4)a_2 + 4a_3 = 6\left[\frac{10 - 9}{4} - \frac{9 - 5}{3}\right] \longrightarrow 14a_2 + 4a_3 = -6.5$$

$i = 2$ $h_2 a_2 + 2(h_2 + h_3)a_3 + h_3 a_4 = 6\left[\frac{y_4 - y_3}{h_3} - \frac{y_3 - y_2}{h_2}\right]$

$$4a_2 + 2(4 + 3)a_3 + 3a_4 = 6\left[\frac{8 - 10}{3} - \frac{10 - 9}{4}\right] \longrightarrow 4a_2 + 14a_3 + 3a_4 = -5.5$$

$i = 3$ $h_3 a_3 + 2(h_3 + h_4)a_4 + h_4 a_5 = 6\left[\frac{y_5 - y_4}{h_4} - \frac{y_4 - y_3}{h_3}\right]$

$$3a_3 + 2(3 + 4)a_4 + 4 \cdot 0 = 6\left[\frac{y_5 - y_4}{h_4} - \frac{y_4 - y_3}{h_3}\right] \longrightarrow 3a_3 + 14a_4 = 2.5$$

The system of three linear equations can be written in a matrix form:

$$\begin{bmatrix} 14 & 4 & 0 \\ 4 & 14 & 3 \\ 0 & 3 & 14 \end{bmatrix} \begin{bmatrix} a_2 \\ a_3 \\ a_4 \end{bmatrix} = \begin{bmatrix} -6.5 \\ -5.5 \\ 2.5 \end{bmatrix} \qquad (5.90)$$

The system in Eq. (5.90) is solved with MATLAB:

```
>> A = [14 4 0; 4 14 3; 0 3 14];
>> B = [-6.5; -5.5; 2.5];
```

```
>> coefficients = (A\B)'
coefficients =
  -0.3665  -0.3421   0.2519
```

$$a_2 \qquad a_3 \qquad a_4$$

With the coefficients known, the polynomials are (from Eq. (5.88)):

$i = 1 \quad f_1(x) = \dfrac{a_1}{6h_1}(x_2 - x)^3 + \dfrac{a_2}{6h_1}(x - x_1)^3 + \left[\dfrac{y_1}{h_1} - \dfrac{a_1 h_1}{6}\right](x_2 - x) + \left[\dfrac{y_2}{h_1} - \dfrac{a_2 h_1}{6}\right](x - x_1)$

$\qquad f_1(x) = \dfrac{0}{6 \cdot 3}(11 - x)^3 + \dfrac{-0.3665}{6 \cdot 3}(x - 8)^3 + \left[\dfrac{5}{3} - \dfrac{0 \cdot 3}{6}\right](11 - x) + \left[\dfrac{9}{3} - \dfrac{-0.3665 \cdot 3}{6}\right](x - 8)$

$\qquad f_1(x) = (-0.02036)(x - 8)^3 + 1.667(11 - x) + 3.183(x - 8) \quad$ for $\ 8 \le x \le 11$

$i = 2 \quad f_2(x) = \dfrac{a_2}{6h_2}(x_3 - x)^3 + \dfrac{a_3}{6h_2}(x - x_2)^3 + \left[\dfrac{y_2}{h_2} - \dfrac{a_2 h_2}{6}\right](x_3 - x) + \left[\dfrac{y_3}{h_2} - \dfrac{a_3 h_2}{6}\right](x - x_2)$

$\qquad f_2(x) = \dfrac{-0.3665}{6 \cdot 4}(15 - x)^3 + \dfrac{-0.3421}{6 \cdot 4}(x - 11)^3 + \left[\dfrac{9}{4} - \dfrac{-0.3665 \cdot 4}{6}\right](15 - x) + \left[\dfrac{10}{4} - \dfrac{-0.3421 \cdot 4}{6}\right](x - 11)$

$\qquad f_2(x) = (-0.01527)(15 - x)^3 + (-0.01427)(x - 11)^3 + 2.494(15 - x) + 2.728(x - 11) \quad$ for $\ 11 \le x \le 15$

$i = 3 \quad f_3(x) = \dfrac{a_3}{6h_3}(x_4 - x)^3 + \dfrac{a_4}{6h_3}(x - x_3)^3 + \left[\dfrac{y_3}{h_3} - \dfrac{a_3 h_3}{6}\right](x_4 - x) + \left[\dfrac{y_4}{h_3} - \dfrac{a_4 h_3}{6}\right](x - x_3)$

$\qquad f_3(x) = \dfrac{-0.3421}{6 \cdot 3}(18 - x)^3 + \dfrac{0.2519}{6 \cdot 3}(x - 15)^3 + \left[\dfrac{10}{3} - \dfrac{-0.3421 \cdot 3}{6}\right](18 - x) + \left[\dfrac{8}{3} - \dfrac{0.2519 \cdot 3}{6}\right](x - 15)$

$\qquad f_3(x) = (-0.019)(18 - x)^3 + 0.014(x - 15)^3 + 3.504(18 - x) + 2.5407(x - 15) \quad$ for $\ 15 \le x \le 18$

$i = 4 \quad f_4(x) = \dfrac{a_4}{6h_4}(x_5 - x)^3 + \dfrac{a_5}{6h_4}(x - x_4)^3 + \left[\dfrac{y_4}{h_4} - \dfrac{a_4 h_4}{6}\right](x_5 - x) + \left[\dfrac{y_5}{h_4} - \dfrac{a_5 h_4}{6}\right](x - x_4)$

$\qquad f_4(x) = \dfrac{0.2519}{6 \cdot 4}(22 - x)^3 + \dfrac{0}{6 \cdot 4}(x - 18)^3 + \left[\dfrac{8}{4} - \dfrac{0.2519 \cdot 4}{6}\right](22 - x) + \left[\dfrac{7}{4} - \dfrac{0 \cdot 4}{6}\right](x - 18)$

$\qquad f_4(x) = 0.0105(22 - x)^3 + 1.832(22 - x) + 1.75(x - 18) \quad$ for $\ 18 \le x \le 22$

(b) The interpolated value of y for $x = 12.7$ is calculated by substituting the value of x in $f_2(x)$:

$\qquad f_2(x) = (-0.01527)(15 - 12.7)^3 + (-0.01427)(12.7 - 11)^3 + 2.494(15 - 12.7) + 2.728(12.7 - 11)$

$\qquad f_2(x) = 10.11$

(c) The plot on the right shows the data points
and the polynomial.

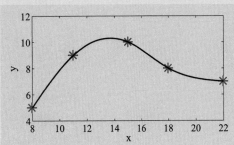

5.7 USE OF MATLAB BUILT-IN FUNCTIONS FOR CURVE FITTING AND INTERPOLATION

MATLAB has built-in functions for curve fitting and interpolation. In addition, MATLAB has an interactive tool for curve fitting, called the basic fitting interface. This section describes how to use the functions `polyfit` (for curve fitting) and `interp1` (for interpolation). Polynomials can be easily used and mathematically manipulated with MATLAB.

The `polyfit` command

The `polyfit` command can be used for curve fitting a given set of n points with polynomials of various degrees and for determining the polynomial of order $n - 1$ that passes through all the points. The form of the command is:

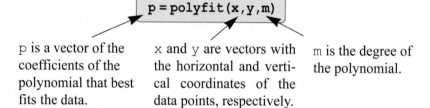

$$p = polyfit(x,y,m)$$

| p is a vector of the coefficients of the polynomial that best fits the data. | x and y are vectors with the horizontal and vertical coordinates of the data points, respectively. | m is the degree of the polynomial. |

The `interp1` command

The `interp1` (the last character is the number one) command executes one-dimensional interpolation at one point. The format of the command is:

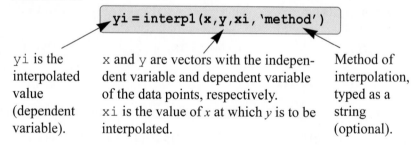

$$yi = interp1(x,y,xi, \text{'method'})$$

| yi is the interpolated value (dependent variable). | x and y are vectors with the independent variable and dependent variable of the data points, respectively. xi is the value of x at which y is to be interpolated. | Method of interpolation, typed as a string (optional). |

- The vector x must be monotonic (the elements must be in ascending or descending order).

- xi can be a scalar (interpolation at one point) or a vector (interpolation at several points). yi is a scalar or a vector with the corresponding interpolated values at the point(s) xi.

- MATLAB can interpolate using one of several methods that can be specified. These methods include:

'nearest'	returns the value of the data point that is nearest to the interpolated point.
'linear'	uses linear spline interpolation.

`'spline'`	uses cubic spline interpolation with "not-a-knot" conditions where the third derivatives at the second and second to last points are continuous. This is not the natural spline presented in this chapter.
`'pchip'`	also called `'cubic'`, uses piecewise cubic Hermite interpolation.

- When the `'nearest'` and the `'linear'` methods are used, the value(s) of xi must be within the domain of x. If the `'spline'` or the `'pchip'` methods are used, xi can have values outside the domain of x and the function `interp1` performs extrapolation.

- The `'spline'` method can give large errors if the input data points are nonuniform such that some points are much closer together than others.

Specification of the method is optional. If no method is specified, the default is `'linear'`.

Two examples of using MATLAB's built-in functions for curve fitting and interpolation are shown. First, the `polyfit` function is used for determining the fourth order polynomial that curve-fits the data points in Example 5-3:

```
>> x = 0:0.4:6;
>> y = [0 3 4.5 5.8 5.9 5.8 6.2 7.4 9.6 15.6 20.7 26.7 31.1 35.6 39.3 41.5];
>> p = polyfit(x,y,4)

p =

  -0.2644   3.1185  -10.1927   12.8780   -0.2746
```

> The polynomial that corresponds to these coefficients is:
> $$f(x) = (-0.2644)x^4 + 3.1185x^3 - 10.1927x^2 + 12.878x - 0.2746 .$$

In the second example, the `interp1` command is used for the interpolation in Example 5-8:

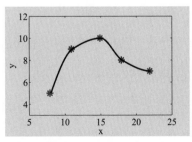

Figure 5-22: Interpolation using MATLAB's `interp1` function.

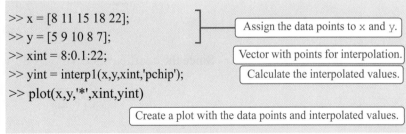

```
>> x = [8 11 15 18 22];          Assign the data points to x and y.
>> y = [5 9 10 8 7];
>> xint = 8:0.1:22;              Vector with points for interpolation.
>> yint = interp1(x,y,xint,'pchip');    Calculate the interpolated values.
>> plot(x,y,'*',xint,yint)

                                Create a plot with the data points and interpolated values.
```

The resulting plot is shown in Fig. 5-22.

MATLAB also has an interactive tool for curve fitting and interpolation, called the basic fitting interface. To activate the interface, the user first has to make a plot of the data points and then in the Figure Window select **Basic Fitting** in the **Tools** menu. (A detailed description of the basic fitting interface is available in MATLAB, An Introduction with Applications, by Amos Gilat, Wiley, 2005.)

5.8 CURVE FITTING WITH A LINEAR COMBINATION OF NONLINEAR FUNCTIONS

The method of least squares that was applied for curve fitting with a linear function in Section 5.2, and with quadratic and higher-order polynomials in Section 5.4, can be generalized in terms of curve fitting with a linear combination of nonlinear functions. A linear combination of m nonlinear functions can be written as:

$$F(x) = C_1 f_1(x) + C_2 f_2(x) + C_3 f_3(x) + \dots + C_m f_m(x) = \sum_{j=1}^{m} C_j f_j(x) \quad (5.91)$$

where $f_1, f_2, \dots, f_m$ are prescribed functions, and $C_1, C_2, \dots, C_m$ are unknown coefficients. Using least-squares regression, Eq. (5.91) is used to fit a given set of n points $(x_1, y_1), (x_2, y_2), \dots, (x_n, y_n)$ by minimizing the total error that is given by the sum of the squares of the residuals:

$$E = \sum_{i=1}^{n} \left[y_i - \sum_{j=1}^{m} C_j f_j(x_i) \right]^2 \quad (5.92)$$

The function E in Eq. (5.92) has a minimum for those values of the coefficients $C_1, C_2, \dots, C_m$ where the partial derivative of E with respect to each of the coefficients is equal to zero:

$$\frac{\partial E}{\partial C_k} = 0 \quad \text{for } k = 1, 2, \dots, m \quad (5.93)$$

Substituting Eq. (5.92) into Eq. (5.93) gives:

$$\frac{\partial E}{\partial C_k} = \sum_{i=1}^{n} 2 \left[y_i - \sum_{j=1}^{m} C_j f_j(x_i) \right] \left[(-1) \frac{\partial}{\partial C_k} \left(\sum_{j=1}^{m} C_j f_j(x_i) \right) \right] = 0$$

$$\text{for } k = 1, 2, \dots, m \quad (5.94)$$

Since the coefficients $C_1, C_2, \dots, C_m$ are independent of each other,

$$\frac{\partial}{\partial C_k} \left(\sum_{j=1}^{m} C_j f_j(x_i) \right) = f_k(x_i) \quad (5.95)$$

and Eq. (5.94) becomes:

$$\frac{\partial E}{\partial C_k} = -\sum_{i=1}^{n} 2 \left[y_i - \sum_{j=1}^{m} C_j f_j(x_i) \right] f_k(x_i) = 0 \quad (5.96)$$

The last equation can be rewritten in the form:

$$\sum_{i=1}^{n} \sum_{j=1}^{m} C_j f_j(x_i) f_k(x_i) = \sum_{i=1}^{n} y_i f_k(x_i) \quad \text{for } k = 1, 2, ..., m \quad (5.97)$$

In Eq. (5.97), x_i, y_i, and $f_k(x_i)$ are all known quantities, and the $C_1, C_2, ..., C_m$ are the unknowns. The set of Eqs. (5.97) is a system of m linear equations for the unknown coefficients $C_1, C_2, ..., C_m$.

The functions $f_k(x)$ can be any functions. For example, if $F(x) = C_1 f_1(x) + C_2 f_2(x)$ such that $f_1(x) = 1$ and $f_2(x) = x$, then Eqs. (5.97) reduce to Eqs. (5.9) and (5.10). If the functions $f_k(x)$ are chosen such that $F(x)$ is quadratic (i.e., $f_1(x) = 1$, $f_2(x) = x$, and $f_3(x) = x^2$), then Eqs. (5.97) reduce to Eqs. (5.23)–(5.25). In general, the functions $f_k(x)$ are chosen because there is a guiding theory that predicts the trend of the data. Example 5-9 shows how the method is used for curve fitting data points with nonlinear approximating functions.

Example 5-9: Curve fitting with linear combination of nonlinear functions.

The following data is obtained from wind-tunnel tests, for the variation of the ratio of the tangential velocity of a vortex to the free stream flow velocity $y = V_\theta / V_\infty$ versus the ratio of the distance from the vortex core to the chord of an aircraft wing, $x = R/C$:

x	0.6	0.8	0.85	0.95	1.0	1.1	1.2	1.3	1.45	1.6	1.8
y	0.08	0.06	0.07	0.07	0.07	0.06	0.06	0.06	0.05	0.05	0.04

Theory predicts that the relationship between x and y should be of the form $y = \dfrac{A}{x} + \dfrac{B e^{-2x^2}}{x}$. Find the values of A and B using the least-squares method to fit the above data.

SOLUTION

In the notation of Eq. (5.91) the approximating function is $F(x) = C_1 f_1(x) + C_2 f_2(x)$ with $F(x) = y$, $C_1 = A$, $C_2 = B$, $f_1(x) = \dfrac{1}{x}$, and $f_2(x) = \dfrac{e^{-2x^2}}{x}$. The equation has two terms, which means that $m = 2$, and since there are 11 data points, $n = 11$. Substituting this information in Eq. (5.97) gives the following system of two linear equations for A and B.

$$\sum_{i=1}^{11} A \frac{1}{x_i} \frac{1}{x_i} + \sum_{i=1}^{11} B \frac{e^{-2x_i^2}}{x_i} \frac{1}{x_i} = \sum_{i=1}^{11} y_i \frac{1}{x_i} \quad \text{for } k = 1$$

$$\sum_{i=1}^{11} A \frac{1}{x_i} \frac{e^{-2x_i^2}}{x_i} + \sum_{i=1}^{11} B \frac{e^{-2x_i^2}}{x_i} \frac{e^{-2x_i^2}}{x_i} = \sum_{i=1}^{11} y_i \frac{e^{-2x_i^2}}{x_i} \quad \text{for } k = 2$$

These two equations can be rewritten as:

$$A\sum_{i=1}^{11}\frac{1}{x_i^2} + B\sum_{i=1}^{11}\frac{e^{-2x_i^2}}{x_i^2} = \sum_{i=1}^{11}y_i\frac{1}{x_i}$$

$$A\sum_{i=1}^{11}\frac{e^{-2x_i^2}}{x_i^2} + B\sum_{i=1}^{11}\frac{e^{-4x_i^2}}{x_i^2} = \sum_{i=1}^{11}y_i\frac{e^{-2x_i^2}}{x_i}$$

The system can be written in a matrix form:

$$\begin{bmatrix} \sum_{i=1}^{11}\frac{1}{x_i^2} & \sum_{i=1}^{11}\frac{e^{-2x_i^2}}{x_i^2} \\ \sum_{i=1}^{11}\frac{e^{-2x_i^2}}{x_i^2} & \sum_{i=1}^{11}\frac{e^{-4x_i^2}}{x_i^2} \end{bmatrix}\begin{bmatrix} A \\ B \end{bmatrix} = \begin{bmatrix} \sum_{i=1}^{11}y_i\frac{1}{x_i} \\ \sum_{i=1}^{11}y_i\frac{e^{-2x_i^2}}{x_i} \end{bmatrix}$$

The system is solved by using MATLAB. The following MATLAB program in a script file solves the system and then makes a plot of the data points and the curve-fitted function.

```
x = [0.6 0.8 0.85 0.95 1.0 1.1 1.2 1.3 1.45 1.6 1.8];
y = [0.08 0.06 0.07 0.07 0.07 0.06 0.06 0.06 0.05 0.05 0.04];
a(1,1) = sum(1./x.^2);
a(1,2) = sum(exp(-2*x.^2)./x.^2);
a(2,1) = a(1,2);
a(2,2) = sum(exp(-4*x.^2)./x.^2);
b(1,1) = sum(y./x);
b(2,1) = sum((y.*exp(-2*x.^2))./x);
AB = a\b
xfit = 0.6:0.02:1.8;
yfit = AB(1)./xfit + AB(2)*exp(-2*xfit.^2)./xfit;
plot(x,y,'o',xfit,yfit)
```

When the program is executed, the solution for the coefficients is displayed in the Command Window (the two elements of the vector AB), and a plot with the data points and the curve-fitted function is created.

Command Window:

```
AB =
    0.0743          ←   The coefficient A.
   -0.0597          ←   The coefficient B.
```

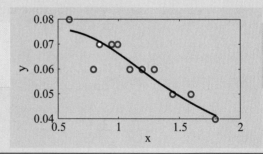

5.9 PROBLEMS

Problems to be solved by hand
Solve the following problems by hand. When needed, use a calculator, or write a MATLAB script file to carry out calculations. If using MATLAB, do not use built-in functions for curve fitting and interpolation.

5.1   The following data is given:

x	2	5	6	8	9	13	15
y	7	8	10	11	12	14	15

(*a*) Use linear least-squares regression to determine the coefficients m and b in the function $y = mx + b$ that best fits the data.
(*b*) Use Eq. (5.5) to determine the overall error.

5.2   The following data is given:

x	-7	-5	-1	0	2	5	6
y	15	12	5	2	0	-5	-9

(*a*) Use linear least-squares regression to determine the coefficients m and b in the function $y = mx + b$ that best fits the data.
(*b*) Use Eq. (5.5) to determine the overall error.

5.3   The following data gives the approximate population of the world for selected years from 1850 until 2000.

Year	1850	1900	1950	1980	2000
Population (billions)	1.3	1.6	3	4.4	6

Assume that the population growth can be modeled with an exponential function $p = be^{mx}$ where x is the year and p is the population in billions. Write the equation in a linear form (Section 5.3), and use linear least-squares regression to determine the constants b and m for which the function best fits the data. Use the equation to estimate the population in the year 1970.

5.4   The following data is given:

x	-0.2	-0.1	0.2	0.7	1.3
y	5.2	3	0.6	0.4	0.2

Determine the coefficients m and b in the function $y = \dfrac{1}{mx + b}$ that best fit the data. (Write the equation in a linear form (Section 5.3), and use linear least-squares regression to determine the value of the coefficients.)

5.5   The following data is given:

x	1	3	5	7	10
y	2.2	5.0	5.5	6.1	6.6

Determine the coefficients m and b in the function $y = \dfrac{mx}{b+x}$ that best fit the data. (Write the equation in a linear form (Section 5.3), and use linear least-squares regression to determine the value of the coefficients.)

5.6 To measure g (the acceleration due to gravity) the following experiment is carried out. A ball is dropped from the top of a 30-m-tall building. As the object is falling down, its speed v is measured at various heights by sensors that are attached to the building. The data measured in the experiment is given in the table.

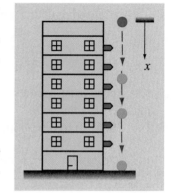

x (m)	0	5	10	15	20	25
v (m/s)	0	9.85	14.32	17.63	19.34	22.41

In terms of the coordinates shown in the figure (positive down), the speed of the ball v as a function of the distance x is given by $v^2 = 2gx$. Using linear regression, determine the experimental value of g.

5.7 The atmospheric pressure, p, as a function of height, h, can be modeled by an exponential function of the form $p = be^{-mh}$. The following are values of pressure measured at different heights. Using linear regression, determine the constants m and h that best fit the data. Use the equation to estimate the atmospheric pressure at a height of 7000 m.

h (m)	0	5,000	10,000	15,000	20,000
p (Pa)	100,000	47,500	22.600	10,800	5,100

5.8 In an electrophoretic fiber making process, the diameter of the fiber, d, is related to the current flow, I. The following is measured during production:

I (nA)	300	300	350	400	400	500	500	650	650
d (μm)	22	26	27	30	34	33	33.5	37	42

The relationship between the current and the diameter can be modeled with an equation of the form $d = a + b\sqrt{I}$. Use the data to determine the constants a and b that best fit the data.

5.9 Determine the coefficients of the polynomial $y = a_2x^2 + a_1x + a_0$ that best fit the data given in Problem 5.5.

5.10 Using the least-squares method (section 5.8), fit a linear combination of a straight line, $\sin x$, and e^x, to the following data:

x	0.1	0.4	0.5	0.7	0.7	0.9
y	0.61	0.92	0.99	1.52	1.47	2.03

5.11   The fuel economy of a car (miles per gallon) varies with its speed. In an experiment, the following five measurements are obtained:

Speed (mph)	10	25	40	55	70
Fuel economy (mpg)	12	26	28	30	24

Determine the fourth-order polynomial in the Lagrange form that passes through the points. Use the polynomial to calculate the fuel economy at 65 mph.

5.12   Determine the fourth-order Newton's interpolating polynomial that passes through the data points given in Problem 5.11. Use the polynomial to calculate the fuel economy at 65 mph.

5.13   The following data is given:

x	1	2.5	2	3	4	5
y	1	7	5	8	2	1

(*a*) Write the polynomial in Lagrange form that passes through the points, then use it to calculate the interpolated value of *y* at *x* = 3.5 .
(*a*) Write the polynomial in Newton's form that passes through the points, then use it to calculate the interpolated value of *y* at *x* = 3.5 .

5.14   Use linear splines interpolation with the data in Problem 5.11, to calculate the fuel economy at a speed of
(*a*) 30 mph   (*b*) 65 mph.

5.15   Use quadratic splines interpolation with the data in Problem 5.11, to calculate the fuel economy at a speed of
(*a*) 30 mph   (*b*) 65 mph.

5.16   Use natural cubic splines interpolation (based on Lagrange-form polynomials [Eqs. (5.86)–(5.89)]) with the data in Problem 5.11, to calculate the fuel economy at a speed of
(*a*) 30 mph   (*b*) 65 mph.

Problems to be programmed in MATLAB
Solve the following problems using MATLAB environment. Do not use MATLAB's built-in functions for curve fitting and interpolation.

5.17 Modify the MATLAB user-defined function LinearRegression in Program 5-1. In addition to determining the constants a_1 and a_0 the modified function should also calculate the overall error E according to Eq. (5.6). For function name use [a, Er] = LinReg(x, y). The input arguments x and y are vectors with the coordinates of the data points. The output argument a is a two-element vector with the values of the constants a_1 and a_0. The output argument Er is the value of the overall error.
(*a*) Use the function to solve Example 5-1.
(*b*) Use the function to solve Problem 5.2.

5.18 Write a MATLAB user-defined function that determines the best fit of an exponential function of the form $y = be^{-mx}$ to a given set of data points. Name the function `[b m]= ExpoFit(x,y)`, where the input arguments `x` and `y` are vectors with the coordinates of the data points, and the output arguments `b` and `m` are the values of the coefficients. The function `ExpoFit` should use the approach that is described in Section 5.3 for determining the value of the coefficients. Use the function to solve Problem 5.7.

5.19 Write a MATLAB user-defined function that determines the best fit of a power function of the form $y = bx^m$ to a given set of data points. Name the function `[b m]= PowerFit(x,y)`, where the input arguments `x` and `y` are vectors with the coordinates of the data points, and the output arguments `b` and `m` are the values of the coefficients. The function `PowerFit` should use the approach that is described in Section 5.3 for determining the value of the coefficients. Use the function to solve Problem 5.7.

5.20 Write a MATLAB user-defined function that determines the coefficients of a quadratic polynomial, $f(x) = a_2 x^2 + a_1 x + a_0$, that best fits a given set of data points. Name the function `a = QuadFit(x,y)`, where the input arguments `x` and `y` are vectors with the coordinates of the data points, and the output argument `a` is a three-element vector with the values of the coefficients a_2, a_1 and a_0.
(a) Use the function to find the quadratic polynomial that best fits the data in Example 5-2.
(b) Write a program in a script file that plots the data points and the curve of the quadratic polynomial that best fits the data.

5.21 Write a MATLAB user-defined function that determines the coefficients of a cubic polynomial, $f(x) = a_3 x^3 + a_2 x^2 + a_1 x + a_0$, that best fits a given set of data points. The function should also calculate the overall error E according to Eq. (5.6). Name the function `[a,Er] = CubicPolyFit(x,y)`, where the input arguments `x` and `y` are vectors with the coordinates of the data points, and the output argument `a` is a four-element vector with the values of the coefficients a_3, a_2, a_1 and a_0. The output argument `Er` is the value of the overall error.
(a) Use `CubicPolyFit` to find the quadratic polynomial that best fits the data in Example 5-3.
(b) Write a program in a script file that plots the data points and the curve of the cubic polynomial that best fits the data.

5.22 Write a MATLAB user-defined function for interpolation with natural cubic splines. Name the function `Yint = CubicSplines(x,y,xint)`, where the input arguments `x` and `y` are vectors with the coordinates of the data points, and `xint` is the x coordinate of the interpolated point. The output argument `Yint` is the y value of the interpolated point.
(a) Use the function with the data in Example 5-8 for calculating the interpolated value at $x = 12.7$.
(b) Use the function with the data in Problem 5.27 for calculating the enthalpy per unit mass at $T = 14000$ K and at $T = 24000$ K.

Problems in math, science, and engineering
Solve the following problems using MATLAB environment. As stated, use the MATLAB programs that are presented in the chapter, programs developed in previously solved problems, or MATLAB's built-in functions.

5.23   In a uniaxial tension test, a dog-bone-shaped specimen is pulled in a machine. During the test, the force applied to the specimen, F, and the length of a gage section, L, are measured. The true stress, σ_t, and the true strain, ε_t, are defined by:

$$\sigma_t = \frac{F}{A_0}\frac{L}{L_0} \quad \text{and} \quad \varepsilon_t = \ln\frac{L}{L_0}$$

where A_0 and L_0 are the initial cross-sectional area and gage length, respectively. The true stress–strain curve in the region beyond yielding is often modeled by:

$$\sigma_t = K\varepsilon_t^m$$

The following are values of F and L measured in an experiment. Use the approach from Section 5.3 for determining the value of the coefficients K and m that best fit the data. The initial cross-sectional area and gage length are $A_0 = 1.25 \times 10^{-4}\,\text{m}^2$, and $L_0 = 0.0125$ m.

F (kN)	24.6	29.3	31.5	33.3	34.8	35.7	36.6	37.5	38.8	39.6	40.4
L (mm)	12.58	12.82	12.91	12.95	13.05	13.21	13.35	13.49	14.08	14.21	14.48

5.24 The stress concentration factor k is the ratio between the maximum stress σ_{max} and the average stress σ_{ave}, $k = \sigma_{max}/\sigma_{ave}$. For a plate of width D with a central hole of diameter d loaded with an axial force F (see figure), the maximum stress is at the edge of the hole, and the average stress is given by $\sigma_{ave} = F/[t(D-d)]$, where t is the thickness of the plate. The stress concentration factor measured in five tests with plates with different ratios of d/D is shown in the table.

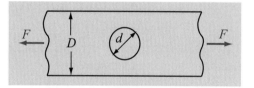

d/D	0.05	0.25	0.45	0.65	0.85
k	2.91	2.40	2.17	2.11	2.03

(a) Use an exponential function $k = be^{m(d/D)}$ to model the relationship between k and d/D. Determine the values of b and m that best-fit the data.
(b) Plot the data points and the curve-fitted model.
(c) Use the model to predict the stress concentration factor for $d/D = 0.15$.

5.25 A hot-wire anemometer is a device for measuring flow velocity, by measuring the cooling effect of the flow on the resistance of a hot wire. The following data are obtained in calibration tests.

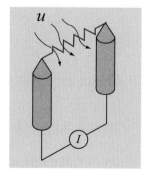

u (ft/s)	4.72	12.49	20.03	28.33	37.47	41.43	48.38	55.06
V (Volt)	7.18	7.3	7.37	7.42	7.47	7.5	7.53	7.55

u (ft/s)	66.77	59.16	54.45	47.21	42.75	32.71	25.43	8.18
V (Volt)	7.58	7.56	7.55	7.53	7.51	7.47	7.44	7.28

Determine the coefficients of the exponential function $u = Ae^{BV}$ that best fit the data,
(a) using the user-defined function ExpoFit developed in Problem 5-18.
(b) using MATLAB built-in functions.
   In each part plot the data points and the fitting function.

5.26 The yield stress of many metals, σ_y, varies with the size of the grains. Often, the relationship between the grain size, d, and the yield stress is modeled with the Hall–Petch equation:

$$\sigma_y = \sigma_0 + kd^{\left(-\frac{1}{2}\right)}$$

The following are results from measurements of average grain size and yield stress.

d (mm)	0.006	0.011	0.017	0.025	0.039	0.060	0.081	0.105
σ_y (MPa)	334	276	249	235	216	197	194	182

(a) Determine the constants σ_0 and d such that the Hall–Petch equation will best-fit the data. Plot the data points (circle markers) and the Hall–Petch equation as a solid line. Use the Hall–Petch equation to estimate the yield stress of a specimen with a grain size of 0.05 mm.

(b) Use the user-defined function QuadFit from Problem 5.20 to find the quadratic function that best fits the data. Plot the data points (circle markers) and the quadratic equation as a solid line. Use the quadratic equation to estimate the yield stress of a specimen with a grain size of 0.05 mm.

5.27 Values of enthalpy per unit mass, h, of an equilibrium Argon plasma (Ar, Ar^+, A^{++}, A^{+++} ions and electrons) versus temperature are:

$T \times 10^3$ (K)	5	7.5	10	12.5	15	17.5	20	22.5	25	27.5	30
h (MJ/kg)	3.3	7.5	41.8	51.8	61	101.1	132.9	145.5	171.4	225.8	260.9

Write a program in a script file that uses interpolation to calculate h at temperatures ranging from 5000 K to 30000 K in increments of 500 K. The program should generate a plot that shows the interpolated points, and the data points from the table (use an asterisk marker).

(a) For interpolation use the user-defined function CubicSplines from Problem 5.22.

(b) For interpolation use MATLAB's built-in function interp1 with the spline option.

5.28 The following are measurements of the rate coefficient, k, for the reaction $CH_4 + O \rightarrow CH_3 + OH$ at different temperatures, T.

T (K)	595	623	761	849	989	1076	1146	1202	1382	1445	1562
$k \times 10^{20}$ (m³/s)	2.12	3.12	14.4	30.6	80.3	131	186	240	489	604	868

(a) Use the method of least-squares to best-fit a function of the form $\ln(k) = C + b\ln(T) - \dfrac{D}{T}$ to the data. Determine the constants C, b, and D by curve fitting a linear combination of the functions $f_1(T) = 1$, $f_2(T) = \ln(T)$, and $f_3(T) = \dfrac{1}{T}$ to the given data (Section 5.8).

(b) Usually, the rate coefficient is expressed in the form of an Arrhenius equation $k = AT^b e^{-E_a/(RT)}$, where A and b are constants, $R = 8.314$ J/mole/K is the universal gas constant, and E_a is the activation energy for the reaction. Having determined the constants C, b, and D in part (a), deduce the values of A (m³/s) and E_a (J/mole) in the Arrhenius expression.

Chapter 6

Numerical Differentiation

6.1 BACKGROUND

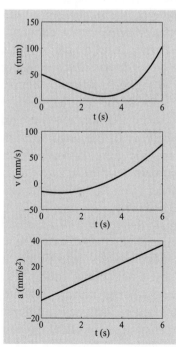

Figure 6-1: Position, velocity, and acceleration as a function of time.

Differentiation gives a measure of the rate at which a quantity changes. Rates of change of quantities appear in many disciplines, especially science and engineering. One of the more fundamental of these rates is the relationship between position, velocity, and acceleration. If the position, x of an object that is moving along a straight line is known as a function of time, t, (the top curve in Fig. 6-1):

$$x = f(t) \tag{6.1}$$

the object's velocity, $v(t)$, is the derivative of the position with respect to time (the middle curve in Fig. 6-1):

$$v = \frac{df(t)}{dt} \tag{6.2}$$

The velocity v is the slope of the position–time curve. Similarly, the object's acceleration, $a(t)$, is the derivative of the velocity with respect to time (the bottom curve in Fig. 6-1):

$$a = \frac{dv(t)}{dt} \tag{6.3}$$

The acceleration a is the slope of the velocity–time curve.

Many models in physics and engineering are expressed in terms of rates. In an electrical circuit, the current in a capacitor is related to the time derivative of the voltage. In analyzing conduction of heat, the amount of heat flow is determined from the derivative of the tempera-

233

ture. Differentiation is also used for finding the maximum and minimum values of functions.

The need for numerical differentiation

The function to be differentiated can be given as an analytical expression or as a set of discrete points (tabulated data). When the function is given as a simple mathematical expression, the derivative can be determined analytically. When analytical differentiation of the expression is difficult or not possible, numerical differentiation has to be used. When the function is specified as a set of discrete points, differentiation is done by using a numerical method.

Numerical differentiation also plays an important role in some of the numerical methods used for solving differential equations, as shown in Chapters 8 and 9.

Approaches to numerical differentiation

Numerical differentiation is carried out on data that are specified as a set of discrete points. In many cases the data are measured or are recorded in experiments, or they may be the result of large-scale numerical calculations. If there is a need to calculate the numerical derivative of a function that is given in an analytical form, then the differentiation is done by using discrete points of the function. This means that in all cases numerical integration is done by using the values of points.

For a given set of points, two approaches can be used to calculate a numerical approximation of the derivative at one of the points. One approach is to use a *finite difference approximation* for the derivative. A finite difference approximation of a derivative at a point x_i is an approximate calculation based on the value of points in the neighborhood of x_i. This approach is illustrated in Fig. 6-2a where the derivative at point x_i is approximated by the slope of the line that connects the point before x_i with the point after x_i. The accuracy of a finite difference approximation depends on the accuracy of the data points, the spacing between the points, and the specific formula used for the approximation. The simplest formula approximates the derivative as the slope of the line that connects two adjacent points. Finite difference approximation is covered in Sections 6.2 and 6.3.

The second approach is to approximate the points with an analytical expression that can be easily differentiated, and then to calculate the derivative by differentiating the analytical expression. The approximate analytical expression can be derived by using curve fitting. This approach is illustrated in Fig. 6-2b, where the points are curve fitted by $f(x)$, and the derivative at point x_i is obtained by analytically differentiating the approximating function and evaluating the result at the point x_i. This approach for numerical differentiation is described in Section 6.6.

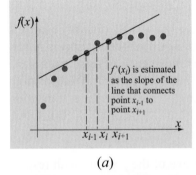

(a)

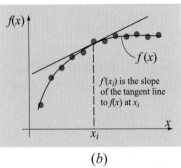

(b)

Figure 6-2: Numerical differentiation using (a) finite difference approximation and (b) function approximation.

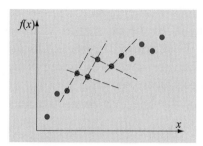

Figure 6-3: Numerical differentiation of data with scatter.

Noise and scatter in the data points

When the data to be differentiated is obtained from experimental measurements, usually there is scatter in the data because of the experimental errors or uncertainties in the measurement (e.g., electrical noise). A set of data points that contains scatter is shown schematically in Fig. 6-3. If this data set is differentiated using a two-point finite difference approximation, which is the simplest form of finite difference approximation (slope of the line that connects two adjacent points), then large variations (positive and negative values) will be seen in the value of the derivative from point to point. It is obvious from the data in the figure that the value of y generally increases with increasing x, which means that the derivative of y w.r.t x is positive. Better results can be obtained by using higher-order formulas of finite difference approximation that take into account the values from more than two points. For example, (see the formulas in Section 6.4) there are four, five, and seven-point finite difference formulas. As mentioned before, the differentiation can also be done by curve fitting the data with an analytical function that is then differentiated. In this case, the data is smoothed out before it is differentiated, eliminating the problem of wrongly amplified slopes between successive points.

6.2 FINITE DIFFERENCE APPROXIMATION OF THE DERIVATIVE

The derivative $f'(x)$ of a function $f(x)$ at the point $x = a$ is defined by:

$$\frac{df(x)}{dx}\bigg|_{x=a} = f'(a) = \lim_{x \to a}\frac{f(x)-f(a)}{x-a} \tag{6.4}$$

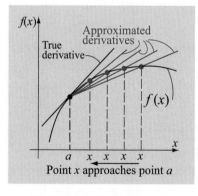

Figure 6-4: Definition of derivative.

Graphically, the definition is illustrated in Fig. 6-4. The derivative is the value of the slope of the tangent line to the function at $x = a$. The derivative is obtained by taking a point x near $x = a$ and calculating the slope of the line that connects the two points. The accuracy of calculating the derivative in this way increases as point x is closer to point a. In the limit as point x approaches point a, the derivative is the slope of the line that is tangent to $f(x)$ at $x = a$. In Calculus, application of the limit condition in Eq. (6.4), which means that point x approaches point a, is used for deriving rules of differentiation that give an analytic expression for the derivative.

In finite difference approximations of the derivative, values of the function at different points in the neighborhood of the point $x = a$ are used for estimating the slope. It should be remembered that the function that is being differentiated is prescribed by a set of discrete points. Various finite difference approximation formulas exist. Three such formulas, where the derivative is calculated from the values of two points, are presented in this section.

Forward, backward, and central difference formulas for the first derivative

The forward, backward, and central finite difference formulas are the simplest finite difference approximations of the derivative. In these approximations, illustrated in Fig. 6-5, the derivative at point (x_i) is calculated from the values of two points. The derivative is estimated as the value of the slope of the line that connects the two points.

- *Forward difference* is the slope of the line that connects points $(x_i, f(x_i))$ and $(x_{i+1}, f(x_{i+1}))$:

$$\left.\frac{df}{dx}\right|_{x=x_i} = \frac{f(x_{i+1}) - f(x_i)}{x_{i+1} - x_i} \tag{6.5}$$

- *Backward difference* is the slope of the line that connects points $(x_{i-1}, f(x_{i-1}))$ and $(x_i, f(x_i))$:

$$\left.\frac{df}{dx}\right|_{x=x_i} = \frac{f(x_i) - f(x_{i-1})}{x_i - x_{i-1}} \tag{6.6}$$

- *Central difference* is the slope of the line that connects points $(x_{i-1}, f(x_{i-1}))$ and $(x_{i+1}, f(x_{i+1}))$:

$$\left.\frac{df}{dx}\right|_{x=x_i} = \frac{f(x_{i+1}) - f(x_{i-1})}{x_{i+1} - x_{i-1}} \tag{6.7}$$

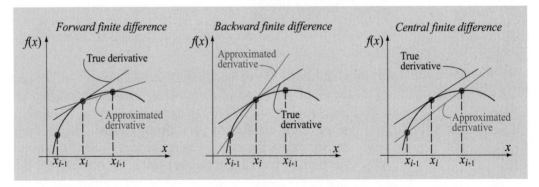

Figure 6-5: Finite difference approximation of derivative.

The first two examples show applications of the forward, backward, and central finite difference formulas. Example 6-1 compares numerical differentiation with analytical differentiation, and in Example 6-2 the formulas are used for differentiation of discrete data.

Example 6-1: Comparing numerical and analytical differentiation.

Consider the function $f(x) = x^3$. Calculate its first derivative at point $x = 3$ numerically with the forward, backward, and central finite difference formulas and using:

(a) Points $x = 2$, $x = 3$, and $x = 4$.

(b) Points $x = 2.75$, $x = 3$, and $x = 3.25$.

Compare the results with the exact (analytical) derivative.

SOLUTION

Analytical differentiation: The derivative of the function is $f'(x) = 3x^2$, and the value of the derivative at $x = 3$ is $f'(3) = 3 \cdot 3^2 = 27$.

Numerical differentiation

(a) The points used for numerical differentiation are:

x: 2   3   4

$f(x)$:   8   27   64

Using Eqs. (6.5) through (6.7), the derivatives using the forward, backward, and central finite difference formulas are:

Forward finite difference:

$$\left.\frac{df}{dx}\right|_{x=3} = \frac{f(4)-f(3)}{4-3} = \frac{64-27}{1} = 37 \quad error = \left|\frac{37-27}{27} \cdot 100\right| = 37.04\,\%$$

Backward finite difference:

$$\left.\frac{df}{dx}\right|_{x=3} = \frac{f(3)-f(2)}{3-2} = \frac{27-8}{1} = 19 \quad error = \left|\frac{19-27}{27} \cdot 100\right| = 29.63\,\%$$

Central finite difference:

$$\left.\frac{df}{dx}\right|_{x=3} = \frac{f(4)-f(2)}{4-2} = \frac{64-8}{1} = 28 \quad error = \left|\frac{28-27}{27} \cdot 100\right| = 3.704\,\%$$

(b) The points used for numerical differentiation are:

x: 2.75   3   3.25

$f(x)$: 2.75^3 3^3 3.25^3

Using Eqs. (6.5) through (6.7), the derivatives using the forward, backward, and central finite difference formulas are:

Forward finite difference:

$$\left.\frac{df}{dx}\right|_{x=3} = \frac{f(3.25)-f(3)}{3.25-3} = \frac{3.25^3-27}{0.25} = 29.3125 \quad error = \left|\frac{29.3125-27}{27} \cdot 100\right| = 8.565\,\%$$

Backward finite difference:

$$\left.\frac{df}{dx}\right|_{x=3} = \frac{f(3)-f(2.75)}{3-2.75} = \frac{27-2.75^3}{3-2.75} = 24.8125 \quad error = \left|\frac{24.8125-27}{27} \cdot 100\right| = 8.102\,\%$$

Central finite difference:

$$\left.\frac{df}{dx}\right|_{x=3} = \frac{f(3.25)-f(2.75)}{3.25-2.75} = \frac{3.25^3-2.75^3}{3.25-2.75} = 27.0625 \quad error = \left|\frac{27.0625-27}{27} \cdot 100\right| = 0.2315\,\%$$

The results show that the central finite difference formula gives a more accurate approximation. This will be discussed further in the next section. In addition, smaller separation between the points gives a significantly more accurate approximation.

Example 6-2: Damped vibrations.

In a vibration experiment, a block of mass m is attached to a spring of stiffness k, and a dashpot with damping coefficient c, as shown in the figure. To start the experiment the block is moved from the equilibrium position and then released from rest. The position of the block as a function of time is recorded at a frequency of 5 Hz (5 times a second). The recorded data for the first 10 s is shown in the figure. The data points for $4 \le t \le 8$ s are given in the table below.

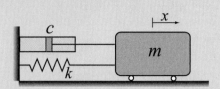

(*a*) The velocity of the block is the derivative of the position w.r.t. time. Use the central finite difference formula to calculate the velocity at time $t = 5$ and $t = 6$ s.
(*b*) Write a user-defined MATLAB function that calculates the derivative of a function that is given by a set of

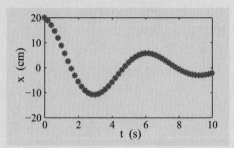

discrete points. Name the function dx=deriva-
tive(x,y) where x and y are vectors with the coordinates of the points, and dx is a vector with the value of the derivative $\frac{dy}{dx}$ at each point. The function should calculate the derivative at the *first* and *last* points using the *forward* and *backward finite difference formulas*, respectively, and using the central finite difference formula for all of the other points.

Use the given data points to calculate the velocity of the block for $4 \le t \le 8$ s. Calculate the acceleration of the block by differentiating the velocity. Make a plot of the displacement, velocity, and acceleration, versus time for $4 \le t \le 8$ s.

t (s)	4.0	4.2	4.4	4.6	4.8	5.0	5.2	5.4	5.6	5.8	6.0	6.2	6.4	6.6
x (cm)	-5.87	-4.23	-2.55	-0.89	0.67	2.09	3.31	4.31	5.06	5.55	5.78	5.77	5.52	5.08

t (s)	6.8	7.0	7.2	7.4	7.6	7.8	8.0
x (cm)	4.46	3.72	2.88	2.00	1.10	0.23	-0.59

SOLUTION

(*a*) The velocity is calculated by using Eq. (6.7):

$$\text{for } t = 5 \text{ s:} \qquad \left.\frac{dx}{dt}\right|_{x=5} = \frac{f(5.2) - f(4.8)}{5.2 - 4.8} = \frac{3.31 - 0.67}{0.4} = 6.6 \text{ cm/s}$$

$$\text{for } t = 6 \text{ s:} \qquad \left.\frac{dx}{dt}\right|_{x=5} = \frac{f(6.2) - f(5.8)}{6.2 - 5.8} = \frac{5.77 - 5.55}{0.4} = 0.55 \text{ cm/s}$$

(*b*) The user-defined function dx=derivative(x,y) that is listed next calculates the derivative of a function that is given by a set of discrete points.

Program 6-1: Function file. Derivative of a function given by points.

```
function dx = derivative(x,y)
% derivative calculates the derivative of a function that is given by a set
% of points. The derivatives at the first and last points are calculated by
% using the forward and backward finite difference formula, respectively.
% The derivative at all the other points is calculated by the central
```

```
% finite difference formula.
% Input variables:
% x  A vector with the coordinates x of the data points.
% y  A vector with the coordinates y of the data points.
% Output variable:
% dx  A vector with the value of the derivative at each point.

n = length(x);
dx(1) = (y(2) - y(1))/(x(2) - x(1));
for i = 2:n - 1
    dx(i) = (y(i + 1) - y(i - 1))/(x(i + 1) - x(i - 1));
end
dx(n) = (y(n) - y(n - 1))/(x(n) - x(n - 1));
```

The user-defined function `derivative` is used in the following script file. The program determines the velocity (the derivative of the given data points) and the acceleration (the derivative of the velocity) and then displays three plots.

```
t = 4:0.2:8;
x = [-5.87 -4.23 -2.55 -0.89 0.67 2.09 3.31 4.31 5.06 5.55 5.78 5.77 5.52 5.08 4.46 3.72 2.88 2.00 1.10 0.23 -0.59];
vel = derivative(t,x)
acc = derivative(t,vel);
subplot (3,1,1)
plot(t,x)
subplot (3,1,2)
plot(t,vel)
subplot (3,1,3)
plot(t,acc)
```

When the script file is executed, the following plots are displayed (the plots were formatted in the Figure Window):

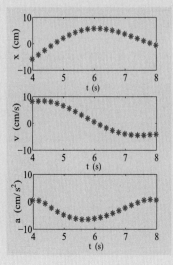

6.3 FINITE DIFFERENCE FORMULAS USING TAYLOR SERIES EXPANSION

The forward, backward, and central difference formulas, as well as many other finite difference formulas for approximating derivatives, can be derived by using Taylor series expansion. The formulas give an estimate of the derivative at a point from the values of points in its neighborhood. The number of points used in the calculation varies with the formula, and the points can be ahead, behind, or on both sides of the point at which the derivative is calculated. One advantage of using Taylor series expansion for deriving the formulas is that it also provides an estimate for the truncation error in the approximation.

In this section, several finite difference formulas are derived. Although the formulas can be derived for points that are not evenly spaced, the derivation here is for points are equally spaced. Section 6.3.1 gives formulas for approximating the first derivative, and Section 6.3.2 deals with finite difference formulas for the second derivative. The methods used for deriving the formulas can also be used for obtaining finite difference formulas for approximating higher-order derivatives. A summary of finite difference formulas for evaluating derivatives up to the fourth derivative is presented in Section 6.4.

6.3.1 Finite Difference Formulas of First Derivative

Several formulas for approximating the first derivative at point x_i based on the values of the points near x_i are derived by using the Taylor series expansion. All the formulas derived in this section are for the case where the points are equally spaced.

Two-point forward difference formula for first derivative

The value of a function at point x_{i+1} can be approximated by a Taylor series in terms of the value of the function and its derivatives at point x_i:

$$f(x_{i+1}) = f(x_i) + f'(x_i)h + \frac{f''(x_i)}{2!}h^2 + \frac{f'''(x_i)}{3!}h^3 + \frac{f^{(4)}(x_i)}{4!}h^4 + \ldots \quad (6.8)$$

where $h = x_{i+1} - x_i$ is the spacing between the points. By using two-term Taylor series expansion with a remainder (see Chapter 2), Eq. (6.8) can be rewritten as:

$$f(x_{i+1}) = f(x_i) + f'(x_i)h + \frac{f''(\xi)}{2!}h^2 \quad (6.9)$$

where ξ is a value of x between x_i and x_{i+1}.
Solving Eq. (6.9) for $f'(x_i)$ yields:

$$f'(x_i) = \frac{f(x_{i+1}) - f(x_i)}{h} - \frac{f''(\xi)}{2!}h \quad (6.10)$$

An approximate value of the derivative $f'(x_i)$ can now be calculated if the second term on the right-hand side of Eq. (6.10) is ignored. Ignoring this second term introduces a truncation (discretization) error. Since this term is proportional to h, the truncation error is said to be on the order of h (written as $O(h)$):

$$truncation \;\; error \; = \; -\frac{f''(\xi)}{2!}h \; = \; O(h) \tag{6.11}$$

It should be pointed out here that the magnitude of the truncation error is not really known since the value of $f''(\xi)$ is not known. Nevertheless, Eq. (6.11) is valuable since it implies that smaller h gives a smaller error. Moreover, as will be shown later in this chapter, it provides a means for comparing the size of the error in different finite difference formulas.

Using the notation of Eq. (6.11), the approximated value of the first derivative is:

$$f'(x_i) \; = \; \frac{f(x_{i+1}) - f(x_i)}{h} + O(h) \tag{6.12}$$

The approximation in Eq. (6.12) is the same as the forward difference formula in Eq. (6.5).

Two-point backward difference formula for first derivative

The backward difference formula can also be derived by application of Taylor series expansion. The value of the function at point x_{i-1} is approximated by a Taylor series in terms of the value of the function and its derivatives at point x_i:

$$f(x_{i-1}) \; = \; f(x_i) - f'(x_i)h + \frac{f''(x_i)}{2!}h^2 - \frac{f'''(x_i)}{3!}h^3 + \frac{f^{(4)}(x_i)}{4!}h^4 + \dots \tag{6.13}$$

where $h = x_i - x_{i-1}$. By using a two-term Taylor series expansion with a remainder (see Chapter 2), Eq. (6.13) can be rewritten as:

$$f(x_{i-1}) \; = \; f(x_i) - f'(x_i)h + \frac{f''(\xi)}{2!}h^2 \tag{6.14}$$

where ξ is a value of x between x_{i-1} and x_i. Solving Eq. (6.14) for $f'(x_i)$ yields:

$$f'(x_i) \; = \; \frac{f(x_i) - f(x_{i-1})}{h} + \frac{f''(\xi)}{2!}h \tag{6.15}$$

An approximate value of the derivative, $f'(x_i)$, can be calculated if the second term on the right-hand side of Eq. (6.15) is ignored. This yields:

$$f'(x_i) \; = \; \frac{f(x_i) - f(x_{i-1})}{h} + O(h) \tag{6.16}$$

The approximation in Eq. (6.16) is the same as the backward difference formula in Eq. (6.6).

Two-point central difference formula for first derivative

The central difference formula can be derived by using three terms in the Taylor series expansion and a remainder. The value of the function at point x_{i+1} in terms of the value of the function and its derivatives at point x_i is given by:

$$f(x_{i+1}) = f(x_i) + f'(x_i)h + \frac{f''(x_i)}{2!}h^2 + \frac{f'''(\xi_1)}{3!}h^3 \tag{6.17}$$

where ξ_1 is a value of x between x_i and x_{i+1}. The value of the function at point x_{i-1} in terms of the value of the function and its derivatives at point x_i is given by:

$$f(x_{i-1}) = f(x_i) - f'(x_i)h + \frac{f''(x_i)}{2!}h^2 - \frac{f'''(\xi_2)}{3!}h^3 \tag{6.18}$$

where ξ_2 is a value of x between x_{i-1} and x_i. In the last two equations, the spacing of the intervals is taken to be equal so that $h = x_{i+1} - x_i = x_i - x_{i-1}$. Subtracting Eq. (6.18) from Eq. (6.17) gives:

$$f(x_{i+1}) - f(x_{i-1}) = 2f'(x_i)h + \frac{f'''(\xi_1)}{3!}h^3 + \frac{f'''(\xi_2)}{3!}h^3 \tag{6.19}$$

An estimate for the first derivative is obtained by solving Eq. (6.19) for $f'(x_i)$ while neglecting the remainder terms, which introduces a truncation error, which is of the order of h^2:

$$f'(x_i) = \frac{f(x_{i+1}) - f(x_{i-1})}{2h} + O(h^2) \tag{6.20}$$

The approximation in Eq. (6.20) is the same as the central difference formula Eq. (6.7) for equally spaced intervals. A comparison of Eqs. (6.12), (6.16), and (6.20) shows that in the forward and backward difference approximation the truncation error is of the order of h, while in the central difference approximation the truncation error is of the order of h^2. This indicates that the central difference approximation gives a more accurate approximation of the derivative. This can be observed schematically in Fig. 6-5, where the slope of the line that represents the approximated derivative in the central difference approximation appears to be closer to the slope of the tangent line than the lines from the forward and backward approximations.

Three-point forward and backward difference formulas for the first derivative

The forward and backward difference formulas, Eqs. (6.12) and (6.16), give an estimate for the first derivative with a truncation error of $O(h)$. The forward difference formula evaluates the derivative at point x_i based on the values at that point and the point immediately to the right of it x_{i+1}. The backward difference formula evaluates the derivative at point x_i based on the values at that point and the one immediately to the left of it, x_{i-1}. Clearly, the forward difference formula can be useful for evaluating the first derivative at the first point x_1 and at all interior points, while the backward difference formula is useful for evaluating the first derivative at the last point and all interior points. The central difference formula, Eq. (6.20), gives an estimate for the first derivative with an error of $O(h^2)$. The central difference formula evaluates the first derivative at a given point x_i by using the points x_{i-1} and x_{i+1}. Consequently, for a function that is given by a discrete set of n points, the central difference formula is useful only for **interior points** and not for the endpoints (x_1 or x_n). An estimate for the first derivative at the endpoints, with error of $O(h^2)$, can be calculated with three-point forward and backward difference formulas, which are derived next.

The **three-point forward difference** formula calculates the derivative at point x_i from the value at that point and the next two points, x_{i+1} and x_{i+2}. It is assumed that the points are equally spaced such that $h = x_{i+2} - x_{i+1} = x_{i+1} - x_i$. (The procedure can be applied to unequally spaced points.) The derivation of the formula starts by using three terms of the Taylor series expansion with a remainder, for writing the value of the function at point x_{i+1} and at point x_{i+2} in terms of the value of the function and its derivatives at point x_i:

$$f(x_{i+1}) = f(x_i) + f'(x_i)h + \frac{f''(x_i)}{2!}h^2 + \frac{f'''(\xi_1)}{3!}h^3 \qquad (6.21)$$

$$f(x_{i+2}) = f(x_i) + f'(x_i)2h + \frac{f''(x_i)}{2!}(2h)^2 + \frac{f'''(\xi_2)}{3!}(2h)^3 \qquad (6.22)$$

where ξ_1 is a value of x between x_i and x_{i+1}, and ξ_2 is a value of x between x_i and x_{i+2}. Equations (6.21) and (6.22) are next combined such that the terms with the second derivative vanish. This is done by multiplying Eq. (6.21) by 4 and subtracting Eq. (6.22):

$$4f(x_{i+1}) - f(x_{i+2}) = 3f(x_i) + 2f'(x_i)h + \frac{4f'''(\xi_1)}{3!}h^3 - \frac{f'''(\xi_2)}{3!}(2h)^3 \qquad (6.23)$$

An estimate for the first derivative is obtained by solving Eq. (6.23) for $f'(x_i)$ while neglecting the remainder terms, which introduces a truncation error of the order of h^2:

$$f'(x_i) = \frac{-3f(x_i) + 4f(x_{i+1}) - f(x_{i+2})}{2h} + O(h^2) \qquad (6.24)$$

Equation (6.24) is the three-point forward difference formula that estimates the first derivative at point x_i from the value of the function at that point and at the next two points, x_{i+1} and x_{i+2}, with an error of $O(h^2)$. The formula can be used for calculating the derivative at the first point of a function that is given by a discrete set of n points.

The ***three-point backward difference*** formula yields the derivative at point x_i from the value of the function at that point and at the previous two points, x_{i-1} and x_{i-2}. The formula is derived in the same way that Eq. (6.24) was derived. The three-term Taylor series expansion with a remainder is written for the value of the function at point x_{i-1}, and at point x_{i-2} in terms of the value of the function and its derivatives at point x_i. The equations are then manipulated to obtain an equation without the second derivative terms, which is then solved for $f'(x_i)$. The formula that is obtained is:

$$f'(x_i) = \frac{f(x_{i-2}) - 4f(x_{i-1}) + 3f(x_i)}{2h} + O(h^2) \qquad (6.25)$$

where $h = x_i - x_{i-1} = x_{i-1} - x_{i-2}$ is the distance between the points.

Example 6-3 shows application of the three-point forward difference formula for the first derivative.

Example 6-3: Comparing numerical and analytical differentiation.

Consider the function $f(x) = x^3$. Calculate the first derivative at point $x = 3$ numerically with the three-point forward difference formula, using:

(*a*) Points $x = 3$, $x = 4$, and $x = 5$.

(*b*) Points $x = 3$, $x = 3.25$, and $x = 3.5$.

Compare the results with the exact value of the derivative, obtained analytically.

SOLUTION

Analytical differentiation: The derivative of the function is $f'(x) = 3x^2$, and the value of the derivative at $x = 3$ is $f'(3) = 3 \cdot 3^2 = 27$.

Numerical differentiation

(*a*) The points used for numerical differentiation are:

x: 3   4   5

$f(x)$:   27   64   125

Using Eq. (6.24), the derivative using the three-point forward difference formula is:

$$f'(3) = \frac{-3f(3) + 4f(4) - f(5)}{2 \cdot 1} = \frac{-3 \cdot 27 + 4 \cdot 64 - 125}{2} = 25 \quad error = \left|\frac{25 - 27}{27}\right| \cdot 100 = 7.41\%$$

(*b*) The points used for numerical differentiation are:

x: 3   3.25   3.5

$f(x)$:   27 3.25^3 3.5^3

Using Eq. (6.24), the derivative using the three points forward finite difference formula is:

$$f'(3) = \frac{-3f(3) + 4f(3.25) - f(3.5)}{2 \cdot 0.25} = \frac{-3 \cdot 27 + 4 \cdot 3.25^3 - 3.5^3}{0.5} = 26.875$$

$$error = \left|\frac{26.875 - 27}{27}\right| \cdot 100 = 0.46\%$$

The results show that the three-point forward difference formula gives a much more accurate value for the first derivative than the two-point forward finite difference formula in Example 6-1. For $h = 1$ the error reduces from 37.04% to 7.4%, and for $h = 0.25$ the error reduces from 8.57% to 0.46%.

6.3.2 Finite Difference Formulas for the Second Derivative

The same approach used in Section 6.3.1 to develop finite difference formulas for the first derivative can be used to develop expressions for higher-order derivatives. In this section, expressions based on central differences, one-sided forward differences, and one-sided backward differences are presented for approximating the second derivative at a point x_i.

Three-point central difference formula for the second derivative

Central difference formulas for the second derivative can be developed using any number of points on either side of the point x_i, where the second derivative is to be evaluated. The formulas are derived by writing the Taylor series expansion with a remainder at points on either side of x_i in terms of the value of the function and its derivatives at point x_i. Then, the equations are combined in such a way that the terms containing the first derivatives are eliminated. For example, for points x_{i+1}, and x_{i-1} the four-term Taylor series expansion with a remainder is:

$$f(x_{i+1}) = f(x_i) + f'(x_i)h + \frac{f''(x_i)}{2!}h^2 + \frac{f'''(x_i)}{3!}h^3 + \frac{f^{(4)}(\xi_1)}{4!}h^4 \quad (6.26)$$

$$f(x_{i-1}) = f(x_i) - f'(x_i)h + \frac{f''(x_i)}{2!}h^2 - \frac{f'''(x_i)}{3!}h^3 + \frac{f^{(4)}(\xi_2)}{4!}h^4 \quad (6.27)$$

where ξ_1 is a value of x between x_i and x_{i+1}, and ξ_2 is a value of x between x_i and x_{i-1}. Adding Eq. (6.26) and Eq. (6.27) gives:

$$f(x_{i+1}) + f(x_{i-1}) = 2f(x_i) + 2\frac{f''(x_1)}{2!}h^2 + \frac{f^{(4)}(\xi_1)}{4!}h^4 + \frac{f^{(4)}(\xi_2)}{4!}h^4 \quad (6.28)$$

An estimate for the second derivative can be obtained by solving Eq. (6.28) for $f''(x_i)$ while neglecting the remainder terms. This introduces a truncation error of the order of h^2.

$$f''(x_i) = \frac{f(x_{i-1}) - 2f(x_i) + f(x_{i+1})}{h^2} + O(h^2) \quad (6.29)$$

Equation (6.29) is the three-point central difference formula that provides an estimate of the second derivative at point x_i from the value of the function at that point, at the previous point, x_{i-1}, and at the next point x_{i+1}, with a truncation error of $O(h^2)$.

The same procedure can be used to develop a higher-order (fourth-order) accurate formula involving the five points x_{i-2}, x_{i-1}, x_i, x_{i+1}, and x_{i+2}:

$$f''(x_i) = \frac{-f(x_{i-2}) + 16f(x_{i-1}) - 30f(x_i) + 16f(x_{i+1}) - f(x_{i+2})}{12h^2} + O(h^4) \quad (6.30)$$

Three-point forward and backward difference formulas for the second derivative

The ***three-point forward difference*** formula that estimates the second derivative at point x_i from the value of that point and the next two points, x_{i+1} and x_{i+2}, is developed by multiplying Eq. (6.21) by 2 and subtracting it from Eq. (6.22). The resulting equation is then solved for $f''(x_i)$:

$$f''(x_i) = \frac{f(x_i) - 2f(x_{i+1}) + f(x_{i+2})}{h^2} + O(h) \quad (6.31)$$

The ***three-point backward difference*** formula that estimates the second derivative at point x_i from the value of that point and the previous two points, x_{i-1} and x_{i-2}, is derived similarly. It is done by writing the three-term Taylor series expansion with a remainder, for the value of the function at point x_{i-1} and at point x_{i-2}, in terms of the value of the function and its derivatives at point x_i. The equations are then manipulated to obtain an equation without the terms that include the first derivative, which is then solved for $f''(x_i)$. The resulting formula is:

$$f''(x_i) = \frac{f(x_{i-2}) - 2f(x_{i-1}) + f(x_i)}{h^2} + O(h) \quad (6.32)$$

Formulas for higher-order derivatives can be derived by using the same methods that are used here for the second derivative. A list of such formulas is given in the next section. Example 6-3 shows application of the three-point forward difference formula for the second derivative.

Example 6-4: Comparing numerical and analytical differentiation.

Consider the function $f(x) = \dfrac{2^x}{x}$. Calculate the second derivative at $x = 2$ numerically with the three-point central difference formula using:
(a) Points $x = 1.8$, $x = 2$, and $x = 2.2$.
(b) Points $x = 1.9$, $x = 2$, and $x = 2.1$.
Compare the results with the exact (analytical) derivative.
SOLUTION

Analytical differentiation: The second derivative of the function $f(x) = \dfrac{2^x}{x}$ is:

$$f''(x) = \frac{2^x[\ln(2)]^2}{x} - \frac{2 \cdot 2^x \ln(2)}{x^2} + \frac{2 \cdot 2^x}{x^3}$$

and the value of the derivative at $x = 2$ is $f''(2) = 0.5746$.

Numerical differentiation

(*a*) The numerical differentiation is done by substituting the values of the points $x = 1.8$, $x = 2$, and $x = 2.2$ in Eq. (6.29). The operations are done with MATLAB, in the Command Window:

```
>> xa = [1.8 2 2.2];
>> ya = 2.^xa./xa;
>> df = (ya(1) - 2*ya(2) + ya(3))/0.2^2
df =
   0.57748177389232
```

(*b*) The numerical differentiation is done by substituting the values of the points $x = 1.9$, $x = 2$, and $x = 2.1$ in Eq. (6.29). The operations are done with MATLAB, in the Command Window:

```
>> xb = [1.9 2 2.1];
>> yb = 2.^xb./xb;
>> dfb = (yb(1) - 2*yb(2) + yb(3))/0.1^2
dfb =
   0.57532441566441
```

Error in part (*a*): $error = \dfrac{0.577482 - 0.5746}{0.5746} \cdot 100 = 0.5016\,\%$

Error in part (*b*): $error = \dfrac{0.575324 - 0.5746}{0.5746} \cdot 100 = 0.126\,\%$

The results show that the three-point central difference formula gives a quite accurate approximation for the value of the second derivative.

6.4 SUMMARY OF FINITE DIFFERENCE FORMULAS FOR NUMERICAL DIFFERENTIATION

Table 6-1 lists difference formulas, of various accuracy, that can be used for numerical evaluation of first, second, third, and fourth derivatives. The formulas can be used when the function that is being differentiated is specified as a set of discrete points with the independent variable equally spaced.

Table 6-1: Finite difference formulas.

First Derivative		
Method	**Formula**	**Truncation Error**
Two-point forward difference	$f'(x_i) = \dfrac{f(x_{i+1}) - f(x_i)}{h}$	$O(h)$
Three-point forward difference	$f'(x_i) = \dfrac{-3f(x_i) + 4f(x_{i+1}) - f(x_{i+2})}{2h}$	$O(h^2)$

Table 6-1: Finite difference formulas.

Two-point backward difference	$f'(x_i) = \dfrac{f(x_i) - f(x_{i-1})}{h}$	$O(h)$
Three-point backward difference	$f'(x_i) = \dfrac{f(x_{i-2}) - 4f(x_{i-1}) + 3f(x_i)}{2h}$	$O(h^2)$
Two-point central difference	$f'(x_i) = \dfrac{f(x_{i+1}) - f(x_{i-1})}{2h}$	$O(h^2)$
Four-point central difference	$f'(x_i) = \dfrac{f(x_{i-2}) - 8f(x_{i-1}) + 8f(x_{i+1}) - f(x_{i+2})}{12h}$	$O(h^4)$

Second Derivative		
Method	**Formula**	**Truncation Error**
Three-point forward difference	$f''(x_i) = \dfrac{f(x_i) - 2f(x_{i+1}) + f(x_{i+2})}{h^2}$	$O(h)$
Four-point forward difference	$f''(x_i) = \dfrac{2f(x_i) - 5f(x_{i+1}) + 4f(x_{i+2}) - f(x_{i+3})}{h^2}$	$O(h^2)$
Three-point backward difference	$f''(x_i) = \dfrac{f(x_{i-2}) - 2f(x_{i-1}) + f(x_i)}{h^2}$	$O(h)$
Four-point backward difference	$f''(x_i) = \dfrac{-f(x_{i-3}) + 4f(x_{i-2}) - 5f(x_{i-1}) + 2f(x_i)}{h^2}$	$O(h^2)$
Three-point central difference	$f''(x_i) = \dfrac{f(x_{i-1}) - 2f(x_i) + f(x_{i+1})}{h^2}$	$O(h^2)$
Five-point central difference	$f''(x_i) = \dfrac{-f(x_{i-2}) + 16f(x_{i-1}) - 30f(x_i) + 16f(x_{i+1}) - f(x_{i+2})}{12h^2}$	$O(h^4)$

Third Derivative		
Method	**Formula**	**Truncation Error**
Four-point forward difference	$f'''(x_i) = \dfrac{-f(x_i) + 3f(x_{i+1}) - 3f(x_{i+2}) + f(x_{i+3})}{h^3}$	$O(h)$
Five-point forward difference	$f'''(x_i) = \dfrac{-5f(x_i) + 18f(x_{i+1}) - 24f(x_{i+2}) + 14f(x_{i+3}) - 3f(x_{i+4})}{2h^3}$	$O(h^2)$
Four-point backward difference	$f'''(x_i) = \dfrac{-f(x_{i-3}) + 3f(x_{i-2}) - 3f(x_{i-1}) + f(x_i)}{h^3}$	$O(h)$
Five-point backward difference	$f'''(x_i) = \dfrac{3f(x_{i-4}) - 14f(x_{i-3}) + 24f(x_{i-2}) - 18f(x_{i-1}) + 5f(x_i)}{2h^3}$	$O(h^2)$
Four-point central difference	$f'''(x_i) = \dfrac{-f(x_{i-2}) + 2f(x_{i-1}) - 2f(x_{i+1}) + f(x_{i+2})}{2h^3}$	$O(h^2)$
Six-point central difference	$f'''(x_i) = \dfrac{f(x_{i-3}) - 8f(x_{i-2}) + 13f(x_{i-1}) - 13f(x_{i+1}) + 8f(x_{i+2}) - f(x_{i+3})}{8h^3}$	$O(h^4)$

Table 6-1: Finite difference formulas.

Fourth Derivative		
Method	Formula	Truncation Error
Five-point forward difference	$f^{\text{iv}}(x_i) = \dfrac{f(x_i) - 4f(x_{i+1}) + 6f(x_{i+2}) - 4f(x_{i+3}) + f(x_{i+4})}{h^4}$	$O(h)$
Six-point forward difference	$f^{\text{iv}}(x_i) = \dfrac{3f(x_i) - 14f(x_{i+1}) + 26f(x_{i+2}) - 24f(x_{i+3}) + 11f(x_{i+4}) - 2f(x_{i+5})}{h^4}$	$O(h^2)$
Five-point backward difference	$f^{\text{iv}}(x_i) = \dfrac{f(x_{i-4}) - 4f(x_{i-3}) + 6f(x_{i-2}) - 4f(x_{i-1}) + f(x_i)}{h^4}$	$O(h)$
Six-point backward difference	$f^{\text{iv}}(x_i) = \dfrac{-2f(x_{i-5}) + 11f(x_{i-4}) - 24f(x_{i-3}) + 26f(x_{i-2}) - 14f(x_{i-1}) + 3f(x_i)}{h^4}$	$O(h^2)$
Five-point central difference	$f^{\text{iv}}(x_i) = \dfrac{f(x_{i-2}) - 4f(x_{i-1}) + 6f(x_i) - 4f(x_{i+1}) + f(x_{i+2})}{h^4}$	$O(h^2)$
Seven-point central difference	$f^{\text{iv}}(x_i) = \dfrac{f(x_{i-3}) + 12f(x_{i-2}) - 39f(x_{i-1}) + 56f(x_i) + 39f(x_{i+1}) + 12f(x_{i+2}) - f(x_{i+3})}{6h^4}$	$O(h^4)$

6.5 DIFFERENTIATION FORMULAS USING LAGRANGE POLYNOMIALS

The differentiation formulas can also be derived by using Lagrange polynomials. For the first derivative, the two-point central, three-point forward, and three-point backward difference formulas are obtained by considering three points (x_i, y_i), (x_{i+1}, y_{i+1}), and (x_{i+2}, y_{i+2}). The polynomial, in Lagrange form, that passes through the points is given by:

$$f(x) = \frac{(x-x_{i+1})(x-x_{i+2})}{(x_i-x_{i+1})(x_i-x_{i+2})} y_i + \frac{(x-x_i)(x-x_{i+2})}{(x_{i+1}-x_i)(x_{i+1}-x_{i+2})} y_{i+1}$$
$$+ \frac{(x-x_i)(x-x_{i+1})}{(x_{i+2}-x_i)(x_{i+2}-x_{i+1})} y_{i+2} \tag{6.33}$$

Differentiating Eq. (6.33) gives:

$$f'(x) = \frac{2x-x_{i+1}-x_{i+2}}{(x_i-x_{i+1})(x_i-x_{i+2})} y_i + \frac{2x-x_i-x_{i+2}}{(x_{i+1}-x_i)(x_{i+1}-x_{i+2})} y_{i+1}$$
$$+ \frac{2x-x_i-x_{i+1}}{(x_{i+2}-x_i)(x_{i+2}-x_{i+1})} y_{i+2} \tag{6.34}$$

The first derivative at either one of the three points is calculated by substituting the corresponding value of x (x_i, x_{i+1} or x_{i+2}) in Eq. (6.34). This gives the following three formulas for the first derivative at the three points.

$$f'(x_i) = \frac{2x_i - x_{i+1} - x_{i+2}}{(x_i - x_{i+1})(x_i - x_{i+2})} y_i + \frac{x_i - x_{i+2}}{(x_{i+1} - x_i)(x_{i+1} - x_{i+2})} y_{i+1}$$
$$+ \frac{x_i - x_{i+1}}{(x_{i+2} - x_i)(x_{i+2} - x_{i+1})} y_{i+2} \tag{6.35}$$

$$f'(x_{i+1}) = \frac{x_{i+1} - x_{i+2}}{(x_i - x_{i+1})(x_i - x_{i+2})} y_i + \frac{2x_{i+1} - x_i - x_{i+2}}{(x_{i+1} - x_i)(x_{i+1} - x_{i+2})} y_{i+1}$$
$$+ \frac{x_{i+1} - x_i}{(x_{i+2} - x_i)(x_{i+2} - x_{i+1})} y_{i+2} \tag{6.36}$$

$$f'(x_{i+2}) = \frac{x_{i+2} - x_{i+1}}{(x_i - x_{i+1})(x_i - x_{i+2})} y_i + \frac{x_{i+2} - x_i}{(x_{i+1} - x_i)(x_{i+1} - x_{i+2})} y_{i+1}$$
$$+ \frac{2x_{i+2} - x_i - x_{i+1}}{(x_{i+2} - x_i)(x_{i+2} - x_{i+1})} y_{i+2} \tag{6.37}$$

When the points are equally spaced, Eq. (6.35) reduces to the three-point forward difference formula (Eq. (6.24)), Eq. (6.36) reduces to the two-point central difference formula (Eq. (6.20)), and Eq. (6.37) reduces to the three-point backward difference formula (Eq. (6.25)).

Equation (6.34) has two other important features. It can be used when the points *are not* spaced equally, and it can be used for calculating the value of the first derivative at any point between x_i and x_{i+2}.

Other difference formulas with more points and for higher-order derivatives can also be derived by using Lagrange polynomials. Use of Lagrange polynomials to derive finite difference formulas is sometimes easier than using the Taylor series. However, the Taylor series provides an estimate of the truncation error.

6.6 DIFFERENTIATION USING CURVE FITTING

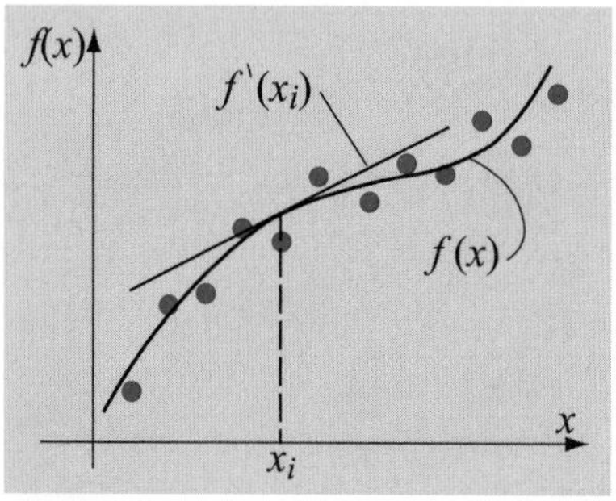

Figure 6-6: Numerical differentiation using curve fitting.

A different approach to differentiation of data specified by a set of discrete points is to first approximate with an analytical function that can be easily differentiated. The approximate function is then differentiated for calculating the derivative at any of the points (Fig. 6-6). Curve fitting is described in Chapter 5. For data that shows a nonlinear relationship, curve fitting is often done by using least squares with an exponential function, a power function, low-order polynomial, or a combination of a nonlinear functions, which are simple to differentiate. This procedure may be preferred when the data contains scatter, or noise, since the curved-fitted function smooths out the noise.

6.7 USE OF MATLAB BUILT-IN FUNCTIONS FOR NUMERICAL DIFFERENTIATION

In general, it is recommended that the techniques described in this chapter be used to develop script files that perform the desired differentiation. MATLAB does not have built-in functions that perform numerical differentiation of an arbitrary function or discrete data. There is, however, a built-in function called $\texttt{diff}$, which can be used to perform numerical differentiation, and another built-in function called $\texttt{poly-der}$, which determines the derivative of polynomial.

The $\texttt{diff}$ command

The built-in function $\texttt{diff}$ calculates the differences between adjacent elements of a vector. The simplest form of the command is:

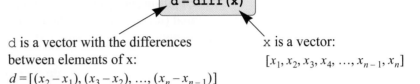

$$d = \texttt{diff(x)}$$

d is a vector with the differences between elements of x:
$$d = [(x_2 - x_1), (x_3 - x_2), ..., (x_n - x_{n-1})]$$

x is a vector:
$$[x_1, x_2, x_3, x_4, ..., x_{n-1}, x_n]$$

The vector $\texttt{d}$ is one element shorter than the vector $\texttt{x}$.

For a function represented by a discrete set of n points (x_1, y_1), (x_2, y_2), (x_3, y_3), ..., (x_n, y_n), the first derivative with the two-point forward difference formula, Eq. (6.5), can be calculated using the $\texttt{diff}$ command by entering $\texttt{diff(y)./diff(x)}$. The result is a vector whose elements are:

$$\left[\frac{(y_2 - y_1)}{(x_2 - x_1)}, \frac{(y_3 - y_2)}{(x_3 - x_2)}, \frac{(y_4 - y_3)}{(x_4 - x_3)}, ..., \frac{(y_n - y_{n-1})}{(x_n - x_{n-1})}\right]$$

When the spacing between the points is the same such that $h = (x_2 - x_1) = (x_3 - x_2) = ... = (x_n - x_{n-1})$, then the first derivative with the two-point forward difference formula, Eq. (6.12), can be calculated using the $\texttt{diff}$ command by entering $\texttt{diff(y)/h}$.

The $\texttt{diff}$ command has an additional optional input argument that can be used for calculating higher-order derivatives. Its form is:

$$d = \texttt{diff(x,n)}$$

where $\texttt{n}$ is a number (integer) that specifies the number of times that $\texttt{diff}$ is applied recursively. For example, $\texttt{diff(x,2)}$ is the same as $\texttt{diff(diff(x))}$. In other words, for an n-element vector $x_1, ..., x_n$ $\texttt{diff(x)}$ calculates a vector with $n - 1$ elements:

$$x_{i+1} - x_i \quad \text{for} \quad i = 1, ..., n-1 \tag{6.38}$$

and $\texttt{diff(x,2)}$ gives the vector with $n - 2$ elements:

$$((x_{i+2} - x_{i+1}) - (x_{i+1} - x_i)) = x_i - 2x_{i+1} + x_{i+2} \quad \text{for} \quad i = 1, ..., n-1 \tag{6.39}$$

The right-hand side of Eq. (6.39) is the same as the numerator of the three-point forward difference formula for the second derivative at $x = x_i$, Eq. (6.31). Consequently, for a function represented by a discrete set of n points (x_i, y_i), where the distance, h, between the points is the same, an estimate of the second derivative according to the three-point forward difference formula can be calculated with MATLAB by entering `diff(y,2)/h^2`.

Similarly, `diff(y,3)` yields the numerator of the third derivative in the four-point forward difference formula (see Table 6-1). In general, `diff(y,n)` gives the numerator in the forward difference formula of the nth derivative.

The `polyder` *command*

The built-in function `polyder` can calculate the derivative of a polynomial (it can also calculate the derivative of a product and quotient of two polynomials). The simplest form of the command is:

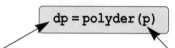

dp is a vector with the coefficients of the polynomial that is the derivative of the polynomial p.

p is a vector with the coefficients of the polynomial that is differentiated.

In MATLAB, polynomials are represented by a row vector in which the elements are the coefficients of the polynomial in order from the coefficient of the highest order term to the zeroth order term. If p is a vector of length n, then dp will be a vector of length $n - 1$. For example, to find the derivative of the polynomial $f(x) = 4x^3 + 5x + 7$, define a vector p = [4 0 5 7], and type df = polyder(p). The output will be df = [12 0 5], representing $12x^2 + 5$, which is the derivative of $f(x)$.

```
>> p = [4 0 5 7];
>> dp = polyder(p)
dp =
   12   0   5
```

The `polyder` command can be useful for calculating the derivative when a function represented by a set of discrete data points is approximated by a curve-fitted polynomial.

6.8 RICHARDSON'S EXTRAPOLATION

Richardson's extrapolation is a method for calculating a more accurate approximation of a derivative from two less accurate approximations of that derivative.

In general terms, consider the value, D, of a derivative (unknown) that is calculated by the difference formula:

$$D = D(h) + k_2 h^2 + k_4 h^4 \qquad (6.40)$$

where $D(h)$ is a function that approximates the value of the derivative and k_2h^2 and k_4h^4 are error terms in which the coefficients, k_2 and k_4 are independent of the spacing h. Using the same formula for calculating the value of D but using a spacing of $h/2$ gives:

$$D = D\left(\frac{h}{2}\right) + k_2\left(\frac{h}{2}\right)^2 + k_4\left(\frac{h}{2}\right)^4 \tag{6.41}$$

Equation (6.41) can be rewritten (after multiplying by 4) as:

$$4D = 4D\left(\frac{h}{2}\right) + k_2h^2 + k_4\frac{h^4}{4} \tag{6.42}$$

Subtracting Eq. (6.40) from Eq. (6.42) eliminates the terms with h^2, and gives:

$$3D = 4D\left(\frac{h}{2}\right) - D(h) - k_4\frac{3h^4}{4} \tag{6.43}$$

Solving Eq. (6.43) for D yields a new approximation for the derivative:

$$D = \frac{1}{3}\left(4D\left(\frac{h}{2}\right) - D(h)\right) - k_4\frac{h^4}{4} \tag{6.44}$$

The error term in Eq. (6.44) is now $O(h^4)$. The value, D, of the derivative can now be approximated by:

$$D = \frac{1}{3}\left(4D\left(\frac{h}{2}\right) - D(h)\right) + O(h^4) \tag{6.45}$$

This means that an approximated value of D with error $O(h^4)$ is obtained from two lower-order approximations ($D(h)$ and $D\left(\frac{h}{2}\right)$) that were calculated with an error $O(h^2)$. Equation (6.45) can be used for obtaining a more accurate approximation for any formula that calculates the derivative with an error $O(h^2)$. The formula is used for calculating one approximation with a spacing of h and a second approximation with a spacing of $h/2$. The two approximations are then substituted in Eq. (6.45), which gives a new estimate with an error of $O(h^4)$. The procedure is illustrated in Example 6-5.

Equation (6.45) can also be derived directly from a particular finite difference formula. As an example, consider the three-point central difference formula for the first derivative for points with equal spacing of h, such that $x_{i+1} = x_i + h$ and $x_{i-1} = x_i - h$. Writing the five-term Taylor series expansion with a remainder for the value of the function at point x_{i+1} in terms of the value of the function and its derivatives at point x_i gives:

$$f(x_i + h) = f(x_i) + f'(x_i)h + \frac{f''(x_i)}{2!}h^2 + \frac{f'''(x_i)}{3!}h^3 + \frac{f^{IV}(x_i)}{4!}h^4 + \frac{f^V(\xi_1)}{5!}h^5 \tag{6.46}$$

where ξ_1 is a value of x between x_i and $x_i + h$. In the same manner, the value of the function at point x_{i-1} is expressed in terms of the value of the function and its derivatives at point x_i:

$$f(x_i - h) = f(x_i) - f'(x_i)h + \frac{f''(x_i)}{2!}h^2 - \frac{f'''(x_i)}{3!}h^3 + \frac{f^{iv}(x_i)}{4!}h^4 - \frac{f^v(\xi_2)}{5!}h^5 \quad (6.47)$$

where ξ_2 is a value of x between $x_i - h$ and x_i. Subtracting Eq. (6.47) from Eq. (6.46) gives:

$$f(x_i + h) - f(x_i - h) = 2f'(x_i)h + 2\frac{f'''(x_i)}{3!}h^3 + \frac{f^v(\xi_1)}{5!}h^5 + \frac{f^v(\xi_2)}{5!}h^5 \quad (6.48)$$

Assuming that the fifth derivative is continuous in the interval $[x_{i-1}, x_{i+1}]$, the two remainder terms in Eq. (6.48) can be combined and written as $O(h^5)$. Then, solving Eq. (6.48) for $f'(x_i)$ gives:

$$f'(x_i) = \frac{f(x_i + h) - f(x_i - h)}{2h} - \frac{f'''(x_i)}{3!}h^2 + O(h^4) \quad (6.49)$$

which is the approximation for the first derivative with a spacing of h.

The derivation of Eqs. (6.46)–(6.49) can be repeated if the spacing between the points is changed to $h/2$. For this case the equation for the value of the derivative is:

$$f'(x_i) = \frac{f(x_i + h/2) - f(x_i - h/2)}{2(h/2)} - \frac{f'''(x_i)}{3!}\left(\frac{h}{2}\right)^2 + O(h^4) \quad (6.50)$$

or

$$f'(x_i) = \frac{f(x_i + h/2) - f(x_i - h/2)}{h} - \frac{f'''(x_i)}{4 \cdot 3!}h^2 + O(h^4) \quad (6.51)$$

Multiplying Eq. (6.51) by 4 gives:

$$4f'(x_i) = 4\frac{f(x_i + h/2) - f(x_i - h/2)}{h} - \frac{f'''(x_i)}{3!}h^2 + O(h^4) \quad (6.52)$$

Subtracting Eq. (6.49) from Eq. (6.52) and dividing the result by 3 yields an approximation for the first derivative with error $O(h^4)$:

$$f'(x_i) = \frac{1}{3}\left[4\frac{f(x_i + h/2) - f(x_i - h/2)}{h} - \frac{f(x_i + h) - f(x_i - h)}{2h}\right] + O(h^4) \quad (6.53)$$

First derivative calculated with two-point central difference formula, Eq. (6.20), with error $O(h^2)$ for points with spacing of $h/2$.

First derivative calculated with two-point central difference formula, Eq. (6.20), with error $O(h^2)$ for points with spacing of h.

Equation (6.53) is a special case of Eq. (6.45) where the derivatives are calculated with the two-point central difference formula. Equation

(6.45) can be used with any difference formula with an error $O(h^2)$.

Richardson's extrapolation method can also be used with approximations that have errors of higher order. Two approximations with an error $O(h^4)$—one calculated from points with spacing of h and the other from points with spacing of $h/2$—can be used for calculating a more accurate approximation with an error $O(h^6)$. The formula for this case is:

$$D = \frac{1}{15}\left(16D\left(\frac{h}{2}\right) - D(h)\right) + O(h^6) \tag{6.54}$$

Application of Richardson's extrapolation is shown in Example 6-5.

Example 6-5: Using Richardson's extrapolation in differentiation.

Use Richardson's extrapolation with the results in Example 6-4 to calculate a more accurate approximation for the derivative of the function $f(x) = \frac{2^x}{x}$ at the point $x = 2$.

Compare the results with the exact (analytical) derivative.

SOLUTION

In Example 6-4 two approximations of the derivative of the function at $x = 2$ were calculated using the central difference formula in which the error is $O(h^2)$. In one approximation $h = 0.2$, and in the other $h = 0.1$. The results from Example 6-4 are:

for $h = 0.2$, $f'(2) = 0.577482$. The error in this approximation is 0.5016 %.

for $h = 0.1$, $f'(2) = 0.575324$. The error in this approximation is 0.126 %.

Richardson's extrapolation can be used by substituting these results in Eq. (6.45) (or Eq. (6.53)):

$$D = \frac{1}{3}\left(4D\left(\frac{h}{2}\right) - D(h)\right) + O(h^4) = \frac{1}{3}(4 \cdot 0.575324 - 0.577481) = 0.574605$$

The error now is $\quad error = \dfrac{0.574605 - 0.5746}{0.5746} \cdot 100 = 0.00087\ \%$

This result shows that a much more accurate approximation is obtained by using Richardson's extrapolation.

6.9 ERROR IN NUMERICAL DIFFERENTIATION

Throughout this chapter, expressions have been given for the truncation error, also known as the discretization error. These expressions are generated by the particular numerical scheme used for deriving a specific finite difference formula to estimate the derivative. In each case, the truncation error depends on h (the spacing between the points) raised to some power. Clearly, the implication is that as h is made smaller and smaller, the error could be made arbitrarily small. When the function to be differentiated is specified as a set of discrete data points, the spacing is fixed, and the truncation error cannot be reduced by reducing the size of h. In this case, a smaller truncation error can be obtained by using a finite difference formula that has a higher-order truncation error.

When the function that is being differentiated is given by a mathematical expression, the spacing h for the points that are used in the finite difference formulas can be defined by the user. It might appear then that h can be made arbitrarily small and there is no limit to how small the error can be made. This, however, is not true because the total error is composed of two parts. One is the truncation error arising from the numerical method (the specific finite difference formula) that is used. The second part is a round-off error arising from the finite precision of the particular computer used. Therefore, even if the truncation error can be made vanishingly small by choosing smaller and smaller values of h, the round-off error still remains, or can even grow as h is made smaller and smaller. Example 6-6 illustrates this point.

Example 6-6: Comparing numerical and analytical differentiation.

Consider the function $f(x) = e^x$. Write an expression for the first derivative of the function at $x = 0$ using the two-point central difference formula in Eq. (6.20). Investigate the effect that the spacing, h, between the points has on the truncation and round-off errors.

SOLUTION

The two-point central difference formula in Eq. (6.20) is:

$$f'(x_i) = \frac{f(x_{i+1}) - f(x_{i-1})}{2h} - 2\frac{f'''(\xi)}{3!}h^2$$

where ξ is a value of x between x_{i-1} and x_{i+1}

The points used for calculating the derivative of $f(x) = e^x$ at $x = 0$ are $x_{i-1} = -h$ and $x_{i+1} = h$. Substituting these points in the formula gives:

$$f'(0) = \frac{e^h - e^{-h}}{2h} - 2\frac{f'''(\xi)}{3!}h^2 \qquad (6.55)$$

When the computer calculates the values of e^h and e^{-h}, a round-off error is introduced, since the computer has finite precision. Consequently, the terms e^h and e^{-h} in Eq. (6.55) are replaced by $e^h + R_1$ and $e^{-h} + R_2$ where now e^h and e^{-h} are the exact values, and R_1 and R_2 are the round-off errors:

$$f'(0) = \frac{e^h + R_1 - e^{-h} - R_2}{2h} - 2\frac{f'''(\xi)}{3!}h^2 = \frac{e^h - e^{-h}}{2h} + \frac{R_1 - R_2}{2h} - 2\frac{f'''(\xi)}{3!}h^2 \qquad (6.56)$$

In Eq. (6.56) the last term on the right-hand side is the truncation error. In this term, the value of $f'''(\xi)$ is not known, but it is bounded. This means that as h decreases the truncation error decreases. The round-off error is $(R_1 - R_2)/(2h)$. As h decreases the round-off error increases. The total error is the sum of the truncation error and round-off error. Its behavior is shown schematically in the figure on the right. As h decreases, the total error initially decreases, but after a certain value (which depends on the precision of the computer used) the total error increases as h decreases further.

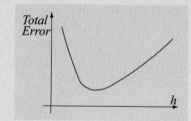

6.10 NUMERICAL PARTIAL DIFFERENTIATION

All the numerical differentiation methods presented so far considered functions with one independent variable. Most problems in engineering or science involve functions of several independent variables since real-life applications are either two or three-dimensional, and in addition may be a function of time. For example, the temperature distribution in an object is a function of the three coordinates used to describe the object: $T(x, y, z)$, or $T(r, \theta, z)$, or $T(r, \theta, \phi)$. The temperature may also be a function of time: $T(x, y, z, t)$. If there is a need for evaluating the amount of heat flow in a given direction, say z, the partial derivative in the z direction is required: $\dfrac{\partial T(x, y, z, t)}{\partial z}$. Another example is the determination of strains from displacements. If two-dimensional displacements are measured on the surface of a structure, the strains are determined from the partial derivatives of the displacements.

For a function of several independent variables, the partial derivative of the function with respect to one of the variables represents the rate of change of the value of the function with respect to this variable, while all the other variables are kept constant (see Section 2.6). For a function $f(x, y)$ with two independent variables, the partial derivatives with respect to x and y at the point (a, b) are defined as:

$$\left.\frac{\partial f(x, y)}{\partial x}\right|_{\substack{x = a \\ y = b}} = \lim_{x \to a} \frac{f(x, b) - f(a, b)}{x - a} \tag{6.57}$$

$$\left.\frac{\partial f(x, y)}{\partial y}\right|_{\substack{x = a \\ y = b}} = \lim_{y \to b} \frac{f(a, y) - f(a, b)}{y - b} \tag{6.58}$$

This means that the finite difference formulas that are used for approximating the derivatives of functions with one independent variable can be adopted for calculating partial derivatives. The formulas are applied for one of the variables, while the other variables are kept constant. For example, consider a function of two independent variables $f(x, y)$ specified as a set of discrete $m \cdot n$ points (x_1, y_1), (x_1, y_2), ..., (x_n, y_m). The spacing between the points in each direction is constant such that $h_x = x_{i+1} - x_i$ and $h_y = y_{i+1} - y_i$. Figure 6-7 shows a case where $n = 5$ and $m = 4$. An approximation for the partial derivative at point (x_i, y_i) with the two-point forward difference formula is:

$$\left.\frac{\partial f}{\partial x}\right|_{\substack{x = x_i \\ y = y_i}} = \frac{f(x_{i+1}, y_i) - f(x_i, y_i)}{h_x} \tag{6.59}$$

$$\left.\frac{\partial f}{\partial y}\right|_{\substack{x = x_i \\ y = y_i}} = \frac{f(x_i, y_{i+1}) - f(x_i, y_i)}{h_y} \tag{6.60}$$

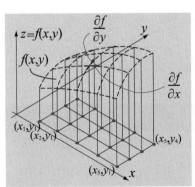

Figure 6-7: A function with two independent variables.

In the same way, the two-point backward and central difference formulas are:

$$\frac{\partial f}{\partial x}\bigg|_{\substack{x=x_i \\ y=y_i}} = \frac{f(x_i, y_i) - f(x_{i-1}, y_i)}{h_x} \qquad \frac{\partial f}{\partial y}\bigg|_{\substack{x=x_i \\ y=y_i}} = \frac{f(x_i, y_i) - f(x_i, y_{i-1})}{h_y} \qquad (6.61)$$

$$\frac{\partial f}{\partial x}\bigg|_{\substack{x=x_i \\ y=y_i}} = \frac{f(x_{i+1}, y_i) - f(x_{i-1}, y_i)}{2h_x} \qquad \frac{\partial f}{\partial y}\bigg|_{\substack{x=x_i \\ y=y_i}} = \frac{f(x_i, y_{i+1}) - f(x_i, y_{i-1})}{2h_y} \qquad (6.62)$$

The second partial derivatives with the three-point central difference formula are:

$$\frac{\partial^2 f}{\partial x^2}\bigg|_{\substack{x=x_i \\ y=y_i}} = \frac{f(x_{i-1}, y_i) - 2f(x_i, y_i) + f(x_{i+1}, y_i)}{h_x^2} \qquad (6.63)$$

$$\frac{\partial^2 f}{\partial y^2}\bigg|_{\substack{x=x_i \\ y=y_i}} = \frac{f(x_i, y_{i-1}) - 2f(x_i, y_i) + f(x_i, y_{i+1})}{h_y^2} \qquad (6.64)$$

Similarly, all the finite difference formulas listed in Section 6.4 can be adapted for calculating partial derivatives of different orders with respect to one of the variables.

A second-order partial derivative can also be mixed $\frac{\partial^2 f}{\partial x \partial y}$. This derivative is carried out successively $\frac{\partial^2 f}{\partial x \partial y} = \frac{\partial}{\partial y}\left(\frac{\partial f}{\partial x}\right) = \frac{\partial}{\partial x}\left(\frac{\partial f}{\partial y}\right)$. A finite difference formula for the mixed derivative can be obtained by using the first-order finite difference formulas for partial derivatives. For example, the second-order mixed four-point central finite difference formula is obtained from Eqs. (6.62):

$$\frac{\partial^2 f}{\partial x \partial y}\bigg|_{\substack{x=x_i \\ y=y_i}} = \frac{[f(x_{i+1}, y_{i+1}) - f(x_{i-1}, y_{i+1})] - [f(x_{i+1}, y_{i-1}) - f(x_{i-1}, y_{i-1})]}{2h_x \cdot 2h_y} \qquad (6.65)$$

Application of finite difference formulas for numerical partial differentiation is shown in Example 6-7.

Example 6-7: Numerical partial differentiation.

The following two-dimensional data for the x component of velocity u as a function of the two coordinates x and y is measured from an experiment:

	x = 1.0	x = 1.5	x = 2.0	x = 2.5	x = 3.0
y = 1.0	163	205	250	298	349
y = 2.0	228	291	361	437	517
y = 3.0	265	350	448	557	676

(a) Using central difference approximations, calculate $\partial u/\partial x$, $\partial u/(\partial y)$, $\partial^2 u/\partial y^2$, and $\partial^2 u/\partial x \partial y$ at the point $(2, 2)$.

(*b*) Using a three-point forward difference approximation, calculate $\partial u / \partial x$ at the point (2,2).

(*c*) Using a three-point forward difference approximation, calculate $\partial u / (\partial y)$ at the point $(2, 1)$.

SOLUTION

(*a*) In this part $x_i = 2$, $y_i = 2$, $x_{i-1} = 1.5$, $x_{i+1} = 2.5$, $y_{i-1} = 1$, $y_{i+1} = 3$, $h_x = 0.5$, $h_y = 1$.

Using Eqs. (6.59) and (6.60), the partial derivatives $\partial f / \partial x$ and $\partial f / (\partial y)$ are:

$$\left. \frac{\partial u}{\partial x} \right|_{\substack{x = x_i \\ y = y_i}} = \frac{u(x_{i+1}, y_i) - u(x_{i-1}, y_i)}{2h_x} = \frac{u(2.5, 2) - u(1.5, 2)}{2 \cdot 0.5} = \frac{437 - 291}{1} = 146$$

$$\left. \frac{\partial u}{\partial y} \right|_{\substack{x = x_i \\ y = y_i}} = \frac{u(x_i, y_{i+1}) - u(x_i, y_{i-1})}{2h_y} = \frac{u(2, 3) - u(2, 1)}{2 \cdot 1} = \frac{448 - 250}{2} = 99$$

The second partial derivative $\partial^2 u / \partial y^2$ is calculated with Eq. (6.64):

$$\left. \frac{\partial^2 u}{\partial y^2} \right|_{\substack{x = x_i \\ y = y_i}} = \frac{u(x_i, y_{i-1}) - 2u(x_i, y_i) + u(x_i, y_{i+1})}{h_y^2} = \frac{250 - (2 \cdot 361) + 448}{1^2} = -24$$

The second mixed derivative $\partial^2 u / \partial x \partial y$ is given by Eq. (6.65):

$$\left. \frac{\partial^2 u}{\partial x \partial y} \right|_{\substack{x = x_i \\ y = y_i}} = \frac{[u(x_{i+1}, y_{i+1}) - u(x_{i-1}, y_{i+1})] - [u(x_{i+1}, y_{i-1}) - u(x_{i-1}, y_{i-1})]}{2h_x \cdot 2h_y}$$

$$= \frac{[u(2.5, 3) - u(1.5, 3)] - [u(2.5, 1) - u(1.5, 1)]}{2 \cdot 0.5 \cdot 2 \cdot 1} = \frac{[557 - 350] - [298 - 205]}{2 \cdot 0.5 \cdot 2 \cdot 1} = 57$$

(*b*) In this part $x_i = 2$, $x_{i+1} = 2.5$, $x_{i+2} = 3.0$, $y_i = 2$, and $h_x = 0.5$. The formula for the partial derivative $\partial u / \partial x$ with the three-points forward finite difference formula can be written from the formula for the first derivative in Section 6.4.

$$\left. \frac{\partial u}{\partial x} \right|_{\substack{x = x_i \\ y = y_i}} = \frac{-3u(x_i, y_i) + 4u(x_{i+1}, y_i) - u(x_{i+2}, y_i)}{2h_x} =$$

$$= \frac{-3u(2, 2) + 4u(2.5, 2) - u(3.0, 2)}{2 \cdot 0.5} = \frac{-3 \cdot 361 + 4 \cdot 437 - 517}{2 \cdot 0.5} = 148$$

(*c*) In this part $y_i = 1$, $y_{i+1} = 2$, $y_{i+2} = 3$, $x_i = 2$, and $h_y = 1.0$. The formula for the partial derivative $\partial u / \partial y$ with the three-points forward difference formula can be written from the formula for the first derivative in Section 6.4.

$$\left. \frac{\partial u}{\partial y} \right|_{\substack{x = x_i \\ y = y_i}} = \frac{-3u(x_i, y_i) + 4u(x_i, y_{i+1}) - u(x_i, y_{i+2})}{2h_y} =$$

$$= \frac{-3u(2, 1) + 4u(2, 2) - u(2, 3)}{2 \cdot 1} = \frac{-3 \cdot 250 + 4 \cdot 361 - 448}{2 \cdot 1} = 123$$

6.11 PROBLEMS

Problems to be solved by hand
Solve the following problems by hand. When needed, use a calculator, or write a MATLAB script file to carry out the calculations. If using MATLAB, do not use built-in functions for differentiation.

6.1   Given the following data,

x	0.398	0.399	0.400	0.401	0.402
$f(x)$	0.408591	0.409671	0.410752	0.411834	0.412915

find the first derivative $f'(x)$ at the point $x = 0.399$.
(*a*) Use the three-point forward difference formula.
(*b*) Use the two-point central difference formula.

6.2   The following data shows the population of Nepal in selected years between 1980 and 2005.

Year	1980	1985	1990	1995	2000	2005
Population (millions)	15	17	19.3	22	24.5	27.1

Calculate the rate of growth of the population in millions per year for 2005.
(*a*) Use two-point backward difference formula.
(*b*) Use three-point backward difference formula.
(*c*) Using the slope in 2005 from part (b), apply the two-point central difference formula to extrapolate and predict the population in the year 2010.

6.3   The following data is given for the stopping distance of a car versus the speed at which it begins braking.

Stopping distance (ft)	20	35	80	110	150
Speed (mph)	20	30	40	50	60

(*a*) Calculate the rate of change of the stopping distance at a speed of 60 mph using (*i*) the two-point backward difference formula, and (*ii*) the three-point backward difference formula.
(*b*) Calculate an estimate for the stopping distance at 70 mph by using the results from part (*a*) for the slope and the two-point central difference formula applied at the speed of 60 mph.

6.4   Given three **unequally** spaced points (x_i, y_i), (x_{i+1}, y_{i+1}), and (x_{i+2}, y_{i+2}), use Taylor series expansion to develop a finite difference formula to evaluate the first derivative dy/dx at the point $x = x_i$. Verify that when the spacing between these points is equal, the three-point forward difference formula is obtained. The answer should involve y_i, y_{i+1}, and y_{i+2}.

6.5   Using a 4-term Taylor series expansion, derive a four-point backward difference formula for evaluating the first derivative of a function given by a set of unequally spaced points. The formula should give the derivative at point $x = x_i$, in terms of x_i, x_{i-1}, x_{i-2}, x_{i-3}, $f(x_i)$, $f(x_{i-1})$, $f(x_{i-2})$, and $f(x_{i-3})$.

6.6   Derive a finite difference approximation formula for $f''(x_i)$ using three points x_{i-1}, x_i, and x_{i+1}, where the spacing is such that $x_i - x_{i-1} = h$ and $x_{i+1} - x_i = 2h$.

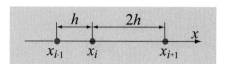

6.7   A particular finite difference formula for the first derivative of a function is:

$$f'(x_i) = \frac{-f(x_{i+3}) + 9f(x_{i+1}) - 8f(x_i)}{6h}$$

where the points x_i, x_{i+1}, x_{i+2}, and x_{i+3} are all equally spaced with step size h. What is the order of the truncation or discretization error?

6.8   The following data shows the number of female and male physicians in the U.S. for various years.

Year	1970	1980	1990	2000	2002	2003	2004
# males	308,627	413,395	511,227	618,182	638,182	646,493	647,347
# females	25,401	54,284	104,194	195,537	215,005	225,042	235,627

(*a*) Calculate the rate of change in the number of male and female physicians in 2002 by using the three-point backward difference formula for the derivative, with unequally spaced points, Eq. (6.37).

(*b*) Use the result from part (*a*) and the three-point central difference formula for the derivative with unequally spaced points, Eq. (6.36) to calculate (predict) the number of male and female physicians in 2003.

6.9   Use the data from Problem 6.8 and the four-point backward difference formula that was derived in Problem 6.5 for evaluating the first derivative of a function specified at unequally spaced points to calculate the following quantities.

(*a*) Evaluate the rate of change in the number of male and female physicians in 2004.

(*b*) Use the data from 2003, 2004, together with the slopes in 2004 from part (*a*) to estimate the year in which the number of female and male physicians will be equal. Use the three-point central difference formula for the derivative (Eq. (6.36)) of a function specified at unequally spaced points.

6.10   Use Lagrange interpolation polynomials to find the finite difference formula for the second derivative at the point $x = x_i$ using the unequally spaced points x_i, x_{i+1}, and x_{i+2}. What is the second derivative at $x = x_{i+1}$ and at $x = x_{i+2}$?

6.11   Given the function $f(x) = \dfrac{2x-1}{(x^4 \sin x + x + 1)^{1/4}}$ find the value of the first derivative at $x = 2$.

(*a*) Use analytical differentiation by hand.

(*b*) Use four-point central difference formula with $x_{i-2} = 1.96$, $x_{i-1} = 1.98$, $x_{i+1} = 2.02$, and $x_{i+2} = 2.04$. (Write a MATLAB program in a script file to carry out the calculations.)

6.12 For the function given in Problem 6.11, find the value of the second derivative at $x = 2$.
(*a*) Use analytical differentiation by hand.
(*b*) Use five-point central difference formula with $x_{i-2} = 1.96$, $x_{i-1} = 1.98$, $x_i = 2$, $x_{i+1} = 2.02$, and
$x_{i+2} = 2.04$. (Write a MATLAB program in a script file to carry out the calculations.)

6.13 The following data for the velocity component in the *x*-direction, *u*, are obtained as a function of the two coordinates *x* and *y*.

	$x = 0$	$x = 1$	$x = 2$	$x = 3$	$x = 4$
$y = 0$	0	2	8	13	15
$y = 1$	3	7	10	15	18
$y = 2$	14	8	14	22	22
$y = 3$	7	9	12	16	17
$y = 4$	5	7	10	9	14

Use the four-point central difference formula for $\dfrac{\partial^2 u}{\partial y \partial x}$ to evaluate this derivative at the point $(2, 3)$.

Problems to be programmed in MATLAB
Solve the following problems using MATLAB environment.

6.14 Write a MATLAB user-defined function that determines the first derivative of a function that is given by a set of points with constant spacing. For function name use yd = FirstDeriv(x,y). The input arguments x and y are vectors with the coordinates of the points, and the output argument yd is a vector with the values of the derivative at each point. At the first and last points, the function should calculate the derivative with the three-point forward and backward difference formulas, respectively. At all the other points FirstDeriv should use the two-point central difference formula. Use FirstDeriv to calculate the derivative of the function that is given in Problem 6.1.

6.15 Write a MATLAB user-defined function that calculates the second derivative of a function that is given by a set of data points with constant spacing. For function name and arguments use ydd=SecDeriv(x,y), where the input arguments x and y are vectors with the coordinates of the points, and ydd is a vector with the values of the second derivative at each point. For calculating the derivative, the function SecDeriv should use the finite difference formulas that have a truncation error of $O(h^2)$. Use SecDeriv for calculating the derivative of the function that is given the following set of points.

x	-1	-0.5	0	0.5	1	1.5	2	2.5	3	3.5	4	4.5
$f(x)$	-3.632	-0.3935	1	0.6487	-1.282	-4.518	-8.611	-12.82	-15.91	-15.88	-9.402	9.017

6.16 Write a MATLAB user-defined function that determines the first and second derivatives of a function that is given by a set of points with constant spacing. For function name use [yd,ydd] = FrstScndDeriv(x,y). The input arguments x and y are vectors with the coordinates of the points, and the output arguments yd and ydd are vectors with the values of the first and second derivatives, respectively, at each point. For calculating both derivatives, the function should use the finite difference formulas that have a truncation error of $O(h^2)$.

(a) Use the function FrstScndDeriv to calculate the derivatives of the function that is given by the data in Problem 6.15.

(b) Modify the function (rename it FrstScndDerivPt) such that it also creates three plots (on the same page in a column). The top plot should be of the function, the second plot of the first derivative, and the third of the second derivative. Apply the function FrstScndDerivPt to the data in Problem 6.15.

6.17 Write a MATLAB user-defined function for differentiation of a function that is given in an analytical form. For function name and arguments use dfx = DiffAnaly('FunName',xi). 'FunName' is a string with the name of a function file that calculates the value of the function to be differentiated at a given value of x, and xi is the value of x where the derivative is calculated. The function should calculate the derivative by using the two-point central difference formula. In the formula, the values of (x_{i+1}) and (x_{i-1}) should be taken to be 5% higher and 5% lower than the value of (x_i), respectively.

(a) Use the function to calculate the first derivative of $f(x) = \dfrac{2^x}{x}$ at $x = 2$.

(b) Use the function to calculate the first derivative of the function that is given in Problem 6.11 at $x = 2$.

6.18 Modify the MATLAB user-defined function in Problem 6.17 to include Richardson's extrapolation. The function should calculate a first estimate for the derivative as described in Problem 6.17, and a second estimate by taking the values of (x_{i+1}) and (x_{i-1}) to be 2.5% higher and 2.5% lower than the value of (x_i), respectively. The two estimates should then be used with Richardson's extrapolation for calculating the derivative. For function name and arguments use dfx=DiffRichardson('FunName',xi).

(a) Use the function to calculate the derivative of $f(x) = \dfrac{2^x}{x}$ at $x = 2$.

(b) Use the function to calculate the first derivative of the function that is given in Problem 6.11 at $x = 2$.

6.19 Write a MATLAB user-defined function that calculates the second derivative of a function that is given in an analytical form. For function name and arguments use ddfx = DDiffAnaly('FunName',xi). 'FunName' is a string with the name of a function file that calculates the value of the function to be differentiated at a given value of x, and xi is the value of x where the second derivative is calculated. The function should calculate the second derivative with the three-point central difference formula. In the formula, the values of (x_{i+1}) and (x_{i-1}) should be taken to be 5% higher and 5% lower than the value of (x_i), respectively.

(a) Use the function to calculate the second derivative of $f(x) = \dfrac{2^x}{x}$ at $x = 2$.

(b) Use the function to calculate the second derivative of the function that is given in Problem 6.11 at $x = 2$.

6.20 Write a MATLAB user-defined function that determines the first derivative of a function that is given by a set of points that are not spaced equally. For function name use yd = FirstDeriv-Lag(x,y). The input arguments x and y are vectors with the coordinates of the points, and the output argument yd is a vector with the values of the derivative at each point. At the first and last points, the function should calculate the derivative with Eqs. (6.35) and (6.37), respectively. At all the other points the function should use Eq. (6.36). Use the function to calculate the derivative of the function that is given by the following set of points.

x	-1	-0.6	-0.3	0	0.5	0.8	1.6	2.5	2.8	3.2	3.5	4
$f(x)$	-3.632	-0.8912	0.3808	1.0	0.6487	-0.3345	-5.287	-12.82	-14.92	-16.43	-15.88	-9.402

Problems in math, science, and engineering
Solve the following problems using MATLAB environment. As stated, use the MATLAB programs that are presented in the chapter, programs developed in previously solved problems, or MATLAB's built-in functions.

6.21 A 1-meter-long uniform beam is simply supported at both ends and is subjected to a load. The deflection of the beam is given by the differential equation:

$$\frac{d^2y}{dx^2} = -\frac{M(x)}{EI}$$

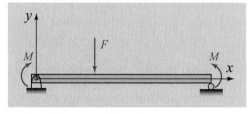

where y is the deflection, x is the coordinate measured along the length of the beam, $M(x)$ is the bending moment, and $EI = 1.2 \times 10^7$ N-m^2 is the flexural rigidity of the beam. The following data is obtained from measuring the deflection of the beam versus position:

x (m)	0	0.2	0.4	0.6	0.8	1.0
y (cm)	0	7.78	10.68	8.38	3.97	0

Find the bending moment $M(x)$ at each location x from this given data. Use the central difference approximation for the internal points ($O(h^2)$). At the endpoints use four-point forward and backward difference approximation ($O(h^2)$).

6.22 The distribution of the x-component of the velocity u of a fluid near a flat surface is measured as a function of the distance y from the surface:

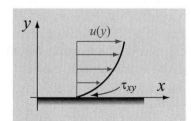

y (m)	0	0.002	0.004	0.006	0.008
u (m/s)	0	0.00618	0.011756	0.01618	0.019021

The shear stress τ_{yx} in the fluid is described by Newton's equation:

$$\tau_{yx} = \mu\frac{\partial u}{\partial y}$$

where μ is the coefficient of dynamic viscosity. The viscosity can be thought of as a measure of the internal friction within the fluid. Fluids that obey Newton's constitutive equation are called Newtonian fluids. Calculate the shear stress at $y = 0$ using (*i*) the two-point forward, and (*ii*) the three-points forward approximations for the derivative. Take $\mu = 0.002$ N-s/m^2.

6.23 A fin is an extended surface used to transfer heat from a base material (at $x = 0$) to an ambient. Heat flows from the base material through the base of the fin, through its outer surface, and through the tip. Measurement of the temperature distribution along a pin fin gives the following data:

x (cm)	0	1	2	3	4	5	6	7	8	9	10
T (K)	473	446.3	422.6	401.2	382	364.3	348.0	332.7	318.1	304.0	290.1

The fin has a length $L = 10$ cm, constant cross-sectional area of 1.6×10^{-5} m^2, and thermal conductivity $k = 240$ W/m/K. The heat flux (W/m^2) is given by $q_x = -k\dfrac{dT}{dx}$

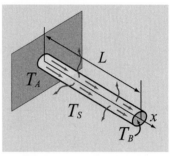

(*a*) Determine the heat flux at $x = 0$. Use the three-point forward difference formula for calculating the derivative.

(*b*) Determine the heat flux at $x = L$. Use the three-point backward difference formula for calculating the derivative.

(*c*) Determine the amount of heat (in W) lost between $x = 0$ and $x = L$.

(The heat flow per unit time in Watts is the heat flux multiplied by the cross-sectional area of the fin.)

6.24 The charge on the capacitor in the RLC circuit shown at various times after the switch is closed at time $t = 0$ is given in the following table. The current, I, as a function of time is given by $I(t) = \dfrac{dQ}{dt}$.

Determine the current as a function of time by numerically differentiating the data.

(*a*) Use the user-defined function `FirstDerivLag` that was written in Problem 6.20.

(*b*) Use the MATLAB built-in function `diff`.

In both parts plot I vs. t.

$Q \times 10^3$ (C)	0	6.67	16.93	23.38	27.44	29.02	29.5	29.29	27.17	24.58	21.43	18.56	16.61	16.01
t (ms)	0	6.9	12.84	17.23	21.24	24.05	27.01	28.49	32.95	36.33	40.1	44.06	48.21	51.94
$Q \times 10^3$ (C)	16.87	18.06	19.51	21.42	22.99	24.05	24.12	23.35	22.5	21.53	20.33	19.44	18.98	19.36
t (ms)	56.94	59.9	62.86	66.69	70.52	75.3	78.48	83.26	86.29	89.33	93.21	97.09	101.6	107.7

6.25 A radar station is tracking the motion of an aircraft. The recorded distance to the aircraft, r, and the angle θ during a period of 60 s is given in the following table. The magnitude of the instantaneous velocity and acceleration of the aircraft can be calculated by:

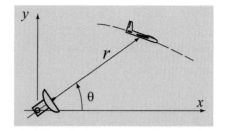

$$v = \sqrt{\left(\frac{dr}{dt}\right)^2 + \left(r\frac{d\theta}{dt}\right)^2} \qquad a = \sqrt{\left[\frac{d^2r}{dt^2} - r\left(\frac{d\theta}{dt}\right)^2\right]^2 + \left[r\frac{d^2\theta}{dt^2} + 2\frac{dr}{dt}\frac{d\theta}{dt}\right]^2}$$

Determine the magnitudes of the velocity and acceleration at the times given in the table. Plot the velocity and acceleration versus time (two separate plots on the same page). Solve the problem by writing a program in a script file. The program evaluates the various derivatives that are required for calculating the velocity and acceleration, and then makes the plots. For calculating the derivatives use:
(*a*) The user-defined function `FrstScndDeriv` that was written in Problem 6.16.
(*b*) MATLAB's built-in function `diff`.

t (s)	0	4	8	12	16	20	24	28
r (km)	18.803	18.861	18.946	19.042	19.148	19.260	19.376	19.495
θ(rad)	0.7854	0.7792	0.7701	0.7594	0.7477	0.7350	0.7215	0.7073
t (s)	32	36	40	44	48	52	56	60
r (km)	19.617	19.741	19.865	19.990	20.115	20.239	20.362	20.484
θ(rad)	0.6925	0.6771	0.6612	0.6448	0.6280	0.6107	0.5931	0.5750

6.26 A projectile is shot from a 200 m tall cliff as shown. Its position (x and y coordinates) as a function of time, t, is given in the table that follows. The velocity of the projectile, v, is given by $v = \sqrt{v_x^2 + v_y^2}$ where the horizontal and vertical components, v_x and v_y are given by: $v_x = \dfrac{\partial x}{\partial t}$ and $v_y = \dfrac{\partial y}{\partial t}$. Write a MAT-LAB program in a script file that:

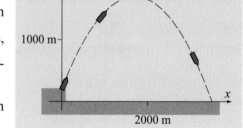

(*a*) Determines v_x and v_y by using the user-defined function `FirstDeriv` that was written in Problem 6.14.
(*b*) Determines the velocity v of the projectile.
(*c*) Displays a figure with plots of v_x, v_y and v as a function of time (three plots in one figure).

t (s)	0	2	4	6	8	10	12	14	16	18
x (m)	0	198	395	593	790	988	1185	1383	1580	1778
y (m)	200	523	806	1050	1254	1420	1546	1633	1681	1690
t (s)	20	22	24	26	28	30	32	34	36	
x (m)	1975	2173	2370	2568	2765	2963	3160	3358	3555	
y (m)	1659	1589	1480	1331	1144	917	651	345	1	

Chapter 7

Numerical Integration

Core Topics	Complementary Topics
Rectangle and midpoint methods (7.2).	Estimation of error (7.8).
Trapezoidal method (7.3).	Richardson's extrapolation (7.9).
Simpson's methods (7.4).	Romberg integration (7.10).
Gauss quadrature (7.5).	Improper integrals (7.11).
Evaluation of multiple integrals (7.6).	
Use of MATLAB built-in functions for integration (7.7).	

7.1 BACKGROUND

Integration is frequently encountered when solving problems and calculating quantities in engineering and science. Integration and integrals are also used when solving differential equations. One of the simplest examples for the application of integration is the calculation of the length of a curve (Fig. 7-1). When a curve in the x-y plane is given by the equation $y = f(x)$, the length L of the curve between the points $x = a$ and $x = b$ is given by:

$$L = \int_a^b \sqrt{1 + [f'(x)]^2}\, dx$$

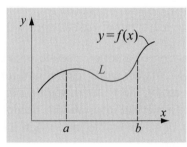

Figure 7-1: Length of a curve.

In engineering there are circumstances that involve experimental or test data, where a physical quantity that has to be determined may be expressed as an integral of other quantities that are measured. For example, the total rate of heat flow through a cross section of width W and height $(b - a)$ is related to the local heat flux via an integral (see Fig. 7-2):

$$\dot{Q} = \int_{y=a}^{y=b} \dot{q}'' W\, dy$$

where $\dot{q}''$ is the heat flux and $\dot{Q}$ is the heat flow rate. Experimental measurements may yield discrete values for the heat flux along the surface as a function of y, but the quantity to be determined may be the total heat flow rate. In this instance, the integrand may be specified as a known set of values for each value of y.

Figure 7-2: Heat flux through a rectangular cross section.

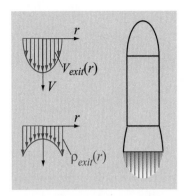

Figure 7-3: Exhaust of a rocket engine.

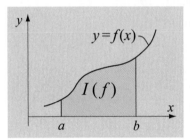

Figure 7-4: Definite integral of *f(x)* between *a* and *b*.

As yet another illustration, consider the exhaust of a rocket engine generating thrust. As shown in Fig. 7-3, the velocity and density of the flow exiting the rocket nozzle are not uniform over the cross-sectional area. For a circular cross section, both will vary with the radial coordinate r. The resulting expression for the magnitude of the thrust can be obtained from conservation of momentum at steady state:

$$T = \int_0^R 2\pi\rho(r)V_{exit}^2(r)r\,dr$$

where T is the thrust, $\rho(r)$ is the mass density of the fluid, $V_{exit}(r)$ is the velocity profile at the exit plane of the engine, r is the radial coordinate, and R is the radius of the rocket nozzle at the exit plane. Computational fluid dynamics calculations can yield $\rho(r)$ and $V_{exit}(r)$, but the thrust (a quantity that can be measured in experiments or tests) must be obtained by integration.

The general form of a definite integral (also called an antiderivative) is:

$$I(f) = \int_a^b f(x)\,dx \tag{7.1}$$

where $f(x)$, called the integrand, is a function of the independent variable x, and a and b are the limits of the integration. The value of the integral $I(f)$ is a number when a and b are numbers. Graphically, as shown in Fig. 7-4, the value of the integral corresponds to the shaded area under the curve of $f(x)$ between a and b.

The need for numerical integration

The integrand can be an analytical function or a set of discrete points (tabulated data). When the integrand is a mathematical expression for which the antiderivative can be found easily, the value of the definite integral can be determined analytically. Numerical integration is needed when analytical integration is difficult or not possible, and when the integrand is given as a set of discrete points.

7.1.1 Overview of Approaches in Numerical Integration

Numerical evaluation of a single integral deals with estimating the number $I(f)$ that is the integral of a function $f(x)$ over an interval from a to b. If the integrand $f(x)$ is an analytical function, the numerical integration is done by using a finite number of points at which the integrand is evaluated (Fig. 7-5). One strategy is to use only the endpoints of the interval, $(a, f(a))$ and $(b, f(b))$. This, however, might not give an accurate enough result, especially if the interval is wide and/or the integrand varies significantly within the interval. Higher accuracy can be achieved by using a composite method where the interval $[a, b]$ is divided into smaller subintervals. The integral over each subinterval is calculated, and the results are added together to give the value of the

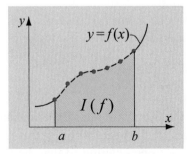

Figure 7-5: Finite number of points are used in numerical integration.

whole integral. If the integrand $f(x)$ is given as a set of discrete points (tabulated data), the numerical integration is done by using these points.

In all cases the numerical integration is carried out by using a set of discrete points for the integrand. When the integrand is an analytical function, the location of the points within the interval $[a, b]$ can be defined by the user or is defined by the integration method. When the integrand is a given set of tabulated points (like data measured in an experiment), the location of the points is fixed and cannot be changed.

Various methods have been developed for carrying out numerical integration. In each of these methods, a formula is derived for calculating an approximate value of the integral from discrete values of the integrand. The methods can be divided into groups called open methods and closed methods.

Closed and open methods

In closed integration methods, the endpoints of the interval (and the integrand) are used in the formula that estimates the value of the integral. In open integration methods, the interval of integration extends beyond the range of the endpoints that are actually used for calculating the value of the integral (Fig. 7-6). The trapezoidal (Section 7.3) and Simpson's (Section 7.4) methods are closed methods, whereas the midpoint method (Section 7.2) and Gauss quadrature (Section 7.5) are open methods.

There are various methods for calculating the value of an integral from the set of discrete points of the integrand. Most commonly, it is done by using Newton–Cotes integration formulas.

Newton–Cotes integration formulas

In numerical integration methods that use Newton–Cotes integration formulas, the value of the integrand between the discrete points is estimated using a function that can be easily integrated. The value of the integral is then obtained by integration. When the original integrand is an analytical function, the Newton–Cotes formula replaces it with a simpler function. When the original integrand is in the form of data points, the Newton–Cotes formula interpolates the integrand between the given points. Most commonly, as with the trapezoidal method (Section 7.3) and Simpson's methods (Section 7.4), the Newton–Cotes integration formulas are polynomials of different degrees.

A different option for integration, once the integrand $f(x)$ is specified as discrete points, is to curve-fit the points with a function $F(x)$ that best fits the points. In other words, as shown in Fig. 7-7, $f(x) \approx F(x)$, where $F(x)$ is a polynomial or a simple function whose antiderivative can be found easily. Then, the integral

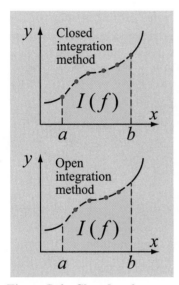

Figure 7-6: Closed and open integration methods.

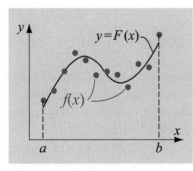

Figure 7-7: Integrating $f(x)$ using an integrable $F(x)$ function.

$$I(f) = \int_a^b f(x)dx \approx \int_a^b F(x)dx$$

is evaluated by direct analytical methods from calculus. This procedure requires numerical methods for finding $F(x)$ (Chapters 5 and 6), but may not require a numerical method to evaluate the integral if $F(x)$ is an integrable function.

7.2 RECTANGLE AND MIDPOINT METHODS

Rectangle method

The simplest approximation for $\int_a^b f(x)dx$ is to take $f(x)$ over the interval $x \in [a, b]$ as a constant equal to the value of $f(x)$ at either one of the endpoints (Fig. 7-8).

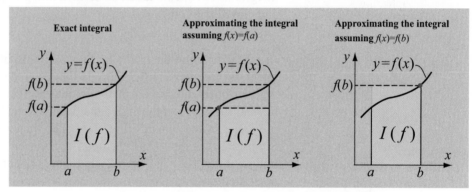

Figure 7-8: Integration using the rectangle method.

The integral can then be calculated in one of two ways:

$$I(f) = \int_a^b f(a)dx = f(a)(b-a) \quad \text{or} \quad I(f) = \int_a^b f(b)dx = f(b)(b-a) \quad (7.2)$$

As Fig. 7-8 shows, the actual integral is approximated by an area of a rectangle. Obviously for the monotonically increasing function shown, the value of the integral is underestimated when $f(x)$ is assumed to be equal to $f(a)$, and overestimated when $f(x)$ is assumed to be equal to $f(b)$. Moreover, the error can be large. When the integrand is an analytical function, the error can be significantly reduced by using the composite rectangle method.

Composite rectangle method

In the composite rectangle method the domain $[a, b]$ is divided into N subintervals. The integral in each subinterval is calculated with the rectangle method, and the value of the whole integral is obtained by adding the values of the integrals in the subintervals. This is shown in Fig. 7-9 where the interval $[a, b]$ is divided into N subintervals by defining the points $x_1, x_2, ..., x_{N+1}$. The first point is $x_1 = a$ and the last point is $x_{N+1} = b$ (it takes $N + 1$ points to define N intervals). Figure 7-9 shows subintervals with the same width, but in general, the subintervals can

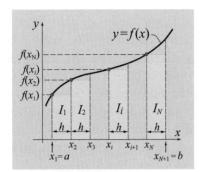

Figure 7-9: The composite rectangle method.

have arbitrary width. In this way smaller intervals can be used in regions where the value of the integrand changes rapidly (large slopes) and larger intervals can be used when the integrand changes more gradually.

In Fig. 7-9, the integrand in each subinterval is assumed to have the value of the integrand at the beginning of the subinterval. By using Eq. (7.2) for each subinterval, the integral over the whole domain can be written as the sum of the integrals in the subintervals:

$$I(f) = \int_a^b f(x)dx \approx \overbrace{f(x_1)(x_2 - x_1)}^{I_1} + \overbrace{f(x_2)(x_3 - x_2)}^{I_2} + \ldots + \overbrace{f(x_i)(x_{i+1} - x_i)}^{I_i}$$

$$+ \ldots + \underbrace{f(x_N)(x_{N+1} - x_N)}_{I_N} = \sum_{i=1}^{N} [f(x_i)(x_{i+1} - x_i)] \qquad (7.3)$$

When the subintervals have the same width, h, Eq. (7.3) can be simplified to:

$$I(f) = \int_a^b f(x)dx \approx h \sum_{i=1}^{N} f(x_i) \qquad (7.4)$$

Equation (7.4) is the formula for the composite rectangle method for the case where the subintervals have identical width h.

Midpoint method

An improvement over the naive rectangle method is the midpoint method. Instead of approximating the integrand by the values of the function at $x = a$ or at $x = b$, the value of the integrand at the middle of the interval, that is, $f\left(\frac{(a+b)}{2}\right)$, is used. Substituting into Eq. (7.1) yields:

$$I(f) = \int_a^b f(x)dx \approx \int_a^b f\left(\frac{a+b}{2}\right)dx = f\left(\frac{a+b}{2}\right)(b - a) \qquad (7.5)$$

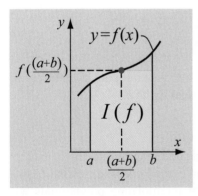

Figure 7-10: The midpoint method.

This method is depicted graphically in Fig. 7-10. As can be seen, the value of the integral is still approximated as the area of a rectangle, but with an important difference—the area is that of an **equivalent** rectangle. This turns out to be more accurate than the rectangle method because for a monotonic function as shown in the figure, the regions of the area under the curve that are ignored may be approximately offset by those regions above the curve that are included. However, this is not true for all cases, so that this method may still not be accurate enough. As in the rectangle method, the accuracy can be increased using a composite midpoint method.

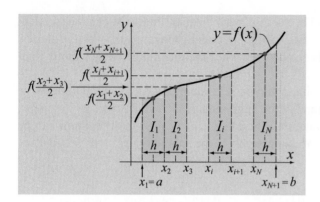

Figure 7-11: The composite midpoint method.

Composite midpoint method

In the composite midpoint method, the domain $[a, b]$ is divided into N subintervals. The integral in each subinterval is calculated with the midpoint method, and the value of the whole integral is obtained by adding the values of the integrals in the subintervals. This is shown in Fig. 7-11 where the interval $[a, b]$ is divided into N subintervals by defining the points $x_1, x_2, ..., x_{N+1}$. The first point is $x_1 = a$ and the last point is $x_{N+1} = b$ (it takes $N + 1$ points to define N intervals). Figure 7-11 shows subintervals with the same width, but in general, the subintervals can have arbitrary width.

By using Eq. (7.5) for each subinterval, the integral over the whole domain can be written as the sum of the integrals in the subintervals:

$$I(f) = \int_a^b f(x)dx \approx \overbrace{f\left(\frac{x_1 + x_2}{2}\right)(x_2 - x_1)}^{I_1} + \overbrace{f\left(\frac{x_2 + x_3}{2}\right)(x_3 - x_2)}^{I_2} + ...$$

$$+ \overbrace{f\left(\frac{x_i + x_{i+1}}{2}\right)(x_{i+1} - x_i)}^{I_i} + ... + \overbrace{f\left(\frac{x_N + x_{N+1}}{2}\right)(x_{N+1} - x_N)}^{I_N}$$

$$= \sum_{i=1}^{N}\left[f\left(\frac{x_i + x_{i+1}}{2}\right)(x_{i+1} - x_i)\right] \tag{7.6}$$

When the subintervals have the same width, h, Eq. (7.6) can be simplified to:

$$I(f) = \int_a^b f(x)dx \approx h\sum_{i=1}^{N}f\left(\frac{x_i + x_{i+1}}{2}\right) \tag{7.7}$$

Equation (7.7) is the formula for the composite midpoint method for the case where the subintervals have identical width h.

7.3 TRAPEZOIDAL METHOD

A refinement over the simple rectangle and midpoint methods is to use a linear function to approximate the integrand over the interval of integration (Fig. 7-12). Newton's form of interpolating polynomials with two points $x = a$ and $x = b$, yields:

$$f(x) \approx f(a) + (x - a)f[a, b] = f(a) + (x - a)\frac{[f(b) - f(a)]}{b - a} \tag{7.8}$$

Substituting Eq. (7.8) into Eq. (7.1) and integrating analytically gives:

$$I(f) \approx \int_a^b \left(f(a) + (x-a)\frac{[f(b)-f(a)]}{b-a} \right) dx$$

$$= f(a)(b-a) + \frac{1}{2}[f(b)-f(a)](b-a) \quad (7.9)$$

Simplifying the result gives an approximate formula popularly known as the trapezoidal rule or trapezoidal method:

$$I(f) \approx \frac{[f(a)+f(b)]}{2}(b-a) \quad (7.10)$$

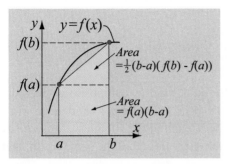

Figure 7-12: The trapezoidal method.

Examining the result before the simplification, that is, the right-hand side of Eq. (7.9), shows that the first term, $f(a)(b-a)$, represents the area of a rectangle of height $f(a)$ and length $(b-a)$. The second term, $\frac{1}{2}[f(b)-f(a)](b-a)$, is the area of the triangle whose base is $(b-a)$ and whose height is $[f(b)-f(a)]$. These are shown in Fig. 7-12 and serve to reinforce the notion that in this method the area under the curve $f(x)$ is approximated by the area of the trapezoid (rectangle + triangle). As shown in Fig. 7-12, this is more accurate than using a rectangle to approximate the shape of the region under $f(x)$.

As with the rectangle and midpoint methods, the trapezoidal method can be easily extended to yield any desired level of accuracy by subdividing the interval $[a, b]$ into subintervals.

7.3.1 Composite Trapezoidal Method

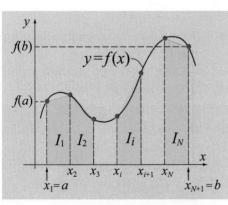

Figure 7-13: The composite trapezoidal method.

The integral over the interval $[a, b]$ can be evaluated more accurately by dividing the interval into subintervals, evaluating the integral for each subintervals (with the trapezoidal method), and adding the results. As shown in Fig. 7-13, the interval $[a, b]$ is divided into N subintervals by defining the points $x_1, x_2, ..., x_{N+1}$ where the first point is $x_1 = a$ and the last point is $x_{N+1} = b$ (it takes $N+1$ points to define N intervals).

The integral over the whole interval can be written as the sum of the integrals in the subintervals:

$$I(f) = \int_a^b f(x)dx = \overbrace{\int_{x_1=a}^{x_2} f(x)dx}^{I_1} + \overbrace{\int_{x_2}^{x_3} f(x)dx}^{I_2} + ... + \overbrace{\int_{x_i}^{x_{i+1}} f(x)dx}^{I_i}$$

$$+ ... + \underbrace{\int_{x_N}^{x_{N+1}} f(x)dx}_{I_N} = \sum_{i=1}^N \int_{x_i}^{x_{i+1}} f(x)dx \quad (7.11)$$

Applying the trapezoidal method to each subinterval $[x_i, x_{i+1}]$ yields:

$$I_i(f) = \int_{x_i}^{x_{i+1}} f(x)dx \approx \frac{[f(x_i) + f(x_{i+1})]}{2}(x_{i+1} - x_i)$$

Substituting the trapezoidal approximation in the right side of Eq. (7.11) gives:

$$I(f) = \int_a^b f(x)dx \approx \frac{1}{2}\sum_{i=1}^{N}[f(x_i) + f(x_{i+1})](x_{i+1} - x_i) \tag{7.12}$$

Equation 7.12 is the general formula for the composite trapezoidal method. Note that the subintervals $[x_i, x_{i+1}]$ need not be identical (i.e., equally spaced) at all. In other words, each of the subintervals can be of different width. If, however, the subintervals are all the same width, that is, if

$$(x_2 - x_1) = (x_3 - x_2) = \dots = (x_{i+1} - x_i) = \dots = (x_N - x_{N-1}) = h$$

then Eq. (7.12) can be simplified to:

$$I(f) \approx \frac{h}{2}\sum_{i=1}^{N}[f(x_{i+1}) + f(x_i)]$$

This can be further reduced to a formula that lends itself to programming by expanding the summation:

$$I(f) \approx \frac{h}{2}[f(a) + 2f(x_2) + 2f(x_3) + \dots + 2f(x_N) + f(b)]$$

or,

$$I(f) \approx \frac{h}{2}[f(a) + f(b)] + h\sum_{i=2}^{N-1} f(x_i) \tag{7.13}$$

Equation (7.13) is the formula for the composite trapezoidal method for the case where the subintervals have identical width h.

Example 7-1 shows how the composite trapezoidal method is programmed in MATLAB and then used for solving a problem.

Example 7-1: Distance traveled by a decelerating airplane.

A Boeing 727-200 airplane of mass $m = 97000$ kg lands at a speed of 93 m/s (about 181 knots) and applies its thrust reversers at $t = 0$. The force F that is applied to the airplane, as it decelerates, is given by $F = -5v^2 - 570000$, where v is the airplane's velocity. Using Newton's second law of motion and flow 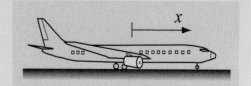 dynamics, the relationship between the velocity and the position x of the airplane can be written as:

$$mv\frac{dv}{dx} = -5v^2 - 570000$$

where x is the distance measured from the location of the jet at $t = 0$.

Determine how far the airplane travels before its speed is reduced to 40 m/s (about 78 knots) by using the composite trapezoidal method to evaluate the integral resulting from the governing differential equation.

SOLUTION

Even though the governing equation is an ODE, it can be expressed as an integral in this case. This is done by separating the variables such that the speed v appears on one side of the equation and x appears on the other.

$$\frac{97000v\,dv}{(-5v^2 - 570000)} = dx$$

Next, both sides are integrated. For x the limits of integration are from 0 to an arbitrary location x, and for v the limits are from 93 m/s to 40 m/s.

$$\int_0^x dx = -\int_{93}^{40} \frac{97000v}{(5v^2 + 570000)}\,dv = \int_{40}^{93} \frac{97000v}{(5v^2 + 570000)}\,dv \qquad (7.14)$$

The objective of this example is to show how the definite integral on the right-hand side of the equation can be determined numerically using the composite trapezoidal method. In this problem, however, the integration can also be carried out analytically. For comparison, the integration is done both ways.

Analytical Integration

The integration can be carried out analytically by using substitution. By substituting $z = 5u^2 + 570000$, the integration can be performed to obtain the value $x = 574.1494$ m.

Numerical Integration

To carry out the numerical integration, the following user-defined function, named `trapezoidal`, is created.

Program 7-1: Function file, integration trapezoidal method.

```
function I = trapezoidal(integrand,a,b,N)
% trapezoidal numerically integrate using the composite trapezoidal method.
% Input Variables:
% integrand  The function to be integrated typed as a string.
% a  Lower limit of integration.
% b  Upper limit of integration.
% N  Number of subintervals.
% Output Variable:
% I  Value of the integral.


h = (b - a)/N;                          Calculate the width h of the subintervals.
func = inline(integrand);               Define the integrand as an inline function.
x = a:h:b;                              Create a vector x with the coordinates of the subintervals.
for i=1:N + 1
    F(i) = func(x(i));                  Create a vector F with the values of
end                                    the integrand at each point x.
I = h*(F(1) + F(N + 1))/2 + h*sum(F(2:N));   Calculate the value of the integral according to Eq. (7.13).
```

The function `trapezoidal` is used next in the Command Window to determine the value of the integral in Eq. (7.14). To examine the effect of the number of subintervals on the result, the function is used three times using $N = 10, 100$, and 1000. The display in the Command Window is:

```
>> format long g
>> distance = trapezoidal('97000*v/(5*v^2+570000)',40,93,10)
distance =

        574.085485133712
>> distance = trapezoidal('97000*v/(5*v^2+570000)',40,93,100)
distance =

        574.148773931409
>> distance=trapezoidal('97000*v/(5*v^2+570000)',40,93,1000)
distance =

        574.149406775129
```

As expected, the results show that the integral is evaluated more accurately as the number of subintervals is increased. When $N = 1000$, the answer is the same as that calculated analytically to four decimal places.

Example 7-1 reveals two key points:

- It is important to check results from numerical computations (performed either by hand or by computer), against known analytical solutions. In the event an analytical solution is not available, it is necessary to check the answer by another numerical method and to compare the two results.

- In most problems involving numerical integration, it is possible to improve on the accuracy of an answer by taking more subintervals, that is, by reducing the size of the subinterval.

7.4 SIMPSON'S METHODS

The trapezoidal method described in the last section relies on approximating the integrand by a straight line. A better approximation can possibly be obtained by approximating the integrand with a nonlinear function that can be easily integrated. One class of such methods, called Simpson's rules or Simpson's methods, uses quadratic (Simpson's 1/3 method) and cubic (Simpson's 3/8 method) polynomials to approximate the integrand.

7.4.1 Simpson's 1/3 Method

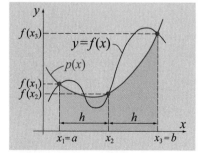

Figure 7-14: Simpson's 1/3 Method.

In this method, a quadratic (second-order) polynomial is used to approximate the integrand (Fig. 7-14). The coefficients of a quadratic polynomial can be determined from three points. For an integral over the domain $[a, b]$, the three points used are the two endpoints $x_1 = a$,

$x_3 = b$, and the midpoint $x_2 = (a+b)/2$. The polynomial can be written in the form:

$$p(x) = \alpha + \beta(x-x_1) + \gamma(x-x_1)(x-x_2) \qquad (7.15)$$

where α, β, and γ are unknown constants evaluated from the condition that the polynomial passes through the points, $p(x_1) = f(x_1)$, $p(x_2) = f(x_2)$, and $p(x_3) = f(x_3)$. These conditions yield:

$$\alpha = f(x_1), \quad \beta = [f(x_2) - f(x_1)]/(x_2 - x_1), \text{ and } \gamma = \frac{f(x_3) - 2f(x_2) + f(x_1)}{2(h)^2}$$

where $h = (b-a)/2$. Substituting the constants back in Eq. (7.15) and integrating $p(x)$ over the interval $[a, b]$ gives:

$$I = \int_{x_1}^{x_3} f(x)dx \approx \int_{x_1}^{x_3} p(x)dx = \frac{h}{3}[f(x_1) + 4f(x_2) + f(x_3)]$$
$$= \frac{h}{3}\left[f(a) + 4f\left(\frac{a+b}{2}\right) + f(b)\right] \qquad (7.16)$$

The value of the integral is shown in Fig. 7-14 as the shaded area between the curve of $p(x)$ and the x axis. The name 1/3 in the method comes from the fact that there is a factor of 1/3 multiplying the expression in the brackets in Eq. (7.16).

As with the rectangular and trapezoidal methods, a more accurate evaluation of the integral can be done with a composite Simpson's 1/3 method. The whole interval is divided into small subintervals. Simpson's 1/3 method is used to calculate the value of the integral in each subinterval, and the values are added together.

Composite Simpson's 1/3 method

In the composite Simpson's 1/3 method (Fig. 7-15) the whole interval $[a, b]$ is divided into N subintervals. In general, the subintervals can have arbitrary width. The derivation here, however, is limited to the case where the subintervals have equal width h, where $h = (b-a)/N$. Since three points are needed for defining a quadratic polynomial, the Simpson's 1/3 method is applied to two adjacent subintervals at a time (the first two, the third and fourth together, and so on). Consequently, the whole interval has to be divided into an **even number** of subintervals.

The integral over the whole interval can be written as the sum of the integrals of couples of adjacent subintervals.

$$I(f) = \int_a^b f(x)dx = \overbrace{\int_{x_1=a}^{x_3} f(x)dx}^{I_2} + \overbrace{\int_{x_3}^{x_5} f(x)dx}^{I_4} + \ldots + \overbrace{\int_{x_{i-1}}^{x_{i+1}} f(x)dx}^{I_i} + \ldots$$

$$+ \underbrace{\int_{x_{N-1}}^{x_{N+1}} f(x)dx}_{I_N} = \sum_{i=2,4,6}^{N} \int_{x_{i-1}}^{x_{i+1}} f(x)dx \qquad (7.17)$$

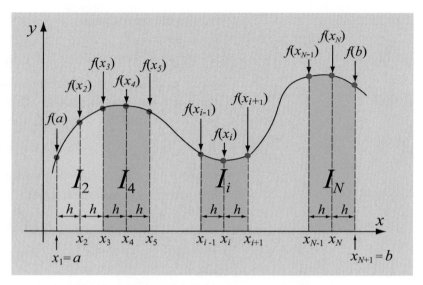

Figure 7-15: Composite Simpson's 1/3 method.

By using Eq. (7.16), the integral over two adjacent intervals $[x_{i-1}, x_i]$ and $[x_i, x_{i+1}]$ can be written in terms of the Simpson's 1/3 method by:

$$I_i(f) = \int_{x_{i-1}}^{x_{i+1}} f(x)dx \approx \frac{h}{3}[f(x_{i-1}) + 4f(x_i) + f(x_{i+1})] \qquad (7.18)$$

where $h = x_{i+1} - x_i = x_i - x_{i-1}$. Substituting Eq. (7.18) in Eq. (7.17) for each of the integrals gives:

$$I(f) \approx \frac{h}{3}[f(a) + 4f(x_2) + f(x_3) + f(x_3) + 4f(x_4) + f(x_5) + f(x_5) + 4f(x_6)$$
$$+ f(x_7) + \ldots + f(x_{N-1}) + 4f(x_N) + f(b)]$$

By collecting similar terms, the right side of the last equation can be simplified to give the general equation for the composite Simpson's 1/3 method for equally spaced subintervals:

$$I(f) \approx \frac{h}{3}\left[f(a) + 4\sum_{i=2,4,6}^{N} f(x_i) + 2\sum_{j=3,5,7}^{N-1} f(x_j) + f(b)\right] \qquad (7.19)$$

where $h = (b-a)/N$.

Equation (7.19) is the composite ***Simpson's 1/3 formula*** for numerical integration. It is important to point out that Eq. (7.19) can be used only if two conditions are satisfied:

- The subintervals must be ***equally spaced***.

- The ***number of subintervals*** within $[a, b]$ ***must be an <u>even</u> number***.

   Equation (7.19) is a weighted addition of the value of the function at the points that define the subintervals. The weight is 4 at all the points x_i with an even index. These are the middle points of each set of two adjacent subintervals (see Eq. (7.18)). The weight is 2 at all the points x_i with an odd index (except the first and last points). These points are at the interface between adjacent pairs of subintervals. Each point is used once as the right endpoint of a pair of subinterval and once as the left endpoint of the next pair of subintervals. The endpoints are used only once. Figure 7-16 illustrates the weighted addition according

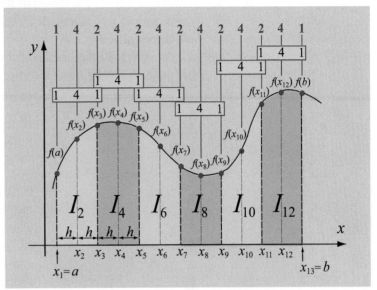

Figure 7-16: Weighted addition with the composite Simpson's 1/3 method.

to Eq. (7.19) for a domain $[a, b]$ that is divided into 12 subintervals. Applying Eq. (7.19) to this illustration gives:

$$I(f) \approx \frac{h}{3}\{f(a) + 4[f(x_2) + f(x_4) + f(x_6) + f(x_8) + f(x_{10}) + f(x_{12})]$$
$$+ 2[f(x_3) + f(x_5) + f(x_7) + f(x_9) + f(x_{11})] + f(b)\}$$

7.4.2 Simpson's 3/8 Method

In this method a cubic (third-order) polynomial is used to approximate the integrand (Fig. 7-17). A third-order polynomial can be determined from four points. For an integral over the domain $[a, b]$, the four points used are the two endpoints $x_1 = a$ and $x_4 = b$, and two points x_2 and x_3 that divide the interval into three equal sections. The polynomial can be written in the form:

$$p(x) = c_3 x^3 + c_2 x^2 + c_1 x + c_0$$

where c_3, c_2, c_1, and c_0 are constants evaluated from the conditions that the polynomial passes through the points, $p(x_1) = f(x_1)$, $p(x_2) = f(x_2)$, $p(x_3) = f(x_3)$, and $p(x_4) = f(x_4)$. Once the constants are determined, the polynomial can be easily integrated to give:

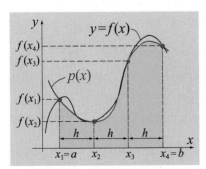

Figure 7-17: Simpson's 3/8 method.

$$I = \int_a^b f(x)dx \approx \int_a^b p(x)dx = \frac{3}{8}h[f(a) + 3f(x_2) + 3f(x_3) + f(b)] \quad (7.20)$$

The value of the integral is shown in Fig. 7-17 as the shaded area between the curve of $p(x)$ and the x axis. The name 3/8 method comes from the 3/8 factor in the expression in Eq. (7.20). Notice that Eq. (7.20) is a weighted addition of the values of $f(x)$ at the two endpoints $x_1 = a$ and $x_4 = b$, and the two points x_2 and x_3 that divide the interval into three equal sections.

As with the other methods, a more accurate evaluation of the integral can be done by using a composite Simpson's 3/8 method.

Composite Simpson's 3/8 Method

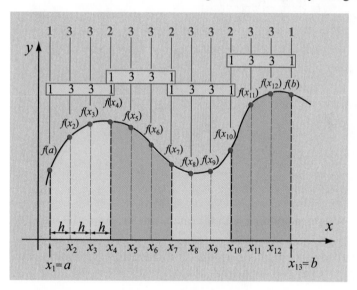

Figure 7-18: Weighted addition with the composite Simpson's 3/8 method.

In the composite Simpson's 3/8 method, the whole interval $[a, b]$ is divided into N subintervals. In general, the subintervals can have arbitrary width. The derivation here, however, is limited to the case where the subintervals have an equal width h, where $h = (b-a)/N$. Since four points are needed for constructing a cubic polynomial, the Simpson's 3/8 method is applied to three adjacent subintervals at a time (the first three, the fourth, fifth, and sixth intervals together, and so on). Consequently, the whole interval has to be divided into a number of subintervals that is divisible by 3.

The integration in each group of three adjacent subintervals is evaluated by using Eq. (7.20). The integral over the whole domain is obtained by adding the integrals in the subinterval groups. The process is illustrated in Fig. 7-18 where the whole domain $[a, b]$ is divided into 12 subintervals that are grouped in four groups of three subintervals. Using Eq. (7.20) for each group and adding the four equations gives:

$$I(f) \approx \frac{3h}{8}\{ f(a) + 3[f(x_2) + f(x_3) + f(x_5) + f(x_6) + f(x_8) + f(x_9)$$
$$+ f(x_{11}) + f(x_{12})] + 2[f(x_4) + f(x_7) + f(x_{10})] + f(b)\} \quad (7.21)$$

For the general case when the domain $[a, b]$ is divided into N subintervals (where N is at least 6 and divisible by 3), Eq. (7.21) can be generalized to:

$$I(f) \approx \frac{3h}{8}\left[f(a) + 3\sum_{i=2,5,8}^{N-1}[f(x_i) + f(x_{i+1})] + 2\sum_{j=4,7,10}^{N-2} f(x_j) + f(b)\right] \quad (7.22)$$

Equation (7.22) is **Simpson's 3/8 method** for numerical integration. Simpson's 3/8 method can be used if the following two conditions are met:

- The subintervals are **equally spaced.**

- The **number of subintervals** within $[a, b]$ **must be divisible by 3**.

Since Simpson's 1/3 method is only valid for an <u>even</u> number of subintervals and Simpson's 3/8 method is only valid for a number of subintervals that is divisible by 3, a combination of both can be used for integration when there are any odd number of intervals. This is done by using Simpson's 3/8 method for the first three subintervals ($[a, x_2]$, $[x_2, x_3]$, and $[x_3, x_4]$) <u>or</u> for the last three subintervals ($[x_{N-2}, x_{N-1}]$, $[x_{N-1}, x_N]$, and $[x_N, x_b]$), and using Simpson's 1/3 method for the remaining even number of subintervals. Such a combined strategy works because the order of the numerical error is the same for both methods (see Section 7.8).

7.5 GAUSS QUADRATURE

Background

In all the integration methods that have been presented so far, the integral of $f(x)$ over the interval $[a, b]$ was evaluated by approximating $f(x)$ with a polynomial that could be easily integrated. Depending on the integration method, the approximating polynomial and $f(x)$ have the same value at one (rectangular and midpoint methods), two (trapezoidal method), or more points (Simpson's methods) within the interval. The integral is evaluated from the value of $f(x)$ at the common points with the approximating polynomial. When two or more points are used, the value of the integral is calculated from weighted addition of the values of $f(x)$ at the different points. The location of the common points is predetermined in each of the integration methods. All the methods have been illustrated so far, using points that are equally spaced. The various methods are summarized in the following table.

Integration Method	Values of the function used in evaluating the integral.
Rectangle Equation (7.2)	$f(a)$ or $f(b)$ (Either one of the endpoints.)
Midpoint Equation (7.5)	$f((a + b)/2)$ (The middle point.)
Trapezoidal Equation (7.9)	$f(a)$ and $f(b)$ (Both endpoints.)

Integration Method	Values of the function used in evaluating the integral.
Simpson's 1/3 Equation (7.16)	$f(a)$, $f(b)$, and $f((a+b)/2)$ (Both endpoints and the middle point.)
Simpson's 3/8 Equation (7.20)	$f(a)$, $f(b)$, $f\left(a+\frac{1}{3}(a+b)\right)$, and $f\left(a+\frac{2}{3}(a+b)\right)$ (Both endpoints and two points that divide the interval into three equal-width subintervals.)

In Gauss quadrature, the integral is also evaluated by using weighted addition of the values of $f(x)$ at different points (called Gauss points) within the interval $[a, b]$. The Gauss points, however, are not equally spaced and do not include the endpoints. The location of the points and the corresponding weights of $f(x)$ are determined in such a way as to minimize the error.

General form of Gauss quadrature

The general form of Gauss quadrature is:

$$\int_a^b f(x)dx \approx \sum_{i=1}^{n} C_i f(x_i) \tag{7.23}$$

where the coefficients C_i are the weights and the x_i are points (Gauss points) within the interval $[a, b]$. For example, for $n = 2$ and $n = 3$ Eq. (7.23) has the form:

$$\int_a^b f(x)dx \approx C_1 f(x_1) + C_2 f(x_2), \quad \int_a^b f(x)dx \approx C_1 f(x_1) + C_2 f(x_2) + C_3 f(x_3)$$

The value of the coefficients C_i and the location of the points x_i depend on the values of n, a, and b, and are determined such that the right side of Eq. (7.23) is exactly equal to the left side for specified functions $f(x)$.

Gauss quadrature integration of $\int_{-1}^{1} f(x)dx$

For the domain $[-1, 1]$ the form of Gauss quadrature is:

$$\int_{-1}^{1} f(x)dx \approx \sum_{i=1}^{n} C_i f(x_i) \tag{7.24}$$

The coefficients C_i and the location of the Gauss points x_i are determined by enforcing Eq. (7.24) to be exact for the cases when $f(x) = 1, x, x^2, x^3, \dots$. The number of cases that have to be considered depends on the value of n. For example, when $n = 2$:

$$\int_{-1}^{1} f(x)dx \approx C_1 f(x_1) + C_2 f(x_2) \tag{7.25}$$

The four constants C_1, C_2, x_1, and x_2 are determined by enforcing Eq. (7.25) to be exact when applied to the following four cases:

Case 1: $f(x) = 1$ $\qquad \int_{-1}^{1} (1)dx = 2 = C_1 + C_2$

Case 2: $f(x) = x$ $\qquad \int_{-1}^{1} xdx = 0 = C_1 x_1 + C_2 x_2$

Case 3: $f(x) = x^2$ $\qquad \int_{-1}^{1} x^2 dx = \frac{2}{3} = C_1 x_1^2 + C_2 x_2^2$

Case 4: $f(x) = x^3$ $\qquad \int_{-1}^{1} x^3 dx = 0 = C_1 x_1^3 + C_2 x_2^3$

The four cases provide a set of four equations for the four unknowns. The equations are nonlinear, which means that multiple solutions can exist. One particular solution can be obtained by imposing an additional requirement. Here the requirement is that the points x_1, and x_2 should be symmetrically located about $x = 0$ ($x_1 = -x_2$). From the second equation, this requirement implies that $C_1 = C_2$. With these requirements, solving the equations gives:

$$C_1 = 1, \quad C_2 = 1, \quad x_1 = -\frac{1}{\sqrt{3}} = -0.57735027, \quad x_2 = \frac{1}{\sqrt{3}} = 0.57735027$$

Substituting the constants back in Eq. (7.25) gives (for $n = 2$):

$$\int_{-1}^{1} f(x)dx \approx f\left(-\frac{1}{\sqrt{3}}\right) + f\left(\frac{1}{\sqrt{3}}\right) \qquad (7.26)$$

The right-hand side of Eq. (7.26) gives the exact value for the integral on the left hand side of the equation when $f(x) = 1$, $f(x) = x$, $f(x) = x^2$, or $f(x) = x^3$. This is illustrated in Fig. 7-19 for the case where $f(x) = x^2$. In this case:

$$\int_{-1}^{1} x^2 dx = \frac{2}{3} = 1\left(-\frac{1}{\sqrt{3}}\right)^2 + 1\left(\frac{1}{\sqrt{3}}\right)^2 \qquad (7.27)$$

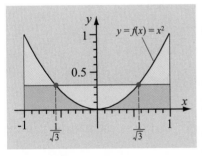

Figure 7-19: Gauss quadrature integration of $f(x) = x^2$.

The value of the integral $\int_{-1}^{1} x^2 dx$ is the area under the curve $f(x) = x^2$. The right-hand side of Eq. (7.27) is the area under the red horizontal line. The two areas are identical since the area between $f(x)$ and the red line for $|x| > 1/(\sqrt{3})$ is the same as the area that is between the red line and $f(x)$ for $|x| < 1/(\sqrt{3})$ (the light shaded areas that are above and below the red line have the same area).

When $f(x)$ is a function that is different from $f(x) = 1$, $f(x) = x$, $f(x) = x^2$, or $f(x) = x^3$, or any linear combination of these, Gauss quadrature gives an approximate value for the integral. For example, if $f(x) = \cos(x)$, the exact value of the integral (the left-hand side of Eq. (7.26) is:

$$\int_{-1}^{1} \cos(x)dx = \sin(x)\Big|_{-1}^{1} = \sin(1) - \sin(-1) = 1.68294197$$

The approximate value of the integral according to Gauss quadrature (the right-hand side of Eq. (7.26)) is:

$$\cos\left(\frac{-1}{\sqrt{3}}\right) + \cos\left(\frac{1}{\sqrt{3}}\right) = 1.67582366$$

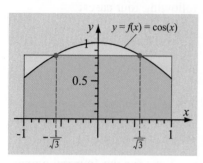

Figure 7-20: Gauss quadrature integration of $f(x) = \cos(x)$.

These results show that Gauss quadrature gives a very good approximation (error of a 4.2%) for the integral, but not the exact value. The last integration is illustrated in Fig. 7-20, where the exact integration is the area under the curve $f(x) = \cos(x)$ and the approximate value of the integral according to Gauss quadrature is the area under the red line. In this case the two areas are not exactly identical. The light shaded area under the red line is a little bit (4.2%) smaller than the light shaded area that is above the red line.

The accuracy of Gauss quadrature can be increased by using a higher value for n in Eq. (7.24). For $n = 3$ the equation has the form:

$$\int_{-1}^{1} f(x)dx \approx C_1 f(x_1) + C_2 f(x_2) + C_3 f(x_3) \tag{7.28}$$

In this case there are six constants: C_1, C_2, C_3, x_1, x_2, and x_3. The constants are determined by enforcing Eq. (7.28) to be exact when $f(x) = 1$, $f(x) = x$, $f(x) = x^2$, $f(x) = x^3$, $f(x) = x^4$, and $f(x) = x^5$. This gives a set of six equations with six unknowns. (The process of finding the unknowns is the same as was done when the value of n was 2.) The constants that are determined are:

$$C_1 = 0.5555556, \quad C_2 = 0.8888889, \quad C_3 = 0.5555556$$

$$x_1 = -0.77459667, \quad x_2 = 0, \quad x_3 = 0.77459667$$

The Gauss quadrature equation for $n = 3$ is then:

$$\int_{-1}^{1} f(x)dx \approx 0.5555556 f(-0.77459667) + 0.8888889 f(0)$$
$$+ 0.5555556 f(0.77459667) \tag{7.29}$$

As an example, the integral when $f(x) = \cos(x)$ is estimated again by using Eq. (7.29):

$$\int_{-1}^{1} \cos(x)dx \approx 0.5555556 \cos(-0.77459667) + 0.8888889 \cos(0)$$
$$+ 0.5555556 \cos(0.77459667) = 1.68285982$$

This value is almost identical to the exact value that was calculated earlier.

The accuracy of Gauss quadrature can be increased even more by using higher values for n. The general equation for estimating the value of an integral is:

$$\int_{-1}^{1} f(x)dx \approx C_1 f(x_1) + C_2 f(x_2) + C_3 f(x_3) + \ldots + C_n f(x_n) \qquad (7.30)$$

Table 7-1 lists the values of the coefficients C_i and the location of the Gauss points x_i for $n = 2, 3, 4, 5, and\ 6$.

Table 7-1: Weight coefficients and Gauss points coordinates.

n (Number of points)	Coefficients C_i (weights)	Gauss points x_i
2	$C_1 = 1$	$x_1 = -0.57735027$
	$C_2 = 1$	$x_2 = 0.57735027$
3	$C_1 = 0.5555556$	$x_1 = -0.77459667$
	$C_2 = 0.8888889$	$x_2 = 0$
	$C_3 = 0.5555556$	$x_3 = 0.77459667$
4	$C_1 = 0.3478548$	$x_1 = -0.86113631$
	$C_2 = 0.6521452$	$x_2 = -0.33998104$
	$C_3 = 0.6521452$	$x_3 = 0.33998104$
	$C_4 = 0.3478548$	$x_4 = 0.86113631$
5	$C_1 = 0.2369269$	$x_1 = -0.90617985$
	$C_2 = 0.4786287$	$x_2 = -0.53846931$
	$C_3 = 0.5688889$	$x_3 = 0$
	$C_4 = 0.4786287$	$x_4 = 0.53846931$
	$C_5 = 0.2369269$	$x_5 = 0.90617985$
6	$C_1 = 0.1713245$	$x_1 = -0.93246951$
	$C_2 = 0.3607616$	$x_2 = -0.66120938$
	$C_3 = 0.4679139$	$x_3 = -0.23861919$
	$C_4 = 0.4679139$	$x_4 = 0.23861919$
	$C_5 = 0.3607616$	$x_5 = 0.66120938$
	$C_6 = 0.1713245$	$x_6 = 0.93246951$

Gauss quadrature integration of $\int_{a}^{b} f(x)dx$

The weight coefficients and the coordinates of the Gauss points given in Table 7-1 are valid only when the interval of the integration is $[-1, 1]$. In general, however, the interval can have any domain $[a, b]$. Gauss quadrature with the coefficients and Gauss points determined for the $[-1, 1]$ interval can still be used for a general domain. This is done by using a transformation. The integral $\int_{a}^{b} f(x)dx$ is transformed into an integral in the form $\int_{-1}^{1} f(t)dt$. This is done by changing variables:

$$x = \frac{1}{2}[t(b-a)+a+b] \quad \text{and} \quad dx = \frac{1}{2}(b-a)dt \qquad (7.31)$$

The integration then has the form:

$$\int_{a}^{b} f(x)dx = \int_{-1}^{1} f\left(\frac{(b-a)t+a+b}{2}\right)\frac{(b-a)}{2}dt$$

Example 7-2 shows how to use the transformation.

Example 7-2: Evaluation of a single definite integral using fourth-order Gauss quadrature.

Evaluate $\int_{0}^{3} e^{-x^2} dx$ using fourth-order Gauss quadrature.

SOLUTION

Step 1: Since the limits of integration are $[0, 3]$, the integral has to be transformed to the form $\int_{-1}^{1} f(t)dt$. In the present problem $a = 0$ and $b = 3$. Substituting these values in Eq. (7.31) gives:

$$x = \frac{1}{2}[t(b-a)+a+b] = \frac{1}{2}[t(3-0)+0+3] = \frac{3}{2}(t+1) \quad \text{and} \quad dx = \frac{1}{2}(b-a)dt = \frac{1}{2}(3-0)dt = \frac{3}{2}dt$$

Substituting these values in the integral gives:

$$I = \int_{0}^{3} e^{-x^2} dx = \int_{-1}^{1} f(t)dt = \int_{-1}^{1} \frac{3}{2}e^{-\left[\frac{3}{2}(t+1)\right]^2} dt$$

Step 2: Use fourth-order Gauss quadrature to evaluate the integral. From Eq. (7.30), and using Table 7-1:

$$I = \int_{-1}^{1} f(t)dt \approx C_1 f(t_1) + C_2 f(t_2) + C_3 f(t_3) + C_4 f(t_4) = 0.3478548 f(-0.86113631)$$

$$+ 0.6521452 f(-0.33998104) + 0.6521452 f(0.33998104) + 0.3478548 f(0.86113631)$$

Evaluating $f(t) = \frac{3}{2}e^{-\left[\frac{3}{2}(t+1)\right]^2}$ gives:

$$I = \frac{3}{2}\left(0.3478548 \frac{3}{2}e^{-\left[\frac{3}{2}((-0.86113631)+1)\right]^2} + 0.6521452 \frac{3}{2}e^{-\left[\frac{3}{2}((-0.33998104)+1)\right]^2}\right.$$

$$\left. + 0.6521452 \frac{3}{2}e^{-\left[\frac{3}{2}((0.33998104)+1)\right]^2} + 0.3478548 \frac{3}{2}e^{-\left[\frac{3}{2}((0.86113631)+1)\right]^2}\right) = 0.8841359$$

The exact value of the integral (when carried out analytically) is 0.8862073. The error is only about 1%.

7.6 EVALUATION OF MULTIPLE INTEGRALS

Double and triple integrals often arise in two-dimensional and three-dimensional problems. A two-dimensional (double integral) has the form:

$$I = \int_A f(x, y)\, dA = \int_a^b \left[\int_{y=g(x)}^{y=p(x)} f(x, y)\, dy \right] dx \qquad (7.32)$$

The integrand $f(x, y)$ is a function of the independent variables x and y. The limits of integration of the inner integral may be a function of x, as in Eq. (7.32), or may be constants, (When they are constants, the integration is over a rectangular region.) Figure 7-21 schematically shows the surface of the function $f(x, y)$ and the projection of the surface on the x-y plane. In this illustration, the domain $[a, b]$ in the x direction is

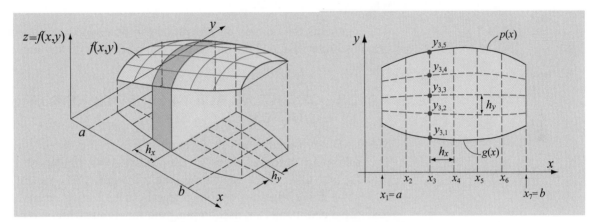

Figure 7-21: Function and domain for double integration.

divided into six equally spaced subintervals. In the y direction the domain $[g(x), p(x)]$ is a function of x, and at every x the y direction is divided into four equally spaced subintervals.

The double integration in Eq. (7.32) can be separated into two parts. The inner integral can be written as:

$$G(x) = \int_{y=g(x)}^{y=p(x)} f(x, y)\, dy \qquad (7.33)$$

and the outer integral can be written with $G(x)$ as its integrand:

$$I = \int_a^b G(x)\, dx \qquad (7.34)$$

The outer integral is evaluated by using one of the numerical methods described in the previous sections. For example, if Simpson's 1/3 method is used, then the outer integral is evaluated by:

$$I(G) \approx \frac{h_x}{3} \{ G(a) + 4[G(x_2) + G(x_4) + G(x_6)] + 2[G(x_3) + G(x_5)] + f(b) \}$$
$$(7.35)$$

where $h_x = \dfrac{b-a}{6}$. Each of the G terms in Eq. (7.35) is an inner integral that has to be integrated according to Eq. (7.33) using the appropriate value of x. In general, the integral $G(x_i)$ can be written as:

$$G(x_i) = \int_{y = g(x_i)}^{y = p(x_i)} f(x_i, y) dy$$

and then integrated numerically. For example, using Simpson's 1/3 method to integrate $G(x_3)$ gives:

$$G(x_3) = \int_{y = g(x_3)}^{y = p(x_3)} f(x_3, y) dy$$

$$= \frac{h_y}{3} \{ f(x_3, y_{3,1}) + 4[f(x_3, y_{3,2}) + f(x_3, y_{3,4})] + 2[f(x_3, y_{3,3})] + f(x_3, y_{3,5}) \}$$

where $h_y = \dfrac{p(x_3) - g(x_3)}{4}$.

In general, the domain of integration can be divided into any number of subintervals, and the integration can be done with any numerical method.

7.7 USE OF MATLAB BUILT-IN FUNCTIONS FOR INTEGRATION

MATLAB has several built-in functions for carrying out integration. The following describes how to use MATLAB's functions `quad`, `quadl`, and `trapz` for evaluating single integrals, and the function `dblquad` for evaluating double integrals. The `quad`, `quadl`, and `dblquad` commands are used to integrate functions, while the `trapz` function is used to integrate tabulated data.

The `quad` *command*

The form of the `quad` command is:

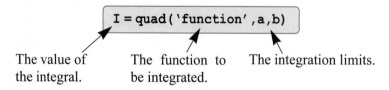

     The value of   The function to   The integration limits.
     the integral.   be integrated.

- The function can be entered as a string, as the name of a function file, or the name of an inline function.

- The function $f(x)$ must be written for an argument x that is a vector (use element-by-element operations), such that it calculates the value of the function for each element of x.

- The user has to make sure that the function does not have a vertical asymptote (singularity) between a and b.

- `quad` calculates the integral with an absolute error that is smaller

than 1.0×10^{-6}. This number can be changed by adding an optional
`tol` argument to the command:

    q = quad('function',a,b,tol)

`tol` is a number that defines the maximum error. With larger `tol` the
integral is calculated less accurately but more quickly.

The quad command uses an adaptive Simpson's method of integration. Adaptive methods are integration schemes that selectively refine
the domain of integration, depending on the behavior of the integrand.
If the integrand varies sharply in the neighborhood of a point within the
domain of integration, then the subintervals in this vicinity are divided
into smaller subintervals.

The `quadl` *command:*

The form of the `quadl` (the last letter is a lower case L) command is
exactly the same as the `quad` command:

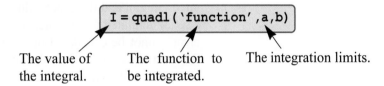

The value of   The function to   The integration limits.
the integral.   be integrated.

All the comments listed above for the `quad` command are valid for the
`quadl` command. The difference between the two commands is in the
numerical method used for calculating the integration. The `quadl` uses
the adaptive Lobatto method.

The `trapz` *command*

The built-in function `trapz` can be used for integrating a function that
is given as discrete data points. It uses the trapezoidal method of numerical integration. The form of the command is:

$$q = trapz(x,y)$$

where `x` and `y` are vectors with the *x* and *y* coordinates of the points,
respectively. The two vectors must be of the same length.

The `dblquad` *command*

The built-in function `dblquad` can be used to evaluate a double integral. The format of the command is:

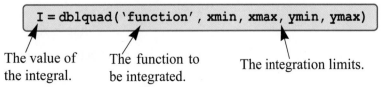

The value of   The function to   The integration limits.
the integral.   be integrated.

• The function can be entered as a string, as the name of a function
  file, or the name of an inline function.

- The function $f(x, y)$ must be written for an argument x that is a vector (use element-by-element operations) and for an argument y that is a scalar.

- The limits of integration are constants.

In the format shown above, the integration is done using the `quad` function and the default tolerance, which is 1.0×10^{-6}. The tolerance can be changed by adding an optional `tol` argument to the command, and the method of integration can be changed to quadl by adding it as an argument:

```
q = dblquad('function',xmin,xmax,ymin,ymax,tol,quadl)
```

7.8 ESTIMATION OF ERROR IN NUMERICAL INTEGRATION

The error is the difference between the value of the numerically calculated integral and the exact value of the integral. When the integrand is a set of tabulated data points, an exact value does not really exist and an error cannot be calculated or even estimated. When the integrand is a function, the error can be calculated if the exact value of the integral can be determined analytically. However, if the value of the integral can be calculated analytically, there is no real need to calculate the value of the integral numerically. A common situation is that the integrand is a mathematical expression, and the integral is evaluated numerically because an exact result obtained by analytical integration is difficult or impossible. In this case, the error can be estimated in some of the numerical integration methods. As an illustration, an estimation of the error in the rectangle method is presented in some detail.

In the rectangle method, the integral of $f(x)$ over the interval $x \in [a, b]$ is calculated by assuming that $f(x) = f(a)$ within the interval:

$$I(f) = \int_a^b f(x)dx \approx \int_a^b f(a)dx = f(a)(b-a)$$

The error E is then:

$$E = \int_a^b f(x)dx - f(a)(b-a) \tag{7.36}$$

An estimate of the error can be obtained by writing the one-term Taylor series expansion with a remainder (see Chapter 2) of $f(x)$ near the point $x = a$:

$$f(x) = f(a) + f'(\xi)(x-a) \tag{7.37}$$

where ξ is a point between a and b. Integrating both sides of Eq. (7.37) gives:

$$\int_a^b f(x)dx = \int_a^b [f(a) + f'(\xi)(x-a)]dx = f(a)(b-a) + \frac{1}{2}f'(\xi)(b-a)^2 \tag{7.38}$$

The error according to Eq. (7.36) can be determined using Eq. (7.38):

$$E = \int_a^b f(x)dx - f(a)(b-a) = \frac{1}{2}f'(\xi)(b-a)^2 \qquad (7.39)$$

Equation (7.39) shows that the error depends on $(b-a)$ and the values of the first derivative of $f(x)$ within the interval $[a, b]$. Obviously, the error can be large if the domain is large and/or the value of the derivatives is large. The error, however, can be reduced significantly if the composite rectangle method is used. The domain is then divided into subintervals of width h where $h = x_{i+1} - x_i$. Equation (7.39) can be used to estimate the error for a subinterval:

$$E = \int_{x_i}^{x_{i+1}} f(x)dx - f(x_i)h = \frac{1}{2}f'(\xi_i)h^2 \qquad (7.40)$$

where ξ_i is a point between x_i and x_{i+1}. Now, the magnitude of the error can be controlled by the size of h. When h is very small (much smaller than 1), the error in the subinterval becomes very small. For the whole interval $[a, b]$, an estimate of the error is obtained by adding the errors from all the subintervals. For the case where h is the same for all subintervals:

$$E = \frac{1}{2}h^2 \sum_{i=1}^{N} f'(\xi_i) \qquad (7.41)$$

If an average value of the derivative $\overline{f'}$ in the interval $[a, b]$ can be estimated by:

$$\overline{f'} \approx \frac{\sum_{i=1}^{N} f'(\xi_i)}{N} \qquad (7.42)$$

then Eq. (7.41) can be simplified by using Eq. (7.42) and recalling that $h = (b-a)/N$:

$$E = \frac{(b-a)}{2}h\overline{f'} = O(h) \qquad (7.43)$$

This equation is an estimate of the error for the composite rectangle method. The error is proportional to h since $\frac{(b-a)}{2}\overline{f'}$ is a constant. It is written as $O(h)$, which means of the order of h.

The error in the composite midpoint, composite trapezoidal, and composite Simpson's methods can be estimated in a similar way. The details are beyond the scope of this book, and the results are as follows.

Composite midpoint method: $\qquad E = \frac{(b-a)}{24}\overline{f''}h^2 = O(h^2) \qquad (7.44)$

Composite trapezoidal method: $\qquad E = -\frac{(b-a)}{12}\overline{f''}h^2 = O(h^2) \qquad (7.45)$

Composite Simpson's 1/3 method: $E = -\frac{(b-a)}{180}\overline{f^{IV}}h^4 = O(h^4)$ (7.46)

Composite Simpson's 3/8 method: $E = -\frac{(b-a)}{80}\overline{f^{IV}}h^4 = O(h^4)$ (7.47)

Note that if the average value of the derivatives in Eqs. (7.43)–(7.47) can be bounded, then bounds can be found for the errors. Unfortunately, such bounds are difficult to find so that the exact magnitude of the error is difficult to calculate in practice.

7.9 RICHARDSON'S EXTRAPOLATION

Richardson's extrapolation is a method for obtaining a more accurate estimate of the value of an integral from two less accurate estimates. For example, two estimates calculated with an error $O(h^2)$ can be used for calculating an estimate with an error $O(h^4)$. This section starts by deriving Richardson's extrapolation formula for this case by considering two initial estimates that are calculated with the composite trapezoidal method (error $O(h^2)$). Next, Richardson's extrapolation formula for obtaining an estimate with an error $O(h^6)$ from two estimates with an error $O(h^4)$ is derived. Finally, a general Richardson's extrapolation formula is presented. This formula uses known estimates of an integral with an error of order h^n, to calculate a new estimate that has an increase of 1 (and possibly 2) in the order of accuracy (i.e., with error $O(h^{n+1})$ or $O(h^{n+2})$).

Richardson's extrapolation from two estimates with an error O(h²)

When an integral $I(f)_h$ is numerically evaluated with a method whose truncation error can be written in terms of even powers of h, starting with h^2, then the true (unknown) value of the integral $I(f)$ can be expressed as the sum of $I(f)_h$ and the error:

$$I(f) = I(f)_h + Ch^2 + Dh^4 + \dots$$ (7.48)

where C, D, ..., are constants. For example, if the composite trapezoidal method is used for calculating $I(f)_h$ (with an error given by Eq. (7.45)), then $I(f)$ can be expressed by:

$$I(f) = I(f)_h - \frac{(b-a)}{12}\overline{f''}h^2$$ (7.49)

Two estimated values of an integral $I(f)_{h1}$ and $I(f)_{h2}$ can be calculated by using a different number of subintervals (in one estimate $h = h_1$ and in the other $h = h_2$). Substituting each of the estimates in Eq. (7.48) gives:

$$I(f) = I(f)_{h1} + Ch_1^2$$ (7.50)

and

$$I(f) = I(f)_{h2} + Ch_2^2 \qquad (7.51)$$

If it is assumed that C is the same (the average value of the second derivative $\overline{f''}$ is independent of the value of h), then Eqs. (7.50) and (7.51) can be solved for $I(f)$ in terms of $I(f)_{h_1}$ and $I(f)_{h_2}$:

$$I(f) = \frac{I(f)_{h_1} - \left(\dfrac{h_1}{h_2}\right)^2 I(f)_{h_2}}{1 - \left(\dfrac{h_1}{h_2}\right)^2} \qquad (7.52)$$

Equation (7.52) gives a new estimate for $I(f)$, which has an error $O(h^4)$, from the values of $I(f)_{h_1}$ and $I(f)_{h_2}$, each of which have an error $O(h^2)$. The proof that Eq. (7.52) has an error $O(h^4)$ is beyond the scope of this book.[1]

A special case is when $h_2 = \frac{1}{2}h_1$. The two estimates of the value of the integral used for the extrapolation are such that the second estimate has double the number of subintervals compared with the first estimate. In this case Eq. (7.52) reduces to:

$$I(f) = \frac{4I(f)_{h_2} - I(f)_{h_1}}{3} \qquad (7.53)$$

Richardson's extrapolation from two estimates with an error $O(h^4)$

When an integral $I(f)_h$ is numerically evaluated with a method whose truncation error can be written in terms of even powers of h, starting with h^4, then the true (unknown) value of the integral $I(f)$ can be expressed as the sum of $I(f)_h$ and the error:

$$I(f) = I(f)_h + Ch^4 + Dh^6 + \dots \qquad (7.54)$$

where C, D, ..., are constants.

Two estimated values of an integral $I(f)_{h1}$ and $I(f)_{h2}$ can be calculated with the same method (which has an error $O(h^4)$) by using a different number of subintervals (in one estimate $h = h_1$ and in the other $h = h_2$) Substituting each of the estimates in Eq. (7.54) gives:

$$I(f) = I(f)_{h1} + Ch_1^4 \qquad (7.55)$$

and

$$I(f) = I(f)_{h2} + Ch_2^4 \qquad (7.56)$$

If it is assumed that C is the same (the average value of the fourth deriv-

---

1. The interested reader is referred to: P. J. Davis, and P. Rabinowitz, *Numerical Integration*, Blaisdell Publishing Company, Waltham, Massachusetts, 1967, pp. 52–55, 166; L. F. Richardson and J. A. Gaunt, Phil. Trans. Roy. Soc. London A, Vol. 226, pp. 299-361, 1927.

ative $\overline{f^{IV}}$ is independent of the value of h), then Eqs. (7.55) and (7.56) can be solved for $I(f)$ in terms of $I(f)_{h_1}$ and $I(f)_{h_2}$:

$$I(f) = \frac{I(f)_{h_1} - \left(\frac{h_1}{h_2}\right)^4 I(f)_{h_2}}{1 - \left(\frac{h_1}{h_2}\right)^4}$$

(7.57)

Equation (7.57) gives an estimate for $I(f)$ with an error $O(h^6)$ from the values of $I(f)_{h_1}$ and $I(f)_{h_2}$, each of which were calculated with an error $O(h^4)$. (The proof is, again, beyond the scope of this book. See footnote on the previous page.)

A special case is when $h_2 = \frac{1}{2}h_1$. The two estimates of the value of the integral used for the extrapolation are such that in the second estimate the number of subintervals is doubled compared with the first estimate. In this case Eq. (7.57) reduces to:

$$I(f) = \frac{16I(f)_{h_2} - I(f)_{h_1}}{15}$$

(7.58)

Richardson's general extrapolation formula

A general extrapolation formula can be derived for the case when the two initial estimates of the value of the integral have the same estimated error of order h^p and are obtained such that in one the number of subintervals is twice the number of subintervals of the other.

If I_n is an estimate of the value of the integral that is obtained by using n subintervals and I_{2n} is an estimate of the value of the integral that is obtained by using $2n$ subintervals, where in both the estimated error is of order h^p, then a new estimate for the value of the integral can be calculated by:

$$I = \frac{2^p I_{2n} - I_n}{2^p - 1}$$

(7.59)

In general, the new estimate of the integral has an estimated error of order $h^{(p+1)}$. The error is of order $h^{(p+2)}$ when the truncation error can be written in terms of even powers of h. Substituting $p = 2$ and $p = 4$ in Eq. (7.59) gives Eqs. (7.53) and (7.58), respectively. In the same way, the extrapolation equation for using two estimates with an error $O(h^6)$ to obtain a new estimate with an error $O(h^8)$ is obtained by substituting $p = 6$ in Eq. (7.59):

$$I = \frac{2^6 I_{2n} - I_n}{2^6 - 1} = \frac{64}{63}I_{2n} - \frac{1}{63}I_n$$

(7.60)

7.10 ROMBERG INTEGRATION

Romberg integration is a scheme for improving the accuracy of the estimate of the value of an integral by successive application of Richardson's extrapolation formula (see Section 7.9). The scheme uses a series of initial estimates of the integral calculated with the composite trapezoidal method by using different numbers of subintervals. The Romberg integration scheme, illustrated in Fig. 7-22, follows these steps:

Step 1: The value of the integral is calculated with the composite trapezoidal method several times. In the first time, the number of subintervals is n, and in each calculation that follows the number of subintervals is doubled. The values obtained are listed in the first (left) column in Fig. 7-22. In the first row $I_{1,1}$ is calculated with the composite trapezoidal method using n subintervals. In the second row $I_{2,1}$ is calculated using $2n$ subintervals, $I_{3,1}$ using $4n$ subintervals, and so on. The error in the calculations of the integrals in the first column is $O(h^2)$.

Step 2: Richardson's extrapolation formula, Eq. (7.53), is used for obtaining improved estimates for the value of the integral from the values listed in the first (left) column in Fig. 7-22. This is the <u>first level</u> of the Romberg integration. The first two values $I_{1,1}$ and $I_{2,1}$ give the estimate $I_{1,2}$:

$$I_{1,2} = \frac{4I_{2,1} - I_{1,1}}{3} \tag{7.61}$$

The second and third values ($I_{2,1}$ and $I_{3,1}$) give the estimate $I_{2,2}$:

$$I_{2,2} = \frac{4I_{3,1} - I_{2,1}}{3} \tag{7.62}$$

and so on. The new improved estimates are listed in the second column in Fig. 7-22. According to Richardson's extrapolation formula they have an error of $O(h^4)$.

Step 3: Richardson's extrapolation formula, Eq. (7.58), is used for obtaining improved estimates for the value of the integral from the values listed in the second column in Fig. 7-22. This is the <u>second level</u> of the Romberg integration. The first two values $I_{1,2}$ and $I_{2,2}$ give the estimate $I_{1,3}$:

$$I_{1,3} = \frac{16I_{2,2} - I_{1,2}}{15} \tag{7.63}$$

The second and third values ($I_{2,2}$ and $I_{3,2}$) give the estimate $I_{2,3}$:

$$I_{2,3} = \frac{16I_{3,2} - I_{2,2}}{15} \tag{7.64}$$

and so on. The new improved estimates are listed in the third column in

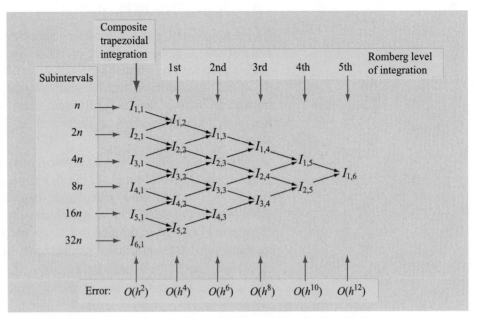

Figure 7-22: Romberg integration method.

Fig. 7-22. According to Richardson's extrapolation formula they have an error of $O(h^6)$.

Step 4 and beyond: The process of calculating improved estimates for the value of the integral can continue where each new column is a higher level of Romberg integration.

The equation for calculating the extrapolated values in each level from the values in the previous level can be written in a general form:

$$I_{i,j} = \frac{4^{j-1}I_{i+1,j-1} - I_{i,j-1}}{4^{j-1} - 1} \tag{7.65}$$

The values of the first column $I_{1,1}$ through $I_{k,1}$ are calculated by using the composite trapezoidal method. Then, the extrapolated values in the rest of the columns are calculated by using Eq. (7.65) for $j = 2, 3, ..., k$ and in each column $i = 1, 2, ..., (k-j+1)$, where k is the number of elements in the first column. The highest level of Romberg integration that can be calculated is $k-1$. The process can continue until there is only one term in the last column (highest level of Romberg integration), or the process can be stopped when the differences between the improved estimated values of the integral are smaller than a predetermined tolerance. Example 7-3 shows an application of Romberg integration.

Example 7-3: Romberg integration with comparison to the composite trapezoidal method.

Evaluate $\int_0^1 \frac{1}{(1+x)} dx$ using three levels of Romberg integration. Use an initial step size of $h = 1$ (one subinterval). Compare your result with the exact answer. What number of subintervals would be required if you were to use the composite trapezoidal method to obtain the same level of accuracy?

SOLUTION

Exact answer: The exact answer to this problem can be obtained analytically. The answer is: $\ln(2) = 0.69314718$.

Romberg Integration

To carry out the numerical integration, a user-defined function, which is listed below, named Romberg is created.

> **Program 7-2: Function file. Romberg integration.**

```
function IR = Romberg(integrand,a,b,Ni,Levels)
% Romberg numerically integrate using the Romberg integration method.
% Input Variables:
% integrand  The function to be integrated typed as a string.
% a  Lower limit of integration.
% b  Upper limit of integration.
% Ni Initial number of subintervals.
% Levels Number of levels of Romberg integration.
% Output Variable:
% IR  A matrix with the estimated values of the integral.

% Creating the first column with the composite trapezoidal method:
for i=1:Levels + 1
    Nsubinter=Ni*2^(i - 1);
    IR(i,1)=trapezoidal(integrand,a,b,Nsubinter);
end
% Calculating the extrapolated values using Eq. (7.65):
for j=2:Levels + 1
    for i=1:(Levels - j + 2)
        IR(i,j)=(4^(j - 1)*IR(i + 1,j - 1) - IR(i,j - 1))/(4^(j - 1) - 1);
    end
end
end
```

> Create the first column of Fig. 7-22 by using the user-defined function **trapezoidal** (listed in Section 7.3).

> Calculate the extrapolated values, level after level, using Eq. (7.65).

The function Romberg is next used in the Command Window to determine the value of the integral $\int_0^1 \frac{1}{(1+x)} dx$. The initial number of subintervals (for the first estimate with the composite trapezoidal method) is 1.

```
>> format long
>> IntVal = Romberg('1/(1+x)',0,1,1,3)
```

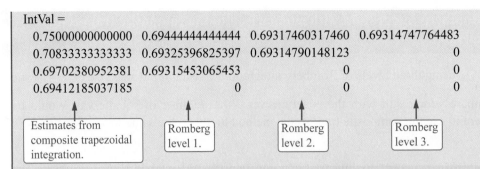

```
IntVal =
   0.75000000000000   0.69444444444444   0.69317460317460   0.69314747764483
   0.70833333333333   0.69325396825397   0.69314790148123              0
   0.69702380952381   0.69315453065453              0              0
   0.69412185037185              0              0              0
```

| Estimates from composite trapezoidal integration. | Romberg level 1. | Romberg level 2. | Romberg level 3. |

The results show that with the composite trapezoidal method (first column) the most accurate value that is obtained (using eight subintervals) is accurate to two decimal places. The first-level Romberg integration (second column) increases the accuracy to four decimal places. The second-level Romberg integration (third column) increases the accuracy to six decimal places. The result from the third-level Romberg integration (fourth column) is also accurate to six decimal places, but the value is closer to the exact answer.

The number of calculations that was executed is 10 (4 in the composite trapezoidal method and 6 in the Romberg integration procedure). To obtain an estimate for the integral with an accuracy of six decimal places by applying only the composite trapezoidal method, the method has to be applied with 276 subintervals. This is shown below where `trapezoidal` is used in the Command Window.

```
>> ITrap = trapezoidal('1/(1+x)',0,1,277)

ITrap =

    0.69314799511374
```

7.11 IMPROPER INTEGRALS

In all integrals $\int_a^b f(x)dx$ that have been considered so far in this chapter, the limits of integration a and b are finite, and the integrand $f(x)$ is finite and continuous in the domain of integration. There are, however, situations in science and engineering where one, or both integration limits, are infinite, and cases where the integrand is not continuous within the range of integration. For example, in statistics the integral

$$\int_{-\infty}^b \frac{1}{\sqrt{2\pi}} e^{\left(\frac{-x^2}{2}\right)} dx$$ is used to calculate the cumulative probability that a

quantity will have a value of b or smaller.

7.11.1 Integrals with Singularities

An integral $\int_a^b f(x)dx$ has a singularity when there is a point c within the domain, $a \le c \le b$, where the value of the integrand $f(c)$ is not defined ($|f(x)| \to \infty$ as $x \to c$). If the singularity is not at one of the endpoints, the integral can always be written as a sum of two integrals. One over $[a, c]$ and one over $[c, b]$. Mathematically, integrals that have a singularity at one of the endpoints might or might not have a finite value. For example, the function $1/(\sqrt{x})$ has a singular point at $x = 0$, but the integral of this function over $[0, 2]$ has a value of 2, $\int_0^1 \frac{1}{\sqrt{x}}\,dx = 2$. On the other hand, the integral $\int_0^1 \frac{1}{x}\,dx$ does not have a finite value.

Numerically, there are several ways of integrating an integral that has a finite value when the integrand has a singularity at one of the endpoints. One possibility is to use an open integration method where the endpoints are not used for determining the integral. Two such methods presented in this chapter are the composite midpoint method (Section 7.2) and Gauss quadrature (Section 7.5). Another possibility is to use a numerical method that uses the value of the integrand at the endpoint, but instead of using the endpoint itself, for example, $x = a$, the integration starts at a point that is very close to the end point $x = a + \varepsilon$ where $\varepsilon \ll |a|$.

In some cases it is also possible to eliminate a singularity analytically. This can be done by using a change of variable or transformation. Subsequently, the transformed integral can be integrated numerically.

7.11.2 Integrals with Unbounded Limits

Integrals with one or two unbounded limits can have one of the following forms:

$$I = \int_{-\infty}^b f(x)\,dx, \quad I = \int_a^\infty f(x)\,dx, \quad I = \int_{-\infty}^\infty f(x)\,dx \qquad (7.66)$$

In general, integrals with unbounded limits might have a finite value (converge) or might not have a finite value (diverge). When the integral has a finite value, it is possible to carry out the integration numerically. Typically, the integrand of such an integral has a finite value over a small range of the domain of integration and a value close to zero everywhere else. The numerical integration can then be done by replacing the unbounded limit (or limits) with a finite limit (or limits) where the value of the integrand is close to zero. Then, the numerical integration can be carried out with any of the methods described in this chapter. The integration is done successively, where in each the absolute value of the limit is increased. The calculations stop when the value of the integral does not change much with successive integrations.

In some cases it is also possible to use a change of variable to transform the integral such that the transformed integral will have bounded limits. Subsequently, the transformed integral can be integrated numerically.

7.12 PROBLEMS

Problems to be solved by hand

Solve the following problems by hand. When needed, use a calculator, or write a MATLAB script file to carry out the calculations. If using MATLAB, do not use built-in functions for integration.

7.1   The function $f(x)$ is given in the following tabulated form. Compute $\int_0^1 f(x)dx$ with $h = 0.25$ and with $h = 0.5$, using:

(*a*) The composite rectangle method.
(*b*) The composite midpoint method.
(*c*) The composite trapezoidal method.

x	0	0.25	0.5	0.75	1.0
$f(x)$	0.9162	0.8109	0.6931	0.5596	0.4055

7.2   To estimate the surface area and volume of a football, the diameter of the ball is measured at different points along the ball. The surface area, S, and volume, V, can be determined by:

$$S = 2\pi \int_0^R r\,dz \quad \text{and} \quad V = \pi \int_0^R r^2\,dz$$

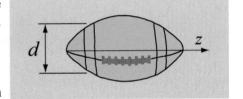

Use the composite Simpson's 1/3 method, with the data given below to determine the volume and surface area of the ball.

z (in.)	0	1.0	2.0	3.0	4.0	5.0	6.0	7.0	8.0	9.0	10.0	11.0	12.0
d (in.)	0	2.6	3.2	4.8	5.6	6	6.2	6.0	5.6	4.8	3.3	2.6	0

7.3   An approximate map of the state of Texas is shown in the figure. For determining the area of the state, the map is divided into two parts (one above and one below the *x* axis). Determine the area of the state by numerically integrating the two areas. For each part make a list of the coordinate *y* of the border as a function of x. Start with $x = 0$ and use increments of 50 mi, such that the last point is $x = 750$.

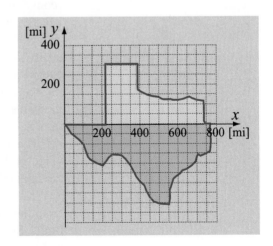

Once the tabulated data is available, determine the integrals once with the composite midpoint method and once with the composite Simpson's 3/8 method.

7.4   Evaluate the integrals:

$$I_1 = \int_0^\pi \sin^2 x\,dx \quad \text{and} \quad I_2 = \int_0^{2\pi} \sin^2 x\,dx$$

using the following methods:

(*a*) Simpson's 1/3 method. Divide the whole interval into six subintervals.

(*b*) Simpson's 3/8 method. Divide the whole interval into six subintervals.

(*c*) Second-order Gauss quadrature.

Compare the results and discuss the reasons for the differences. The exact value of the integrals is $I_1 = \pi/2$, and $I_2 = \pi$.

7.5 The equation of a circle with a radius of 1 (unit circle) is given by $x^2 + y^2 = 1$, and its area is $A = \pi$. Consequently:

$$\int_{-1}^{1} \sqrt{1 - x^2}\, dx = \frac{\pi}{2}$$

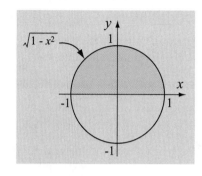

Evaluate the integral using the following methods:

(*a*) Simpson's 1/3 method. Divide the whole interval into eight subintervals.

(*b*) Simpson's 3/8 method. Divide the whole interval into nine subintervals.

(*c*) Second order Gauss quadrature.

Compare the results and discuss the reasons for the differences.

7.6 The shape of the centroid line of the Gateway Arch in St. Louis can be modeled approximately with the equation:

$$f(x) = 693.9 - 68.8\cosh\left(\frac{x}{99.7}\right) \quad \text{for} \quad -299.25 \le x \le 299.25 \text{ ft.}$$

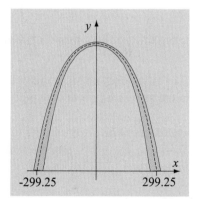

By using the equation $L = \int_a^b \sqrt{1 + [f'(x)]^2}\, dx$ determine the length of the arch with the following integration methods:

(*a*) Simpson's 1/3 method. Divide the whole interval into eight subintervals.

(*b*) Simpson's 3/8 method. Divide the whole interval into nine subintervals.

(*c*) Second order Gauss quadrature.

7.7 The roof of a silo is made by revolving the curve $y = 10\cos\left(\frac{\pi}{10}x\right)$ from $x = -5$ m to $x = 5$ m about the y axis, as shown in the figure to the right.

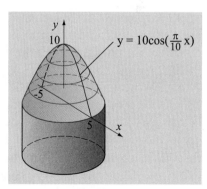

The surface area, S, that is obtained by revolving a curve $y = f(x)$ in the domain from a to b around the y axis can be calculated by:

$$S = 2\pi \int_a^b x\sqrt{1 + [f'(x)]^2}\, dx$$

Calculate the surface area of the roof with the following integration methods:

(*a*) Simpson's 1/3 method. Divide the whole interval into eight subintervals.

(*b*) Simpson's 3/8 method. Divide the whole interval into nine subintervals.

(*c*) Second order Gauss quadrature.

7.8 In the standard Simpson's 1/3 method (Eq. (7.16)), the points used for the integration are the endpoints of the domain, a and b, and the middle point $(a+b)/2$. Derive a new formula for Simpson's 1/3 method in which the points used for the integration are $x = a$, $x = b$, and $x = (a+b)/3$.

7.9 Evaluate the integral $-\int_{-1}^{1} \frac{1}{64}(2\cos\theta + 5\cos 3\theta)^2 \sin\theta \, d\theta$ using Simpson's 3/8 method and dividing the whole interval into eight subintervals. This is the orthogonality condition for the third-order Legendre polynomial.

7.10 Evaluate the integral $\int_{-1}^{1} 225x(1-x^2)\sqrt{(1-x^2)^3} \, dx$ using second-level Romberg integration. Use $n = 1$ in the first estimate with the trapezoidal method. The integral is the orthogonality condition for the associated Legendre functions $P_3^2(x)$ and $P_3^3(x)$.

7.11 Evaluate the integral $53.3904 \int_0^{10} (1 - e^{-0.18355x}) \, dx$ by using third-order (two-point) Gauss quadrature.

7.12 Evaluate the integral

$$\int_0^{0.8} (0.2 + 25x - 200x^2 + 675x^3 - 900x^4 + 400x^5) \, dx$$

using third-order (two-point) Gauss quadrature.

7.13 Show that the truncation error for the composite trapezoidal method is of the order of h^2, where h is the step size (width of subinterval).

Problems to be programmed in MATLAB
Solve the following problems using MATLAB environment. Do not use MATLAB's built-in functions for integration.

7.14 Write a user-defined MATLAB function for integration of a function, $I = \int_a^b y(x)dx$, where x and $y(x)$ are given as a set of tabulated points, with the midpoint method. For the function name and arguments use I=IntegrationMidpoint(x,y), where the input arguments x and y are vectors with the coordinates of the points, and I is the value of the integral. Use the function to solve Problem 7.9.

7.15 Write a user-defined MATLAB function for integration of a function with the composite trapezoidal method. For function name and arguments use I=Compzoidal('FunName',a,b). 'FunName' is a string with the name of a function file that calculates the value of the function to be integrated for a given value of x. a and b are the limits of integration, and I is the value of the integral. The integration function calculates the value of the integral in iterations. In the first iteration the interval $[a, b]$ is divided into two subintervals. In every iteration that follows the number of subintervals is doubled. The iterations stop when the difference in the value of the integral between two successive iterations is smaller than 1%. Use the function to solve Problems 7.4 and 7.5.

7.16 Write a user-defined MATLAB function for integration of a function with the composite Simpson's 1/3 method. For function name and arguments use `I=Simpson13('FunName',a,b)`. `'FunName'` is a string with the name of a function file that calculates the value of the function to be integrated for a given value of x. a and b are the limits of integration, and I is the value of the integral. The integration function calculates the value of the integral in iterations. In the first iteration the interval $[a, b]$ is divided into two subintervals. In every iteration that follows, the number of subintervals is doubled. The iterations stop when the difference in the value of the integral between two successive iterations is smaller than 1%. Use the function to solve Problems 7.4 and 7.5.

7.17 Write a user-defined MATLAB function for integration of a function with the composite Simpson's 3/8 method. For function name and arguments use `I=Simpsons38('FunName',a,b)`. `'FunName'` is a string with the name of a function file that calculates the value of the function to be integrated for a given value of x. a and b are the limits of integration, and I is the value of the integral. The integration function calculates the value of the integral in iterations. In the first iteration the interval $[a, b]$ is divided into three subintervals. In every iteration that follows the number of subintervals is doubled. The iterations stop when the difference in the value of the integral between two successive iterations is smaller than 1%. Use the function to solve Problems 7.4 and 7.5.

7.18 Write a user-defined MATLAB function for integration of a function $f(x)$ in the domain $[-1, 1]$ ($\int_{-1}^{1} f(x)dx$) with fourth-order Gauss quadrature. For function name and arguments use `I=GaussQuad4('FunName')`, where `'FunName'` is a string with the name of a function file that calculates the value of the function to be integrated. Use the function to evaluate the integral in Example 7-5.

7.19 Write a user-defined MATLAB function for integration of a function $f(x)$ in the domain $[a, b]$ ($\int_{a}^{b} f(x)dx$) with fourth-order Gauss quadrature. For function name and arguments use `I=GaussQuad4ab('FunName',a,b)`, where `'FunName'` is a string with the name of a function file that calculates the value of the function to be integrated, and a and b are the limits of integration. Use the function to evaluate the integral in Example 7-2.

Problems in math, science, and engineering
Solve the following problems using MATLAB environment. As stated, use the MATLAB programs that are presented in the chapter, programs developed in previously solved problems, or MATLAB's built-in functions.

7.20 The density, ρ, of the Earth varies with the radius, r. The following table gives the approximate density at different radii:

r (km)	0	800	1200	1400	2000	3000	3400	3600	4000	5000	5500	6370
ρ (kg/m³)	13000	12900	12700	12000	11650	10600	9900	5500	5300	4750	4500	3300

The mass of the Earth can be calculated by:

$$m = \int_{0}^{6370} \rho 4\pi r^2 dr$$

(*a*)   Write a user-defined MATLAB function for integration with the composite trapezoidal method of tabulated data with general spacing (the spacing between the points does not have to be the same). For function name and arguments, use `I=TrapTabulated(x,y)`, where the input arguments x and y are vectors with the data points. To carry out the integration, use Eq. (7.8). Use the function to calculate the mass of the Earth using the data given in the table.

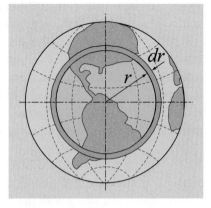

(*b*)   Use MATLAB's built-in `trapz` function to calculate the mass of the Earth using the data given in the table.

(*c*)   Use MATLAB's built-in `interp1` function (with the `spline` option for the interpolation method) to generate a new interpolated data set from the data that is given in the table. For spacing divide the domain [0, 6370] into 50 equal subintervals (use the `linspace` command). Calculate the mass of the Earth by integrating the interpolated data set with MATLAB's built-in `trapz` function.

7.21 The perimeter, *P*, of an ellipse is given by:

$$P = 4a \int_0^{\pi/2} \sqrt{1 - k^2 \sin^2 \theta} \; d\theta$$

where $k = \dfrac{\sqrt{a^2 - b^2}}{a}$.

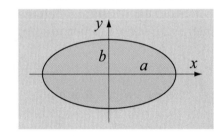

Write a user-defined MATLAB function that calculates the perimeter of an ellipse. For function name and arguments use `P=EllipsePer(a,b)`, where a and b are the major and minor axes, respectively, and P is the perimeter.

   Use the function to calculate the perimeter of ellipses given by:

(*a*) $\dfrac{x^2}{5^2} + \dfrac{y^2}{2^2} = 1$ (*b*) $\dfrac{x^2}{4^2} + \dfrac{y^2}{7^2} = 1$

7.22 A cross-sectional area has the geometry of half an ellipse, as shown in the figure to the right. The coordinate $\bar{x}$ of the centroid of the area can be calculated by:

$$\bar{x} = \frac{M_y}{A}$$

where *A* is the area given by $A = \dfrac{1}{2}\pi ab$, and M_y is the moment of the area about the *y* axis, given by:

$$M_y = \int_A x_c \, dA = 2b \int_0^a x \sqrt{1 - \frac{x^2}{a^2}} \; dx$$

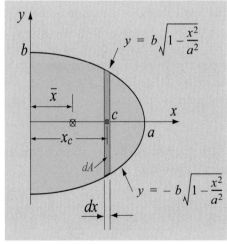

(*a*) Write a MATLAB program in a script file that calculates $\bar{x}$ when *a* = 40 mm and *b* = 15 mm. For the integration first use the user-defined function `Compzoidal` that was created in Problem 7.15.

(*b*) Replace the integration function in the program from part (*a*) with one of MATLAB's built-in integration functions. Repeat the calculation, and compare the results.

7.23 A cylindrical vertical tank of diameter $D = 2$ ft has a hole of diameter $d = 2$ in. near its bottom. Water enters the tank through a pipe at the top at a rate of $Q = 0.35$ ft^3/s. The time, t, that is required for the height of the water level in the tank to change from its initial ($t = 0$) level of $h_0 = 9$ ft to level of h is given by:

$$t = \int_{h_0}^{h} \frac{1}{\left[\dfrac{4Q}{\pi D^2} - \dfrac{a^2}{D^2} \sqrt{2gy} \right]} \, dy$$

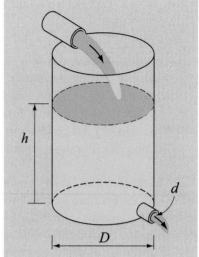

where $g = 32.2$ ft/s^2.

(a) Use the composite trapezoidal method (user-defined function created in Problem 7.15) to determine how long it would take for the height of the water level to change to $h = 5$ ft.

(b) Determine the times that correspond to water levels from $h = 9$ ft to $h = 1$ ft in increments of 0.2 ft, and plot h as a function of t. To carry out the calculations and make the plot, write a MATLAB program in a script file. For integration use one of MATLAB's built-in functions.

7.24 The moment of inertia, I_y, about the y axis of the cross-sectional area of the half ellipse that is shown in Problem 7.22 is given by:

$$I_y = \int_A x_c^2 \, dA = 2b \int_0^a x^2 \sqrt{1 - \frac{x^2}{a^2}} \, dx$$

Calculate I_y when $a = 40$ mm and $b = 15$ mm. For the integration use one of MATLAB's built-in integration functions.

7.25 The moment of inertia, I_x, about the x axis of the cross-sectional area of the half ellipse that is shown in Problem 7.22 is given by:

$$I_x = \int_0^a \frac{1}{12} \left(2b \sqrt{1 - \frac{x^2}{a^2}} \right)^3 dx$$

Calculate I_x when $a = 40$ mm and $b = 15$ mm. For the integration use one of MATLAB's built-in integration functions.

7.26 An aluminum beam with a length, $L = 2$ m, is clamped at one end and is loaded by an axial force $P = 20$ kN at the other end. The cross-sectional area of the beam varies with the length of the beam and is given as a function of x by:

$$A = \frac{W}{2}\left(2 - \frac{x}{L}\right)\frac{H}{2}\left(2 - \frac{x}{L}\right)$$

where $W = 80$ mm and $H = 30$ mm. The normal strain, ε, along the beam is given by:

$$\varepsilon = \frac{du}{dx} = \frac{P}{EA}$$

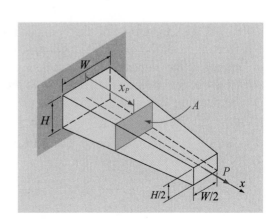

where $E = 70$ GPa is the modulus of elasticity of the beam.

The displacement δ at a point $x = x_P$ along the beam can be calculated by:

$$\delta = \int_0^{x_P} \frac{P}{EA}\, dx$$

(a) Determine the displacement δ at the endpoint of the beam ($x_P = L$). Calculate the displacement twice, once by using the user-defined function from Problem 7.16, and once by using one of MATLAB's built-in integration functions.

(b) Write a MATLAB program in a script file that calculates, and makes a plot, of the displacement δ as a function of x. For the integration use one of MATLAB's built-in integration functions.

7.27 In the design of underground pipes, there is a need to estimate the temperature of the ground. The temperature of the ground at various depths can be estimated by modeling the ground as a semi-infinite solid initially at constant temperature. The temperature at depth, x, and time, t, can be calculated from the expression:

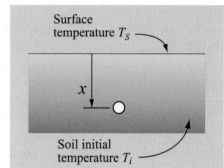

$$\frac{T(x, t) - T_S}{T_i - T_S} = erf\left(\frac{x}{2\sqrt{\alpha t}}\right) = \frac{2}{\pi}\int_0^{\frac{x}{2\sqrt{\alpha t}}} e^{-u^2}\, du$$

where T_S is the surface temperature, T_i is the initial soil temperature, and $\alpha = 0.138 \times 10^{-6}$ m^2/s is the thermal diffusivity of the soil. Answer the following questions taking $T_S = -15\ ^\circ$C and $T_i = 12\ ^\circ$C.

(a) Find the temperature at a depth $x = 1$ m after 30 days ($t = 2.592 \times 10^6$ s).

(b) Write a MATLAB program in a script file that generates a plot that shows the temperature as a function of time at a depth of $x = 0.5$ m for 40 days. Use increments of 1 day.

(c) Write a MATLAB program that generates a three-dimensional plot (T vs x and t) showing the temperature as a function of depth and time for $0 \le x \le 3$ m and $0 \le t \le 2.592 \times 10^7$ s.

7.28 A thermocouple is used to measure the temperature of a flowing gas in a duct. The time-dependence of the temperature of the spherical junction of a thermocouple is given by the implicit integral equation:

$$t = -\int_{T_i}^{T} \frac{\rho V C}{A_S[h(T - T_\infty) + \varepsilon \sigma_{SB}(T^4 - T_{surr}^4)]}\, dT$$

where T is the temperature of the thermocouple junction at time t, ρ is the density of the junction material, V is the volume of the spherical junction, C is the heat capacity of the junction, A_S is the surface area of the junction, h is the convection heat transfer coefficient, T_∞ is the temperature of the flowing gas, ε is the emissivity of the junction material, σ_{SB} is the Stefan–Boltzmann constant, and T_{surr} is the temperature of the surrounding duct wall.

For $T_i = 298$ K, $\varepsilon = 0.9$, $\rho = 8500$ kg/m^3, $C = 400$ J/kg/K, $T_\infty = 473$ K, $h = 400$ W/m^2/K, and $T_{surr} = 673$ K, determine the time, t, it takes for the thermocouple junction temperature, T, to increase to 490 K, using Romberg integration.

Chapter 8

Ordinary Differential Equations: Initial-Value Problems

Core Topics

Euler's methods (explicit, implicit, errors) (8.2).

Modified Euler method (8.3).

Midpoint method (8.4).

Runge–Kutta methods (second, third, fourth order) (8.5).

Multistep methods (8.6).

Predictor–corrector methods (8.7).

Systems of first-order ODEs (8.8).

Solving a higher order initial value ODE (8.9).

Use of MATLAB built-in functions for solving initial-value ODEs (8.10).

Complementary Topics

Local truncation error in second-order Runge–Kutta method (8.11).

Step size for desired accuracy (8.12).

Stability (8.13).

Stiff ODEs (8.14).

8.1 BACKGROUND

A differential equation is an equation that contains derivatives of an unknown function. The solution of the equation is the function that satisfies the differential equation. A differential equation that has one independent variable is called an ***ordinary differential equation*** (ODE). A first-order ODE involves the first derivative of the dependent variable with respect to the independent variable. For example, if x is the independent variable and y is the dependent variable, the equation has combinations of the variables x, y, and $\frac{dy}{dx}$. A first-order ODE is linear, if it is a linear function of y and $\frac{dy}{dx}$ (it can be a nonlinear function of x). Examples of a linear and a nonlinear first-order ODE are given in Eqs. (8.1) and (8.2), respectively.

$$\frac{dy}{dx} + ax^2 + by = 0 \quad \text{(linear)} \tag{8.1}$$

$$\frac{dy}{dx} + ayx + b\sqrt{y} = 0 \quad \text{(nonlinear)} \tag{8.2}$$

where a and b are constants.

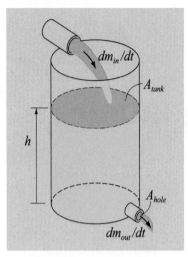

Figure 8-1: Water tank.

Differential equations appear in all branches of science and engineering. They are also encountered in economics, business applications, and social sciences where ideas have to be quantified and predictive models are needed. Differential equations provide detailed information regarding distributions or variations of the dependent variable as a function of the independent variable.

As an example, consider the cylindrical water tank shown in Fig. 8-1. The tank is being filled at the top, and water flows out of the tank through a pipe connected at the bottom. The rate of water flow into the tank varies with time and is given by the equation:

$$\frac{dm_{in}}{dt} = K_1 + K_2\cos(Ct) \tag{8.3}$$

where dm_{in}/dt is mass flow per unit time, and K_1, K_2, and C are constants. The rate that water is flowing out of the tank, dm_{out}/dt, depends on the height, h, of the water in the tank (pressure) and is given by:

$$\frac{dm_{out}}{dt} = \rho A_{pipe}\sqrt{2gh} \tag{8.4}$$

where ρ is the density of water, A_{pipe} is the cross-sectional area of the pipe at the exit, and g is the acceleration due to gravity. The time rate of change of the mass in the tank, $\frac{dm}{dt}$, is the difference between the rate of mass entering the tank and the rate of mass leaving the tank (conservation of mass) per unit time:

$$\frac{dm}{dt} = \frac{dm_{in}}{dt} - \frac{dm_{out}}{dt} \tag{8.5}$$

The mass of the water in the tank can be expressed in terms of the height, h, and the cross-sectional area of the tank, A_{tank}, by $m = \rho A_{tank}h$, which means that $\frac{dm}{dt} = \rho A_{tank}\frac{dh}{dt}$. Substituting this relationship and the expressions from Eqs. (8.3) and (8.4) in Eq. (8.5) gives an equation for the rate of change of the height, h:

$$\rho A_{tank}\frac{dh}{dt} = K_1 + K_2\cos(Ct) - \rho A_{pipe}\sqrt{2gh} \tag{8.6}$$

Equation (8.6) is a first-order ODE, with t as the independent variable and h the dependent variable. The solution of Eq. (8.6) is a function $h(t)$ that satisfies the equation. In general, an infinite number of functions satisfy the equation. To obtain a specific solution, a first-order ODE must have an initial condition or constraint that specifies the value of the dependent variable at a particular value of the independent variable. As discussed in Chapters 1 and 2, a properly formulated physical problem resulting in an ODE of the first order has to have one such constraint. Because such problems are typically time-dependent problems (i.e., problems in which time t is the independent variable), the constraint is called an ***initial condition*** and the problem is called an ***initial-value problem (IVP)***. Such a constraint associated with Eq. (8.6) may be written as:

$$h = h_0 \quad \text{at} \quad t = t_0, \quad \text{or} \quad h(t_0) = h_0$$

First-order ODE problem statement

A first order ODE has the form:

$$\frac{dy}{dx} = f(x, y) \quad \text{with the initial condition: } y(x_1) = y_1 \qquad (8.7)$$

The solution is the function $y(x)$ that satisfies the equation and the initial condition.

In the differential equation, Eq. (8.7), the function $f(x, y)$ gives the slope of $y(x)$ as a function of x and y. For example, consider the following ODE:

$$\frac{dy}{dx} = f(x, y) = -1.2y + 7e^{-0.3x} \qquad (8.8)$$

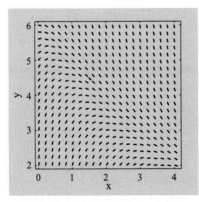

Figure 8-2: Illustration of the function $f(x, y)$.

The value of $f(x, y)$ in Eq. (8.8) for the domain $0 \geq x \geq 4$ and $2 \geq y \geq 6$ is illustrated in Fig. 8-2, where the slope at many points within the domain is plotted. The slopes are tangent to the solution, which means that the slopes are like "flow lines" that show the direction that $y(x)$ follows. It is clear from Fig. 8-2 that, in general, there can be many (infinite) solutions since it is possible to draw different lines that follow the "flow lines." Three possible solutions are shown in Fig. 8-3. The solution to a specific problem is fixed by the initial condition that defines the point where $y(x)$ starts and the prescribed domain for x, which specifies where $y(x)$ ends. The initial conditions for the three solutions that are shown in Fig. 8-3 are $y(0) = 3$, $y(1.5) = 2$, and $y(2) = 6$.

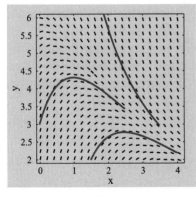

Figure 8-3: Different solutions for the same ODE.

Analytical solution of a first-order ODE

Analytical solution of an ODE is a mathematical expression of the function $y(x)$ that satisfies the differential equation and has the value $y(x_1) = y_1$. Once the function $y(x)$ is known, its value at any x can be calculated. As mentioned in Chapter 2, there are techniques for solving relatively simple first-order ODEs, but in many situations an analytical solution is not possible.

Numerical solution of a first-order ODE

A numerical solution of a first order ODE formulated as Eq. (8.7) is a set of discrete points that approximate the function $y(x)$. When a differential equation is solved numerically, the problem statement also includes the domain of the solution. For example, a solution is required for values of the independent variable from $x = a$ to $x = b$ (the domain is $[a, b]$). Depending on the numerical method used to solve the equation, the number of points between a and b at which the solution is obtained can be set in advance, or it can be decided by the method. For example, the domain can be divided into N subintervals of equal width defined by $N+1$ values of the independent variable from $x_1 = a$ to $x_{N+1} = b$. The solution consists of values of the dependent variable that are determined at each value of the independent variable. The solution then is a set of points (x_1, y_1), (x_2, y_2), ..., (x_{N+1}, y_{N+1}) that define the function $y(x)$.

Overview of numerical methods used for solving a first-order ODE

Numerical solution is a procedure for calculating an estimate of the exact solution at a set of discrete points. The solution process is incremental, which means that it is determined in steps. It starts at the point where the initial value is given. Then, using the known solution at the first point, a solution is determined at a second nearby point. This is followed by a solution at a third point, and so on. There are procedures with a **single-step** and **multistep** approach. In a single-step approach, the solution at the next point, x_{i+1}, is calculated from the already known solution at the present point, x_i. In a multi-step approach, the solution at x_{i+1} is calculated from the known solutions at several previous points. The idea is that the value of the function at several previous points can give a better estimate for the trend of the solution. Also, two types of methods, **explicit**, and **implicit**, can be used for calculating the solution at each step. The difference between the methods is in the way that the solution is calculated at each step.

Explicit methods are those methods that use an <u>explicit</u> formula for calculating the value of the dependent variable at the next value of the independent variable. In an explicit formula, the right-hand side of the equation only has known quantities. In other words, the next unknown value of the dependent variable, y_{i+1}, is calculated by evaluating an expression of the form:

$$y_{i+1} = F(x_i, x_{i+1}, y_i) \tag{8.9}$$

where x_i, y_i, and x_{i+1} are all known quantities.

In **implicit** methods, the equation used for calculating y_{i+1} from the known x_i, y_i, and x_{i+1} has the form:

$$y_{i+1} = F(x_i, x_{i+1}, y_{i+1}) \tag{8.10}$$

Here, the unknown y_{i+1} appears on both sides of the equation. In general, the right-hand side of Eq. (8.10) is nonlinear, and the equation must actually be solved numerically for y_{i+1} using the methods described in Chapter 3. If the function on the right-hand side of Eq. (8.10) depends linearly on y_{i+1}, then it is actually an explicit formula just like (8.9) because Eq. (8.10) can be rewritten and solved for y_{i+1} to obtain an explicit expression in the form of Eq. (8.9). Implicit methods provide improved accuracy over explicit methods, but require more effort at each step.

Errors in numerical solution of ODEs

Two types of errors, **round-off errors** and **truncation errors**, occur when ODEs are solved numerically. Round-off errors are due to the way that computers carry out calculations (see Chapter 1). Truncation errors are due to the approximate nature of the method used to calculate the solution. Since the numerical solution of a differential equation is calculated in increments (steps), the truncation error at each step of the solution consists of two parts. One, called **local truncation error**, is due to the application of the numerical method in a single step. The second part, called **propagated, or accumulated, truncation error**, is due to the accumulation of local truncation errors from previous steps. Together, the two parts are the **global (total) truncation error** in the solution. More details on truncation errors are provided in the following sections where the error in various methods is analyzed (see Sections 8.2.2 and 8.11).

Single-step explicit methods

In a single-step explicit method, illustrated in Fig. 8-4, the approximate numerical solution (x_{i+1}, y_{i+1}) is calculated from the known solution at point (x_i, y_i) by:

$$x_{i+1} = x_i + h \tag{8.11}$$

$$y_{i+1} = y_i + Slope \cdot h \tag{8.12}$$

where h is the step size, and the *Slope* is a constant that estimates the value of $\dfrac{dy}{dx}$ in the interval from x_i to x_{i+1}. The numerical solution starts at the point where the initial value is known. This corresponds to $i = 1$ and point (x_1, y_1). Then i is increased to $i = 2$, and the solution at the next point, (x_2, y_2), is calculated by using Eqs. (8.11) and (8.12). The procedure continues with $i = 3$ and so on until the points cover the whole domain of the solution.

Many single-step explicit methods use the form of Eqs. (8.11) and (8.12), and several are covered in Sections 8.2 through 8.6. The difference between the methods is in the value used for the constant *Slope* in Eq. (8.12) and in the way that it is determined. The simplest is Euler's explicit method, described in Section 8.2, in which *Slope* is equal to the slope of $y(x)$ at (x_i, y_i). In the modified Euler method, described in Section 8.3, the value of *Slope* is an average of the slope of $y(x)$ at (x_i, y_i) and an estimate of the slope of $y(x)$ at the end of the interval, (x_{i+1}, y_{i+1}). In another method, called the midpoint method, Section 8.4, the value of *Slope* is an estimate of the slope of $y(x)$ at the middle of the interval, (i.e., at $(x_i + x_{i+1})/2$). A more sophisticated class of methods, called Runge–Kutta methods, is presented in Section 8.5. In these methods the value of *Slope* is calculated from a weighted average of estimates of the slope of $y(x)$ at several points within an interval.

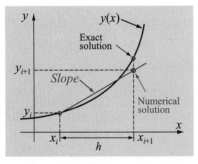

Figure 8-4: Single-step explicit methods.

8.2 EULER'S METHODS

Euler's method is the simplest technique for solving a first-order ODE of the form of Eq. (8.7):

$$\frac{dy}{dx} = f(x, y) \quad \text{with the initial condition: } y(x_1) = y_1$$

The method can be formulated as an explicit or an implicit method. The more commonly used explicit formulation is presented in detail in the next section. The implicit formulation is covered in Section 8.2.3.

8.2.1 Euler's Explicit Method

Euler's explicit method (also called the forward Euler method) is a single-step, numerical technique for solving a first-order ODE. The method uses Eqs. (8.11) and (8.12), where the value of the constant *Slope* in Eq. (8.12) is the slope of $y(x)$ at point (x_i, y_i). This slope is actually calculated from the differential equation:

$$Slope = \left.\frac{dy}{dx}\right|_{x=x_i} = f(x_i, y_i) \tag{8.13}$$

Euler's method assumes that for a short distance h near (x_i, y_i), the function $y(x)$ has a constant slope equal to the slope at (x_i, y_i). With this assumption, the next point of the numerical solution (x_{i+1}, y_{i+1}) is calculated by:

$$x_{i+1} = x_i + h \tag{8.14}$$

$$y_{i+1} = y_i + f(x_i, y_i)h \tag{8.15}$$

Euler's explicit method is illustrated schematically in Fig. 8-5. To simplify the illustration, the step size is exaggerated and the difference (error) between the numerical and exact solutions at x_i is ignored. A more precise illustration is presented in Section 8.2.2. It is obvious from Fig. 8-5 that the error in this method depends on the value of h and is smaller for smaller h. A detailed discussion of the error is given later in Section 8.2.2.

Equation (8.15) of Euler's method can be derived in several ways. Starting with the given differential equation:

$$\frac{dy}{dx} = f(x, y) \tag{8.16}$$

An approximate solution of Eq. (8.16) can be obtained either by numerically integrating the equation or by using a finite difference approximation for the derivative.

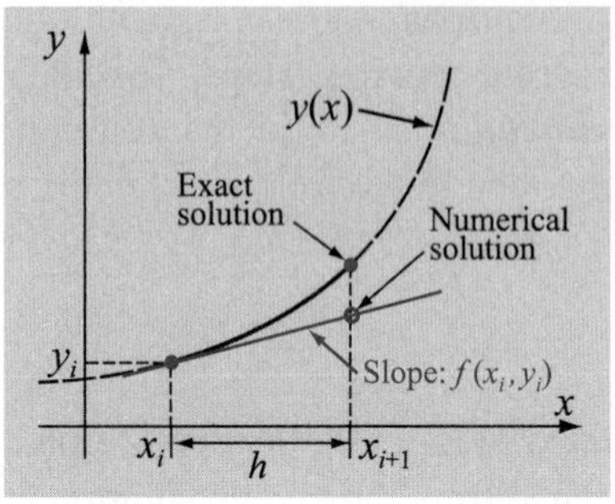

Figure 8-5: Euler's explicit method.

Deriving Euler's method by using numerical integration

Equation (8.16) can be written as an integration problem by multiplying both sides by dx:

$$\int_{y_i}^{y_{i+1}} dy = \int_{x_i}^{x_{i+1}} f(x, y) dx \tag{8.17}$$

Carrying out the integration on the left-hand side and solving for y_{i+1} gives:

$$y_{i+1} = y_i + \int_{x_i}^{x_{i+1}} f(x, y) dx \tag{8.18}$$

The second term on the right-hand side is an integral that has to be evaluated. Chapter 7 describes several methods of numerical integration. The simplest of these is the rectangle method (Section 7.2.1), where the integrand is approximated by the constant value $f(x_i, y_i)$. Using this approach, Eq. (8.18) becomes:

$$y_{i+1} = y_i + \int_{x_i}^{x_{i+1}} f(x, y) dx = y_i + f(x_i, y_i)(x_{i+1} - x_i) \tag{8.19}$$

which is the same as Eq. (8.15) since $h = (x_{i+1} - x_i)$.

Deriving Euler's method by using finite difference approximation for the derivative

Euler's formula, Eq. (8.15), can also be derived by using an approximation for the derivative in the differential equation. The derivative $\dfrac{dy}{dx}$ in Eq. (8.15) can be approximated with the forward difference formula (see Section 6.2) by evaluating the ODE at the point $x = x_i$:

$$\left.\frac{dy}{dx}\right|_{x_i} \approx \frac{y_{i+1} - y_i}{x_{i+1} - x_i} = f(x_i, y_i) \tag{8.20}$$

Solving Eq. (8.20) for y_{i+1} gives Eq. (8.15) of Euler's method. (Because the equation can be derived in this way, the method is also known as the forward Euler method.)

  Example 8-1 shows the application of Euler's explicit method in the solution of first order ODEs. The equation that is solved numerically can be (and also is) solved analytically. This provides an opportunity to compare the numerical and exact solutions, and to examine the effect of the step size, h, on the error.

Example 8-1: Solving a first-order ODE using Euler's explicit method.

Use Euler's explicit method to solve the ODE $\dfrac{dy}{dx} = -1.2y + 7e^{-0.3x}$ from $x = 0$ to $x = 2.5$ with the initial condition $y = 3$ at $x = 0$.

(*a*) Solve by hand using $h = 0.5$.

(*b*) Write a MATLAB program in a script file that solves the equation using $h = 0.1$.

(*c*) Use the program from part (*b*) to solve the equation using $h = 0.01$.

In each part compare the results with the exact (analytical) solution: $y = \dfrac{70}{9}e^{-0.3x} - \dfrac{43}{9}e^{-1.2x}$.

SOLUTION

(*a*) ***Solution by hand:*** The first point of the solution is $(0, 3)$, which is the point where the initial condition is given. For the first point $i = 1$. The values of x and y are $x_1 = 0$ and $y_1 = 3$.

The rest of the solution is determined by using Eqs. (8.14) and (8.15). In the present problem these equations have the form:

$$x_{i+1} = x_i + h = x_i + 0.5 \tag{8.21}$$

$$y_{i+1} = y_i + f(x_i, y_i)h = y_i + (-1.2y_i + 7e^{-0.3x_i})0.5 \tag{8.22}$$

Equations (8.21) and (8.22) are applied five times with $i = 1, 2, 3, 4$, and 5.

First step: For the first step $i = 1$. Equations (8.21) and (8.22) give:

$x_2 = x_1 + 0.5 = 0 + 0.5 = 0.5$

$y_2 = y_1 + (-1.2y_1 + 7e^{-0.3x_1})0.5 = 3 + (-1.2 \cdot 3 + 7e^{-0.3 \cdot 0})0.5 = 4.7$

The second point is $(0.5, 4.7)$.

Second step: For the second step $i = 2$. Equations (8.21) and (8.22) give:

$x_3 = x_2 + 0.5 = 0.5 + 0.5 = 1.0$

$y_3 = y_2 + (-1.2y_2 + 7e^{-0.3x_2})0.5 = 4.7 + (-1.2 \cdot 4.7 + 7e^{-0.3 \cdot 0.5})0.5 = 4.893$

The third point is $(1.0, 4.893)$.

Third step: For the third step $i = 3$. Equations (8.21) and (8.22) give:

$x_4 = x_3 + 0.5 = 1.0 + 0.5 = 1.5$

$y_4 = y_3 + (-1.2y_3 + 7e^{-0.3x_3})0.5 = 4.893 + (-1.2 \cdot 4.893 + 7e^{-0.3 \cdot 1.0})0.5 = 4.550$

The fourth point is $(1.5, 4.550)$.

Fourth step: For the fourth step $i = 4$. Equations (8.21) and (8.22) give:

$x_5 = x_4 + 0.5 = 1.5 + 0.5 = 2.0$

$y_5 = y_4 + (-1.2y_4 + 7e^{-0.3x_4})0.5 = 4.550 + (-1.2 \cdot 4.550 + 7e^{-0.3 \cdot 1.5})0.5 = 4.052$

The fifth point is $(2.0, 4.052)$.

Fifth step: For the fifth step $i = 5$. Equations (8.21) and (8.22) give:

$x_6 = x_5 + 0.5 = 2.0 + 0.5 = 2.5$

$y_6 = y_5 + (-1.2y_5 + 7e^{-0.3x_5})0.5 = 4.052 + (-1.2 \cdot 4.052 + 7e^{-0.3 \cdot 2.0})0.5 = 3.542$

The sixth point is $(2.5, 3.542)$.

The figure on the right shows the calculated numerical solution (red points) and the exact analytical solution (solid line).
The values of the exact and numerical solutions, and the error, which is the difference between the two, are:

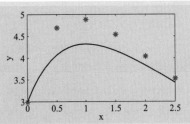

i	1	2	3	4	5	6
x_i	0.0	0.5	1.0	1.5	2.0	2.5
y_i (numerical)	3.0	4.70	4.893	4.55	4.052	3.542
y_i (exact)	3.0	4.072	4.323	4.170	3.835	3.436
Error	0.0	-0.6277	-0.5696	0.3803	-0.2165	-0.1054

(b) To solve the ODE with MATLAB, a user-defined function (called odeEuler) that solves a first-order, initial value problem using Euler's explicit method is written. Then a program in a script file uses the function to solve part (b) of the problem. The script file also creates a plot that shows the numerical solution and the exact solution.

Program 8-1: User-defined function. Solving first-order ODE using Euler's explicit Method.

```
function [x, y] = odeEULER(ODE,a,b,h,yINI)
% odeEULER solves a first-order initial value ODE using Euler's
% explicit method.
% Input variables:
% ODE   Name (string) of a function file that calculates dy/dx.
% a     The first value of x.
% b     The last value of x.
% h     Step size.
% yINI    The value of the solution y at the first point (initial value).
% Output variable:
% x     A vector with the x coordinate of the solution points.
% y     A vector with the y coordinate of the solution points.
```

```
x(1) = a;  y(1) = yINI;
N = (b - a)/h;
for i = 1:N
    x(i + 1) = x(i) + h;
    y(i + 1) = y(i) + feval(ODE,x(i),y(i))*h;
end
```

Assign the initial value to x(1) and y(1).
Determine the number of steps.

Apply Eq. (8.14).
Apply Eq. (8.15).

The following program in a script file uses the function odeEULER for solving the ODE with Euler's explicit method (part (b) of the problem). The program also plots the numerical and the exact solutions.

```
clear all
a = 0; b = 2.5; h = 0.1; yINI = 3;
[x, y] = odeEULER('Chap8Exmp1ODE',a,b,h,yINI);
xp = a:0.1:b;
yp=70/9*exp(-0.3*xp) - 43/9*exp(-1.2*xp);
```

Assign the domain, step size, and initial value to variables.
Use the odeEULER function.
Create vectors for plotting the exact solution.

```
plot(x,y,'--r',xp,yp)
xlabel('x'); ylabel('y')
```

The user-defined function `Chap8Exmp1ODE` (listed below), which is typed as the first argument in the function `odeEULER`, calculates the value of dy/dx.

```
function dydx = Chap8Exmp1ODE(x,y)
dydx = -1.2*y + 7*exp(-0.3*x);
```

The plot that is created when the program is executed is shown below with the numerical (red) and the exact (black) solutions.

(*c*) The program from part (*b*) is executed again with $h = 0.01$. The result is shown in the plot below.

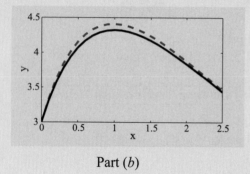

Part (*b*)

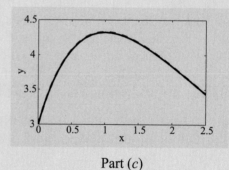

Part (*c*)

As expected, the numerical solution becomes more accurate as h decreases.

8.2.2 Analysis of Truncation Error in Euler's Explicit Method

As mentioned in Section 8.1, when ODEs are solved numerically there are two sources of error, round-off and truncation (see also Chapter 1). The round-off errors are due to the way that computers carry out calculations. The truncation error is due to the approximate nature of the method used for calculating the solution in each increment (step). In addition, since the numerical solution of a differential equation is calculated in increments (steps), the truncation error consists of a local truncation error and propagated truncation error (see Section 8.1). The truncation errors in Euler's explicit method are discussed in this section. The discussion is divided into two parts. First, the local truncation error is analyzed, and then the results are used for determining an estimate of the global truncation error.

A note about notation in this section

In the present section, different values of the dependent variable are calculated and compared at the same value of the independent variable. To clarify the presentation, each quantity is identified with a superscript. A dependent variable that is calculated by a numerical method is written as y^{NS} (in the rest of the chapter it is just y). The true solution of the ODE is written as y^{TS}. In addition, the notation y^{Taylor} is used for the value that is calculated with a Taylor series expansion.

Local truncation error

The local truncation error is the error inherent in the formula used to obtain the numerical solution in a single step (subinterval). It is the difference between the numerical solution and an exact solution for that step. In a general step i, the numerical solution y_i^{NS} at $x = x_i$ is known (previously calculated), and the numerical solution, y_{i+1}^{NS} at $x = x_{i+1}$ is calculated with an approximate formula. The value of the exact solution for this step can be expressed by a two-term Taylor series expansion with a remainder (see Chapter 2):

$$y_{i+1}^{Taylor} = y_i^{NS} + f(x_i, y_i^{NS})h + \frac{d^2 y}{dx^2}\bigg|_{x=\xi_i} \frac{h^2}{2} \tag{8.23}$$

where $f(x_i, y_i^{NS}) = \dfrac{dy}{dx}\bigg|_{x=x_i}$ and ξ_i is a value of x between x_i and x_{i+1}.

With Euler's explicit method the numerical solution y_{i+1}^{NS} at $x = x_{i+1}$ is calculated with:

$$y_{i+1}^{NS} = y_i^{NS} + f(x_i, y_i^{NS})h \tag{8.24}$$

Comparing Eq. (8.23) with Eq. (8.24) shows that the Euler explicit formula is an approximation consisting of the first two terms of the Taylor series expansion. The difference between the two (truncation error) is because the remainder term in Eq. (8.23) was truncated. This error, called the local truncation error in step i, e_i^{TR}, is given by:

$$e_i^{TR} = y_{i+1}^{Taylor} - y_{i+1}^{NS} = \frac{d^2 y}{dx^2}\bigg|_{x=\xi_i} \frac{h^2}{2} = O(h^2) \tag{8.25}$$

This is the truncation error inherent in every step. It should be emphasized, as illustrated in Fig. 8-6, that the value y_i^{NS} that is used in both Eq. (8.23) and Eq. (8.24) is not the true solution. y_i^{TS} is the true solution of the ODE at $x = x_i$. y_i^{NS} is a numerical solution at $x = x_i$ obtained by applying Euler's formula in previous steps. Moreover, the value $f(x_i, y_i^{NS})$ is not the derivative (slope) of the true solution at $x = x_i$ since $y_i^{NS} \neq y_i^{TS}$. The difference between the numerical solution and the true solution at $x = x_{i+1}$ is the total error, part of which is due to the local truncation error and the rest due to the accumulation of truncation errors from previous steps. (The discussion here excludes round-off errors.) The total error due to truncation alone is called the *global truncation error*. An estimate of the global truncation error for Euler's explicit method is given next.

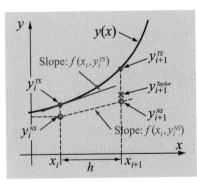

Figure 8-6: Numerical and true solution.

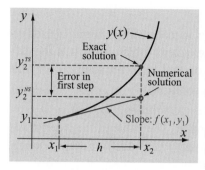

Figure 8-7: Error in the first step.

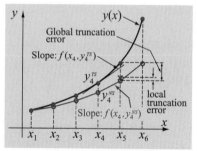

Figure 8-8: Local and global truncation error.

Global truncation error

In the first subinterval (step), $[x_1, x_2]$, the local truncation error is the same as the global truncation error because the true solution at the first point is known (initial condition). Starting with the second subinterval (step), $[x_2, x_3]$, there is an additional error in each step because the value $f(x_i, y_i^{NS})$ (the slope) in Eq. (8.24) is not the true value of the derivative at $x = x_i$. (The true value is $f(x_i, y_i^{TS})$.) In other words, the truncation error is propagated, or accumulated, from the previous subinterval(s) to the next subinterval. The error in the first step is shown in Fig. 8-7, and the accumulation of error in the first few points is illustrated in Fig. 8-8. At some point x_i, the global truncation error is the difference between the true solution, y_i^{TS}, and the numerical solution y_i^{NS}:

$$E_i^{TR} = y_i^{TS} - y_i^{NS} \tag{8.26}$$

Similarly, at x_{i+1} the global truncation error is:

$$E_{i+1}^{TR} = y_{i+1}^{TS} - y_{i+1}^{NS} \tag{8.27}$$

The value of the true solution, y_{i+1}^{TS}, in step i can also be expressed by a two-term Taylor series expansion with a remainder:

$$y_{i+1}^{TS} = y_i^{TS} + f(x_i, y_i^{TS})h + \frac{d^2y}{dx^2}\bigg|_{x=\eta_i} \frac{h^2}{2} \tag{8.28}$$

where $f(x_i, y_i^{TS}) = \dfrac{dy}{dx}\bigg|_{x=x_i}$, and η_i is a value of x between x_i and x_{i+1}.

Subtracting Eq. (8.28) and Eq. (8.24), and using Eqs. (8.26) and (8.27) yields:

$$E_{i+1}^{TR} = E_i^{TR} + [f(x_i, y_i^{TS}) - f(x_i, y_i^{NS})]h + \frac{d^2y}{dx^2}\bigg|_{x=\eta_i} \frac{h^2}{2} \tag{8.29}$$

The mean-value theorem for derivatives (see Chapter 2) gives:

$$[f(x_i, y_i^{TS}) - f(x_i, y_i^{NS})] = \frac{\partial f(x, y)}{\partial y}\bigg|_{\substack{y=\gamma_i \\ x=x_i}} (y_i^{TS} - y_i^{NS}) = E_i^{TR} \frac{\partial f(x, y)}{\partial y}\bigg|_{\substack{y=\gamma_i \\ x=x_i}} \tag{8.30}$$

where γ_i is a value of y between y_i^{TS} and y_i^{NS}. Substituting Eq. (8.30) in Eq. (8.29) gives:

$$E_{i+1}^{TR} = E_i^{TR}[1 + hf_y(x_i, \gamma_i)] + \alpha_i h^2 \tag{8.31}$$

where $f_y(x_i, \gamma_i) = \dfrac{\partial f(x, y)}{\partial y}\bigg|_{\substack{y=\gamma_i \\ x=x_i}}$ and $\alpha_i = \dfrac{1}{2}\dfrac{d^2y}{dx^2}\bigg|_{x=\eta_i}$. Suppose now

that in the domain of the solution, $f_y(x_i, \gamma_i)$ is bounded by a positive number C. Then $|f_y(x_i, \gamma_i)| \le C$ and Eq. (8.31) can be written as:

$$E_{i+1}^{TR} \le E_i^{TR}[1 + hC] + \alpha_i h^2 \tag{8.32}$$

Equation (8.32) can be used for showing how the truncation error propagates (accumulates) as the numerical solution progresses. At the first point of the solution where the initial value is given, $E_1^{TR} = 0$ (since $y_1^{TS} = y_1^{NS}$). Then, from Eq. (8.32) the truncation error at the second point is:

$$E_2^{TR} \leq \alpha_1 h^2 \tag{8.33}$$

At the third point:

$$E_3^{TR} \leq E_2^{TR}[1 + hC] + \alpha_2 h^2 = \alpha_1 h^2[1 + hC] + \alpha_2 h^2 = h^2\{[1 + hC]\alpha_1 + \alpha_2\} \tag{8.34}$$

In the same way, at the fourth point:

$$E_4^{TR} \leq h^2\{[1 + hC]^2\alpha_1 + [1 + hC]\alpha_2 + \alpha_3\} \tag{8.35}$$

At each step a new (local) truncation error is added, and the truncation errors from the previous steps are multiplied (magnified) by $[1 + hC]$. For the ith step the expression for the global truncation error is:

$$E_i^{TR} \leq h^2\{[1 + hC]^{i-2}\alpha_1 + [1 + hC]^{i-3}\alpha_2 + \ldots + [1 + hC]\alpha_{i-2} + \alpha_{i-1}\} \tag{8.36}$$

Suppose now that in the domain of the solution α_i is bounded by a positive number M. Then $|\alpha_i| \leq M$ and Eq. (8.36) can be written as:

$$E_i^{TR} \leq h^2 M\{1 + [1 + hC] + \ldots + [1 + hC]^{i-3} + [1 + hC]^{i-2}\} \tag{8.37}$$

If z is defined as $z = [1 + hC]$, then the expression inside the parentheses $\{\ \}$ in Eq. (8.37) resembles the series:

$$1 + z + z^2 + \ldots + z^m = \frac{z^{m+1} - 1}{z - 1} \tag{8.38}$$

Using Eq. (8.38), Eq. (8.37) can be written as:

$$E_i^{TR} \leq h^2 M\left\{\frac{[1 + hC]^{i-1} - 1}{hC}\right\} = \frac{hM}{C}\{[1 + hC]^{i-1} - 1\} \tag{8.39}$$

It is difficult to determine the order of magnitude of E^{TR} directly from Eq. (8.39). It is possible, however, to determine a bound by considering the Taylor series expansion of the exponential function e^s about $s = 0$:

$$e^s = 1 + s + \frac{s^2}{2} + \frac{s^3}{3!} + \ldots \tag{8.40}$$

Substituting $s = hC$ gives:

$$e^{hC} = 1 + hC + \frac{(hC)^2}{2} + \frac{(hC)^3}{3!} + \ldots \tag{8.41}$$

Equation (8.41) implies that $1 + hC \leq e^{hC}$, or $[1 + hC]^{i-1} \leq e^{hC(i-1)}$. Using this result in Eq. (8.39) gives:

$$E_i^{TR} \leq \frac{hM}{C}[e^{hC(i-1)} - 1] \tag{8.42}$$

which is a bound on the global truncation error in step i. If the solution process continues for N steps, the bound on the global truncation error at the last point $x = x_{N+1}$ is:

$$E_{N+1}^{TR} \le \frac{hM}{C}[e^{hCN} - 1] = \frac{hM}{C}[e^{C(x_{N+1} - x_i)} - 1] \qquad (8.43)$$

since the number of subintervals (steps) is given by $N = (x_{N+1} - x_1)/h$. Equation (8.43) shows that $E_{N+1}^{TR} \le O(h)$, so that the global truncation error is of the order of h, even though the local truncation error is of the order of h^2.

In summary, the local truncation error of Euler's explicit method is $O(h^2)$. The global truncation error is $O(h)$. The total numerical error is the sum of the global truncation error and the round-off error. The truncation error can be reduced by using smaller h (step size). However, if h becomes too small such that round-off errors become significant, the total error might increase. It should be emphasized that Eq. (8.43) cannot be used for calculating an actual value for the bound of global truncation error because the values of the constants M and C are not known. The usefulness of Eq. (8.43) is in the fact that it can be used to compare the accuracy of Euler's method for solving an ODE with other methods.

A numerical example of the errors is included in part (*a*) of Example 8-1. The step size in part (*a*) of the solution is $h = 0.5$, and the list of the errors in each step shows that the errors indeed are of the order of 0.5.

8.2.3 Euler's Implicit Method

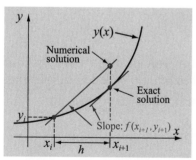

Figure 8-9: Euler's implicit method.

The form of Euler's implicit method is the same as the explicit scheme, except, as illustrated schematically in Fig. 8-9, for a short distance, h, near (x_i, y_i) the slope of the function $y(x)$ is taken to be a constant equal to the slope at the endpoint of the interval (x_{i+1}, y_{i+1}). With this assumption, the next point of the numerical solution (x_{i+1}, y_{i+1}) is calculated by:

$$x_{i+1} = x_i + h \qquad (8.44)$$

$$y_{i+1} = y_i + f(x_{i+1}, y_{i+1})h \qquad (8.45)$$

Now, the unknown y_{i+1} appears on both sides of Eq. (8.45), and unless $f(x_{i+1}, y_{i+1})$ depends on y_{i+1} in a simple linear or quadratic form, it is not easy or even possible to solve the equation for y_{i+1} explicitly. In general, Eq. (8.45) is a nonlinear equation for the unknown y_{i+1}, which must be solved numerically using the methods described in Chapter 3.

The derivation of Eq. (8.45) is similar to the derivation of Eq. (8.15) (see Section 8.2.1), except that in the derivation that uses integration, the approximation of the integral with the rectangle method uses the end point of the interval. In the derivation that uses an approximation of the derivative, the backward difference formula is used. Because of this, Euler's implicit method is also known as the backward Euler method. The local and global truncation errors are the same as those in the explicit method (see Section 8.2.2). Example 8-3 shows a solution of an initial value ODE using Euler's implicit method.

Example 8-2: Solving a first-order ODE using Euler's implicit method.

A chemical compound decays over time when exposed to air, at a rate proportional to its concentration to the power of $3/2$. At the same time, the compound is produced by another process. The differential equation for its instantaneous concentration is:

$$\frac{dn(t)}{dt} = -0.8n^{3/2} + 10n_1(1 - e^{-3t}) \tag{8.46}$$

where $n(t)$ is the instantaneous concentration and $n_1 = 2000$ is the initial concentration at $t = 0$. Solve the differential equation to find the concentration as a function of time from $t = 0$ until $t = 0.5$ s, using Euler's implicit method and Newton's method for solving for the roots of a nonlinear equation. Use a step size of $h = 0.002$ s, and plot n versus time.

SOLUTION

In this problem t is the independent variable, and n is the dependent variable. The function $f(t, n)$ is given by:

$$f(t, n) = -0.8n^{3/2} + 10n_1(1 - e^{-3t}) \tag{8.47}$$

The numerical solution of the differential equation is done incrementally by using Eqs. (8.44) and (8.45):

$$t_{i+1} = t_i + h \tag{8.48}$$

$$n_{i+1} = n_i + [-0.8n_{i+1}^{3/2} + 10n_1(1 - e^{-3t_{i+1}})]h \tag{8.49}$$

where $f(t, n)$ from Eq. (8.47) was substituted in Eq. (8.45).

At each step of the solution, Eq. (8.49) has to be solved for n_{i+1}. Since Eq. (8.49) cannot be solved explicitly, the solution has to be done numerically. To carry out the numerical solution, Eq. (8.49) is written in the form $g(x) = 0$, where $x = n_{i+1}$:

$$g(x) = x + 0.8x^{3/2}h - 10n_1(1 - e^{-3t_{i+1}})h - n_i = 0 \tag{8.50}$$

A numerical solution of Eq. (8.50) with Newton's method (see Section 3.5) requires the derivative of $g(x)$, which is given by:

$$g'(x) = 1 + 0.8 \cdot \frac{3}{2} \cdot x^{1/2} \cdot h \tag{8.51}$$

The iteration equation for solving Eq. (8.50) with Newton's method is obtained by substituting Eqs. (8.50) and (8.51) in Eq. (3.14):

$$x_{j+1} = x_j - \frac{x_j + 0.8x_j^{3/2}h - 10n_1(1 - e^{-3t_{i+1}})h - n_i}{1 + 0.8 \cdot \frac{3}{2} \cdot x_j^{1/2} \cdot h} \qquad (8.52)$$

The MATLAB program (script file) that solves the problem is listed below. The program starts by calculating the number of steps N ($N = (b-a)/h$) in the solution. The program has a main loop for i (from 1 to N), in which the solution at point $i+1$ is calculated. Inside the main loop there is a nested loop (uses the index j) in which at each step the solution of Eq. (8.50) is determined by using Newton's method. The iterations are performed by using Eq. (8.52). The starting value of the iterations is the solution (the value on n) at the previous point. The iterations stop when the estimated relative error (Eq. (3.18)) is smaller than 0.0001. The nested loop is limited to 20 iterations.

Program 8-2: Script file. Solving first-order ODE using Euler's implicit method.

```
% Solving first order ODE with Euler's implicit method.
clear all
a = 0; b = 0.5; h = 0.002;          Assign the domain to variables a and b, and the step size to h.
N = (b - a)/h;                      Calculate the number of steps.
n(1) = 2000; t(1) = a;              Assign the initial condition to the first point of the solution.
for i=1:N
    t(i + 1) = t(i) + h;            Calculate the next value of the independent variable, Eq. (8.48).
    x = n(i);
% Newton's method starts.            Start the solution of Eq. (8.50) using Newton's method.
    for j = 1:20
        num = x + 0.800*x^(3/2)*h - 10.0*n(1)*(1 - exp(-3*t(i + 1)))*h - n(i);
        denom = 1 + 0.800*1.5*x^(1/2)*h;
        xnew = x - num/denom;                              Eq. (8.52).
        if abs((xnew - x)/x) < 0.0001
            break                        Check if the error is small enough to stop the iterations.
        else
            x = xnew;
        end
    end
    if j == 20
        fprintf('Numerical solution could not be calculated at t = %g s', t(i))
        break
    end
% Newton's method ends.
    n(i + 1) = xnew;         The solution from Newton's method is assigned to the solution of the ODE at the next point.
end
plot(t,n)
axis([0 0.5 0 2000]), xlabel('t (s)'), ylabel('n')
```

The plot that displayed when the program is executed is shown in the figure.

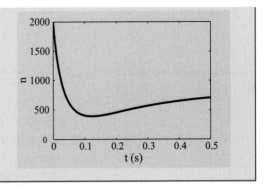

8.3 MODIFIED EULER'S METHOD

The modified Euler method is a single-step, explicit, numerical technique for solving a first-order ODE. The method is a modification of Euler's explicit method. (This method is sometimes called Heun's[1] method.) As discussed in Section 8.2.1, the main assumption in Euler's explicit method is that in each subinterval (step) the derivative (slope) between points (x_i, y_i) and (x_{i+1}, y_{i+1}) is constant and equal to the derivative (slope) of $y(x)$ at point (x_i, y_i). This assumption is the main source of error. In the modified Euler method the slope used for calculating the value of y_{i+1} is modified to include the effect that the slope changes within the subinterval. The slope used in the modified Euler method is the average of the slope at the beginning of the interval and an estimate of the slope at the end of the interval. The slope at the beginning is given by:

$$\left.\frac{dy}{dx}\right|_{x = x_i} = f(x_i, y_i) \tag{8.53}$$

The estimate of the slope at the end of the interval is determined by first calculating an approximate value for y_{i+1}, written as y_{i+1}^{Eu} using Euler's explicit method:

$$y_{i+1}^{Eu} = y_i + f(x_i, y_i)h \tag{8.54}$$

and then estimating the slope at the end of the interval by substituting the point (x_{i+1}, y_{i+1}^{Eu}) in the equation for $\frac{dy}{dx}$:

$$\left.\frac{dy}{dx}\right|_{\substack{y = y_{i+1}^{Eu} \\ x = x_{i+1}}} = f(x_{i+1}, y_{i+1}^{Eu}) \tag{8.55}$$

---

1. There is inconsistency in the literature in the use of the name Heun's method. Sometimes the name is used for the method presented in this section, and at other times for a different method. Since the term "modified (or improved) Euler method" is frequently associated with the method presented in this section, we (as many other authors) use "modified Euler" for the name of the method presented here. A different method presented in Section 8.5.1 is referred to as Heun's method.

Once the two slopes are calculated, a better value of y_{i+1} is calculated using the average of the two slopes:

$$y_{i+1} = y_i + \frac{f(x_i, y_i) + f(x_{i+1}, y_{i+1}^{Eu})}{2}h \qquad (8.56)$$

The modified Euler method is illustrated schematically in Fig. 8-10. Figure 8-10a shows the slope at the beginning of the interval that is given by Eq. (8.53) and the value of y_{i+1}^{Eu} that is calculated with Eq. (8.54). Figure 8-10b shows the estimated slope at the end of the interval that is calculated with Eq. (8.55), and Fig. 8-10c shows the value of y_{i+1} that is obtained with Eq. (8.56) using the average of the slopes. Equation (8.56) can also be derived by integrating the ODE over the interval $[x_i, x_{i+1}]$ using the trapezoidal method.

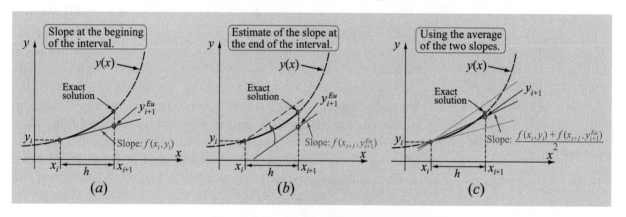

Figure 8-10: The modified Euler method.

The modified Euler method is summarized in the following algorithm.

Algorithm for the modified Euler method

1. Given a solution at point (x_i, y_i), calculate the next value of the independent variable:

$$x_{i+1} = x_i + h$$

2. Calculate $f(x_i, y_i)$.

3. Estimate y_{i+1} using Euler's method:

$$y_{i+1}^{EU} = y_i + f(x_i, y_i)h$$

4. Calculate $f(x_{i+1}, y_{i+1})$.

5. Calculate the numerical solution at $x = x_{i+1}$:

$$y_{i+1} = y_i + \frac{f(x_i, y_i) + f(x_{i+1}, y_{i+1}^{Eu})}{2}h$$

The implementation of the modified Euler method is shown in Example 8-3 where the problem from Example 8-1 is solved again.

Example 8-3: Solving a first-order ODE using the modified Euler method.

Use the modified Euler method to solve the ODE $\frac{dy}{dx} = -1.2y + 7e^{-0.3x}$ from $x = 0$ to $x = 2.5$ with the initial condition $y(0) = 3$. Write a MATLAB program in a script file that solves the equation using $h = 0.5$.

Compare the results with the exact (analytical) solution: $y = \frac{70}{9}e^{-0.3x} - \frac{43}{9}e^{-1.2x}$.

SOLUTION

The following MATLAB program in a script file solves the equation and creates a plot that shows the numerical solution and the exact solution.

Program 8-3: User-defined function. Solving first-order ODE using the modified Euler method.

```
function [x, y] = odeModEuler(ODE,a,b,h,yINI)
% odeModEuler solves a first order ODE using the
% modified Euler method.
% Input variables:
% ODE   Name (string) of a function file that calculates dy/dx.
% a     The first value of x.
% b     The last value of x.
% h     Step size.
% yINI  The value of the solution y at the first point (initial value.
% Output variable:
% x     A vector with the x coordinate of the solution points.
% y     A vector with the y coordinate of the solution points.
x(1) = a;  y(1) = yINI;
N = (b - a)/h;
for i = 1:N
    x(i + 1) = x(i) + h;
    SlopeEu = feval(ODE,x(i),y(i));
    yEu = y(i) + SlopeEu*h;
    SlopeEnd = feval(ODE,x(i + 1),yEu);
    y(i + 1) = y(i) + (SlopeEu + SlopeEnd)*h/2;
end
```

- Assign the initial value to x(1) and y(1).
- Determine the number of steps.
- Apply Eq. (8.14).
- Determine the slope at the beginning of the interval, Eq. (8.53).
- Apply Eq. (8.54).
- Determine the estimated slope at the end of the interval, Eq. (8.55).
- Calculate the numerical solution in step *i*, Eq. (8.56).

The problem is solved with the following program in a script file that uses the function `odeModEuler`. The function ODE in the input argument of `odeModEuler` is the same as in Example 8-1 (`Chap8Exmp1ODE`). The program also creates a plot that shows the numerical and the exact solutions.

```
clear all
a = 0; b = 2.5; h = 0.5; yINI = 3;
[x, y] = odeModEuler('Chap8Exmp1ODE',a,b,h,yINI);
xp = a:0.1:b;
yp = 70/9*exp(-0.3*xp) - 43/9*exp(-1.2*xp);
plot(x,y,'*r',xp,yp)
```

- Assign the domain, step size and initial value to variables.
- Use the `odeModEuler` function.
- Create vectors for plotting the exact solution.

xlabel('x'); ylabel('y')

The plot produced by the program is shown in the figure on the right. The figure shows that the calculated points are much closer to the exact solution than in Example 8-1, part (a), where Euler's explicit method was employed using the same step size.

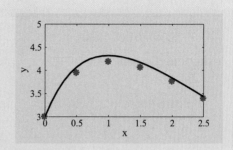

The numerical values of the exact and numerical solutions, and the error, which is the difference between the two, are:

i	1	2	3	4	5	6
x_i	0.0	0.5	1.0	1.5	2.0	2.5
y_i (numerical)	3.0	3.946	4.188	4.063	3.764	3.394
y_i (exact)	3.0	4.072	4.323	4.170	3.835	3.436
Error	0.0	0.1261	0.1351	0.1063	0.0716	0.0425

Comparing the error values here with those in Example 8-1, where the problem was solved with Euler's explicit method using the same size subintervals, shows that the error with the modified Euler method is much smaller.

8.4 MIDPOINT METHOD

The midpoint method is another modification of Euler's explicit method. Here, the slope used for calculating y_{i+1} is an estimate of the slope at the middle point of the interval (step). This estimate is calculated in two steps. First, Euler's method is used to calculate an approximate value of y at the middle point of the interval $x_m = x_i + h/2$, written as y_m:

$$y_m = y_i + f(x_i, y_i)\frac{h}{2} \qquad (8.57)$$

Then, the estimated slope at the midpoint is calculated by substituting (x_m, y_m) in the differential equation for $\frac{dy}{dx}$:

$$\left.\frac{dy}{dx}\right|_{x = x_m} = f(x_m, y_m) \qquad (8.58)$$

The slope from Eq. (8.58) is then used for calculating the numerical solution, y_{i+1}:

$$y_{i+1} = y_i + f(x_m, y_m)h \qquad (8.59)$$

The midpoint method is illustrated schematically in Fig. 8-11. Figure 8-11a shows the determination of the midpoint with Euler's explicit method using Eq. (8.57). Figure 8-11b shows the estimated slope that is calculated with Eq. (8.58), and Fig. 8-11c shows the value of y_{i+1} obtained with Eq. (8.59). Equation (8.59) can also be derived by integrating the ODE over the interval $[x_i, x_{i+1}]$ using the rectangle method applied at the midpoint of the interval.

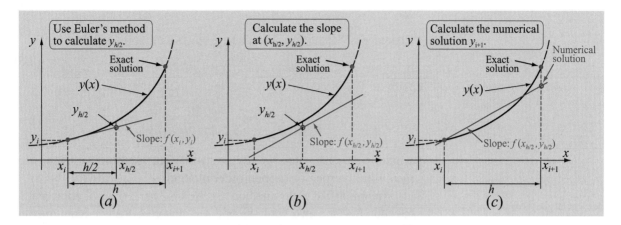

Figure 8-11: The midpoint method.

8.5 RUNGE–KUTTA METHODS

Runge–Kutta methods are a family of single-step, explicit, numerical techniques for solving a first-order ODE. As was stated in Section 8.1, for a subinterval (step) defined by $[x_i, x_{i+1}]$, where $h = x_{i+1} - x_i$, the value of y_{i+1} is calculated by:

$$y_{i+1} = y_i + Slope \cdot h \tag{8.60}$$

where *Slope* is a constant. The value of *Slope* in Eq. (8.60) is obtained by considering the slope at several points within the subinterval. Various types of Runge–Kutta methods are classified according to their order. The order identifies the number of points within the subinterval that are used for determining the value of *Slope* in Eq. (8.60). Second-order Runge–Kutta methods use the slope at two points, third-order methods use three points, and so on. The so-called classical Runge–Kutta method is of fourth order and uses four points. The order of the method is also related to the global truncation error of each method. For example, the second-order Runge–Kutta method is second-order accurate globally; that is, it has a local truncation error of $O(h^3)$ and a global truncation error of $O(h^2)$. For each order there are several methods. The differences between the methods is in the location of the points within the subinterval that are used for determining the slopes and in the way that the constant *Slope* in Eq. (8.60) is determined from the slopes at the different points within the subinterval.

Runge–Kutta methods give a more accurate solution compared to the simpler Euler's explicit method. The accuracy increases (i.e., the truncation error decreases) with increasing order. In each step, however, they require several evaluations (depending on the order) of the function for the derivative $f(x, y)$.

8.5.1 Second-Order Runge–Kutta Methods

The general form of second-order Runge–Kutta methods is:

$$y_{i+1} = y_i + (c_1 K_1 + c_2 K_2)h \tag{8.61}$$

with

$$\begin{aligned} K_1 &= f(x_i, y_i) \\ K_2 &= f(x_i + a_2 h, y_i + b_{21} K_1 h) \end{aligned} \tag{8.62}$$

where c_1, c_2, a_2, and b_{21} are constants. The values of these constants vary with the specific second-order method.

The modified Euler method (Section 8.3) and the midpoint method (Section 8.4) are two versions of a second-order Runge–Kutta method. These two versions and an additional version called Heun's method are presented next, using the general form of Eqs. (8.61) and (8.62).

Modified Euler method in the form of a second-order Runge–Kutta method

For the modified Euler method, the constants in Eqs. (8.61) and (8.62) are:

$$c_1 = \frac{1}{2}, \quad c_2 = \frac{1}{2}, \quad a_2 = 1, \quad \text{and} \quad b_{21} = 1$$

Substituting these constants in Eqs. (8.61) and (8.62) yields:

$$y_{i+1} = y_i + \frac{1}{2}(K_1 + K_2)h \tag{8.63}$$

where

$$\begin{aligned} K_1 &= f(x_i, y_i) \\ K_2 &= f(x_i + h, y_i + K_1 h) \end{aligned} \tag{8.64}$$

Equations (8.63) and (8.64) can also be derived by integrating the ODE over the interval $[x_i, x_{i+1}]$ using the trapezoidal method.

Midpoint method in the form of a second-order Runge–Kutta method

For the midpoint method, the constants in Eqs. (8.61) and (8.62) are:

$$c_1 = 0, \quad c_2 = 1, \quad a_2 = \frac{1}{2}, \quad \text{and} \quad b_{21} = \frac{1}{2}$$

Substituting these constants in Eqs. (8.61) and (8.62) yields:

$$y_{i+1} = y_i + K_2 h \tag{8.65}$$

where

$$\begin{aligned} K_1 &= f(x_i, y_i) \\ K_2 &= f\left(x_i + \frac{1}{2}h, y_i + \frac{1}{2}K_1 h\right) \end{aligned} \tag{8.66}$$

Heun's method

In Heun's method the constants in Eqs. (8.61) and (8.62) are:

$$c_1 = \frac{1}{4}, \quad c_2 = \frac{3}{4}, \quad a_2 = \frac{2}{3}, \quad \text{and} \quad b_{21} = \frac{2}{3}$$

Substituting these constants in Eqs. (8.61) and (8.62) yields:

$$y_{i+1} = y_i + \left(\frac{1}{4}K_1 + \frac{3}{4}K_2\right)h \tag{8.67}$$

where

$$K_1 = f(x_i, y_i)$$
$$K_2 = f\left(x_i + \frac{2}{3}h, y_i + \frac{2}{3}K_1 h\right) \tag{8.68}$$

Truncation error in second-order Runge–Kutta methods

The local truncation error in second-order Runge–Kutta methods is $O(h^3)$, and the global truncation error is $O(h^2)$. This is smaller by a factor of h than the truncation errors in Euler's explicit method. This means that, for the same accuracy, a larger step size can be used. In each step, however, the function $f(x, y)$ in the second-order Runge–Kutta methods is calculated twice. The derivation of an estimate of the local truncation error for the modified Euler's version of the second-order Runge–Kutta method is presented in Section 8.11.

Second-order Runge–Kutta methods and Taylor series expansion

Second-order Runge–Kutta methods can be associated with Taylor series expansion. For the interval defined by $[x_i, x_{i+1}]$, where $h = x_{i+1} - x_i$ and the value of y_i is known, the value of y_{i+1} can be approximated with three terms of a Taylor series expansion:

$$y_{i+1} = y_i + \frac{dy}{dx}\bigg|_{x_i} h + \frac{1}{2}\frac{d^2y}{dx^2}\bigg|_{x_i} h^2 + O(h^3) \tag{8.69}$$

The first derivative $\dfrac{dy}{dx}\bigg|_{x_i}$ is given by the right-hand side of the differential equation evaluated at $x = x_i$:

$$\frac{dy}{dx}\bigg|_{x_i} = f(x_i, y_i) \tag{8.70}$$

and the second derivative $\dfrac{d^2y}{dx^2}\bigg|_{x_i}$ can be determined from the first derivative by using the chain rule (see Chapter 2):

$$\frac{d^2y}{dx^2}\bigg|_{x_i} = \frac{\partial f(x, y)}{\partial x}\bigg|_{x_i, y_i} + \frac{\partial f(x, y)}{\partial y}\bigg|_{x_i, y_i} \frac{dy}{dx}\bigg|_{x_i} \tag{8.71}$$

Substituting Eqs. (8.70) and (8.71) in Eq. (8.69) yields:

$$y_{i+1} = y_i + f(x_i, y_i)h + \frac{1}{2}\frac{\partial f(x, y)}{\partial x}\bigg|_{x_i, y_i} h^2 + \frac{1}{2}\frac{\partial f(x, y)}{\partial y}f(x_i, y_i)h^2 + O(h^3) \quad (8.72)$$

In the general form of second-order Runge–Kutta methods (see Eqs. (8.61) and (8.62)), K_2 is defined as:

$$K_2 = f(x_i + a_2 h, y_i + b_{21} K_1 h) \quad (8.73)$$

Using Taylor series expansion for a function of two variables (see Chapter 2), the right-hand side of the last equation can be expanded to give:

$$K_2 = f(x_i + a_2 h, y_i + b_{21} K_1 h) =$$
$$= f(x_i, y_i) + a_2 h \frac{\partial f(x, y)}{\partial x}\bigg|_{x_i, y_i} + b_{21} h \frac{\partial f(x, y)}{\partial y}\bigg|_{x_i, y_i} K_1 + O(h^2) \quad (8.74)$$

Substituting $K_1 = f(x_i, y_i)$ and K_2 from Eq. (8.74) in Eq. (8.61) gives:

$$y_{i+1} = y_i + c_1 f(x_i, y_i)h + c_2 f(x_i, y_i)h + c_2 a_2 h^2 \frac{\partial f(x, y)}{\partial x}\bigg|_{x_i, y_i}$$
$$+ c_2 b_{21} h^2 \frac{\partial f(x, y)}{\partial y}\bigg|_{x_i, y_i} f(x_i, y_i) + O(h^3) \quad (8.75)$$

Equations (8.72) and (8.75) are two different equations for calculating the (same) value of y_{i+1}. Since the equations have the same type of terms, the coefficients in both equations have to be equal. Matching coefficients of terms with the same power of h gives the following three equations:

$$c_1 + c_2 = 1, \quad c_2 a_2 = \frac{1}{2}, \quad c_2 b_{21} = \frac{1}{2} \quad (8.76)$$

The three equations have four unknowns, which means that there is no unique solution; instead, many solutions exist. The constants of the modified Euler method, midpoint, and Heun's methods are three examples of such solutions. If the coefficients are chosen to satisfy Eq. (8.76), then the local truncation error of the method can be seen from Eq. (8.72) and (8.75) to be of the order $O(h^3)$.

Example 8-4 shows a solution of the ODE that was solved in Examples 8-1 and 8-3 using the second-order Runge–Kutta method worked out manually.

Example 8-4: Solving by hand a first-order ODE using the second-order Runge–Kutta method.

Use the second-order Runge–Kutta method (modified Euler version) to solve the ODE $\frac{dy}{dx} = -1.2y + 7e^{-0.3x}$ from $x = 0$ to $x = 2.0$ with the initial condition $y = 3$ at $x = 0$. Solve by hand using $h = 0.5$.

SOLUTION

The equation is solved with the modified Euler method in Example 8-3, where the solution is obtained by writing a MATLAB program in a script file. Here, in order to illustrate how the Runge–Kutta method is applied, the calculations are carried out by hand.

The first point of the solution is $(0, 3)$, which is the point where the initial condition is given. The values of x and y at the first point are $x_1 = 0$ and $y_1 = 3$.

The rest of the solution is done by steps. In each step the next value of the independent variable is given by:

$$x_{i+1} = x_i + h = x_i + 0.5 \tag{8.77}$$

The value of the dependent variable y_{i+1} is calculated by first calculating K_1 and K_2 using Eq. (8.64):

$$K_1 = f(x_i, y_i)$$
$$K_2 = f(x_i + h, y_i + K_1 h) \tag{8.78}$$

and then substituting the Ks in Eq. (8.63):

$$y_{i+1} = y_i + \frac{1}{2}(K_1 + K_2)h \tag{8.79}$$

First step: In the first step $i = 1$. Equations (8.77)–(8.79) give:

$x_2 = x_1 + 0.5 = 0 + 0.5 = 0.5$

$K_1 = -1.2y_1 + 7e^{-0.3x_1} = -1.2 \cdot 3 + 7e^{-0.3 \cdot 0} = 3.4$

$y_1 + K_1 h = 3 + 3.4 \cdot 0.5 = 4.7$

$K_2 = -1.2(y_1 + K_1 h) + 7e^{-0.3(x_1 + 0.5)} = -1.2 \cdot 4.7 + 7e^{-0.3 \cdot 0.5} = 0.385$

$y_2 = y_1 + \frac{1}{2}(K_1 + K_2)h = 3 + \frac{1}{2}(3.4 + 0.385) \cdot 0.5 = 3.946$

At the end of the first step: $x_2 = 0.5$, $y_2 = 3.946$

Second step: In the second step $i = 2$. Equations (8.77)–(8.79) give:

$x_3 = x_2 + 0.5 = 0.5 + 0.5 = 1.0$

$K_1 = -1.2y_2 + 7e^{-0.3x_2} = -1.2 \cdot 3.946 + 7e^{-0.3 \cdot 0.5} = 1.290$

$y_2 + K_1 h = 3.946 + 1.290 \cdot 0.5 = 4.591$

$K_2 = -1.2(y_2 + K_1 h) + 7e^{-0.3(x_2 + 0.5)} = -1.2 \cdot 4.591 + 7e^{-0.3 \cdot 1.0} = -0.3223$

$$y_3 = y_2 + \frac{1}{2}(K_1 + K_2)h = 3.946 + \frac{1}{2}(1.290 + (-0.3223)) \cdot 0.5 = 4.188$$

At the end of the second step: $x_3 = 1.0$, $y_3 = 4.188$

Third step: In the third step $i = 3$. Equations (8.77)–(8.79) give:

$$x_4 = x_3 + 0.5 = 1.0 + 0.5 = 1.5$$
$$K_1 = -1.2y_3 + 7e^{-0.3x_3} = -1.2 \cdot 4.188 + 7e^{-0.3 \cdot 1.0} = 0.1601$$
$$y_3 + K_1 h = 4.188 + 0.1601 \cdot 0.5 = 4.268$$
$$K_2 = -1.2(y_3 + K_1 h) + 7e^{-0.3(x_3 + 0.5)} = -1.2 \cdot 4.268 + 7e^{-0.3 \cdot 1.5} = -0.6582$$
$$y_4 = y_3 + \frac{1}{2}(K_1 + K_2)h = 4.188 + \frac{1}{2}(0.1601 + (-0.6582)) \cdot 0.5 = 4.063$$

At the end of the third step: $x_4 = 1.5$, $y_4 = 4.063$

Fourth step: In the third step $i = 4$. Equations (8.77)–(8.79) give:

$$x_5 = x_4 + 0.5 = 1.5 + 0.5 = 2.0$$
$$K_1 = -1.2y_4 + 7e^{-0.3x_4} = -1.2 \cdot 4.063 + 7e^{-0.3 \cdot 1.5} = -0.4122$$
$$y_4 + K_1 h = 4.063 + (-0.4122) \cdot 0.5 = 3.857$$
$$K_2 = -1.2(y_4 + K_1 h) + 7e^{-0.3(x_4 + 0.5)} = -1.2 \cdot 3.857 + 7e^{-0.3 \cdot 2.0} = -0.7867$$
$$y_5 = y_4 + \frac{1}{2}(K_1 + K_2)h = 4.063 + \frac{1}{2}(-0.4122 + (-0.7867)) \cdot 0.5 = 3.763$$

At the end of the fourth step: $x_5 = 2.0$, $y_5 = 3.763$

The solution obtained is obviously identical (except for rounding errors) to the solution in Example 8-3.

8.5.2 Third-Order Runge–Kutta Methods

The general form of third-order Runge–Kutta methods is:

$$y_{i+1} = y_i + (c_1 K_1 + c_2 K_2 + c_3 K_3)h \tag{8.80}$$

with

$$K_1 = f(x_i, y_i)$$
$$K_2 = f(x_i + a_2 h, y_i + b_{21} K_1 h) \tag{8.81}$$
$$K_3 = f(x_i + a_3 h, y_i + b_{31} K_1 h + b_{32} K_2 h)$$

where c_1, c_2, c_3, a_2, a_3, b_{21}, b_{31} and b_{32} are eight constants. Six equations that relate the eight constants can be derived by comparing Eqs. (8.80) and (8.81) with a four-term Taylor series expansion that estimates the value of y_{i+1}. (The derivation is beyond the scope of this book.) This means that, as in the second-order Runge–Kutta methods, it is possible to have many third-order methods that have different sets of constants.

One method is called the ***classical third-order Runge–Kutta method***. The values of the eight constants in this method are:

$$c_1 = \frac{1}{6}, \quad c_2 = \frac{4}{6}, \quad c_3 = \frac{1}{6}, \quad a_2 = \frac{1}{2}, \quad a_3 = 1, \quad b_{21} = \frac{1}{2}, \quad b_{31} = -1, \text{ and } b_{32} = 2$$

With these constants the equations for the classical third-order Runge–Kutta method are:

$$y_{i+1} = y_i + \frac{1}{6}(K_1 + 4K_2 + K_3)h \tag{8.82}$$

where

$$K_1 = f(x_i, y_i)$$

$$K_2 = f\left(x_i + \frac{1}{2}h, y_i + \frac{1}{2}K_1 h\right) \qquad (8.83)$$

$$K_3 = f(x_i + h, y_i - K_1 h + 2K_2 h)$$

Equations (8.82) and (8.83) can also be derived by integrating the ODE over the interval $[x_i, x_{i+1}]$ using Simpson's 1/3 method.

Truncation errors in third-order Runge–Kutta methods

The local truncation error in third-order Runge–Kutta methods is $O(h^4)$, and the global truncation error is $O(h^3)$. The derivation of these error estimates is tedious and beyond the scope of this book.

Other variations of third-order Runge–Kutta method use different combinations of constants in Eqs. (8.80) and (8.81). The constants of three such methods, as well as the constants of the classical third-order Runge–Kutta method, are listed in Table 8-1.

Table 8-1: Constants of third-order Runge–Kutta methods.

Method	c_1	c_2	c_3	a_2	b_{21}	a_3	b_{31}	b_{31}
Classical	1/6	4/6	1/6	1/2	1/2	1	−1	2
Nystrom's	2/8	3/8	3/8	2/3	2/3	2/3	0	2/3
Nearly Optimal	2/9	3/9	4/9	1/2	1/2	3/4	0	3/4
Heun's Third	1/4	0	3/4	1/3	1/3	2/3	0	2/3

8.5.3 Fourth-Order Runge–Kutta Methods

The general form of fourth-order Runge–Kutta methods is:

$$y_{i+1} = y_i + (c_1 K_1 + c_2 K_2 + c_3 K_3 + c_4 K_4)h \qquad (8.84)$$

with

$$K_1 = f(x_i, y_i)$$

$$K_2 = f(x_i + a_2 h, y_i + b_{21} K_1 h)$$

$$K_3 = f(x_i + a_3 h, y_i + b_{31} K_1 h + b_{32} K_2 h) \qquad (8.85)$$

$$K_4 = f(x_i + a_4 h, y_i + b_{41} K_1 h + b_{42} K_2 h + b_{43} K_3 h)$$

where c_1, c_2, c_3, c_4, a_2, a_3, a_4 b_{21}, b_{31}, b_{32}, b_{41}, b_{42} and b_{43} are 13 constants. The values of these constants vary with the specific fourth-order method.

The *classical fourth-order Runge–Kutta method* is among the methods more commonly used. The constants of this method are:

$$c_1 = c_4 = \frac{1}{6}, \quad c_2 = c_3 = \frac{2}{6}, \quad a_2 = a_3 = b_{21} = b_{32} = \frac{1}{2}$$

$$a_4 = b_{43} = 1, \quad b_{31} = b_{41} = b_{42} = 0$$

With these constants the equations for the classical fourth-order Runge–Kutta method are:

$$y_{i+1} = y_i + \frac{1}{6}(K_1 + 2K_2 + 2K_3 + K_4)h \qquad (8.86)$$

where

$$
\begin{aligned}
K_1 &= f(x_i, y_i) \\[4pt]
K_2 &= f\left(x_i + \frac{1}{2}h, y_i + \frac{1}{2}K_1 h\right) \\[4pt]
K_3 &= f\left(x_i + \frac{1}{2}h, y_i + \frac{1}{2}K_2 h\right) \\[4pt]
K_4 &= f(x_i + h, y_i + K_3 h)
\end{aligned}
\qquad (8.87)
$$

The classical fourth-order Runge–Kutta method is illustrated schematically in Fig. 8-12. Figures (*a*) through (*c*) show the determination of the slopes in Eq. (8.87). Figure (*a*) shows the slope K_1 and how it is used to find the slope K_2; figure (*b*) shows how slope K_2 is used to find the slope K_3; figure (*c*) shows how slope K_3 is used to find the slope K_4; and figure (*d*) illustrates the application of Eq. (8.86) where the slope

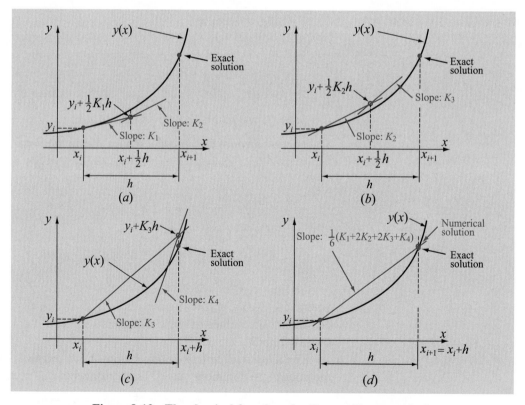

Figure 8-12: The classical fourth-order Runge–Kutta method.

used for calculating y_{i+1} is a weighted average of the slopes K_1, K_2, K_3, and K_4.

Truncation errors in fourth-order Runge–Kutta Methods

The local truncation error in fourth-order Runge–Kutta methods is $O(h^5)$, and the global truncation error is $O(h^4)$. The derivation of these error estimates is tedious and beyond the scope of this book.

Equations (8.86) and (8.87) can also be derived by integrating the ODE over the interval $[x_i, x_{i+1}]$ using Simpson's 1/3 method. However, unlike the third-order Runge–Kutta methods, the function evaluations in the fourth-order methods are combined in such a way as to give a local truncation error of $O(h^5)$.

Examples 8-5 and 8-6 show solutions of first-order initial value problems using the classical fourth-order Runge–Kutta method. In Example 8-5 the first three steps of a solution are calculated by hand, and in Example 8-6, a MATLAB user-defined function is implemented.

Example 8-5: Solving by hand a first-order ODE using the fourth-order Runge–Kutta method.

Use the classical fourth-order Runge–Kutta method to solve the ODE $\dfrac{dy}{dx} = -1.2y + 7e^{-0.3x}$ from $x = 0$ to $x = 1.5$ with the initial condition $y = 3$ at $x = 0$.
Solve by hand using $h = 0.5$.

Compare the results with the exact (analytical) solution: $y = \dfrac{70}{9}e^{-0.3x} - \dfrac{43}{9}e^{-1.2x}$.

SOLUTION

The first point of the solution is $(0, 3)$, which is the point where the initial condition is given. The values of x and y at the first point are $x_1 = 0$ and $y_1 = 3$.

The rest of the solution is done in steps. In each step the next value of the independent variable is calculated by:

$$x_{i+1} = x_i + h = x_i + 0.5 \tag{8.88}$$

The value of the dependent variable y_{i+1} is calculated by first evaluating K_1, K_2, K_3 and K_4 using Eq. (8.87):

$$K_1 = f(x_i, y_i)$$
$$K_2 = f\left(x_i + \frac{1}{2}h, y_i + \frac{1}{2}K_1 h\right)$$
$$K_3 = f\left(x_i + \frac{1}{2}h, y_i + \frac{1}{2}K_2 h\right) \tag{8.89}$$
$$K_4 = f(x_i + h, y_i + K_3 h)$$

and then substituting the Ks in Eq. (8.86):

$$y_{i+1} = y_i + \frac{1}{6}(K_1 + 2K_2 + 2K_3 + K_4)h \tag{8.90}$$

First step: In the first step $i = 1$. Equations (8.88)–(8.90) give:

$$x_2 = x_1 + 0.5 = 0 + 0.5 = 0.5$$

$$K_1 = -1.2y_1 + 7e^{-0.3x_1} = -1.2 \cdot 3 + 7e^{-0.3 \cdot 0} = 3.4$$

$$x_1 + \frac{1}{2}h = 0 + \frac{1}{2} \cdot 0.5 = 0.25 \quad y_1 + \frac{1}{2}K_1h = 3 + \frac{1}{2} \cdot 3.4 \cdot 0.5 = 3.85$$

$$K_2 = -1.2\left(y_1 + \frac{1}{2}K_1h\right) + 7e^{-0.3\left(x_1 + \frac{1}{2}h\right)} = -1.2 \cdot 3.85 + 7e^{-0.3 \cdot 0.25} = 1.874$$

$$y_1 + \frac{1}{2}K_2h = 3 + \frac{1}{2} \cdot 1.874 \cdot 0.5 = 3.469$$

$$K_3 = -1.2\left(y_1 + \frac{1}{2}K_2h\right) + 7e^{-0.3\left(x_1 + \frac{1}{2}h\right)} = -1.2 \cdot 3.469 + 7e^{-0.3 \cdot 0.25} = 2.331$$

$$y_1 + K_3h = 3 + 2.331 \cdot 0.5 = 4.166$$

$$K_4 = -1.2(y_1 + K_3h) + 7e^{-0.3(x_1 + h)} = -1.2 \cdot 4.166 + 7e^{-0.3 \cdot 0.5} = 1.026$$

$$y_2 = y_1 + \frac{1}{6}(K_1 + 2K_2 + 2K_3 + K_4)h = 3 + \frac{1}{6}(3.4 + 2 \cdot 1.874 + 2 \cdot 2.331 + 1.026) \cdot 0.5 = 4.069$$

At the end of the first step: $x_2 = 0.5$, $y_2 = 4.069$

Second step: In the second step $i = 2$. Equations (8.88)–(8.90) give:

$$x_3 = x_2 + 0.5 = 0.5 + 0.5 = 1.0$$

$$K_1 = -1.2y_2 + 7e^{-0.3x_2} = -1.2 \cdot 4.069 + 7e^{-0.3 \cdot 0.5} = 1.142$$

$$x_2 + \frac{1}{2}h = 0.5 + \frac{1}{2} \cdot 0.5 = 0.75 \quad y_2 + \frac{1}{2}K_1h = 4.069 + \frac{1}{2} \cdot 1.142 \cdot 0.5 = 4.355$$

$$K_2 = -1.2\left(y_2 + \frac{1}{2}K_1h\right) + 7e^{-0.3\left(x_2 + \frac{1}{2}h\right)} = -1.2 \cdot 4.355 + 7e^{-0.3 \cdot 0.75} = 0.3636$$

$$y_2 + \frac{1}{2}K_2h = 4.069 + \frac{1}{2} \cdot 0.3636 \cdot 0.5 = 4.16$$

$$K_3 = -1.2\left(y_2 + \frac{1}{2}K_2h\right) + 7e^{-0.3\left(x_2 + \frac{1}{2}h\right)} = -1.2 \cdot 4.16 + 7e^{-0.3 \cdot 0.75} = 0.5976$$

$$y_2 + K_3h = 4.069 + 0.5976 \cdot 0.5 = 4.368$$

$$K_4 = -1.2(y_2 + K_3h) + 7e^{-0.3(x_2 + h)} = -1.2 \cdot 4.368 + 7e^{-0.3 \cdot 1.0} = -0.0559$$

$$y_3 = y_2 + \frac{1}{6}(K_1 + 2K_2 + 2K_3 + K_4)h = 4.069 + \frac{1}{6}[1.142 + 2 \cdot 0.3636 + 2 \cdot 0.5976 + (-0.0559)] \cdot 0.5 = 4.32$$

At the end of the second step: $x_3 = 1.0$, $y_3 = 4.32$

Third step: In the third step $i = 3$. Equations (8.88)–(8.90) give:

$$x_4 = x_3 + 0.5 = 1.0 + 0.5 = 1.5$$

$$K_1 = -1.2y_3 + 7e^{-0.3x_3} = -1.2 \cdot 4.32 + 7e^{-0.3 \cdot 1.0} = 0.001728$$

$$x_3 + \frac{1}{2}h = 1.0 + \frac{1}{2} \cdot 0.5 = 1.25 \quad y_3 + \frac{1}{2}K_1h = 4.32 + \frac{1}{2} \cdot 0.001728 \cdot 0.5 = 4.320$$

$$K_2 = -1.2\left(y_3 + \frac{1}{2}K_1h\right) + 7e^{-0.3\left(x_3 + \frac{1}{2}h\right)} = -1.2 \cdot 4.32 + 7e^{-0.3 \cdot 1.25} = -0.373$$

$$y_3 + \frac{1}{2}K_2h = 4.32 + \frac{1}{2} \cdot (-0.373) \cdot 0.5 = 4.227$$

$$K_3 = -1.2\left(y_3 + \frac{1}{2}K_2h\right) + 7e^{-0.3\left(x_3 + \frac{1}{2}h\right)} = -1.2 \cdot 4.227 + 7e^{-0.3 \cdot 1.25} = -0.2614$$

$$y_3 + K_3h = 4.32 + (-0.2614) \cdot 0.5 = 4.189$$

$$K_4 = -1.2(y_3 + K_3h) + 7e^{-0.3(x_3 + h)} = -1.2 \cdot 4.189 + 7e^{-0.3 \cdot 1.5} = -0.5634$$

$$y_4 = y_3 + \frac{1}{6}(K_1 + 2K_2 + 2K_3 + K_4)h = 4.32 + \frac{1}{6}[0.001728 + 2 \cdot (-0.373) + 2 \cdot (-0.2614) + (-0.5634)] \cdot 0.5 = 4.167$$

At the end of the third step: $x_4 = 1.5$, $y_4 = 4.167$

A comparison between the numerical solution and the exact solution is shown in the following table and figure. The error is 0.003. The global truncation error in the second-order Runge–Kutta method is of the order of h^4. In this problem $h^4 = 0.5^4 = 0.0625$, which is smaller than the actual error. It should be remembered, however, that in the error term the h^4 is multiplied by a constant whose value is not known. This shows that the estimates of truncation errors are good for comparing the accuracy of different methods, but they do not necessarily give an accurate numerical value for the error.

i	1	2	3	4
x_i	0.0	0.5	1.0	1.5
y_i (numerical)	3.0	4.069	4.32	4.167
y_i (exact)	3.0	4.072	4.323	4.170
Error	0.0	0.003	0.003	0.003

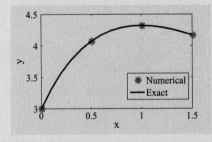

Example 8-6: A user-defined function for solving a first-order ODE using the fourth-order Runge–Kutta method.

Write a user-defined MATLAB function that solves a first-order ODE using the classical fourth-order Runge–Kutta method. Name the function [x,y]=odeRK4(ODE,a,b,h), where ODE is the name of a user-defined function that calculates the derivative dy/dx.

Use the function odeRK4 to solve:

$$\frac{dy}{dx} = -1.2y + 7e^{-0.3x} \text{ from } x = 0 \text{ to } x = 2.5 \text{ with the initial condition } y = 3 \text{ at } x = 0.$$

using $h = 0.5$.

Compare the results with the exact (analytical) solution: $y = \frac{70}{9}e^{-0.3x} - \frac{43}{9}e^{-1.2x}$.

SOLUTION

To solve the problem, a user-defined MATLAB function called odeRK4, which solves a first-order initial value ODE, is written. The function is then used in a script file, which also generates a plot that shows a comparison between the numerical and the exact solutions. The ODE itself is written in a separate user-defined function that is used by the odeRK4 function.

Program 8-4: User-defined function. Solving first-order ODE using Runge–Kutta fourth-order method.

```
function [x, y] = odeRK4(ODE,a,b,h,yIni)
% odeRK4 solves a first order initial value ODE using Runge-Kutta fourth order method.
% Input variables:
% ODE   Name (string) of a function file that calculates dy/dx.
% a     The first value of x.
% b     The last value of x.
% h     Step size.
% yIni    The value of the solution y at the first point (initial value).
% Output variable:
% x     A vector with the x coordinate of the solution points.
% y     A vector with the y coordinate of the solution points.
```

```
x(1) = a;  y(1) = yIni
n = (b - a)/h;
for i = 1:n
   x(i + 1) = x(i) + h;
   K1 = feval(ODE,x(i),y(i));
   xhalf = x(i) + h/2;
   yK1 = y(i) + K1*h/2;
   K2 = feval(ODE,xhalf,yK1);
   yK2 = y(i) + K2*h/2;
   K3 = feval(ODE,xhalf,yK2);
   yK3 = y(i) + K3*h;
   K4 = feval(ODE,x(i + 1),yK3);
   y(i + 1) = y(i) + (K1 + 2*K2 + 2*K3 + K4)*h/6;
end
```

> Assign the initial value to the first point of the solution.
> Calculate the number of steps.
> Calculate the next value of the independent variable.
> Calculate K_1, Eq. (8.87).
> Calculate K_2, Eq. (8.87).
> Calculate K_3, Eq. (8.87).
> Calculate K_4, Eq. (8.87).
> Calculate the next value of the dependent variable, Eq. (8.86).

The following script file uses the function `odeRK4` for solving the ODE.

```
a = 0; b = 2.5;
h = 0.5; yIni = 3;
[x,y] = odeRK4('Chap8Exmp6ODE',a,b,h,yIni)
xp = a:0.1:b;
yp = 70/9*exp(-0.3*xp) - 43/9*exp(-1.2*xp);
plot(x,y,'*r',xp,yp)
yExact = 70/9*exp(-0.3*x) - 43/9*exp(-1.2*x)
error = yExact - y
```

> Use the user-defined `odeRK4` function.
> Create vectors for plotting the exact solution.
> Calculate the exact solution at points of the numerical solution.

The user-defined function `Chap8Exmp6ODE` used in the argument of the function `odeRK4` calculates the value of *dy/dx*:

```
function dydx = Chap8Exmp6ODE(x,y)
dydx = -1.2*y + 7*exp(-0.3*x);
```

When the script file is executed, the following data is displayed in the Command Window. In addition, a plot that shows the points of the numerical solution and the exact solution is displayed in the Figure Window.

```
x =
     0    0.5000    1.0000    1.5000    2.0000    2.5000
y =
   3.0000    4.0698    4.3203    4.1676    3.8338    3.4353
yExact =
   3.0000    4.0723    4.3229    4.1696    3.8351    3.4361
error =
     0    0.0025    0.0026    0.0020    0.0013    0.0008
```

The results (the `error` vector and the figure) show that the numerical solution has a very small error, even though the step size is quite large.

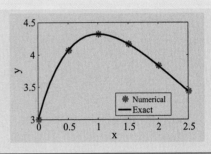

8.6 MULTISTEP METHODS

In single-step methods the solution y_{i+1} at the next point $x = x_{i+1}$ is obtained by using only the value of y_i and x_i at the previous point. In multistep methods, the solution y_{i+1} at the next point $x = x_{i+1}$ is calculated by considering two or more previous points. Multistep methods can be explicit or implicit. In explicit multistep methods, the solution, y_{i+1}, at the next point is calculated from an explicit formula. For example, if three prior points are used, the next unknown value of the dependent variable, y_{i+1}, is calculated by evaluating an expression of the form:

$$y_{i+1} = F(x_{i-2}, y_{i-2}, x_{i-1}, y_{i-1}, x_i, y_i, x_{i+1}) \tag{8.91}$$

This form is explicit since the right-hand side of the equation has only known quantities. Such a method obviously cannot be used for determining the solution at the second point, since only one prior point (the initial condition) is known. Depending on the number of prior points used, the first few points can be determined by single-step methods or by multistep methods that use fewer prior points. In implicit multistep methods, the unknown y_{i+1} appears on both sides of the equation, which is then solved numerically using the methods described in Chapter 3.

8.6.1 Adams–Bashforth Method

The Adams–Bashforth method is an explicit multistep method for solving a first-order ODE. There are several Adams–Bashforth formulas for calculating the value of y_{i+1} by using the known solution at two or more previous points. The formulas are classified according to their order, which is the number of points used in the formula, and is also the order of the global truncation error of the scheme. The second-order formula uses the points (x_i, y_i) and (x_{i-1}, y_{i-1}). The third-order formula uses the three points (x_i, y_i), (x_{i-1}, y_{i-1}), and (x_{i-2}, y_{i-2}), and so on.

The formulas of the Adams–Bashforth method are derived by integrating the differential equation over an arbitrary interval $[x_i, x_{i+1}]$. If the differential equation is given by:

$$\frac{dy}{dx} = f(x, y) \tag{8.92}$$

integration from x_i to x_{i+1} gives:

$$y_{i+1} = y_i + \int_{x_i}^{x_{i+1}} f(x, y)dx \tag{8.93}$$

The integration is carried out by approximating $f(x, y)$ with a polynomial that interpolates the value of $f(x, y)$ at (x_i, y_i) and at previous points.

Second-order Adams–Bashforth method

If $f(x, y)$ in Eq. (8.93) is approximated by a polynomial that interpolates the value of $f(x, y)$ at (x_i, y_i) and at the previous point, (x_{i-1}, y_{i-1}), then the polynomial is of first order and can be written in the form:

$$f(x, y) = f(x_i, y_i) + \frac{f(x_i, y_i) - f(x_{i-1}, y_{i-1})}{h}(x - x_i) \tag{8.94}$$

where $h = x_i - x_{i-1}$. Equation (8.94) is substituted in Eq. (8.93), which is then integrated. Carrying out the integration gives:

$$y_{i+1} = y_i + \frac{h}{2}[3f(x_i, y_i) - f(x_{i-1}, y_{i-1})] \tag{8.95}$$

Equation (8.95) is the second-order Adams–Bashforth method that approximates the solution of the differential equation at x_{i+1} from the previously calculated solutions (x_i, y_i) and (x_{i-1}, y_{i-1}).

Higher-order formulas that include more previous points are derived in the same way. The formulas for the third, fourth, and fifth-order Adams–Bashforth method are given next.

Third-order Adams–Bashforth method

The third-order formula gives the solution y_{i+1} in terms of the known values of the solution at the previous three points:

$$y_{i+1} = y_i + \frac{h}{12}[23f(x_i, y_i) - 16f(x_{i-1}, y_{i-1}) + 5f(x_{i-2}, y_{i-2})] \qquad (8.96)$$

When solving a first-order ODE, the first point is known (initial condition). The solutions at the second and third points have to be determined by other methods, and the formula in Eq. (8.96) can be used starting from the fourth point.

Fourth-order Adams–Bashforth method

The fourth-order formula gives the solution y_{i+1} in terms of the known values of the function in the previous four points:

$$y_{i+1} = y_i + \frac{h}{24}[55f_i - 59f_{i-1} + 37f_{i-2} - 9f_{i-3}] \qquad (8.97)$$

where the notation $f_i = f(x_i, y_i)$, $f_{i-1} = f(x_{i-1}, y_{i-1})$, and so on, is used. When solving a first-order ODE, other methods have to be used for evaluating y at the second, third, and fourth points, and the formula in Eq. (8.97) can be used starting from the fifth point.

8.6.2 Adams–Moulton Method

The Adams–Moulton method is an implicit multistep method for solving first-order ODEs. There are several Adams–Moulton formulas for calculating the value of y_{i+1} by using the previously calculated solutions at two or more points. The formulas are classified according to their order, which refers to the number of points used in the formula and the order of the global truncation error. The second-order formula uses the points (x_i, y_i) and (x_{i-1}, y_{i-1}). The third-order formula uses the three points (x_i, y_i), (x_{i-1}, y_{i-1}), and (x_{i-2}, y_{i-2}), and so on. The approach used in this method is similar to that of the Adams–Bashforth, where the function $f(x, y)$ in Eq. (8.93) is approximated with a polynomial. The difference between the methods is that in the Adams–Moulton method, the points used for determining the interpolation points include (x_i, y_i) and previous points, as well as the point (x_{i+1}, y_{i+1}) where the solution is to be determined. Consequently, the variable y_{i+1} also appears on the right-hand side of the equation, which makes the method implicit. The derivation of the equations is beyond the scope of this book, but the second, third, and fourth-order formulas are given here.

Second-order Adams–Moulton method

The second-order formula is:

$$y_{i+1} = y_i + \frac{h}{2}[f(x_i, y_i) + f(x_{i+1}, y_{i+1})] \tag{8.98}$$

The formula is an implicit form of the modified Euler method.

Third-order Adams–Moulton method

The third-order formula is:

$$y_{i+1} = y_i + \frac{h}{12}[5f(x_{i+1}, y_{i+1}) + 8f(x_i, y_i) - f(x_{i-1}, y_{i-1})] \tag{8.99}$$

Fourth-order Adams–Moulton method

The fourth-order formula is:

$$y_{i+1} = y_i + \frac{h}{24}[9f_{i+1} + 19f_i - 5f_{i-1} + f_{i-2}] \tag{8.100}$$

All of the formulas above can be used only if the solutions at the required number of previous points are already known.

The Adams–Moulton methods can be used in two ways. If they are used by themselves, they have to be solved numerically, since the equations contain the unknown y_{i+1} on both sides of the equation. Usually, however, they are used in conjunction with other equations in methods that are called **predictor–corrector methods**, which are presented in the next section.

8.7 PREDICTOR–CORRECTOR METHODS

Predictor–corrector methods refer to a family of schemes for solving ODEs using two formulas; **predictor** and **corrector** formulas. The predictor is an explicit formula and is used first to determine an estimate of the solution y_{i+1}. Since the predictor is an explicit formula, the value of y_{i+1} is calculated from the known solution at the previous point (x_i, y_i) (single-step method) or several previous points (multistep methods). Once an estimate of y_{i+1} is found, the corrector is applied. The corrector uses the estimated value of y_{i+1} on the right-hand side of an otherwise implicit formula for calculating a new, more accurate, value for y_{i+1} on the left-hand side. Therefore, the corrector equation, which is usually an implicit equation, is being used in an *explicit* manner since no solution of a nonlinear equation is required. This scheme utilizes the benefits of the implicit formula while avoiding the difficulties associated with solving an implicit equation directly. Furthermore, the application of the corrector can be repeated several times such that the new value of y_{i+1} is substituted back on the right-hand side of the corrector formula to obtain a more refined value for y_{i+1}. The predictor–corrector method is summarized in the following algorithm.

Algorithm for the predictor–corrector method

Given a solution at points $(x_1, y_1), (x_2, y_2), ..., (x_i, y_i)$.

1. Calculate y_{i+1} using an explicit method.

2. Substitute y_{i+1} from Step 1, as well as any required values from the already known solution at previous points, in the right-hand side of an implicit formula to obtain a refined value for y_{i+1}.

3. Repeat Step 2 by substituting the refined value of y_{i+1} back in the implicit formula, to obtain an even more refined value for y_{i+1}. Step 2 can be repeated as many times as necessary to produce the desired level of accuracy, that is, until further repetitions do not change the answer for y_{i+1} to a specified number of decimal places.

The simplest example of a predictor–corrector method is the modified Euler method which is presented in Section 8.3. Recall that in this method a first estimate of y_{i+1} is calculated with Euler's explicit formula, Eq. (8.54), which is the predictor. The estimate is then used to calculate a more accurate value of y_{i+1} using Eq. (8.56), which is the corrector. In addition, as mentioned earlier in this section, this method can be modified such that the corrector equation is applied several times at each step. With this modification, the modified Euler predictor–corrector method is presented in the following algorithm.

Algorithm for the modified Euler predictor–corrector method

Given a solution at point (x_i, y_i).

1. Calculate a first estimate for $y_{i+1}^{(1)}$ using Euler's explicit method as a predictor:

$$y_{i+1}^{(1)} = y_i + f(x_i, y_i)h \qquad (8.101)$$

2. Calculate better estimates for y_{i+1} by using Eq. (8.56) repetitively as a corrector:

$$y_{i+1}^{(k)} = y_i + \frac{f(x_i, y_i) + f(x_{i+1}, y_{i+1}^{(k-1)})}{2}h \qquad \text{for} \quad k = 2, 3, ... \quad (8.102)$$

3. Stop the iterations when $\left| \dfrac{y_{i+1}^{(k)} - y_{i+1}^{(k-1)}}{y_{i+1}^{(k-1)}} \right| \leq \varepsilon$

Adams–Bashforth and Adams–Moulton predictor–corrector methods

The Adams–Bashforth method (Section 8.6.1), which is an explicit method, and the Adams–Moulton method (Section 8.6.2), which is an implicit method, can be used together in a predictor–corrector method. For example, with the formulas of the third order, the predictor equation that calculates the first estimate of $y_{i+1}^{(1)}$ is:

$$y_{i+1}^{(1)} = y_i + \frac{h}{12}[23f(x_i, y_i) - 16f(x_{i-1}, y_{i-1}) + 5f(x_{i-2}, y_{i-2})] \qquad (8.103)$$

The corrector equation is:

$$y_{i+1}^{(k)} = y_i + \frac{h}{12}[5f_{i+1}^{(k-1)} + 8f_i - f_{i-1}] \qquad \text{for} \quad k = 2, 3, \dots \qquad (8.104)$$

where $f_{i+1}^{(k-1)} = f(x_{i+1}, y_{i+1}^{(k-1)})$, $f_i = f(x_i, y_i)$, and $f_{i-1} = f(x_{i-1}, y_{i-1})$.

8.8 SYSTEM OF FIRST-ORDER ORDINARY DIFFERENTIAL EQUATIONS

Ordinary differential equations are used to describe, or simulate, processes and systems that are modeled by rates. Frequently, these processes and systems are associated with several dependent variables that are affecting each other. In many of these instances there is a need to solve a system of coupled first-order ODEs. In addition, as will be shown in Section 8.9, initial value problems involving ODEs of second and higher orders are solved by converting the equation into a system of first-order equations.

A simple example of application of a system of two first-order ODEs is in the simulation of growth or decay of two populations that are affecting each other (the so-called predator–prey problem). Suppose a community consists of N_L lions (predators) and N_G gazelles (prey), with b and d representing the birth and death rates of the respective species. Then, the rate of change (growth or decay) of the lion (L) and gazelle populations can be modeled by the equations:

$$\frac{dN_L}{dt} = b_L N_L N_G - d_L N_L \qquad (8.105)$$
$$\frac{dN_G}{dt} = b_G N_G - d_G N_G N_L$$

The dependent variables in Eq. (8.105) are N_L and N_G, and the independent variable is time. The constants b_L and b_G are the birth rates of lions and gazelles, respectively, and d_L and d_G are their respective death rates. The first equation indicates that the rate of change of the population of the lions increases as a function of the product $N_L N_G$, and decreases linearly as a function of the number of lions. For the gazelles, the second equation states that the rate of change of their population increases linearly as a function of the number of gazelles and decreases as a function of the product $N_L N_G$. Once an initial condition is specified (the number of lions and gazelles at $t = 0$), a solution of the system of equations in Eq. (8.105) will give the population of both species as a function of time (see Example 8-11). The differential equations in Eq. (8.105) are coupled since each equation contains both dependent variables.

Chemical reactions are often written as "equations" but with arrows instead of an equal sign. For example, the reaction:

$$H + Br_2 \xrightarrow{k_1} HBr + Br$$

$$HBr + Br \xrightarrow{k_2} H + Br_2$$

represents the physical process of a hydrogen atom (H) colliding with a bromine molecule (Br_2), resulting in the formation of a new molecule (HBr) and a bromine atom (Br). The rate at which this reaction occurs is proportional to the amount of reactants (in this case, (H) and (Br_2)) that are present. Therefore, the rate of production of (HBr) and (Br) is written as:

$$\frac{dn_{Br}}{dt} = k_1 n_H n_{Br_2} - k_2 n_{HBr} n_{Br}$$

$$\frac{dn_H}{dt} = k_2 n_{HBr} n_{Br} - k_1 n_H n_{Br_2}$$

(8.106)

where n_i is the number of atoms or molecules of species i per unit volume, and k_1 and k_2 are the *rate coefficients* for the reactions. It can be seen from Eqs. (8.106) that the number of atoms per unit volume, n_{Br} and n_H, are the dependent variables and time is the independent variable.

General form of a system of first-order ODEs

A system of n first-order ordinary differential equations has the form:

$$\frac{dy_1}{dt} = f_1(t, y_1, y_2, y_3 ..., y_n) \qquad y_1(t_1) = Y_1$$

$$\frac{dy_2}{dt} = f_2(t, y_1, y_2, y_3 ..., y_n) \qquad y^{(2)}(t_1) = Y_2$$

$$...$$

$$\frac{dy_n}{dt} = f_n(t, y_1, y_2, y_3 ..., y_n) \qquad y^{(n)}(t_1) = Y_3$$

(8.107)

where t is the independent variable and $y_1, y_2, y_3 ..., y_n$ are the dependent variables. The right-hand sides of Eq. (8.107) can be nonlinear and of arbitrary complexity.

Some systems of first-order ODEs may be solved with any of the previously discussed explicit methods. For single-step methods, the general form is the same as in Eqs. (8.11) and (8.12), except that the second equation is applied for each of the dependent variables.

$$t_{i+1} = t_i + h$$

(8.108)

$$y_{1, i+1} = y_{1, i} + (Slope)_1 \cdot h$$

$$y_{2, i+1} = y_{2, i} + (Slope)_2 \cdot h$$

$$...$$

$$y_{n, i} = y_{n, i} + (Slope)_n \cdot h$$

(8.109)

where h is the step size and $(Slope)_i$ is a quantity that estimates the value of $\dfrac{dy_i}{dt}$ in the interval from t_i to t_{i+1}.

The next three sections present the details of solving a system of ODEs with Euler's explicit method, The modified Euler method, and the fourth-order Runge–Kutta method. Other explicit methods can be used in the same way.

8.8.1 Solving a System of First-Order ODEs Using Euler's Explicit Method

The application of Euler's method for solving a system of ODEs is shown first for the case of two ODEs. A system of two first-order ODEs, with y and z as the dependent variables and x as the independent variable, has the form:

$$\frac{dy}{dx} = f_1(x, y, z) \tag{8.110}$$

$$\frac{dz}{dx} = f_2(x, y, z) \tag{8.111}$$

for the domain $[a, b]$ with the initial conditions: $y(a) = y_1$ and $z(a) = z_1$.

For a system of two ODEs, Euler's explicit method is given by:

$$x_{i+1} = x_i + h \tag{8.112}$$

$$y_{i+1} = y_i + f_1(x_i, y_i, z_i)h \tag{8.113}$$

$$z_{i+1} = z_i + f_2(x_i, y_i, z_i)h \tag{8.114}$$

The solution process begins with $i = 1$ at the first point, x_1, where the values y_1 and z_1 are known. Then, once h is assigned a value, Eqs. (8.112)–(8.114) are used to calculate the second (next) point of the solution (since all the quantities on the right-hand side of the equations are known). The process then continues with $i = 2, 3, \ldots$ all the way to the end of the domain of the solution. This approach can be readily extended to solve a system of n ODEs.

8.8.2 Solving a System of First-Order ODEs Using Second-Order Runge–Kutta Method (Modified Euler Version)

The application of the second-order Runge–Kutta method for solving a system of ODEs is shown for the case of three ODEs. A system of three first-order ODEs, with y, z, and w as the dependent variables and x as the independent variable, has the form:

$$\frac{dy}{dx} = f_1(x, y, z, w) \tag{8.115}$$

$$\frac{dz}{dx} = f_2(x, y, z, w) \tag{8.116}$$

$$\frac{dw}{dx} = f_3(x, y, z, w) \tag{8.117}$$

for the domain $[a, b]$ with the initial conditions: $y(a) = y_1$, $z(a) = z_1$, and $w(a) = w_1$.

The system is solved by using the modified Euler version of the second-order Runge–Kutta method; see Eqs. (8.63) and (8.64) in Section 8.5.1. When this formulation is used to solve one equation, the value of y_{i+1} at each interval is calculated in three steps. In the first step the value of K_1 is calculated; in the second step the value of K_2 is calculated by using K_1 from the first step; and finally in the third step both Ks are used for calculating y_{i+1}. When solving a system of equations, this process is applied in parallel for each of the equations. This means that the K_1 for all of the ODEs is calculated first. Then, the K_1s are used for calculating the K_2s for all of the equations, and once the two Ks for each equation are known, the value of each dependent variable is calculated at x_{i+1}.

For the system of three ODEs given in Eqs. (8.115)–(8.117), the solution process in each step starts by calculating the value of the independent variable at the end of the step:

$$x_{i+1} = x_i + h \tag{8.118}$$

Then, the K_1s associated with each of the ODEs are calculated:

$$
\begin{aligned}
K_{y,1} &= f_1(x_i, y_i, z_i, w_i) \\
K_{z,1} &= f_2(x_i, y_i, z_i, w_i) \\
K_{w,1} &= f_3(x_i, y_i, z_i, w_i)
\end{aligned}
\tag{8.119}
$$

where $K_{y,1}$, $K_{z,1}$, and $K_{w,1}$ are the K_1 of the first, second, and third ODEs, respectively. Next, the K_2s associated with each of the ODEs are calculated:

$$
\begin{aligned}
K_{y,2} &= f_1(x_i + h, y_i + K_{y,1}h, z_i + K_{z,1}h, w_i + K_{w,1}h) \\
K_{z,2} &= f_2(x_i + h, y_i + K_{y,1}h, z_i + K_{z,1}h, w_i + K_{w,1}h) \\
K_{w,2} &= f_3(x_i + h, y_i + K_{y,1}h, z_i + K_{z,1}h, w_i + K_{w,1}h)
\end{aligned}
\tag{8.120}
$$

where $K_{y,2}$, $K_{z,2}$, and $K_{w,2}$ are the K_2 of the first, second, and third ODEs, respectively. Finally, the values of the three dependent variables at x_{i+1} are calculated by:

$$
\begin{aligned}
y_{i+1} &= y_i + \frac{1}{2}(K_{y,1} + K_{y,2})h \\
z_{i+1} &= z_i + \frac{1}{2}(K_{z,1} + K_{z,2})h \\
w_{i+1} &= w_i + \frac{1}{2}(K_{w,1} + K_{w,2})h
\end{aligned}
\tag{8.121}
$$

The generalization to a system of n equations is presented in the following algorithm.

Algorithm for solving a system of first-order ODEs with second-order Runge–Kutta method (modified Euler version)

A system of n ODEs with the independent variable x and dependent variables $y_1, y_2, ..., y_n$ is written as:

$$\frac{dy_1}{dx} = f_1(x, y_1, ..., y_n)$$
$$...$$
$$\frac{dy_n}{dx} = f_n(x, y_1, ..., y_n)$$
(8.122)

over the interval $[a, b]$ with the initial conditions:

$$y_1\big|_{x=a} = Y_1$$
$$...$$
$$y_n\big|_{x=a} = Y_n$$

1. Choose a step size $h = (b-a)/N$, where N is the number of steps.

2. For $i = 1, ..., N$,

   - Calculate the next value of the independent variable:

   $$x_{i+1} = x_i + h$$
   (8.123)

   - Using x_i and the known $y_{1,i}, ..., y_{n,i}$, calculate K_1 for each ODE:

   $$K_{1,1} = f_1(x_i, y_{1,i}, ..., y_{n,i})$$
   $$...$$
   $$K_{n,1} = f_n(x_i, y_{1,i}, ..., y_{n,i})$$
   (8.124)

   - Using x_i, the known $y_{1,i}, ..., y_{n,i}$, and the known K_1s, calculate the K_2s for each ODE:

   $$K_{1,2} = f_1(x_i + h, y_{1,i} + K_{1,1}h, ..., y_{n,i} + K_{n,1}h)$$
   $$...$$
   $$K_{n,2} = f_n(x_i + h, y_{1,i} + K_{1,1}h, ..., y_{n,i} + K_{n,1}h)$$
   (8.125)

   - Using the known $y_{1,i}, ..., y_{n,i}$, $K_{1,1}, ..., K_{n,1}$ and $K_{1,2}, ..., K_{n,2}$, calculate the solution $y_{1,i+1}, ..., y_{n,i+1}$:

   $$y_{1,i+1} = y_{1,i} + \frac{1}{2}(K_{1,1} + K_{1,2})h$$
   $$...$$
   $$y_{n,i+1} = y_{n,i} + \frac{1}{2}(K_{n,1} + K_{n,2})h$$
   (8.126)

Example 8-7 shows the application of Euler's explicit method and Runge–Kutta second-order method (modified Euler version) for solving a system of equations.

Example 8-7: Solving a system of two first-order ODEs using Euler's explicit method and second-order Runge–Kutta method.

Consider the following initial value problem consisting of two first-order ODEs:

$$\frac{dy}{dx} = (-y+z)e^{(1-x)} + 0.5y \quad \text{with the initial condition} \quad y(0) = 3 \tag{8.127}$$

$$\frac{dz}{dx} = y - z^2 \quad \text{with the initial condition} \quad z(0) = 0.2 \tag{8.128}$$

on the domain from $x = 0$ to $x = 3$.
(a) Solve the system for the first three steps by hand with Euler's explicit method using $h = 0.25$.
(b) Solve the system for the first two steps by hand with the second-order Runge–Kutta method (modified Euler version) using $h = 0.25$.
(c) Write a MATLAB program in a script file that solves the system with the second-order Runge–Kutta method (modified Euler version) using $h = 0.1$.
Show the results from the three parts in a plot.

SOLUTION

(a) ***Solution by hand with Euler's method:*** The first point of the solution is $x_1 = 0$, $y_1 = 3$, and $z_1 = 0.2$, which is the point where the initial conditions are given. For the first point $i = 1$.
The rest of the solution is determined by using Eqs. (8.112)–(8.114). In the present problem these equations have the form:

$$x_{i+1} = x_i + h = x_i + 0.25 \tag{8.129}$$

$$y_{i+1} = y_i + f_1(x_i, y_i, z_i)h = y_i + [(-y_i + z_i)e^{(1-x_i)} + 0.5y_i]h \tag{8.130}$$

$$z_{i+1} = z_i + f_2(x_i, y_i, z_i)h = z_i + [y_i - z_i^2]h \tag{8.131}$$

Equations (8.129)–(8.131) are next applied three times with $i = 1, 2, 3$.
First step: For the first subinterval $i = 1$. Equations (8.129)–(8.131) give:
$x_2 = x_1 + 0.25 = 0 + 0.25 = 0.25$
$y_2 = y_1 + [(-y_1 + z_1)e^{(1-x_1)} + 0.5y_1]h = 3 + [(-3 + 0.2)e^{(1-0)} + 0.5 \cdot 3]0.25 = 1.472$
$z_2 = z_1 + [y_1 - z_1^2]h = 0.2 + [3 - 0.2^2]0.25 = 0.94$
The second point of the solution is: $x_2 = 0.25$, $y_2 = 1.472$, and $z_2 = 0.94$.
Second step: For the second step $i = 2$. Equations (8.129)–(8.131) give:
$x_3 = x_2 + 0.25 = 0.25 + 0.25 = 0.5$
$y_3 = y_2 + [(-y_2 + z_2)e^{(1-x_2)} + 0.5y_2]h = 1.472 + [(-1.472 + 0.94)e^{(1-0.25)} + 0.5 \cdot 1.472]0.25 = 1.374$
$z_3 = z_2 + [y_2 - z_2^2]h = 0.94 + [1.472 - 0.94^2]0.25 = 1.0871$
The third point of the solution is: $x_3 = 0.5$, $y_3 = 1.374$, and $z_3 = 1.087$.
Third step: For the third step $i = 3$. Equations (8.129)–(8.131) give:
$x_4 = x_3 + 0.25 = 0.5 + 0.25 = 0.75$
$y_4 = y_3 + [(-y_3 + z_3)e^{(1-x_3)} + 0.5y_3]h = 1.374 + [(-1.374 + 1.087)e^{(1-0.5)} + 0.5 \cdot 1.374]0.25 = 1.427$
$z_4 = z_3 + [y_3 - z_3^2]h = 1.087 + [1.374 - 1.087^2]0.25 = 1.135$

The fourth point of the solution is: $x_3 = 0.75$, $y_3 = 1.427$, and $z_3 = 1.135$.

The solution obtained with Euler's explicit method is shown in the figure at the end of the solution of part (c).

(b) ***Solution by hand with second-order Runge–Kutta method:*** The first point of the solution is $x_1 = 0$, $y_1 = 3$, and $z_1 = 0.2$, which is the point where the initial conditions are given. For the first point $i = 1$.

The rest of the solution is determined by using Eqs. (8.124)–(8.126). In the present problem these equations have the form:

$$x_{i+1} = x_i + h = x_i + 0.25 \tag{8.132}$$

$$K_{y,1} = (-y_i + z_i)e^{(1-x_i)} + 0.5y_i$$
$$K_{z,1} = y_i - z_i^2 \tag{8.133}$$

$$K_{y,2} = (-y_{est} + z_{est})e^{(1-x_{i+1})} + 0.5y_{est}$$
$$K_{z,2} = y_{est} - z_{est}^2 \tag{8.134}$$

where: $y_{est} = y_i + K_{y,1}h$, and $z_{est} = z_i + K_{z,1}h$.

> Note: To simplify the equations, the quantities y_{est} and z_{est} are introduced. They are calculated after $K_{y,1}$ and $K_{z,1}$ are determined, and are used for evaluating $K_{y,2}$ and $K_{z,2}$.

$$y_{i+1} = y_i + \frac{1}{2}(K_{y,1} + K_{y,2})h$$

$$z_{i+1} = z_i + \frac{1}{2}(K_{z,1} + K_{z,2})h \tag{8.135}$$

Equations (8.132)–(8.135) are next applied three times with $i = 1, 2, 3$.

First step: For the first interval $i = 1$. Equations (8.132)–(8.135) give:

$x_2 = x_1 + 0.25 = 0 + 0.25 = 0.25$

$K_{y,1} = (-y_1 + z_1)e^{(1-x_1)} + 0.5y_1 = (-3 + 0.2)e^{(1-0)} + 0.5 \cdot 3 = -6.111$

$K_{z,1} = y_1 - z_1^2 = 3 - 0.2^2 = 2.96$

$y_{est} = y_1 + K_{y,1}h = 3 + (-6.111)0.25 = 1.472$

$z_{est} = z_1 + K_{z,1}h = 0.2 + 2.96 \cdot 0.25 = 0.94$

$K_{y,2} = (-y_{est} + z_{est})e^{(1-x_{i+1})} + 0.5y_{est} = (-1.472 + 0.94)e^{(1-0.25)} + 0.5 \cdot 1.472 = -0.3902$

$K_{z,2} = y_{est} - z_{est}^2 = 1.472 - 0.94^2 = 0.5884$

$y_2 = y_1 + \frac{1}{2}(K_{y,1} + K_{y,2})h = 3 + \frac{1}{2}[-6.111 + (-0.3902)]0.25 = 2.187$

$z_2 = z_1 + \frac{1}{2}(K_{z,1} + K_{z,2})h = 0.2 + \frac{1}{2}[2.96 + 0.5884]0.25 = 0.6436$

The second point of the solution is: $x_2 = 0.25$, $y_2 = 2.187$, and $z_2 = 0.6436$.

Second step: For the second interval $i = 2$. Equations (8.132)–(8.135) give:

$x_3 = x_2 + 0.25 = 0.25 + 0.25 = 0.5$

$K_{y,1} = (-y_2 + z_2)e^{(1-x_2)} + 0.5y_2 = (-2.187 + 0.6436)e^{(1-0.25)} + 0.5 \cdot 2.187 = -2.173$

$K_{z,1} = y_2 - z_2^2 = 2.187 - 0.6436^2 = 1.773$

$$y_{est} = y_2 + K_{y,1}h = 2.187 + (-2.173)0.25 = 1.644$$

$$z_{est} = z_2 + K_{z,1}h = 0.6436 + 1.773 \cdot 0.25 = 1.087$$

$$K_{y,2} = (-y_{est} + z_{est})e^{(1-x_{i+1})} + 0.5y_{est} = (-1.644 + 1.087)e^{(1-0.5)} + 0.5 \cdot 1.644 = -0.09634$$

$$K_{z,2} = y_{est} - z_{est}^2 = 1.644 - 1.087^2 = 0.4624$$

$$y_3 = y_2 + \frac{1}{2}(K_{y,1} + K_{y,2})h = 2.187 + \frac{1}{2}[-2.173 + (-0.09634)]0.25 = 1.903$$

$$z_3 = z_2 + \frac{1}{2}(K_{z,1} + K_{z,2})h = 0.6436 + \frac{1}{2}[1.773 + 0.4624]0.25 = 0.9230$$

The third point of the solution is: $x_3 = 0.5$, $y_3 = 1.903$, and $z_3 = 0.9230$.

The solution points that were obtained with the second-order Runge–Kutta method are shown in the figure at the end of the solution of part (c).

c) *Solution with a computer program that uses the second-order Runge–Kutta method.* First, a user-defined function, named Sys2ODEsRK2, that solves a system of two ODEs using the second-order Runge–Kutta method is written. This function uses two other user-defined functions (listed in the function definition line as ODE1 and ODE2) that calculate the value of *dy/dx* and *dz/dx* in each step. The functions are named odeExample7dydx and odeExample7dzdx. The solution itself is done in a MATLAB program in a script file. The program also produces a plot that shows the solution from part (c) and the solutions calculated in parts (a) and (b).

> **Program 8-5: User-defined function. Solving a system of two first-order ODEs using a second-order Runge–Kutta method (modified Euler version).**

```
function [x, y, z] = Sys2ODEsRK2(ODE1,ODE2,a,b,h,yINI,zINI)
% Sys2ODEsRK2 solves a system of two first-order initial value ODEs using
% second-order Runge-Kutta method.
% The independent variable is x, and the dependent variables are y and z.
% Input variables:
% ODE1   Name (string) of a function file that calculates dy/dx.
% ODE2   Name (string) of a function file that calculates dz/dx.
% a      The first value of x.
% b      The last value of x.
% h      The size of a increment.
% yINI   The initial value of y.
% zINI   The initial value of z.
% Output variable:
% x   A vector with the x coordinate of the solution points.
% y   A vector with the y coordinate of the solution points.
% z   A vector with the z coordinate of the solution points.
```

x(1) = a; y(1) = yINI; z(1) = zINI;   `Assign the initial value to x(1), y(1), and z(1).`

N = (b - a)/h;   `Determine the number of steps.`

for i = 1:N

x(i + 1) = x(i) + h;   `Calculate the next value of the independent variable.`

Ky1 = feval(ODE1,x(i),y(i),z(i));   `Calculate the $K_1$s, Eq. (8.133).`

Kz1 = feval(ODE2,x(i),y(i),z(i));

```
    Ky2 = feval(ODE1,x(i + 1),y(i) + Ky1*h,z(i) + Kz1*h);
    Kz2 = feval(ODE2,x(i + 1),y(i) + Ky1*h,z(i) + Kz1*h);
    y(i + 1) = y(i) + (Ky1 + Ky2)*h/2;
    z(i + 1) = z(i) + (Kz1 + Kz2)*h/2;
end
```

> Calculate the K_2 s, Eq. (8.134).

> Calculate the next value of the dependent variables, Eq. (8.135).

Listed next are two user-defined functions that calculate the value of dy/dx, Eq. (8.127), and dz/dx, Eq. (8.128). The functions are named `odeExample7dydx` and `odeExample7dzdx`.

```
function dydx = odeExample7dydx(x,y,z)
dydx = (-y + z)*exp(1 - x) + 0.5*y;
```

```
function dzdx = odeExample7dzdx(x,y,z)
dzdx = y - z^2;
```

The MATLAB program in a script file that uses the above functions for solving the problem is listed next. The program also displays a plot that shows the solution from part (*c*) and the solutions that were obtained in parts (*a*) and (*b*).

```
% Solving Chapter 8 Example 7
clear all
a = 0; b = 3; yINI = 3; zINI = 0.2; h = 0.1;
[x, y, z] = Sys2ODEsRK2('odeExample7dydx','odeExample7dzdx',a,b,h,yINI,zINI);
% Data from part (a)
xa = [0 0.25 0.5 0.75];
ya = [3 1.472 1.374 1.427];
za = [0.2 0.94 1.087 1.135];
% Data from part (b)
xb = [0 0.25 0.5];
yb = [3 2.187 1.903];
zb = [0.2 0.6436 0.9230];
plot(x,y,'-k',x,z,'-r',xa,ya,'*k',xa,za,'*r',xb,yb,'ok',xb,zb,'or')
```

> Use the user-defined function `Sys2ODEsRK2` for solving the system.

The figure on the right shows the numerical solution obtained for the whole domain in part (*c*) (solid lines), the first four points from part (*a*) (star markers), and the three points calculated in part (*b*) (circle markers).

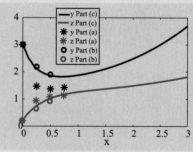

8.8.3 Solving a System of First-Order ODEs Using the Classical Fourth-Order Runge–Kutta Method

Application of the fourth-order Runge–Kutta method for solving a system of ODEs is in principle the same as the second-order method described in the previous section. The only difference is that in the fourth-order Runge–Kutta method (see Eqs. (8.86) and (8.87) in Section 8.5.3) there are four quantities, K_1, K_2, K_3, and K_4 for each ODE that have to be calculated prior to calculating the values of the dependent variables at the next value of the independent variable, x_{i+1}. Again since the equations are coupled, the Ks have to be calculated in parallel, which means that the values of all the K_1s are calculated first. This is followed by calculating all of the K_2s, then all of the K_3s, and finally all of the K_4s. Once the four Ks for each ODE are known, the values of the dependent variables are calculated.

To avoid writing a large number of general equations, application of the fourth-order Runge–Kutta method is presented here for the case of a system of three first-order ODEs. The approach can easily be extended for solving a system with more ODEs.

A system of three first-order ODEs, with y, z, and w as the dependent variables and x as the independent variable, has the form:

$$\frac{dy}{dx} = f_1(x, y, z, w) \tag{8.136}$$

$$\frac{dz}{dx} = f_2(x, y, z, w) \tag{8.137}$$

$$\frac{dw}{dx} = f_3(x, y, z, w) \tag{8.138}$$

for the domain $[a, b]$ with the initial conditions: $y(a) = y_1$, $z(a) = z_1$, and $w(a) = w_1$.

The solution process starts by calculating the value of K_1 for each ODE:

$$
\begin{aligned}
K_{y,1} &= f_1(x_i, y_i, z_i, w_i)\\
K_{z,1} &= f_2(x_i, y_i, z_i, w_i)\\
K_{w,1} &= f_3(x_i, y_i, z_i, w_i)
\end{aligned}
\tag{8.139}
$$

Next, the value of K_2 is calculated for each of the equations:

$$K_{y,2} = f_1\left(x_i + \frac{1}{2}h,\, y_i + \frac{1}{2}K_{y,1}h,\, z_i + \frac{1}{2}K_{z,1}h,\, w_i + \frac{1}{2}K_{w,1}h\right)$$

$$K_{z,2} = f_2\left(x_i + \frac{1}{2}h,\, y_i + \frac{1}{2}K_{y,1}h,\, z_i + \frac{1}{2}K_{z,1}h,\, w_i + \frac{1}{2}K_{w,1}h\right) \tag{8.140}$$

$$K_{w,2} = f_3\left(x_i + \frac{1}{2}h,\, y_i + \frac{1}{2}K_{y,1}h,\, z_i + \frac{1}{2}K_{z,1}h,\, w_i + \frac{1}{2}K_{w,1}h\right)$$

This is followed by the calculation of K_3:

$$K_{y,3} = f_1\left(x_i + \frac{1}{2}h, y_i + \frac{1}{2}K_{y,2}h, z_i + \frac{1}{2}K_{z,2}h, w_i + \frac{1}{2}K_{w,2}h\right)$$

$$K_{z,3} = f_2\left(x_i + \frac{1}{2}h, y_i + \frac{1}{2}K_{y,2}h, z_i + \frac{1}{2}K_{z,2}h, w_i + \frac{1}{2}K_{w,2}h\right) \qquad (8.141)$$

$$K_{w,3} = f_3\left(x_i + \frac{1}{2}h, y_i + \frac{1}{2}K_{y,2}h, z_i + \frac{1}{2}K_{z,2}h, w_i + \frac{1}{2}K_{w,2}h\right)$$

The last quantity to be calculated is K_4:

$$K_{y,4} = f_1(x_i + h, y_i + K_{y,3}h, z_i + K_{z,3}h, w_i + K_{w,3}h)$$
$$K_{z,4} = f_2(x_i + h, y_i + K_{y,3}h, z_i + K_{z,3}h, w_i + K_{w,3}h) \qquad (8.142)$$
$$K_{w,4} = f_3(x_i + h, y_i + K_{y,3}h, z_i + K_{z,3}h, w_i + K_{w,3}h)$$

Once the four Ks for each ODE are determined, the value of the dependent variables (solution) at $x = x_{i+1}$ is calculated by:

$$y_{i+1} = y_i + \frac{1}{6}(K_{y,1} + 2K_{y,2} + 2K_{y,3} + K_{y,4})h$$

$$z_{i+1} = z_i + \frac{1}{6}(K_{z,1} + 2K_{z,2} + 2K_{z,3} + K_{z,4})h \qquad (8.143)$$

$$w_{i+1} = w_i + \frac{1}{6}(K_{w,1} + 2K_{w,2} + 2K_{w,3} + K_{w,4})h$$

A user-defined MATLAB function for solving a system of two first-order ODEs using the fourth-order Runge–Kutta method is given in Example 8-8.

8.9 SOLVING A HIGHER-ORDER INITIAL VALUE PROBLEM

Second-order IVP

A second order ODE with x and y as the independent and dependent variables, respectively, can be written in the form:

$$\frac{d^2y}{dx^2} = f\left(x, y, \frac{dy}{dx}\right) \qquad (8.144)$$

over the domain $[a, b]$. Such an equation can be solved if two constraints are specified. When the two constraints are specified at one value of x, the problem is classified as an initial-value problem (IVP). The two initial conditions are the value of y and the value of the first derivative $\frac{dy}{dx}$ at the first point, a, of the solution domain. If A and B are these values, the initial conditions can be written as:

$$y(a) = A \quad \text{and} \quad \left.\frac{dy}{dx}\right|_{x=a} = B \qquad (8.145)$$

This type of second-order ODE can be transformed into a system of two first-order ODEs that can be solved numerically with the methods presented in Section 8.8. The solution of second-order (and higher-order) equations that are not initial value problems is presented in Chapter 9.

Transforming a second-order ODE into a system of two first-order ODEs is done by introducing a new dependent variable, w, such that:

$$w = \frac{dy}{dx} \tag{8.146}$$

and

$$\frac{dw}{dx} = \frac{d^2y}{dx^2} \tag{8.147}$$

With these definitions, the second-order ODE in Eq. (8.144) with the initial conditions Eq. (8.145) can be written as the following system of two first-order ODEs:

$$\frac{dy}{dx} = w \text{ with } y(a) = A \tag{8.148}$$

$$\frac{dw}{dx} = f(x, y, w) \text{ with } w(a) = B \tag{8.149}$$

Once written in this form, the system can be solved numerically with any of the previously described methods.

As a specific example, consider the second-order ODE:

$$\frac{d^2y}{dt^2} = -0.2\frac{dy}{dt} - 2y + 3\sin(t) \text{ with } y(0) = 3 \text{ and } \frac{dy}{dt}\bigg|_{t=0} = 1.5 \tag{8.150}$$

By defining $w = \frac{dy}{dt}$, such that $\frac{dw}{dt} = \frac{d^2y}{dt^2}$, Eq. (8.150) can be written as the following system of two first-order ODEs:

$$\frac{dy}{dt} = w \quad \text{with} \quad y(0) = 3 \tag{8.151}$$

$$\frac{dw}{dt} = -0.2w - 2y + 3\sin(t) \quad \text{with} \quad w(0) = 1.5 \tag{8.152}$$

The methods of Section 8.8 may now be used to solve the system of Eq. (8.151) and (8.152) numerically.

Solving a second-order initial value problem is illustrated in Example 8-8, where a numerical solution is obtained for the motion of a pendulum. The problem is solved for the case where damping and large displacements of the pendulum are present. This problem can be linearized and solved analytically for small displacements. The nonlinear case of large displacements, however, can only be solved numerically.

Example 8-8: Damped, nonlinear motion of a pendulum.

A pendulum is modeled by a mass that is attached to a weightless rigid rod. According to Newton's second law, as the pendulum swings back and forth, the sum of the forces that are acting on the mass equals the mass times acceleration (see the free body diagram in the figure). Writing the equilibrium equation in the tangential direction gives:

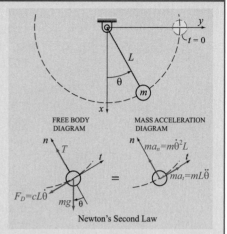

$$\Sigma F_t = -cL\frac{d\theta}{dt} - mg\sin\theta = mL\frac{d^2\theta}{dt^2} \qquad (8.153)$$

where θ is the angle of the pendulum (with respect to the vertical axis, as shown in the figure), $c = 0.16$ (N·s)/m is the damping coefficient, $m = 0.5$ kg is the mass, $L = 1.2$ m is the length, and $g = 9.81$ m/s^2 is the acceleration due to gravity.

Equation (8.153) can be rewritten as the following second-order differential equation:

$$\frac{d^2\theta}{dt^2} = -\frac{c}{m}\frac{d\theta}{dt} - \frac{g}{L}\sin\theta \qquad (8.154)$$

The pendulum is initially displaced such that $\theta = 90°$, and then at $t = 0$ it is released from rest, $\frac{d\theta}{dt} = 0$ (zero initial velocity). Determine the angle of the pendulum as a function of time, $\theta(t)$, for the first 18 seconds after it is released.

SOLUTION

To solve Eq. (8.154), a new dependent variable, w, is introduced, such that:

$$w = \frac{d\theta}{dt} \quad \text{and} \quad \frac{dw}{dt} = \frac{d^2\theta}{dt^2}$$

With these definitions Eq. (8.154) can be rewritten as the following system of two first-order ODEs:

$$\frac{d\theta}{dt} = w \text{ with the initial condition } \theta(0) = \frac{\pi}{2} \qquad (8.155)$$

$$\frac{dw}{dt} = -\frac{c}{m}w - \frac{g}{L}\sin\theta \text{ with the initial condition } w(0) = 0 \qquad (8.156)$$

The system of ODEs is solved with a user-defined MATLAB function named `[t, x, y] = Sys2ODEsRK4(ODE1,ODE2,a,b,h,x1,y1)`, which is listed below. The function solves a system of two ODEs using the fourth-order Runge–Kutta method. The function uses two additional user-defined functions that calculate the value of the right-hand side of each ODE.

> **Program 8-6: User-defined function. Solving a system of two first-order ODEs using fourth-order Runge–Kutta method.**

```
function [t, x, y] = Sys2ODEsRK4(ODE1,ODE2,a,b,h,x1,y1)
% Sys2ODEsRK4 solves a system of two first-order initial value ODEs using
% fourth-order Runge-Kutta method.
% The independent variable is t, and the dependent variables are x and y.
% Input variables:
```

```
% ODE1   Name (string) of a function file that calculates dx/dt.
% ODE2   Name (string) of a function file that calculates dy/dt.
% a      The first value of t.
% b      The last value of t.
% h      The size of a increment.
% x1     The initial value of x.
% y1     The initial value of y.
% Output variable:
% t      A vector with the t coordinate of the solution points.
% x      A vector with the x coordinate of the solution points.
% y      A vector with the y coordinate of the solution points.
```

t(1) = a; x(1) = x1; y(1) = y1;   ┤ Assign the initial values to the first point of the solution. ├

n = (b - a)/h;   ┤ Calculate the number of steps. ├

```
for i = 1:n
   t(i+1) = t(i) + h;
   tm = t(i) + h/2;
   Kx1 = feval(ODE1,t(i),x(i),y(i));
   Ky1 = feval(ODE2,t(i),x(i),y(i));
   Kx2 = feval(ODE1,tm,x(i) + Kx1*h/2,y(i) + Ky1*h/2);
   Ky2 = feval(ODE2,tm,x(i) + Kx1*h/2,y(i) + Ky1*h/2);
   Kx3 = feval(ODE1,tm,x(i) + Kx2*h/2,y(i) + Ky2*h/2);
   Ky3 = feval(ODE2,tm,x(i) + Kx2*h/2,y(i) + Ky2*h/2);
   Kx4 = feval(ODE1,t(i + 1),x(i) + Kx3*h,y(i) + Ky3*h);
   Ky4 = feval(ODE2,t(i + 1),x(i) + Kx3*h,y(i) + Ky3*h);
   x(i+1) = x(i) + (Kx1 + 2*Kx2 + 2*Kx3 + Kx4)*h/6;
   y(i+1) = y(i) + (Ky1 + 2*Ky2 + 2*Ky3 + Ky4)*h/6;
end
```

To solve the system of ODEs in Eqs. (8.155) and (8.156), two user-defined functions are written.

One, named `PendulumDthethaDt`, calculates $\dfrac{d\theta}{dt}$, and the other, named `PendulumDtheth-aDt`, calculates $\dfrac{dw}{dt}$:

```
function dxdt = PendulumDthethaDt(t,x,y)
dxdt = y;                                          ┤ Eq. (8.155). ├
```

```
function dydt = PendulumDwDt(t,x,y)
c = 0.16; m = 0.5; g = 9.81; L = 1.2;
dydt = -(c/m)*y - (g/L)*sin(x);                    ┤ Eq. (8.156). ├
```

The user-defined functions are used in the following MATLAB program (script file) for solving the pendulum problem. The program also displays a plot (θ versus time) of the solution.

```
[t, x, y] = Sys2ODEsRK4('PendulumDthethaDt','PendulumDwDt',0,18,0.1,pi/2,0);
plot(t,x)
xlabel('Time (s)')
ylabel('Angle thetha (rad)')
```

When the script file is executed, the following figure is displayed:

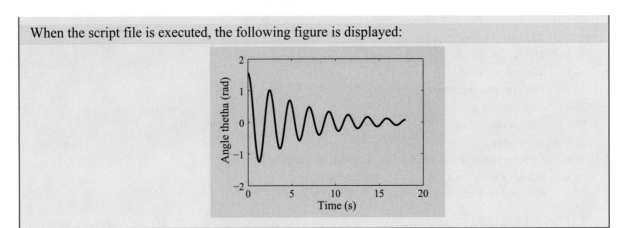

Higher-order IVP

Third and higher-order ODEs can be converted into systems of first-order ODEs in a similar way as second-order ODEs. An nth-order IVP has the form:

$$\frac{d^n y}{dx^2} = f\left(x, y, \frac{dy}{dx}, \frac{d^2 y}{dx^2}, \dots, \frac{d^{(n-1)} y}{dx^{(n-1)}}\right) \quad \text{for} \quad a \leq x \leq b \qquad (8.157)$$

with initial conditions:

$$y(a) = A_1, \quad \left.\frac{dy}{dx}\right|_{x=a} = A_2, \quad \left.\frac{d^2 y}{dx^2}\right|_{x=a} = A_3, \quad \dots, \quad \left.\frac{d^{(n-1)} y}{dx^{(n-1)}}\right|_{x=a} = A_{n-1} \quad (8.158)$$

The nth-order ODE can be transformed into a system of n first-order ODEs by introducing $n-1$ new dependent variables $w_1, w_2, w_3, \dots, w_{n-1}$ such that:

$$w_1 = \frac{dy}{dx}, \quad w_2 = \frac{dw_1}{dx} = \frac{d^2 y}{dx^2}, \quad w_3 = \frac{dw_2}{dx} = \frac{d^3 y}{dx^3}, \quad \dots,$$

$$w_{n-1} = \frac{dw_{n-2}}{dx} = \frac{d^{(n-1)} y}{dx^{(n-1)}} \qquad (8.159)$$

With these definitions, the nth-order ODE in Eq. (8.157) with the initial conditions Eq. (8.158) can be written as the following system of n first-order ODEs, each with its own initial condition:

$$\frac{dy}{dx} = w_1 \quad \text{with} \quad y(a) = A_1 \qquad (8.160)$$

$$\frac{dw_1}{dx} = w_2 \quad \text{with} \quad w_1(a) = A_2 \qquad (8.161)$$

$$\frac{dw_2}{dx} = w_3 \quad \text{with} \quad w_2(a) = A_3 \qquad (8.162)$$

$$\dots$$

$$\frac{dw_{n-2}}{dx} = w_{n-1} \quad \text{with} \quad w_{n-2}(a) = A_{n-1} \qquad (8.163)$$

For example, the third-order IVP:

$$\frac{d^3y}{dx^3} = 2x - 3y + 4\frac{dy}{dx} + x\frac{d^2y}{dx^2} \qquad (8.164)$$

with the initial conditions:

$$y(0) = 3, \quad \frac{dy}{dx}\bigg|_{x=0} = 2, \quad \frac{d^2y}{dx^2}\bigg|_{x=0} = 7 \qquad (8.165)$$

can be transformed into the following system of three first-order ODEs:

$$\frac{dy}{dx} = w_1 \quad \text{with} \quad y(0) = 3 \qquad (8.166)$$

$$\frac{dw_1}{dx} = w_2 \quad \text{with} \quad w_1(0) = 2 \qquad (8.167)$$

$$\frac{dw_2}{dx} = 2x - 3y + 4w_1 + xw_2 \quad \text{with} \quad w_2(0) = 7 \qquad (8.168)$$

Systems of higher-order IVP

Any coupled system of higher-order ODEs can also be rewritten as a system of first-order ODEs using the previously described approach. For example, consider a system of two second-order ODEs:

$$\frac{d^2x}{dt^2} = f\left(x, y, t, \frac{dx}{dt}, \frac{dy}{dt}\right)$$
$$\frac{d^2y}{dt^2} = f\left(x, y, t, \frac{dx}{dt}, \frac{dy}{dt}\right) \qquad (8.169)$$

By defining $u = \dfrac{dx}{dt}$ and $w = \dfrac{dy}{dt}$, the system in Eq. (8.169) can be rewritten as a system of four first-order ODEs:

$$\frac{dx}{dt} = u \qquad\qquad \frac{dy}{dt} = w$$
$$\frac{du}{dt} = f(x, y, t, u, w) \qquad \frac{dw}{dt} = f(x, y, t, u, w) \qquad (8.170)$$

This system can be solved with the methods described in Section 8.8, provided that the initial conditions are prescribed.

8.10 USE OF MATLAB BUILT-IN FUNCTIONS FOR SOLVING INITIAL-VALUE PROBLEMS

MATLAB has a variety of built-in functions that can be used for solving a single, first-order ODE and systems of first-order ODEs. A number of different numerical methods are available for solving ODEs, and most use advanced techniques that optimize the step size automatically, minimize the error per step, or even adapt the step size to the local behavior of the solution. This section introduces the use of these built-in functions. Solution of single first order ODEs is addressed first in Section 8.10.1. Solution of systems of first-order ODEs with MATLAB built-in functions is presented in Section 8.10.2.

For the remainder of this section the independent variable is taken as t (time), and the dependent variable as y. This is done in order to be consistent with the information provided in the **Help** menu of MATLAB, and because in many applications time is the independent variable.

8.10.1 Solving a Single First-Order ODE Using MATLAB

MATLAB has several built-in functions that can be used for solving a single first-order ODE. The differences between the various built-in functions are in the numerical methods used for the solution. The procedure for solving a first-order ODE with MATLAB is summarized in the following list.

• **Problem statement**

The ODE has to be written in the form:

$$\frac{dy}{dt} = f(t, y) \quad \text{with} \quad y(t_1) = y_1$$

Three pieces of information are needed for solving a first-order ODE: an equation that gives an expression for the derivative of y with respect to t, the interval of the independent variable, and the initial value of y. An example is:

$$\frac{dy}{dt} = \frac{t^3 - 2y}{t} \quad \text{for} \quad 1 \leq t \leq 3 \quad \text{with} \quad y = 4.2 \text{ at } t = 1 \quad (8.171)$$

• **User-defined function for calculating the function** $f(t, y)$

The function $f(t, y)$, which gives the value of $\frac{dy}{dt}$ for arbitrary values of t and y, has to be supplied by a user-defined MATLAB function. For the ODE given in Eq. (8.171) the function is:

```
function dydt = DiffEq (t,y)
dydt = (t^3 - 2*y)/t
```

This user-defined MATLAB function is named `DiffEq` in the above example but in general can have any name. As shown later, the name is used in the MATLAB built-in function that actually solves the ODE. The output argument is the value of $\frac{dy}{dt}$ and in its simplest form, the function has two input arguments (t and y). The user-defined function for $f(t, y)$ can have other input arguments in addition to t and y. For example, if f is also a function of temperature, T (a constant), then the function definition line will include T as an input argument as well:

```
function dydt = DiffEq (t, y, T)
```

As will be explained later, a numerical value for T is transferred to the function when it is used in the built-in MATLAB function that actually solves the ODE.

- **Method of solution**

Table 8-2 lists several of the built-in functions available in MATLAB for solving a first-order ODE. A short description of the numerical method used in each built-in function is included in the table.

Table 8-2: MATLAB Built-in functions for solving first-order ODE.

Solver Name	Description
ode45	For nonstiff problems, best to apply as a first try for most problems. Single-step method based on fourth and fifth-order explicit Runge–Kutta methods.
ode23	For nonstiff problems. Single-step method based on second and third-order explicit Runge–Kutta methods. Often quicker but less accurate than ode45.
ode113	For nonstiff problems. Multistep method based on Adams–Bashforth–Moulton methods.
ode15s	For stiff problems. Multistep method that uses a variable-order method. Low to medium accuracy.
ode23s	For stiff problems. One-step solver. Can solve some problems that ode15 cannot. Low accuracy.
ode23t	For moderately stiff problems. Low accuracy.
ode23tb	For stiff problems. Uses an implicit Runge–Kutta method. Often more efficient than ode15s.

The various functions have internal parameters (control parameters) that control the details of the integration with each method, such as step size and error. The default values of these parameters are selected such that the functions perform well when solving common problems. As described later, if needed the user can change the values of the control parameters.

- **Solving the ODE**

The form of the command that is actually used for solving the ODE is the same for all the solvers. The command has a simple form that is suitable for solving common ODEs and a more advanced form that includes additional optional arguments. The simplest form of the command with an example is presented first. The command with the additional optional arguments is presented after the example.

The simple form of MATLAB's built-in function for solving a first-order ODE is:

$$[\texttt{t,y}] = \texttt{SolverName('DiffEq',tspan,yIni)}$$

where:

SolverName   The name of the solver (numerical method) from Table 8-1 that is used (e.g., ode45 or ode23s).

DiffEq   The name of the user-defined function (function file) that calculates $\dfrac{dy}{dt} = f(t, y)$ for given values of t and y. DiffEq can be typed as a string (i.e. 'DiffEq'), or by using a handle (i.e. @DiffEq).

tspan   A vector that specifies the domain of the independent variable. The vector must have at least two elements but can have more. If the vector has only two elements, the elements must be [ta tb], which define the domain of the independent variable. The vector tspan can, however, have additional values between the first and last points. The number of elements in tspan affects the output from the command. See [t,y] below.

yIni   The initial value of y (the value of y at the first point of the domain).

[t,y]   The output, which is the solution of the ODE. t and y are column vectors. The first and the last points are the first and last points of the solution domain. The spacing and number of points in between depend on the input vector tspan. If tspan has two elements (the beginning and end points), the vectors t and y contain the solution at every integration step calculated by the solver. If tspan has more than two elements (additional points between the first and the last), then the vectors t and y contain the solution only at these points. The number of points in tspan does not affect the time steps actually used for the solution by the built-in solver.

A solution of the ODE that was solved in Example 8-6 is shown next in Example 8-9, using MATLAB's ode45 function.

Example 8-9: Using MATLAB's built-in function to solve a first-order ODE.

Use MATLAB's built-in function `ode45`, to solve the ODE:

$\frac{dy}{dx} = -1.2y + 7e^{-0.3x}$ from $x = 0$ to $x = 2.5$ with the initial condition $y = 3$ at $x = 0$.

Compare the results with the exact (analytical) solution: $y = \frac{70}{9}e^{-0.3x} - \frac{43}{9}e^{-1.2x}$.

SOLUTION

The following script file demonstrates the use of MATLAB's `ode45` ODE solver.

```
tspan = [0:0.5:2.5];                               Domain of independent variable.
yIni = 3;                                          Initial value of the dependent variable.
[x,y] = ode45('DiffEqExp8',tspan,yIni)             Use the ode45 function.
yExact = 70/9*exp(-0.3*x) - 43/9*exp(-1.2*x)       Calculate the exact solution at same points as numerical solution.
error = yExact - y
```

The user-defined function `DiffEqExp8` that is used in the argument of the function `ode45` calculates the value of dy/dx:

```
function dydx = DiffEqExp8(x,y)
dydx = -1.2*y + 7*exp(-0.3*x);
```

Since `tspan` is a vector with six elements, the solution is displayed at the six points. When the script file is executed, the following data is displayed in the Command Window.

```
x =
        0
   0.5000
   1.0000
   1.5000
   2.0000
   2.5000
y =
   3.0000
   4.0723
   4.3229
   4.1696
   3.8351
   3.4361
error =
   1.0e-005 *
        0
  -0.2400
  -0.0374
   0.0310
   0.0452
   0.0476
```

The results (the `error` vector) show that the numerical solution has an extremely small error.

As mentioned before, MATLAB's built-in functions for solving a first-order ODE can include additional optional input arguments that can change the default values of the internal parameters that control the integration properties within the solvers. These parameters can be used to transfer values of arguments to the user-defined `DiffEq` function. When these optional input arguments are used, the form of the command is:

$$[\texttt{t,y}]=\texttt{SolverName(DiffEq,tspan,yIni,options,arg1,arg2)}$$

The argument `options` can be used for changing the default values of the internal parameters (called properties) inside the ODE solvers. `options` is a name of a structure array that contains the new values of the properties. (Structure array in MATLAB is an array of fields that can contain different types of data. For example, one field can contain a string, and the next field can contain a number.) The `options` structure is created by a MATLAB built-in function called `odeset`, which has the following form:

$$\texttt{options = odeset('name1',value1,'name2',value2,...)}$$

where `name1`, `name2`, ..., are names of the properties, and `value1`, `value2`, ..., are the corresponding new values of the properties. For example, to set the initial step size to 0.0002 and the relative error to 0.001, the command is written as:

```
options = odeset('InitialStep',0.0002,'RelTol',0.001)
```

Many other properties are listed in the MATLAB help documentation. This feature should be used by experienced users.

The arguments `arg1`, `arg2`, ..., can be used, when needed, for transferring values of arguments to the user-defined function `DiffEq` that calculates $\dfrac{dy}{dt} = f$. In this case f is a function of y, t, and additional parameters, and the function definition line in the user-defined function `DiffEq` has the form (for the case of two additional arguments):

```
dydt = DiffEq(t, y, arg1,arg2)
```

When the function of the ODE solver is executed, the values of the additional arguments are transferred to function `DiffEq`.

Important note

In the function of the solver, the `options` argument has to be present if optional arguments `arg1`, `arg2`, ..., are used for transferring values of arguments to the user-defined function `DiffEq`. If the user does not wish to change any internal parameters, an empty (null) vector (typed as `[]`) should be entered for `options`.

The use of optional arguments in MATLAB's ODE solvers is illustrated in Example 8-10.

Example 8-10: Cooling of a hot plate. (Using a MATLAB's built-in function to solve a first-order ODE.)

When a thin hot plate is suddenly taken out of an oven and is exposed to the surrounding air, it cools due to heat loss by convection and radiation. The rate at which the plate's temperature, T, is changing with time is given by:

$$\frac{dT}{dt} = -\frac{A_s}{\rho V C_v}[\sigma_{SB}\varepsilon(T^4 - T_\infty^4) + h(T - T_\infty)] \tag{8.172}$$

where A_s is the plate's surface area, $\rho = 300\ \text{kg/m}^3$ is its mass density, V is its volume, $C_V = 900$ J/kg/K is its specific heat at constant volume, and $\varepsilon = 0.8$ is its radiative emissivity. Also, $\sigma_{SB} = 5.67 \times 10^{-8}\ \text{W/m}^2/\text{K}^4$ is the Stefan–Boltzmann constant, $h = 30\ \text{W/m}^2/\text{K}$ is the heat transfer coefficient, and T_∞ is the ambient air temperature.

Write a user-defined MATLAB function that calculates the temperature of the plate as a function of time for the first 180 s after the plate is taken out of the oven, and display the result in a figure. Name the function `PlateTemp(T1,Vol,Area,Tamb)` where `T1` is the initial temperature of the plate, `Vol`, and `Area` are the volume and surface area of the plate, respectively, and `Tamb` is the ambient temperature.

Use the function to make a plot that shows the variation of temperature with time for a plate with $V = 0.003\ \text{m}^3$, $A_s = 0.25\ \text{m}^2$, which has an initial temperature of 673 K, when the ambient temperature is 298 K.

SOLUTION

To solve the problem, two user-defined functions are written. One is the function `PlateTemp(T1,Vol,Area,Tamb)`, and the other is a function named `TempRate(t,T,Vol,As,Tamb)` that calculates the value of $\frac{dT}{dt}$ in Eq. (8.172). The ODE is solved inside `PlateTemp` by using MATLAB's built-in function `ode45`. (`ode45` uses `TempRate` when solving the equation). The constants ρ, C_V, ε, and σ_{SB}, are defined inside `TempRate`. The input arguments of `TempRate` are `t,T,Vol,As`, and `Tamb`. The first two arguments, `t` and `T`, are the standard independent and dependent variables. The last three arguments, `Vol,As`, and `Tamb`, are additional arguments that provide the values of the volume, surface area of the plate, and the ambient temperature. The value of these arguments is entered in the function `PlateTemp` and is transferred to `TempRate` through the additional optional arguments in the function `ode45`.

The listing of the user-defined function `PlateTemp` is:

```
function PlateTemp(T1,Vol,Area,Tamb)
% The function PlateTemp calculates the temperature of a plate.
% Input variables:
% T1  The initial temperature in degrees K.
% Vol Volume of the plate in m cube.
% Area Area of the plate in m square.
% Tamb The ambient temperature in degrees K.
```

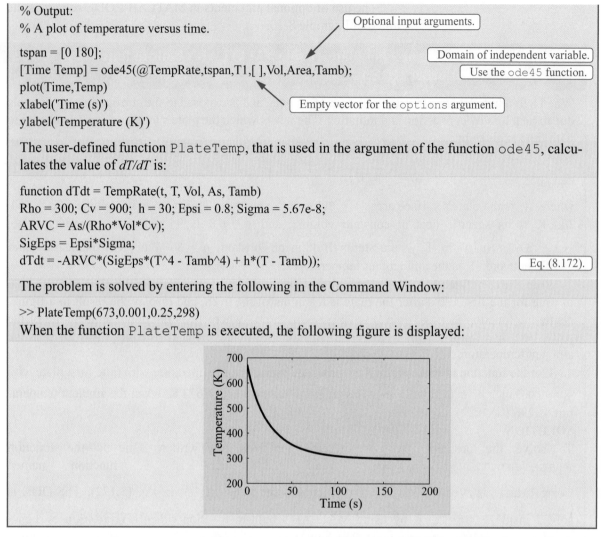

% Output:
% A plot of temperature versus time.

tspan = [0 180];
[Time Temp] = ode45(@TempRate,tspan,T1,[ ],Vol,Area,Tamb);
plot(Time,Temp)
xlabel('Time (s)')
ylabel('Temperature (K)')

Optional input arguments.

Domain of independent variable.

Use the ode45 function.

Empty vector for the options argument.

The user-defined function PlateTemp, that is used in the argument of the function ode45, calculates the value of dT/dT is:

function dTdt = TempRate(t, T, Vol, As, Tamb)
Rho = 300; Cv = 900; h = 30; Epsi = 0.8; Sigma = 5.67e-8;
ARVC = As/(Rho*Vol*Cv);
SigEps = Epsi*Sigma;
dTdt = -ARVC*(SigEps*(T^4 - Tamb^4) + h*(T - Tamb));

Eq. (8.172).

The problem is solved by entering the following in the Command Window:

>> PlateTemp(673,0.001,0.25,298)
When the function PlateTemp is executed, the following figure is displayed:

8.10.2 Solving a System of First-Order ODEs Using MATLAB

Systems of first-order ODEs can be solved with the built-in solvers listed in Table 8-1. The procedures and commands for solving a system are the same as those for a single equation. When solving a system, however, the equations of the system and the initial values are entered as vectors, and the solution that MATLAB returns is in the form of a matrix, where each column contains the solution for one dependent variable.

• **User-defined function for calculating** $\dfrac{dy_i}{dt} = f_i(t, y_1, y_2, \ldots)$

When solving a single ODE, the user-defined function DiffEq calculates the right-hand side of one differential equation, $\dfrac{dy}{dt} = f(t, y)$. The function definition line has the form: dydt = DiffEq(t,y), where the input arguments `t` and `y` are scalars used for calculating the value of $f(t, y)$, which is assigned to `dydt`.

When a system of ODEs is solved (see Eq. (8.107)), there are several dependent variables, $y_1, y_2, \ldots$, and the `DiffEq` function calculates the value of the right-hand side of several differential equations at a specified instant, t:

$$\frac{dy_1}{dt} = f_1(t, y_1, y_2, \ldots), \quad \frac{dy_2}{dt} = f_2(t, y_1, y_2, \ldots), \ldots$$

The function definition line has the same form as when solving a single ODE: `dydt = DiffEq(t, y)`. When used for solving systems, however, `t` is a scalar and `y` is a vector of the dependent variables $[y_1, y_2, \ldots]$. The components of `y` and the value of `t` are used for calculating the values of $f_1(t, y_1, y_2, \ldots), f_2(t, y_1, y_2, \ldots), \ldots$, which have to be assigned as elements in a ***column vector*** to the output argument `dydt`.

- ***Solving a system of ODEs***

The command of the built-in function of the solver is `[t, y] = SolverName(DiffEq, tspan, yIni)`. When used for solving a system of ODEs, the input argument `yIni` is a vector whose components are the initial values of the various dependent variables, $[Y_1, Y_2, \ldots]$ (see Eq. (8.107)). The output arguments are `[t, y]`. When solving one ODE, `t` and `y` are each a column vector. When solving a system of ODEs `t` is a column vector and `y` is a matrix, where each column of `y` is a solution for one dependent variable. The first column is the solution of y_1, the second of y_2, and so on.

Example 8-11 shows how MATLAB built-in functions are used for solving a system of ODEs.

Example 8-11: The predator–prey problem. (Using MATLAB's built-in function to solve a system of two first-order ODEs.)

The relationship between the population of lions (predators), N_L, and the population of gazelles (prey), N_G, that reside in the same area can be modeled by a system of two ODEs. Suppose a community consists of N_L lions (predators) and N_G gazelles (prey), with b and d representing the birth and death rates of the respective species. The rate of change (growth or decay) of the lion (L) and gazelle populations can be modeled by the equations:

$$\frac{dN_L}{dt} = b_L N_L N_G - d_L N_L \tag{8.173}$$

$$\frac{dN_G}{dt} = b_G N_G - d_G N_G N_L$$

Determine the population of the lions and gazelles as a function of time from $t = 0$ to $t = 25$ years, if at $t = 0$, $N_G = 3000$, and $N_L = 500$. The coefficients in the model are: $b_G = 1.1 \text{ yr}^{-1}$, $b_L = 0.00025 \text{ yr}^{-1}$, $d_G = 0.0005 \text{ yr}^{-1}$, and $d_L = 0.7 \text{ yr}^{-1}$.

SOLUTION

In this problem N_L and N_G are the dependent variables, and t is the independent variable. To solve the problem a user-defined function named PopRate(t,N) is written. The function, listed below, calculates the values of $\dfrac{dN_L}{dt}$ and $\dfrac{dN_G}{dt}$ in Eq. (8.173). The input argument N is a vector of the dependent variables, where $N(1) = N_L$ and $N(2) = N_G$. Notice how these components are used when the values of the differential equations are calculated. The output argument dNdt is a column vector where the first element is the value of the derivative $\dfrac{dN_L}{dt}$, and the second is the value of the derivative $\dfrac{dN_G}{dt}$.

```
function dNdt = PopRate(t, N)
bG = 1.1; bL = 0.00025;  dG = 0.0005; dL = 0.7;
f1 = bL*N(1)*N(2) - dL*N(1);
f2 = bG*N(2) - dG*N(2)*N(1);
dNdt = [f1; f2];
```
Calculate Eq. (8.173).

The system of ODEs in Eq. (8.173) is solved with MATLAB's solver ode45. The solver uses the user-defined function PopRate for calculating the right-hand sides of the differential equations. The following MATLAB program in a script file shows the details. Notice that Nini is a vector in which the first element is the initial value of N_L and the second element is the initial value of N_G.

```
tspan = [0 25];                                        Domain of independent variable.
Nini = [500 3000];            Initial values of the dependent variable (notice that Nini is a vector).
[Time Pop] = ode45(@PopRate,tspan,Nini);              Solve the system of ODEs.
plot(Time,Pop(:,1),'-',Time,Pop(:,2),'--')
xlabel('Time (yr)')
ylabel('Population')
legend('Lions','Gazelles')
```

In the output arguments of the solver ode45, the variable Time is a column vector, and the variable Pop is a two-column array, with the solution of N_L and N_G in the first and second columns, respectively. The first few elements of [Time Pop] that are displayed in the Command Window (if the semicolon at the end of the command is removed) are shown below. The program generates a plot that shows the population of the lions and gazelles as a function of time.

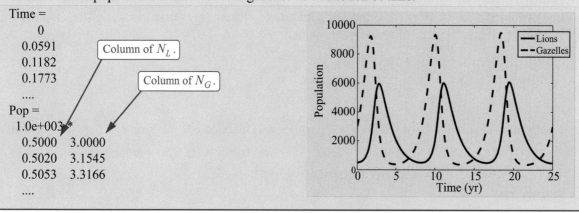

```
Time =
     0
     0.0591        Column of $N_L$.
     0.1182
     0.1773
     ....
Pop =
  1.0e+003 *
     0.5000    3.0000        Column of $N_G$.
     0.5020    3.1545
     0.5053    3.3166
     ....
```

8.11 LOCAL TRUNCATION ERROR IN SECOND-ORDER RANGE–KUTTA METHOD

The second-order Runge–Kutta method was presented in Section 8.5.1. As was mentioned, the local truncation error in this method is $O(h^3)$, and the global truncation error is $O(h^2)$. This section presents the derivation of an estimate of the local truncation error for the modified Euler version of the second-order Runge–Kutta method. In general, the local truncation error at each step of the solution is due to the approximate formula used for calculating the solution.

A note about notation in this section

In the present section, different values of the dependent variable are calculated and compared at the same value of the independent variable. To clarify the presentation, each quantity is identified with a superscript. A dependent variable that is calculated by a numerical method is written as y^{NS} (in the rest of the chapter it is just y). The true solution of the ODE is written as y^{TS}. In addition, the notation y^{Taylor} is used for the value that is calculated with a Taylor series expansion.

The local truncation error is the error inherent in the formula used to obtain the numerical solution in a single step (interval). It is the difference between the numerical solution and an exact solution for that step. In a general step i the numerical solution y_i^{NS} at $x = x_i$ is known (previously calculated), and the numerical solution y_{i+1}^{NS} at $x = x_{i+1}$ is calculated with an approximate formula. The value of the exact solution for this step with y_i presumed known can be expressed by a three-term Taylor series expansion with a remainder (see Chapter 2):

$$y_{i+1}^{Taylor} = y_i^{NS} + \frac{dy}{dx}\bigg|_{x_i, y_i^{NS}} h + \frac{d^2y}{dx^2}\bigg|_{x_i, y_i^{NS}} \frac{h^2}{2} + \alpha_i h^3 \tag{8.174}$$

where the fourth term on the right is the remainder term. Since $\frac{dy}{dx} = f(x, y)$, the second derivative $\frac{d^2y}{dx^2}\bigg|_{x_i, y_i^{NS}}$ can be expressed in terms of $f(x, y)$ and its partial derivatives:

$$\frac{d^2y}{dx^2} = \frac{d}{dx}f(x, y) = \frac{\partial f}{\partial x} + \frac{\partial f}{\partial y}\frac{dy}{dx} \tag{8.175}$$

Substituting Eq. (8.175) in Eq. (8.174) gives:

$$y_{i+1}^{Taylor} = y_i^{NS} + f(x_i, y_i^{NS}) h + \left(\frac{\partial f}{\partial x}\bigg|_{x_i, y_i^{NS}} + \frac{\partial f}{\partial y}\bigg|_{x_i, y_i^{NS}} f(x_i, y_i^{NS})\right)\frac{h^2}{2} + \alpha_i h^3 \tag{8.176}$$

The numerical solution with the modified Euler version of the second-order Runge–Kutta method is given by Eqs. (8.63)–(8.64), which are repeated here:

$$y_{i+1}^{NS} = y_i^{NS} + \frac{1}{2}(K_1 + K_2)h \tag{8.177}$$

$$K_1 = f(x_i, y_i^{NS})$$
$$K_2 = f(x_i + h, y_i^{NS} + K_1 h) \tag{8.178}$$

In the equation for K_2, the function $f(x_i + h, y_i^{NS} + K_1 h)$ can be expanded in a Taylor series expansion (three terms) about $x = x_i$ and $y = y_i^{NS}$ (using Taylor expansion for a function of two variables, see Section 2.7.2, since f is a function of x and y):

$$K_2 = f(x_i + h, y_i^{NS} + K_1 h) = f(x_i, y_i^{NS}) + h\frac{\partial f}{\partial x}\Big|_{x_i, y_i^{NS}} + K_1 h\frac{\partial f}{\partial y}\Big|_{x_i, y_i^{NS}} + \beta_i h^2 \tag{8.179}$$

where the fourth term on the right is the remainder term. Substituting $K_1 = f(x_i, y_i^{NS})$ in Eq. (8.179) and then substituting Eq. (8.179) and $K_1 = f(x_i, y_i^{NS})$ in Eq. (8.177) yields:

$$y_{i+1}^{NS} = y_i^{NS} + f(x_i, y_i^{NS})h + \frac{1}{2}h^2\left(\frac{\partial f}{\partial x}\Big|_{x_i, y_i^{NS}} + f(x_i, y_i^{NS})\frac{\partial f}{\partial y}\Big|_{x_i, y_i^{NS}}\right) + \frac{1}{2}\beta_i h^3 \tag{8.180}$$

The local truncation error, e_i^{TR}, is given by the difference between Eq. (8.180) and Eq. (8.176), which is of the order of h^3:

$$e_{i+1}^{TR} = y_{i+1}^{NS} - y_{i+1}^{Taylor} = \frac{1}{2}\beta_i h^3 - \alpha_i h^3 = O(h^3) \tag{8.181}$$

Equation (8.181) shows that the local truncation error is of the order of $(O(h^3))$. When an interval of solution $[a, b]$ is divided into N subintervals of width h, the local truncation errors accumulate and the global truncation error is $O(h^2)$.

8.12 STEP SIZE FOR DESIRED ACCURACY

The order of magnitude of the truncation errors for the various numerical methods for solving first-order ODEs can be used for determining the step size, h, that should used for solving the ODE with a desired accuracy. Recall from Example 8-1 that, in general, as the step size is made smaller, the more accurate is the solution. If one desires the solution to be of a certain accuracy, for example, correct to five decimal places, information about the truncation error of a particular numerical scheme can help determine the step size necessary to get the desired accuracy. This section describes a method of determining this step size.

The method used to find a step size for desired accuracy is illustrated here using the third-order Runge–Kutta method. As mentioned in Section 8.5.2, the third-order Runge–Kutta method has a local truncation error of the order of h^4. Therefore, the local truncation error for a chosen h can be written as:

$$e_h^{TR} = Ah^4 \tag{8.182}$$

where A is a problem-dependent constant. Note that if h is halved, then the local error is $e_{h/2}^{TR} = \dfrac{Ah^4}{16}$. This means that the local truncation error incurred in advancing the numerical solution over an extent h for the independent variable by using two steps of $h/2$ is approximately equal (neglecting accumulated truncation error) to $2e_{h/2}^{TR}$ or:

$$2e_{h/2}^{TR} = \frac{Ah^4}{8} \tag{8.183}$$

Subtracting Eq. (8.183) from Eq. (8.182) gives:

$$e_h^{TR} - 2e_{h/2}^{TR} = Ah^4 - \frac{Ah^4}{8} = \frac{7}{8}Ah^4 \tag{8.184}$$

Now, suppose that an ODE is solved with the third-order Runge–Kutta method twice: once, a solution is calculated for a single step of size h yielding a numerical solution y_h, and a second time another solution is calculated for two steps of size $h/2$ yielding a numerical solution $y_{h/2}$. Using Eq. (8.181) the numerical solutions can be expressed in terms of Taylor series expansion and the local truncation error:

$$y_h^{NS} = y_h^{Taylor}(x_i + h) + e_h^{TR} \tag{8.185}$$

and

$$y_{h/2}^{NS} = y_h^{Taylor}(x_i + h) + 2e_{h/2}^{TR} \tag{8.186}$$

The factor 2 on the right-hand side of Eq. (8.186) is necessary because in the solution with a step size of $h/2$ two steps are required for a solution at $x_i + h$. Subtracting Eq. (8.185) from Eq. (8.186) and using Eq. (8.184) yields:

$$y_h^{NS} - y_{h/2}^{NS} = \frac{7}{8}Ah^4 \tag{8.187}$$

The significance of Eq. (8.187) is that by running a code already written to solve an ODE using the third-order Runge–Kutta method twice with the different step sizes of h and $h/2$, the problem–dependent constant A can be determined by a numerical experiment. In other words, by running the code once (for one step) with an arbitrarily chosen (but reasonable) step size of h, y_h^{NS} can be found. Then, by running the same code (for two steps) with a step size of $h/2$, $y_{h/2}^{NS}$ can be found. Since h was chosen, it is a known quantity and Eq. (8.187) can be used to determine the parameter A. Once A is determined, Eq. (8.182) can be used for calculating a value of h for a desired level of accuracy. For example, if the desired error is ξ, then the value of h such that $e_h^{TR} \le \xi$ is given by:

$$h = \left(\frac{\xi}{A}\right)^{1/4} \tag{8.188}$$

Similar expressions can be derived for Runge–Kutta methods of different orders or other methods.

Criteria such as Eq. (8.188) establish a guideline and are therefore inexact. The reason is that in this discussion of numerical error, only truncation error (which is method-dependent) has been considered. The total error is the sum of the global truncation error and the round-off error, and (8.186) does not contain any information regarding the machine-dependent round-off error (see Chapter 1). Consequently, the optimum step size provided by (8.186) is approximate and must be viewed as such. Example 8-12 shows how an optimum step size may be found for a particular problem when using the fourth-order Runge–Kutta method.

Example 8-12: Determining the step size for given accuracy.

(a) Develop the formula that can be used for calculating the step size for a required accuracy for the fourth-order Runge–Kutta method.

(b) For the given ODE, apply the formula from part (a) to find the magnitude of the step size such that the local truncation error will be less than 10^{-6}.

$$\frac{dy}{dx} = -\frac{y}{1+x^2}, \quad \text{with} \quad y(0) = 1, \quad \text{for} \quad 0 < x < 1 \tag{8.189}$$

SOLUTION

(a) Since the local truncation error for the fourth-order Runge–Kutta method is $O(h^5)$, the truncation error in a solution with a step size of h can be expressed by (analogous to Eq. (8.182)):

$$e_h^{TR} = Ah^5 \tag{8.190}$$

When the step size is halved, the local truncation error when the independent variable advances two steps is:

$$2e_{h/2}^{TR} = \frac{Ah^5}{16} \tag{8.191}$$

Subtracting the last two equations gives a relation analogous to Eq. (8.184):

$$e_h^{TR} - 2e_{h/2}^{TR} = Ah^5 - \frac{Ah^5}{16} = \frac{15}{16}Ah^5 \tag{8.192}$$

By using Eqs. (8.185) and (8.186), the difference in the truncation errors in Eq. (8.192) can be written in terms of the difference in the numerical solutions that are obtained with step sizes of h and $h/2$:

$$y_h^{NS} - y_{h/2}^{NS} = \frac{15}{16}Ah^5, \quad \text{or} \quad A = \frac{16(y_h^{NS} - y_{h/2}^{NS})}{15} \frac{1}{h^5} \tag{8.193}$$

Once the two numerical solutions are obtained, the constant A in Eq. (8.193) can be determined. With A known, the step size for the required accuracy is calculated by using Eq. (8.190):

$$h = \left(\frac{\xi}{A}\right)^{1/5} \tag{8.194}$$

where ξ is the desired error.

(b) To find the magnitude of the step size such that the local truncation error will be less than 10^{-6}, Eq. (8.189) is solved twice with the fourth-order Runge–Kutta method: once with $h = 1$ and once with $h = 1/2$. The solution of Eq. (8.189) is performed in the Command Window of MATLAB by using the user-defined function `odeRK4` that was developed in Example 8-6:

```
>> [x, y] = odeRK4('Chap8Exmp12ODE',0,1,1,1)
x =
   0   1
y =
   1.00000000000000   0.45666666666667
```

Solve the ODE using $h = 1$.

From this solution $y_h^{NS} = 0.45666666666667$.

```
>> [x, y] = odeRK4('Chap8Exmp12ODE',0,1,0.5,1)
x =
        0   0.50000000000000   1.00000000000000
y =
   1.00000000000000   0.62900807381776   0.45599730642061
```

Solve the ODE using $h = 1/2$.

From this solution $y_{h/2}^{NS} = 0.45599730642061$.

The argument `Chap8Exmp12ODE` in the user-defined function `odeRK4` is the name of the following user-defined function that calculates the value of dy/dx:

```
function dydx = Chap8Exmp12ODE(x,y)
dydx = -y/(1 + x^2);
```

The results from the numerical solutions are used in Eq. (8.193) for determining the value of the constant A:

$$A = \frac{16(0.45666667 - 0.45599730)}{15} \cdot \frac{1}{1^5} = 7.14 \times 10^{-4}$$

Now that A is known, the magnitude of the step size such that the local truncation error will be less than 10^{-6} can be calculated with Eq. (8.194):

$$h = \left(\frac{10^{-6}}{7.14 \times 10^{-4}}\right)^{1/5} = 0.2687$$

Thus, the step size must be smaller than $h = 0.2687$ for solving Eq. (8.189) with the fourth-order Runge–Kutta method for the local truncation error to be smaller than 10^{-6}.

To check the result, Eq. (8.189) is solved again using $h = 0.2$, which is smaller than the value obtained above.

```
>> [x,y]=odeRK4('Chap8Exmp12ODE',0,1,0.2,1);
>> y(6)
ans =
   0.45594036941322
```

Solve the ODE using $h = 0.2$.

Display the last value of y.

This numerical solution gives $y = 0.45594036941322$.

The exact solution of Eq. (8.189) is $y = 0.45593812776600$.

The difference between the two solutions is 2.24×10^{-6} which is larger than but on the order of 10^{-6}. However, the step size of h was calculated for a required local truncation error, and the last solution also contains a global truncation error (round-off errors are likely negligible since MATLAB uses double precision).

8.13 STABILITY

The total numerical error consists of the method-dependent truncation error and the machine-dependent round-off error. When solving an ODE numerically, the error that is introduced in each step is ideally expected not to increase as the solution progresses. The total error is also generally expected to decrease (or at least not increase) as the step size, h, is reduced. In other words, the numerical solution is expected to become more accurate as the step size is reduced. However, in some situations the numerical error grows without bound either as the solution progresses in an initial value problem, or even as the step size is reduced. These situations are symptomatic of *instability*, and the solution is said to become *unstable*. The stability of a solution depends in general on three factors:

1. The particular numerical method used.
2. The step size, h, used in the numerical solution.
3. The specific differential equation being solved.

There are several ways to analyze stability. One way is to perform the calculation first with single-precision and then with double-precision (see Chapter 1 for further discussion) and compare the answers (MATLAB by default uses double-precision). This can help to examine the accumulation of the round-off error part of the total error. The second way is to vary the step size, h, and observe the behavior of the answer as the step size is reduced. This enables control over the truncation error portion of the total error. Very small step size can also result in larger round-off errors. The third way is to use a higher-order numerical method (with a lower truncation error, such as, for example, the fourth-order Runge–Kutta method) and compare the results with the answers obtained with a lower-order method (such as, for example, the second or third-order Runge–Kutta methods). Yet another way to check for stability is to test a method on a differential equation that has a known analytical solution. If the method is unstable when solving the equation with the known solution, it can be expected to be unstable when solving other equations.

Euler's explicit method and higher order single-step Runge–Kutta methods are numerically stable if the step size, h, is sufficiently small. This is illustrated by considering an ODE of the simple form:

$$\frac{dy}{dx} = -\alpha y, \quad \text{with} \quad y(0) = 1 \tag{8.195}$$

where $\alpha > 0$ is a constant. The exact solution of this equation is $y(x) = e^{-\alpha x}$. For an interval $[x_i, x_{i+1}]$, (where $h = x_{i+1} - x_i$) the exact solution at point x_{i+1} is:

$$y^{TS}(x_{i+1}) = e^{-\alpha x_{i+1}} = e^{-\alpha(x_i + h)} = e^{-\alpha x_i}e^{-\alpha h} = y^{TS}(x_i)e^{-\alpha h} \qquad (8.196)$$

The numerical solution with Euler's explicit method gives:

$$y_{i+1} = y_i + (-\alpha y_i)h = (1 - \alpha h)y_i = \gamma y_i \qquad (8.197)$$

where $\gamma = (1 - \alpha h)$.

Comparing Eqs. (8.196) and (8.197) shows that the factor γ in the numerical solution is just an approximation for the factor $e^{-\alpha h}$ in the exact solution. In fact, γ consists of the first two terms of the Taylor series expansion of $e^{-\alpha h}$ for small αh. The two factors are plotted in Fig. 8-13. This factor γ is the source of the error and instability. It can be seen from Eq. (8.197) that when $|\gamma| < 1$ the error will not be amplified. Thus, for stability, $|\gamma| < 1 \Rightarrow -1 < 1 - \alpha h < 1$, which can be written as:

$$0 < \alpha h < 2 \qquad (8.198)$$

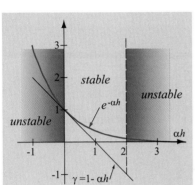

Figure 8-13: Stability criterion.

Equation (8.198) defines a ***stability criterion***. It states that the error for an ODE of the form $\dfrac{dy}{dx} = -\alpha y$ will not grow without bound provided $\alpha h < 2$. Since α is dependent on the problem, this stability criterion sets an upper limit on the step size h for a particular problem.

Example 8-13 is a numerical example of the effect step size has on the stability of the solution of an ODE with the form $\dfrac{dy}{dx} = -\alpha y$ when solved using Euler's explicit method.

Example 8-13: Stability of Euler's explicit method.

Consider the solution of the ODE:

$$\frac{dy}{dx} = -2.5y, \quad \text{with} \quad y(0) = 1, \quad \text{for} \quad 0 < x < 3.4 \qquad (8.199)$$

(a) Solve with Euler's explicit method using $h = 0.2$.
(b) Solve with Euler's explicit method using $h = 0.83$.
The exact (analytical) solution is: $y = e^{-2.5x}$. Show the results from parts (a) and (b) in a plot together with the exact solution.

SOLUTION

Equation (8.199) is in the form of Eq. (8.195) with $\alpha = 2.5$. According to Eq. (8.198), a stable solution will be obtained for $h < \dfrac{2}{2.5} = 0.8$. Consequently, it can be expected that a stable solution will be obtained in part (a), while the solution in part (b) will be unstable. The solutions are carried out by writing the following MATLAB program in a script file.

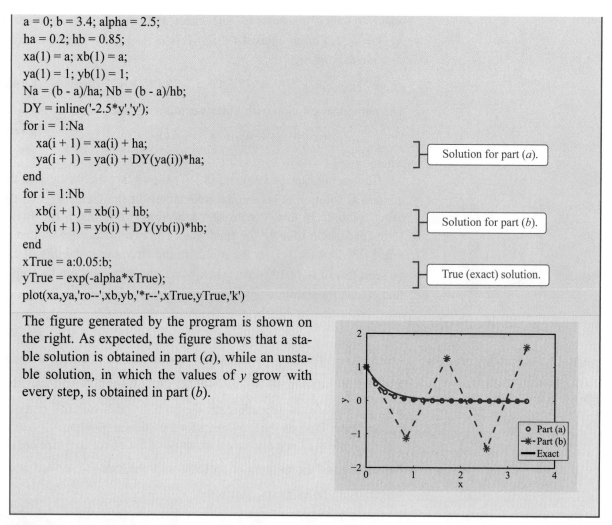

```
a = 0; b = 3.4; alpha = 2.5;
ha = 0.2; hb = 0.85;
xa(1) = a; xb(1) = a;
ya(1) = 1; yb(1) = 1;
Na = (b - a)/ha; Nb = (b - a)/hb;
DY = inline('-2.5*y','y');
for i = 1:Na
    xa(i + 1) = xa(i) + ha;
    ya(i + 1) = ya(i) + DY(ya(i))*ha;
end
for i = 1:Nb
    xb(i + 1) = xb(i) + hb;
    yb(i + 1) = yb(i) + DY(yb(i))*hb;
end
xTrue = a:0.05:b;
yTrue = exp(-alpha*xTrue);
plot(xa,ya,'ro--',xb,yb,'*r--',xTrue,yTrue,'k')
```

Solution for part (*a*).

Solution for part (*b*).

True (exact) solution.

The figure generated by the program is shown on the right. As expected, the figure shows that a stable solution is obtained in part (*a*), while an unstable solution, in which the values of y grow with every step, is obtained in part (*b*).

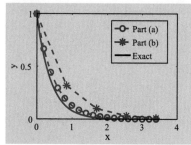

Figure 8-14: Solution of Example 8-13 with Euler's implicit method.

In contrast to Euler's explicit method, Euler's implicit method is unconditionally stable. This means that truncation errors are not magnified as the numerical solution proceeds, regardless of the step size. The solution of Example 8-13 with Euler's implicit method (see Problem 8.17) is shown in Fig. 8-14. Note that although the numerical solution becomes less accurate as h is increased, it does not become unstable. In general, stability criteria such as Eq. (8.198) cannot be derived for ODEs of arbitrary complexity. Nevertheless, stability issues such as those depicted in the figure in Example 8-13 can be observed as the step size is varied.

8.14 STIFF ORDINARY DIFFERENTIAL EQUATIONS

Certain applications in science and engineering involve competing physical phenomena with widely different time scales or spatial (length) scales. For example, widely varying time scales are encountered in problems involving combustion, chemical reactions, electronic networking, and control. These applications frequently lead to systems of ODEs whose solutions include several terms with magnitudes varying with time at a significantly different rate. Such ODE systems are called stiff and are difficult to solve. For example, the solution of a stiff ODE can have sums or differences of exponential terms such as e^{-at} and e^{-bt} (t is the independent variable), where there is a large difference between the magnitude of a and b.

As an illustrative example of stiff equations, consider the following system of two ODEs:

$$\frac{dx}{dt} = 998x - 1998y \tag{8.200}$$

$$\frac{dy}{dt} = 1000x - 2000y \tag{8.201}$$

with $x(0) = 1$ and $y(0) = 2$. An analytical solution to the system can be obtained by subtracting the two ODEs and treating $(x-y)$ as the dependent variable to yield:

$$y - x = e^{-2t} \tag{8.202}$$

Substituting for y from Eq. (8.202) into Eq. (8.200) yields:

$$\frac{dx}{dt} = 998x - 1998(x + e^{-2t}) = -1000x - 1998e^{-2t} \tag{8.203}$$

which has the solution:

$$x(t) = -\frac{1998}{998}e^{-2t} + \frac{2996}{998}e^{-1000t} \tag{8.204}$$

and from (8.200), y is:

$$y(t) = x(t) + e^{-2t} = -\frac{1000}{998}e^{-2t} + \frac{2996}{998}e^{-1000t} \tag{8.205}$$

Evaluating x and y at $t = 0.1$, the exact answers are: $x(0.1) = -1.63910225$, and $y(0.1) = -0.82037150$. Note that in the solutions (8.202) and (8.203), the terms containing e^{-1000t} contribute negligibly to the answer. However, it is precisely these terms that force taking small step sizes for accuracy, when the equations are solved numerically.

Using Euler's explicit method for solving the system of Eqs. (8.200) and (8.201) with a step size of $h = 0.1$ gives the following results for $x(0.1)$ and $y(0.1)$, which are the values of the dependent variables after the first step:

$$x(0.1) = x_1 + (998x_1 - 1998y_1)h = 1 + (998 \cdot 1 - (1998 \cdot 2))0.1 = -298.8$$

$$y(0.1) = y_1 + (1000x_1 - 2000y_1)h = 2 + (1000 \cdot 1 - (2000 \cdot 2))0.1 = -298$$

Clearly, these answers of $x(0.1) = -298.8$ and $y(0.1) = -298$ are completely wrong. As discussed in Section 8.13, stability considerations require that a step size smaller than a critical value must be used in order to get an accurate answer. In this problem, step sizes $h = 0.1$, or smaller, must be chosen. For instance, if a step size of $h = 0.0001$ is used, the numerical solutions for the first step are $x(0.1) = -1.63906946$ and $y(0.1) = -0.820355090$, which are close to the exact solution. As smaller and smaller step sizes are used, the answers will only get slightly more accurate until finally round-off errors begin to grow. Thus, it can be seen that there is a limit to the accuracy that can be attained for stiff equations when an explicit method such as the Euler's method is used.

Solution with implicit methods can give more accurate results for stiff equations. Using Euler's explicit method for solving the system of Eqs. (8.200) and (8.201) with a step size of $h = 0.1$ gives the following equations for the first step:

$$x_2 = x_1 + (998x_2 - 1998y_2)h = 1 + (998x_2 - 1998y_2)0.1$$
$$y_2 = y_1 + (1000x_2 - 2000y_2)h = 2 + (1000x_2 - 2000y_2)0.1$$

which leads to the following system of two simultaneous equations:

$$\begin{bmatrix} 98.8 & -199.8 \\ 100 & -201 \end{bmatrix} \begin{bmatrix} x_2 \\ y_2 \end{bmatrix} = \begin{bmatrix} -1 \\ -2 \end{bmatrix} \tag{8.206}$$

Solving the system in Eq. (8.206) for x_2 and y_2, gives the solution of the ODEs at the end of the first step: $x(0.1) = -1.63861386$ and $y(0.1) = -0.80528053$. This is far more accurate than the answers obtained with Euler's explicit method for a comparable step size. These results, summarized in Table 8-3, show that when solving stiff ODEs, the step size that must be used depends essentially on the stability of the numerical method used, and not on the desired accuracy.

Table 8-3: Solutions of a system of two stiff ODEs.

	Exact solution	Euler's explicit method $h = 0.1$.	Euler's explicit method $h = 0.0001$	Euler's implicit method $h = 0.1$
$x(0.1)$	-1.63910225	-298.8	-1.63906946	-1.63861386
$y(0.1)$	-0.82037150	-298	-0.82035509	-0.80528053

Since Euler's implicit method is unconditionally stable, it is very effective in solving stiff ODEs compared to Euler's explicit method, which is constrained by stability limits.

MATLAB has several built-in ODE solvers specifically intended for stiff systems. These are usually identified by a suffix s that appears after the solver's name. For example, ode15s and ode23s are examples of stiff solvers. In addition, ode23t and ode23tb can also be used for moderately stiff ODEs.

In summary, stiff ODEs are ODEs that have widely varying time (or length) scales. In the course of solving an initial value problem with a set of stiff ODEs, explicit methods require vanishingly small step sizes based on stability requirements so that it is (1) either impractical to obtain a solution because it takes so long to solve the problem or (2) the accuracy of the answer is limited because the round-off error grows with ever smaller step sizes. In contrast, implicit methods are free of stability constraints and are the preferred methods for solving stiff ODEs. Predictor–corrector methods can also be useful in solving some moderately stiff problems.

8.15 PROBLEMS

Problems to be solved by hand
Solve the following problems by hand. When needed, use a calculator, or write a MATLAB script file to carry out the calculations. If using MATLAB, do not use built-in functions for numerical solutions

8.1   Consider the following first-order ODE:
$$\frac{dy}{dx} = yx - x^3 \quad \text{from } x = 0 \text{ to } x = 1.8 \text{ with } y(0) = 1$$
(*a*) Solve by hand with Euler's explicit method using $h = 0.6$.
(*b*) Solve by hand with the modified Euler method using $h = 0.6$.
(*c*) Solve by hand with the classical fourth-order Runge–Kutta method using $h = 0.6$.

The analytical solution of the ODE is: $y = x^2 - e^{\frac{1}{2}x^2} + 2$. In each part, calculate the error between the true solution and the numerical solution at the points where the numerical solution is determined.

8.2   Consider the following first-order ODE:
$$\frac{dy}{dx} = x - y \quad \text{from } x = 0 \text{ to } x = 1.5 \text{ with } y(0) = 1$$
(*a*) Solve by hand with Euler's explicit method using $h = 0.5$.
(*b*) Solve by hand with the modified Euler method using $h = 0.5$.
(*c*) Solve by hand with the classical fourth-order Runge–Kutta method using $h = 0.5$.
The analytical solution of the ODE is: $y = x + 2e^{-x} - 1$. In each part, calculate the error between the true solution and the numerical solution at the points where the numerical solution is determined.

8.3   Consider the following first-order ODE:
$$\frac{dy}{dt} = y + t^3 \quad \text{from } t = 0.5 \text{ to } t = 2 \text{ with } y(0.5) = -1$$
(*a*) Solve by hand with Euler's explicit method using $h = 0.5$.
(*b*) Solve by hand with the midpoint method using $h = 0.5$.
(*c*) Solve by hand with the classical fourth-order Runge–Kutta method using $h = 0.5$.
The analytical solution of the ODE is: $y = -t^3 - 3t^2 - 6t - 6 + (71e^t)/(8e^{0.5})$. In each part, calculate the error between the true solution and the numerical solution at the points where the numerical solution is determined.

8.4   Consider the following system of two ODEs:

$$\frac{dx}{dt} = x+y \qquad \frac{dy}{dt} = y-x \quad \text{from } t = 0 \text{ to } t = 2 \text{ with } x(0) = 1 \text{, and } y(0) = 3.$$

(a) Solve by hand with Euler's explicit method using $h = 0.5$.

(b) Solve by hand with the modified Euler method using $h = 0.5$.

The analytical solution of the system is: $x = e^t[3\sin(t) + \cos(t)]$, $y = e^t[3\cos(t) - \sin(t)]$. In each part, calculate the error between the true solution and the numerical solution at the points where the numerical solution is determined.

8.5   Consider the following system of two ODEs:

$$\frac{dx}{dt} = xt - y \qquad \frac{dy}{dt} = yt + x \qquad \text{from } t = 0 \text{ to } t = 1.2 \text{ with } x(0) = 1 \text{, and } y(0) = 0.5.$$

(a) Solve by hand with Euler's explicit method using $h = 0.4$.

(b) Solve by hand with the classical fourth-order Runge–Kutta method using $h = 0.4$.

The analytical solution of the system is: $x = e^{\frac{1}{2}t^2}\left(\cos t - \frac{1}{2}\sin t\right)$, $y = \left(-e^{\frac{1}{2}t^2}\right)\left(-\sin t - \frac{1}{2}\cos t\right)$. In each part, calculate the error between the true solution and the numerical solution at the points where the numerical solution is determined.

8.6   Write the following second-order ODEs as a system of two first-order ODEs:

(a) $EI\dfrac{d^2y}{dx^2} = -Py + \dfrac{QL}{2}x - \dfrac{Q}{2}x^2$, where E, I, P, Q, and L are constants.

(b) $EI\dfrac{d^2y}{dx^2} = M\left[1 + \left(\dfrac{dy}{dx}\right)^2\right]^{3/2}$, where E, I, and M are constants.

8.7   Write the following second-order ODEs as a system of two first-order ODEs:

(a) $\dfrac{1}{g}\dfrac{d^2h}{dt^2} = \dfrac{T}{w} - 1 - \dfrac{0.008}{w}\left(\dfrac{dh}{dt}\right)^2$, where g, T, and w are constants.

(b) $\dfrac{d^2Q}{dt^2} + \dfrac{500}{15}\dfrac{dQ}{dt} + \dfrac{250}{15}\left(\dfrac{dQ}{dt}\right)^3 + \dfrac{Q}{15 \cdot 4.2 \times 10^{-6}} = \dfrac{1000}{15}$.

8.8   Write the following system of two second-order ODEs as a system of four first-order ODEs:

$$\frac{d^2x}{dt^2} = -\frac{\gamma}{m}\left(\frac{dx}{dt}\right)\sqrt{\left(\frac{dx}{dt}\right)^2 + \left(\frac{dy}{dt}\right)^2} \qquad \frac{d^2y}{dt^2} = -g - \frac{\gamma}{m}\left(\frac{dy}{dt}\right)\sqrt{\left(\frac{dx}{dt}\right)^2 + \left(\frac{dy}{dt}\right)^2}$$

Problems to be programmed in MATLAB
Solve the following problems using the MATLAB environment. Do not use MATLAB's built-in functions for solving differential equations.

8.9   Use the user-defined MATLAB function `odeEULER` (Program 8-1 which is listed in Example 8-1) to solve Problem 8.1. Write a MATLAB program is a script file that solves the ODE in Problem 8.1 three times. Once by using $h = 0.6$, once by using $h = 0.3$, and once by using $h = 0.1$. The program should also plot the exact solution (given in Problem 8.1) and plot the three numerical solutions (all in the same figure).

8.10 Use the user-defined MATLAB function `odeEULER` (Program 8-1 which is listed in Example 8-1) to solve Problem 8.2. Write a MATLAB program is a script file that solves the ODE in Problem 8.2 three times. Once by using $h = 0.5$, once by using $h = 0.25$, and once by using $h = 0.1$. The program should also plot the exact solution (given in Problem 8.2) and plot the three numerical solutions (all in the same figure).

8.11 Write a user-defined MATLAB function that solves a first-order ODE by applying the midpoint method (use the form of second-order Runge–Kutta method, Eqs. (8.65), (8.66)). For function name and arguments use `[x,y]=odeMIDPOINT(ODE,a,b,h,yINI)`, where the input argument `ODE` is the name (string) of the user-defined function that calculates $\frac{dy}{dx}$, a and b define the domain of the solution, h is the step size, and `yINI` is the initial value. The output arguments, x and y, are vectors with the x and y coordinates of the solution.

Use the function `odeMIDPOINT` to solve the ODE in Problem 8.2. Write a MATLAB program in a script file that solves the ODE twice, once by using $h = 0.5$ and once by using $h = 0.1$. The program should also plot the exact solution (given in Problem 8.2) and plot the two numerical solutions (all in the same figure).

8.12 Write a user-defined MATLAB function that solves a first-order ODE by applying the classical third-order Runge–Kutta method, Eqs. (8.82), (8.83). For function name and arguments use `[x,y]=odeRK3(ODE,a,b,h,yINI)`, where the input argument `ODE` is the name (string) of the user-defined function that calculates $\frac{dy}{dx}$, a and b define the domain of the solution, h is the step size, and `yINI` is the initial value. The output arguments, x and y, are vectors with the x and y coordinates of the solution.

Use the function `odeRK3` to solve Problem 8.2. Write a MATLAB program in a script file that solves the ODE in Problem 8.2 twice, once by using $h = 0.5$ and once by using $h = 0.1$. The program should also plot the exact solution (given in Problem 8.2) and plot the two numerical solutions (all in the same figure).

8.13 Write a user-defined MATLAB function that solves a first-order ODE by using the second-order Adams–Bashforth method. The function should use the modified Euler method for calculating the solution at the second point, and Eq. (8.95) for calculating the solution at the rest of the points. For function name and arguments use `[x,y]=odeAdams2(ODE,a,b,h,yINI)`, where the input argument `ODE` is the name (string) of the user-defined function that calculates $\frac{dy}{dx}$, a and b define the domain of the solution, h is the step size, and yINI is the initial value. The output arguments, x and y, are vectors with the x and y coordinates of the solution.

Use the function odeAdams2 to solve the ODE in Problem 8.3. Write a MATLAB program in a script file that solves the ODE by using $h = 0.1$. The program should also plot the exact solution (given in Problem 8.3) and the numerical solution (both in the same figure).

8.14 Write a user-defined MATLAB function that solves a first-order ODE by using the modified Euler predictor–corrector method (see algorithm in Section 8.7). In each step the iterations should continue until the estimated relative error is smaller than 0.001, $\left| \dfrac{y^{(k)}_{i+1} - y^{(k-1)}_{i+1}}{y^{(k-1)}_{i+1}} \right| \le \varepsilon = 0.001$. For function name and arguments use [x,y]=odeEulerPreCor(ODE,a,b,h,yINI), where the input argument ODE is the name (string) of the user-defined function that calculates $\dfrac{dy}{dx}$, a and b define the domain of the solution, h is the step size, and yINI is the initial value. The output arguments, x and y, are vectors with the x and y coordinates of the solution.

Use the function odeAdams2 to solve the ODE in Problem 8.3. Write a MATLAB program in a script file that solves the ODE by using $h = 0.1$. The program should also plot the exact solution (given in Problem 8.3) and the numerical solution (both in the same figure).

8.15 The user-defined MATLAB function Sys2ODEsRK2(ODE1,ODE2,a,b,h,yINI,zINI) (Program 8-5), that is listed in the solution of Example 8-7, solves a system of two ODEs by using the second-order Runge–Kutta method (modified Euler's version). Modify the function such that the two ODEs are entered in one input argument. Similarly, the domain should be entered by using one input argument, and the two initial conditions entered in one input argument. For function name and arguments use [t,x,y]=Sys2ODEsModEu(ODEs,ab,h,INI). The input argument ODE is the name (string) of the user-defined function that calculates $\dfrac{dx}{dt}$ and $\dfrac{dy}{dt}$ (the input arguments of this user-defined function are t,x,y, and the output argument is a two-element vector with the values of $\dfrac{dx}{dt}$ and $\dfrac{dy}{dt}$). ab is a two-element vector that defines the domain of the solution, h is the step size, and INI is a two-element vector with the initial values. The output arguments, t, x, and y, are vectors of the solution.

Use the function Sys2ODEsModEu to solve the system of ODEs in Problem 8.5. Write a MATLAB program in a script file that solves the system by using $h = 0.1$. The program should also plot the exact solution (given in Problem 8.5) and the numerical solution (both in the same figure).

8.16 The user-defined MATLAB function Sys2ODEsRK4(ODE1,ODE2,a,b,h,x1,y1) (Program 8-6) that is listed in the solution of Example 8-8 solves a system of two ODEs by using fourth-order Runge–Kutta method. Modify the function such that the two ODEs are entered in one input argument. Similarly, the domain is entered by using one input argument, and the two initial conditions are entered in one input argument. For function name and arguments use [t,x,y]=Sys2ODEsRKclas(ODEs,ab,h,INI). The input argument ODE is the name (string) of the user-defined function that calculates $\dfrac{dx}{dt}$ and $\dfrac{dy}{dt}$ (the input arguments of this user-defined function are t,x,y, and the output argument is a two-element vector with the values of $\dfrac{dx}{dt}$ and $\dfrac{dy}{dt}$). ab is a two-element vector that defines the domain of the solution, h is the step size, and INI is a two-element vector with the initial values. The output arguments, t, x, and y, are vectors of the solution.

Use the function Sys2ODEsRKclas to solve the system of ODEs in Problem 8.5. Write a MAT-LAB program in a script file that solves the system by using $h = 0.1$. The program should also plot the exact solution and the numerical solution (both in the same figure).

8.17 Write a program in a script file that solves the IVP in Example 8-13, using Euler's implicit method with $h = 0.1$. Plot the numerical and analytical solutions.

Problems in math, science, and engineering
Solve the following problems using the MATLAB environment. As stated, use the MATLAB programs that are presented in the chapter, programs developed in previously solved problems, or MATLAB's built-in functions.

8.18 An inductor and a nonlinear resistor of resistance $R = 500 + 250I^2 \, \Omega$ are connected in series with a DC power source and a switch, as shown in the figure. The switch is initially open and then is closed at time $t = 0$. The current I in the circuit for $t > 0$ is determined from the solution of the equation:

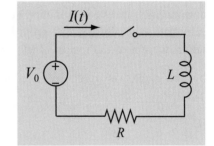

$$\frac{dI}{dt} = \frac{V_0}{L} - \frac{R}{L}I$$

For $V_0 = 1000$ V and $L = 15$ Henries, determine and plot the current as a function of time for $0 \geq t \geq 0.1$ s.
(*a*) Solve the problem with the function odeMIDPOINT that was written in Problem 8.11. For step size use 0.005 s.
(*b*) Solve the problem using one of MATLAB's built-in functions for solving an ODE.

8.19 Consider the cylindrical water tank that is shown in Fig. 8-1 (shown also on the right). The tank is being filled at the top, and water flows out of the tank through a pipe that is connected at the bottom. The rate of change of the height, h, of the water is given by Eq. (8.6):

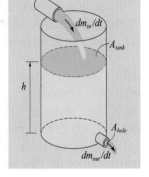

$$\rho A_{tank}\frac{dh}{dt} = K_1 + K_2\cos\left(\frac{\pi}{12}t\right) - \rho A_{pipe}\sqrt{2gh}$$

For the given tank, $A_{tank} = 3.13 \, \text{m}^2$, $A_{pipe} = 0.06 \, \text{m}^2$, $K_1 = 300 \, \text{kg/h}$, $K_2 = 200 \, \text{kg/h}$. Also, $\rho = 1000 \, \text{kg/m}^3$, and $g = 9.81 \, \text{m/s}^2$. Determine, and plot the height of the water as a function of time for $0 \leq t \leq 150$ s, if at $t = 0$, $h = 3$ m.
(*a*) Use the user-defined function odeRK3 that was written in Problem 8.12. For step size use 0.1s.
(*b*) Use one of MATLAB's built-in functions for solving an ODE.

8.20 A spherical water tank of radius $R = 4$ m is emptied through a small circular hole of radius $r = 0.02$ m at the bottom. The top of the tank is open to the atmosphere. The instantaneous water level, h (measured from the bottom of the tank, at the drain), in the tank can be determined from the solution of the following ODE:

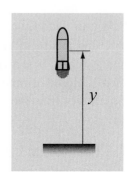

$$\frac{dh}{dt} = -\frac{r^2\sqrt{2gh}}{2hR - h^2}$$

where $g = 9.81$ m/s^2. If the initial ($t = 0$) water level is $h = 6$ m, find the time required to drain the tank to a level of $h = 0.5$ m.

(*a*) Use the fourth-order Runge–Kutta method (use the user-defined function `odeRK` (Program 8-4) that was developed in Example 8-6).

(*b*) Use one of MATLAB's built-in functions for solving an ODE.

8.21 A small rocket having an initial weight of 3000 lb (including 2400 lb of fuel) and initially at rest, is launched vertically upward. The rocket burns fuel at a constant rate of 80 lb/s, which provides a constant thrust, T, of 7000 lb. The instantaneous weight of the rocket is $w(t) = 3000 - 80t$ lb. The drag force, D, experienced by the rocket is given by $D = 0.008g\left(\dfrac{dy}{dt}\right)^2$ lb, where y is distance in ft, and $g = 32.2$ ft/s^2. Using Newton's law, the equation of motion for the rocket is given by:

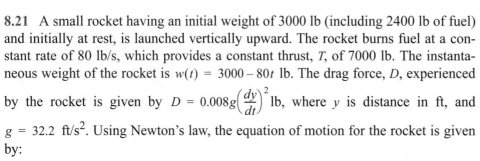

$$\frac{w}{g}\frac{d^2y}{dt^2} = T - w - D$$

Determine and plot the position, velocity, and acceleration of the rocket (three separate figures on one page) as a function of time from $t = 0$ when the rocket start moving upward from rest, until $t = 3$. Reduce the second-order ODE to a system of two first-order ODEs.

(*a*) Use the fourth-order Runge–Kutta method. Use either the user-defined function `Sys2ODEsRK4` (Program 8-6) that was developed in Example 8-8, or the user-defined function `Sys2ODEsRKclas` that was written in Problem 8.16. For step size use 0.05 s.

(*b*) Use one of MATLAB's built-in functions for solving ODEs.

8.22 A U-tube manometer (used to measure pressure) is initially filled with water, but is exposed to a pressure difference such that the water level on the left side of the U-tube is 0.05 m higher than the water level on the water level on the right. At $t = 0$ the pressure difference is suddenly removed. When friction is neglected, the height of the water level on the left side, y, measured from the midplane between the two initial water levels is given by the solution of the equation:

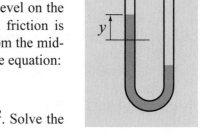

$$L\frac{d^2y}{dt^2} = -2gy$$

where $L = 0.2$ m is the total length of the U-tube, and $g = 9.81$ m/s^2. Solve the ODE and plot y, and $\dfrac{dy}{dt}$ (two separate figures on one page) as a function of t for the first 10 seconds. Reduce the second-order ODE to a system of two first-order ODEs and solve the system by:

(a) using the user-defined function `Sys2ODEsModEu` that was written in Problem 8.15. For step size use 0.02 s.

(b) using one of MATLAB's built-in functions for solving ODEs.

8.23 If the effect of friction is included in the analysis of the U-tube manometer in Problem 8.22, the height of the water level, y, is given by the solution of the equation:

$$L\frac{d^2y}{dt^2} = -0.5\frac{dy}{dt} - 2gy$$

Solve the ODE and plot y, and $\frac{dy}{dt}$ (two separate figures on one page) as a function of t for the first 10 seconds. Reduce the second-order ODE to a system of two first-order ODEs and solve the system by:

(a) using the user-defined function `Sys2ODEsRKclas` that was written in Problem 8.16. For step size use 0.02 s.

(b) using one of MATLAB's built-in functions for solving ODEs.

8.24 Consider forced vibration of a mass–spring system that is shown in the figure. The position x of the mass as a function of time is given by the solution of the equation:

$$\frac{d^2x}{dt^2} = -\frac{k}{m}x + \frac{F_0}{m}\cos\omega t$$

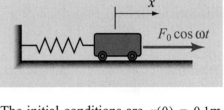

where $m = 2$ kg is the mass, $k = 800$ N/m is the spring constant, $F_0 = 50$ N is the amplitude of the applied harmonic force, and $\omega = 3$ rad/s is the frequency of the applied harmonic force. The initial conditions are $x(0) = 0.1$m, $\frac{dx}{dt}\Big|_{t=0} = 0.1$ m/s. Solve the ODE for $0 \le t \le 10$ s, and plot x and $\frac{dx}{dt}$ as a function of t (two separate plots on the same page).

(a) Use the user-defined function `Sys2ODEsRKclas` that was written in Problem 8.16. For step size use 0.01 s.

(b) Use one of MATLAB's built-in functions for solving ODEs.

8.25 A capacitor of $C = 4.2 \times 10^{-6}$ Farads is added in series to the circuit in Problem 8.18. As shown in the figure, the circuit includes a DC voltage source, $V_0 = 1000$ V, an inductor, $L = 15$ Henries, and a nonlinear resistor of resistance $R = R_0 + R_1 I^2\ \Omega$, where $R_0 = 500\Omega$ and $R_1 = 250\Omega/A^2$. The switch is initially open and then is closed at time $t = 0$. The charge, Q, in the capacitor for $t > 0$ is determined from the solution of the equation:

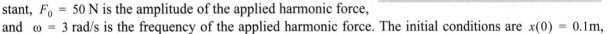

$$\frac{d^2Q}{dt^2} + \frac{R_0}{L}\frac{dQ}{dt} + \frac{R_1}{L}\left(\frac{dQ}{dt}\right)^3 + \frac{Q}{LC} = \frac{V_0}{L}$$

Initially, $Q = 0$ and $\frac{dQ}{dt} = 0$.

(a) Reduce the second-order ODE to a system of two first-order ODEs, and determine the charge, Q, as a function of time for $0 \ge t \ge 0.3$s by solving the system using the fourth-order Runge–Kutta method.

Use either the user-defined function `Sys2ODEsRK4` (Program 8-6) that was developed in Example 8-8, or the user-defined function `Sys2ODEsRKclas` that was written in Problem 8.16. Use $h = 0.002\,\text{s}$, and plot of Q versus time.

(b) Use the results from part (a) to plot the current in the circuit. The current is given by the derivative w.r.t. time of the charge, $I = \dfrac{dQ}{dt}$.

(c) Solve the problem (parts (a) and (b)) using MATLAB's built-in functions.

8.26 A paratrooper jumps from an aircraft in straight and level flight. The motion of the paratrooper is approximately described by the following system of equations:

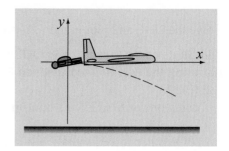

$$\frac{d^2x}{dt^2} = -\frac{\gamma}{m}\left(\frac{dx}{dt}\right)\sqrt{\left(\frac{dx}{dt}\right)^2 + \left(\frac{dy}{dt}\right)^2}$$

$$\frac{d^2y}{dt^2} = -g - \frac{\gamma}{m}\left(\frac{dy}{dt}\right)\sqrt{\left(\frac{dx}{dt}\right)^2 + \left(\frac{dy}{dt}\right)^2}$$

where x and y are the paratrooper's position according to the coordinate system shown in the figure. For $m = 80\,\text{kg}$, $g = 9.81\ \text{m/s}^2$, $\gamma = 5.38\ \text{Ns}^2/\text{m}^2$, and initial conditions:

$$x(0) = 0, \quad y(0) = 0, \quad \left.\frac{dx}{dt}\right|_{t\,=\,0} = 134\,\text{m/s}, \quad \left.\frac{dy}{dt}\right|_{t\,=\,0} = 0$$

Determine and plot the trajectory of the paratrooper for the first 5s. Reduce the system of two second-order ODE to a system of four first-order ODEs and solve using MATLAB's built-in functions.

Chapter 9

Ordinary Differential Equations: Boundary-Value Problems

Core Topics

The shooting method (9.2).

The finite difference method (9.3).

Use of MATLAB built-in functions for solving boundary value ODEs (9.4).

Complementary Topics

Error and stability in numerical solution of boundary value problems (9.5).

9.1 BACKGROUND

A specific solution of a differential equation can be determined if the constraints of the problem are known. A first-order ODE can be solved if one constraint, the value of the dependent variable (initial value) at one point is known. To solve an nth-order equation, n constraints must be known. The constraints can be the value of the dependent variable (solution) and its derivative(s) at certain values of the independent variable. When all the constraints are specified at one value of the independent variable, the problem is called an initial value problem (IVP). Solution of initial value problems is discussed in Chapter 8. In many cases there is a need to solve differential equations of second and higher order that have constraints specified at different values of the independent variable. These problems are called **boundary value problems** (BVP), and the constraints are called boundary conditions because the constraints are often specified at the endpoints or boundaries of the domain of the solution.

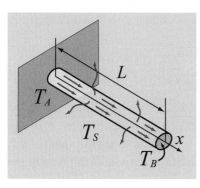

Figure 9-1: Heat flow in a pin fin.

As an example, consider the modeling of temperature distribution in a pin fin used as a heat sink for cooling an object (Fig. 9-1). If convection and radiation are included in the analysis, the steady-state temperature distribution, $T(x)$, along the fin can be obtained from the solution of an equation of the form:

$$\frac{d^2T}{dx^2} - \alpha_1(T - T_S) - \alpha_2(T^4 - T_S^4) = 0 \qquad (9.1)$$

where T_S is the temperature of the surrounding air, and α_1 and α_2 are coefficients. Equation (9.1) is a second-order ODE and can be solved once two boundary conditions are specified. Two such boundary conditions can be the temperatures at the ends of the fin, T_A and T_B.

Problem statement of a second-order boundary value problem

With an independent variable x and dependent variable y, a second-order boundary value problem statement consists of a differential equation:

$$\frac{d^2 y}{dx^2} = f\left(x, y, \frac{dy}{dx}\right) \tag{9.2}$$

a domain of the solution $a \le x \le b$, and the boundary conditions. The two boundary conditions required for a solution are typically given at the endpoints of the domain. Since a boundary condition can be a value of y, or a value of the derivative $\frac{dy}{dx}$, the boundary conditions can be specified in different ways. Common forms of boundary conditions are:

- Two values of y are given—one at $x = a$ and one at $x = b$:

$$y(a) = Y_a \quad \text{and} \quad y(b) = Y_b \tag{9.3}$$

These conditions are called Dirichlet boundary conditions. A second-order ODE with these boundary conditions is called a two-point BVP.

- Two values of $\frac{dy}{dx}$ are given—one at $x = a$ and one at $x = b$:

$$\frac{dy}{dx}\bigg|_{x = a} = D_a \quad \text{and} \quad \frac{dy}{dx}\bigg|_{x = b} = D_b \tag{9.4}$$

These conditions are called Neumann boundary conditions.

- The third possibility involves mixed boundary conditions, which can be written in the form:

$$c_1 \frac{dy}{dx}\bigg|_{x = a} + c_2 y(a) = C_a \quad \text{and} \quad c_3 \frac{dy}{dx}\bigg|_{x = b} + c_4 y(b) = C_b \tag{9.5}$$

where c_1, c_2, c_3, and c_4 are constants. These conditions are called mixed boundary conditions. Special cases are when $c_1 = 0$ and $c_4 = 0$, or when $c_2 = 0$ and $c_3 = 0$. In these cases the value of y is given at one endpoint, and the value of the derivative is given at the other endpoint.

It is also possible to have nonlinear boundary conditions where D_a, D_b, c_1, c_2, C_a, or C_b are nonlinear functions of y or its derivatives.

Boundary value problems with ODEs of order higher than second-order require additional boundary conditions, which are typically the values of higher derivatives of y. For example, the differential equation that relates the deflection of a beam, y, due to the application of a distributed load, $p(x)$, is:

$$\frac{d^4 y}{dx^4} = \frac{1}{EI} p(x) \tag{9.6}$$

where E and I are the elastic modulus of the beam's material and the area moment of inertia of the beam's cross-sectional area, respectively. This fourth-order ODE can be solved if four boundary conditions are specified. The boundary conditions depend on the way that the beam is supported. The beam shown in Fig. 9-2 is clamped at both ends, which means that the deflection and slope of the deflection curve are equal to zero at the ends. This gives the following four boundary conditions:

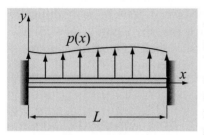

Figure 9-2: Clamped beam under distributed load.

$$y(0) = 0; \qquad \left.\frac{dy}{dx}\right|_{x=0} = 0$$

$$y(L) = 0; \qquad \left.\frac{dy}{dx}\right|_{x=L} = 0 \tag{9.7}$$

Overview of numerical methods used for solving boundary value problems

This chapter presents two approaches for solving boundary value problems: **shooting methods** and **finite differences methods**. Shooting methods reduce the second-order (or higher order) ordinary differential equation to an initial value problem. This is done, as was explained in Section 8.9, by transforming the equation into a system of first-order ODEs. The boundary value at the first point of the domain is known and is used as one initial value for the system. The additional initial values needed for solving the system are guessed. The system is then solved, and the solution at the end of the interval is compared with the specified boundary condition(s) there. If the two disagree to a required accuracy, the guessed initial values are changed and the system is solved again. The calculations are repeated until all the specified boundary conditions are satisfied. The difference between various shooting methods is in the way that the values of the assumed initial conditions are modified after each calculation.

In finite difference methods, the derivatives in the differential equation are approximated with finite difference formulas (see Chapter 6). The domain of the solution is divided into N subintervals that are defined by $(N+1)$ points (mesh points), and the differential equation is approximated at each mesh point of the domain. This results in a system of linear (or nonlinear) algebraic equations. The solution of the system is the numerical solution of the differential equation. The difference between various finite difference methods is in the finite difference formulas used for approximating the differential equation.

Both approaches have their respective advantages and disadvantages. In the finite difference methods, there is no need to solve the differential equation several times in order to match the prescribed boundary conditions at the endpoint of the domain. On the other hand, the solution of nonlinear ODEs using finite difference methods results in the need to solve a system of simultaneous nonlinear equations (usually iteratively), which can be tedious and fraught with difficulty. Shooting methods have the advantage that the solution of nonlinear ODEs is fairly straightforward. The disadvantage of the shooting methods is that the ODE has to be solved several times.

9.2 THE SHOOTING METHOD

In the shooting method a boundary value problem (BVP) is transformed into a system of initial value problems (IVPs). A BVP involving an ODE of second-order can be transformed, as was described in Section 8.9, into a system of two first-order ODEs. This system of ODEs can be solved numerically if the initial condition for each ODE is known. A problem statement of a BVP with an nth-order ODE includes n boundary conditions, where some of the boundary conditions are given at the first point of the domain and others at the end-point. When the nth-order ODE is transformed to a system of n first-order ODEs, the boundary conditions given at the first point of the domain are used as initial conditions for the system. The additional initial conditions required for solving the system are guessed. The system is then solved, and the solution obtained at the end point of the domain is compared with the boundary conditions there. If the numerical solution is not accurate enough, the guessed initial values are changed, and the system is solved again. This process is repeated until the numerical solution agrees with the prescribed boundary condition(s) at the endpoint of the domain. A detailed description of the shooting method for the case of a second-order BVP is given next.

Shooting method for a two-point BVP

Consider a BVP with a second-order ODE where the boundary conditions are the values of the dependent variable at the endpoints:

$$\frac{d^2y}{dx^2} = f\left(x, y, \frac{dy}{dx}\right) \quad \text{for} \quad a \leq x \leq b \quad \text{with} \quad y(a) = Y_a \quad \text{and} \quad y(b) = Y_b \quad (9.8)$$

A numerical solution with the shooting method can be obtained by the following procedure:

Step 1: The ODE is transformed into a system of two first-order IVPs (see Section 8.9). The two equations have the form:

$$\frac{dy}{dx} = w \quad \text{with the initial condition:} \quad y(a) = Y_a \quad (9.9)$$

and

$$\frac{dw}{dx} = f(x, y, w) \qquad (9.10)$$

The problem statement does not include an initial condition for Eq. (9.10).

Step 2: A first estimate (guess) is made for the initial value of Eq. (9.10):

$$w(a) = \left.\frac{dy}{dx}\right|_{x=a} = W_1 \qquad (9.11)$$

This is actually a guess for the slope at $x = a$. With this estimate, the system of Eqs. (9.9) and (9.10) is solved numerically. The numerical solution at $x = b$ (the end of the interval) is y_{b1}, as shown in Fig. 9-3. If the numerical solution is close enough to the boundary condition Y_b (i.e., the error between the two is acceptable), then a solution has been obtained. Otherwise the solution process continues in Step 3.

Step 3: A second estimate (guess) is made for the initial value of Eq. (9.10):

$$w(a) = \left.\frac{dy}{dx}\right|_{x=a} = W_2 \qquad (9.12)$$

With this estimate, the system of Eqs. (9.9) and (9.10) is solved numerically again. The numerical solution at $x = b$ (the end of the interval) is y_{b2} (Fig. 9-3). If the numerical solution is close enough to the boundary condition Y_b (i.e., the error between the two is acceptable), then a solution has been obtained. Otherwise the solution process continues in Step 4.

Step 4: A new estimate for the initial value of Eq. (9.10) is determined by using the results of the previous two solutions:

$$w(a) = \left.\frac{dy}{dx}\right|_{x=a} = W_3 \qquad (9.13)$$

Several methods can be used in this step. For example, if the value of the boundary condition, Y_b, is between y_{b1} and y_{b2} ($y_{b1} < Y_b < y_{b2}$), as illustrated in Fig. 9-3, interpolation can be used for finding a value of W_3 that is between W_1 and W_2. Additional details about how to do this and about other methods are given later in this section.

Step 5: Using $w(a) = W_3$ as the initial value in Eq. (9.10), the system of Eqs. (9.9) and (9.10) is solved numerically again. If the numerical solution at $x = b$ is equal to the boundary condition, Y_b (or the error between the two is acceptable), then a solution has been obtained. Otherwise, as was done in Step 4, a new estimate for the initial value is determined.

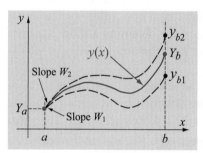

Figure 9-3: Shooting method.

Steps 4 and 5 are repeated until the numerical solution at $x = b$ agrees to a required accuracy with the boundary condition. In each calculation a new estimate for the initial condition for Eq. (9.10) is determined by using the results from the previous calculations. Several methods that can be used for this purpose are presented next.

Estimating the slope (initial value) at $x = a$

As was described earlier, the process of solving a second-order BVP with the shooting method starts by guessing two values for the slope of $y(x)$ at the first point of the domain. "Intelligent" guesses can be made in many situations when the differential equation is associated with a real application. Then, using each of the guesses, the system of equations is solved to give two solutions at the endpoint of the domain. These solutions are then used for estimating a new value for the initial slope that is used in the next calculation for obtaining a more accurate solution. The simplest method is to use linear interpolation.

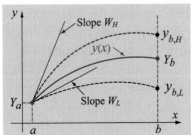

Figure 9-4: Two iterations in a solution with shooting method.

Linear interpolation: Consider two numerical solutions of a second-order BVP, shown in Fig. 9-4, that are obtained by assuming W_H and W_L for the slope (initial condition for Eq. (9.10)) at $x = a$. The solutions at $x = b$ are $y_{b,H}$ and $y_{b,L}$, respectively. Recall that the boundary condition at this point is $y(b) = Y_b$, and suppose that $y_{b,H}$ and $y_{b,L}$ have been found such that $Y_b \leq y_{b,H}$ and $y_{b,L} \leq Y_b$. Using linear interpolation, illustrated in Fig. 9-5, a new value for the slope W_N that corresponds to Y_b can be calculated by:

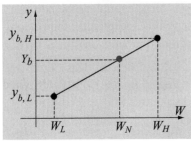

Figure 9-5: Linear interpolation for determining W_N.

$$W_N = W_L + (Y_b - y_{b,L}) \frac{W_H - W_L}{y_{b,H} - y_{b,L}} \qquad (9.14)$$

The new value W_N is used in the next calculation with a new solution at the endpoint. Then, the new solution with either $y_{b,H}$ or $y_{b,L}$ (depending on whether the new solution is above or below the boundary condition) can be used in linear interpolation for calculating the next estimate for the initial slope. Example 9-1 illustrates application of the shooting method with linear interpolation. Two initial solutions are first obtained by using two guesses for the slope at the first point of the interval. Interpolation is then used for determining a third guess for the slope, which gives a much more accurate solution when the problem is solved for the third time.

Example 9-1: Temperature distribution in a pin fin. Solving a second-order ODE (BVP) using the shooting method.

A pin fin is a slender extension attached to a surface in order to increase the surface area and enable greater heat transfer. When convection and radiation are included in the analysis, the steady-state temperature distribution, $T(x)$, along a pin fin can be calculated from the solution of the equation:

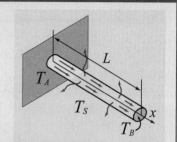

$$\frac{d^2 T}{dx^2} - \frac{h_c P}{k A_c}(T - T_S) - \frac{\varepsilon \sigma_{SB} P}{k A_c}(T^4 - T_S^4) = 0, \ 0 \le x \le L \quad (9.15)$$

with the boundary conditions: $T(0) = T_A$ and $T(L) = T_B$.

In Eq. (9.15), h_c is the convective heat transfer coefficient, P is the perimeter bounding the cross section of the fin, ε is the radiative emissivity of the surface of the fin, k is the thermal conductivity of the fin material, A_c is the cross-sectional area of the fin, T_S is the temperature of the surrounding air, and $\sigma_{SB} = 5.67 \times 10^{-8}$ W/(m^{2}K^4) is the Stefan–Boltzmann constant.

Determine the temperature distribution if $L = 0.1$ m, $T(0) = 473$ K, $T(0.1) = 293$ K, and $T_S = 293$ K. Use the following values for the parameters in Eq. (9.15): $h_c = 40$ W/m^2/K, $P = 0.016$ m^2, $\varepsilon = 0.4$, $k = 240$ W/m/K, and $A_c = 1.6 \times 10^{-5}$ m^2.

SOLUTION

Equation (9.15) is a second-order nonlinear BVP. To solve, it is transformed into a system of two first-order ODEs. The transformation is done by introducing a new variable $w = \frac{dT}{dx}$. With this definition, the system is:

$$\frac{dT}{dx} = w \quad (9.16)$$

$$\frac{dw}{dx} = \frac{h_c P}{k A_c}(T - T_S) + \frac{\varepsilon \sigma_{SB} P}{k A_c}(T^4 - T_S^4) \quad (9.17)$$

The initial value for Eq. (9.16) is $T(0) = 473$. The initial condition for Eq. (9.17) is not known.

To illustrate the shooting method, the system in Eqs. (9.16) and (9.17) is solved three times with three different values for the initial condition of Eq. (9.17). This is executed in a MATLAB program (script file) that uses the user-defined function Sys2ODEsRK2 that was created in Section 8.8.2. (The function Sys2ODEsRK2 solves a system of two first-order ODEs using the second-order Runge–Kutta method.)

The operations in the program, which is listed below, are:

- Solving the system assuming $w(0) = -1000$.

- Solving the system assuming $w(0) = -3500$.

- Using interpolation, Eq. (9.14), calculating a third value for $w(0)$ from the results of the first two solutions.

- Solving the system using the interpolated value of $w(0)$.

For each solution, the program lists the calculated temperature at the endpoint. In addition, the program displays a figure that shows the three solutions.

Program 9-1: Script file. Solving second-order ODE using the shooting method.

% Solving Chapter 9 Example 1
clear all
a = 0; b = 0.1; TINI = 473; wINI1 = -1000; h = 0.001;   Solve the system assuming $w(0) = -1000$.
[x, T1, w] = Sys2ODEsRK2('odeChap9Exmp1dTdx','odeChap9Exmp1dwdx',a,b,h,TINI,wINI1);
n = length(x);
fprintf('The temperature at x=0.1 is %5.3f, for initial value of dt/dx= %4.1f\n',T1(n),wINI1)
wINI2 = -3500;   Solve the system assuming $w(0) = -3500$.
[x, T2, w] = Sys2ODEsRK2('odeChap9Exmp1dTdx','odeChap9Exmp1dwdx',a,b,h,TINI,wINI2);
fprintf('The temperature at x=0.1 is %5.3f, for initial value of dt/dx= %4.1f\n',T2(n),wINI2)
wINI3 = wINI1 + (293 - T1(n))*(wINI2 - wINI1)/(T2(n) - T1(n));   Interpolation using Eq. (9.14).
[x, T3, w] = Sys2ODEsRK2('odeChap9Exmp1dTdx','odeChap9Exmp1dwdx',a,b,h,TINI,wINI3);
fprintf('The temperature at x = 0.1 is %5.3f, for initial value of dt/dx = %4.1f\n',T3(n),wINI3)
plot(x,T1,'-k',x,T2,'-k',x,T3,'-r')   Solve the system with the interpolated value of $w(0)$.
xlabel('Distance (m)'); ylabel('Temperature (K)')

The first two input arguments in the function Sys2ODEsRK2 are the names of user-defined functions that calculate the values of dT/dx, Eq. (9.16), and dw/dx, Eq. (9.17). The two functions, odeChap9Exmp1dTdx and odeChap9Exmp1dwdx, are:

function dTdx = odeChap9Exmp1dTdx(x,T,w)
dTdx = w;

function dwdx = odeChap9Exmp1dwdx(x,T,w)
hc = 40; P = 0.016; eps = 0.4; k = 240; Ac = 1.6E-5; Seg = 5.67E-8;
Ts = 293;
kAc = k*Ac;
A1 = hc*P/kAc; A2 = eps*Seg*P/kAc;
dwdx = A1*(T - Ts) + A2*(T^4 - Ts^4);

When the script file is executed, the following is displayed in the Command Window:

The temperature at x = 0.1 is 536.502, for initial value of dt/dx = -1000.0
The temperature at x = 0.1 is 198.431, for initial value of dt/dx = -3500.0
The temperature at x = 0.1 is 291.835, for initial value of dt/dx = -2800.7

In addition, the program displays a figure (shown on the right) with plots of the three solutions.

The results show that with the first assumption for the slope, $w(0) = -1000$, the temperature at the endpoint is higher than the prescribed boundary condition. The second assumption, $w(0) = -3500$, gives a lower value. Solution with the interpolated value of $w(0) = -2800.7$ gives $T(0.1) = 291.835$ K, which is lower, but close to prescribed boundary condition of 293 K.

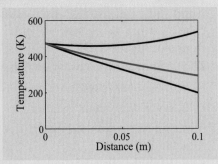

It can be expected that a more accurate solution can be obtained by executing additional calculations with better guesses for $w(0)$.

A more sophisticated method is to consider the error (difference between the numerical solution and the given boundary condition at the endpoint of the solution domain) as a function of the assumed initial slope, and use the methods from Chapter 3 for finding a solution (the zero) of this function. Using this approach with the bisection method and with the secant method is described next.

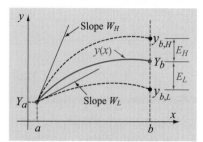

Figure 9-6: Two iterations in a solution with shooting method.

Shooting method using the bisection method: Consider the BVP in Eq. (9.8) which is solved by converting the problem into the system of two IVPs in Eqs. (9.9) and (9.10). Two numerical solutions that have been obtained in previous calculations are shown in Fig. 9-6. In one solution, $y_{b,H}$, the value at $x = b$ is larger than the boundary condition $y(b) = Y_b$, and in the other solution, $y_{b,L}$ the value is lower. The initial values of the slopes at $x = a$ that correspond to these solutions are W_H and W_L (recall that $W = \dfrac{dy}{dx}\Big|_{x=a}$). The error in each numerical solution is $E_H = y_{b,H} - Y_b$ (a positive number) and $E_L = y_{b,L} - Y_b$ (a negative number). With this notation, the error, E, can be considered as a function of the slope, W, and the objective is to find the value of W where $E = 0$.

With the bisection method, illustrated in Fig. 9-7, the initial value for the slope, W_N, in the next calculation is

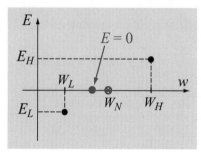

Figure 9-7: Bisection method for determining W_N.

$$W_N = \frac{1}{2}(W_H + W_L) \tag{9.18}$$

When the method is implemented in a computer code, the values of $y_{b,H}$, $y_{b,L}$, W_H, and W_L should be specified. Then after each iteration, the program should select the new values such that the boundary condition will be bounded between $y_{b,H}$ and $y_{b,L}$. Implementation of the bisection method in the shooting method is shown in Example 9-2.

Example 9-2: Temperature distribution in a pin fin. Solving a second-order ODE (BVP) using the shooting method in conjunction with the bisection method.

Write a MATLAB program (script file) that solves the BVP in Example 9-1 by using the bisection method within the shooting method for determining a new estimate for the initial slope $w(0)$. Start by solving the problem twice using $w(0) = -1000$ and $w(0) = -3500$, which, as shown in Example 9-1, give solutions that are higher and lower, respectively, than the boundary condition at $x = 0.1$, and thus can be the starting values for the bisection method. Stop the iterations when the difference between the calculated temperature from the numerical solution and the prescribed boundary condition is less than 0.01 K.

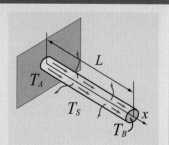

SOLUTION

Recall from Example 9-1 that the second-order BVP to be solved is:

$$\frac{d^2T}{dx^2} - \frac{h_c P}{kA_c}(T - T_S) - \frac{\varepsilon\sigma_{SB}P}{kA_c}(T^4 - T_S^4) = 0, \ 0 \le x \le 0.1 \tag{9.19}$$

with the boundary conditions: $T(0) = 473$ and $T(L) = 293$.

The solution is accomplished by converting Eq. (9.19) into the following system of two first-order ODEs:

$$\frac{dT}{dx} = w \tag{9.20}$$

$$\frac{dw}{dx} = \frac{h_c P}{kA_c}(T - T_S) + \frac{\varepsilon\sigma_{SB}P}{kA_c}(T^4 - T_S^4) \tag{9.21}$$

The initial value for Eq. (9.20) is $T(0) = 473$. The initial condition for Eq. (9.21) is not known. The MATLAB program listed below incorporates the bisection method in the shooting method. The program uses the user-defined function Sys2ODEsRK2 that was written in Section 8.8.2 for solving the system of Eqs. (9.20) and (9.21). The order of operations in the program is:

- Solve the system assuming $w(0) = -1000$.
- Solve the system assuming $w(0) = -3500$.
- Use the first two solutions and start calculating new values for $w(0)$ using the bisection method, Eq. (9.18).
- display a plot of the solution and output the numerical solution at $x = 0.1$ where the boundary value is prescribed.

Program 9-2: Script file. Applying the shooting method in conjunction with the bisection method.

```
% Solving Chapter 9 Example 2
clear all
a = 0; b = 0.1; TINI = 473; h = 0.001; Yb = 293;
tol = 0.01; imax = 15;
wH = -1000;                                       Solve the system assuming w(0) = -1000.
[x, T, w] = Sys2ODEsRK2('odeChap9Exmp1dTdx','odeChap9Exmp1dwdx',a,b,h,TINI,wH);
n = length(x);
wL = -3500;                                       Solve the system assuming w(0) = -3500.
[x, T, w] = Sys2ODEsRK2('odeChap9Exmp1dTdx','odeChap9Exmp1dwdx',a,b,h,TINI,wL);
for i = 1:imax + 1                                The start of the loop of the iterations.
   wi = (wH + wL)/2;                              Calculate a new value for w(0) using bisection, Eq. (9.18).
   [x, T, w] = Sys2ODEsRK2('odeChap9Exmp1dTdx','odeChap9Exmp1dwdx',a,b,h,TINI,wi);
   E = T(n) - Yb;                                 Calculate the error between the new
   if abs(E) < tol                                solution and the boundary condition.
      break                                                                              Solve the system with
   end                                            Stop if the error is smaller than specified.   the new value of w(0).
   if  E > 0
      wH = wi;
   else                                           Assign a new value to W_H, or to W_L.
      wL = wi;
   end
```

```
    end                                          The end of the loop of the iterations.
    if  i > imax
        fprintf('Solution was not obtained in %i iterations.',imax)
    else
        plot(x,T)
        xlabel('Distance (m)'); ylabel('Temperature (K)')
        fprintf('The calculated temperature at x = 0.1 is %5.3f K.\n',T(n))
        fprintf('The solution was obtained in %2.0f iterations.\n',i)
    end
```

The first two input arguments in the function `Sys2ODEsRK2` are the names of user-defined functions that calculate the values of dT/dx, Eq. (9.20), and dw/dx, Eq. (9.21). The two functions, `odeChap9Exmp1dTdx` and `odeChap9Exmp1dwdx`, are listed in Example 9-1.
When the script file is executed, the following is displayed in the Command Window:

The calculated temperature at x = 0.1 is 292.999 K.
The solution was obtained in 9 calculations.

In addition, the program produces a figure (shown on the right) with a plot of the solution.

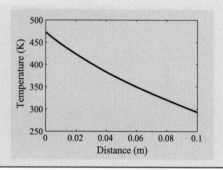

Shooting method using the secant method: The BVP in Eq. (9.8) is solved by converting the problem into the system of two IVPs in Eqs. (9.9) and (9.10). Shown in Fig. 9-8 are two numerical solutions that have been obtained in two iterations, i and $i-1$. The solutions at $x = b$ are $y_{b,i-1}$ and $y_{b,i}$. Recall that the boundary condition at this point is $y(b) = Y_b$. The initial values of the slopes at $x = a$ that correspond to these solutions are W_{i-1} and W_i, respectively (recall that $W = \dfrac{dy}{dx}\Big|_{x=a}$).

The errors in each numerical solution are $E_{i-1} = y_{b,i-1} - Y_b$ and $E_i = y_{b,i} - Y_b$. With this notation, the error, E, can be considered as a function of the slope, W, and the objective is to find the value of W where $E = 0$.

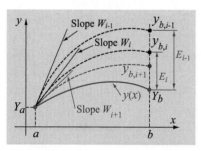

Figure 9-8: Two iterations in a solution with shooting method.

In an iterative process, the secant method determines an estimate for the zero of a function from the value of function at two points near the solution (see Section 3.6). When used in combination with the shooting method, the next estimate for the slope, W_{i+1}, is calculated from the points (W_i, E_i), and (W_{i-1}, E_{i-1}). This is illustrated in Fig. 9-9. The new value of the slope, W_{i+1}, is given by (see Eq. (3.26)):

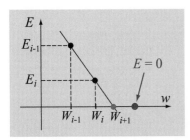

Figure 9-9: Secant method for determining W_{i+1}.

$$W_{i+1} = W_i - \frac{(W_{i-1} - W_i)}{E_{i-1} - E_i} E_i \qquad (9.22)$$

In this method the Es from the previous two iterations can be both positive, negative, or they can have opposite signs.

The shooting method can also be used for solving boundary value problems with derivative, or mixed, boundary conditions. The overall approach is the same as with two-point boundary value problems. The second-order (or higher) ODE is converted into a system of first-order ODEs that is solved as an initial value problem by assuming the unknown initial values needed for the solution. The solution is then compared with the prescribed boundary conditions, and if the results are not accurate enough, the assumed initial values are modified and the system is solved again. When derivative boundary conditions are prescribed at the endpoint, the calculated value of the derivative must be evaluated numerically.

9.3 FINITE DIFFERENCE METHOD

In finite difference methods, the derivatives in the differential equation are replaced with finite difference approximations. As shown in Fig. 9-10, the domain of the solution $[a, b]$ is divided into N subintervals of equal length h, that are defined by $(N + 1)$ points called **grid points**. (In general, subintervals can have unequal length.) The length of each subinterval is then $h = (b - a)/N$. Points a and b are the **endpoints**, and the rest of the points are the **interior points**. The differential equation is then written at each of the interior points of the domain. This results in a system of linear algebraic equations when the differential equation is linear, or in a system of nonlinear algebraic equations when the differential equation is nonlinear. The solution of the system is the numerical solution of the differential equation.

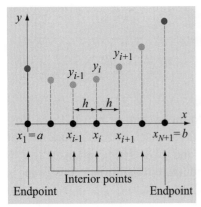

Figure 9-10: Finite difference method.

Many finite difference formulas are listed in Chapter 6 (see Table 6-1). Frequently, the central difference formulas are used in finite difference methods since they give better accuracy. Recall that for a function $y(x)$ that is given at points $(x_1, y_1), ..., (x_i, y_i), ..., (x_{N+1}, y_{N+1})$ that are equally spaced ($h = x_{i+1} - x_i$ for $i = 1...N$), the finite difference approximation of the first and the second derivatives at the interior points, with the central difference formulas, are given by:

$$\frac{dy}{dx} = \frac{y_{i+1} - y_{i-1}}{2h} \quad \text{and} \quad \frac{d^2y}{dx^2} = \frac{y_{i-1} - 2y_i + y_{i+1}}{h^2} \qquad (9.23)$$

Finite difference solution of a linear two-point BVP

The finite difference approximation for a linear second-order differential equation of the form:

$$\frac{d^2y}{dx^2} + f(x)\frac{dy}{dx} + g(x)y = h(x) \tag{9.24}$$

is:

$$\frac{y_{i-1} - 2y_i + y_{i+1}}{h^2} + f(x_i)\frac{y_{i+1} - y_{i-1}}{2h} + g(x_i)y_i = h(x_i) \tag{9.25}$$

The process of converting the differential equation, Eq. (9.24), into the algebraic form, Eq. (9.25) at each point x_i is called **discretization**. For a two-point BVP, the value of the solution at the endpoints, y_1 and y_{N+1} are known. Equation (9.25) is written $N-1$ times for $i = 2, ..., N$. This gives a system of $N-1$ linear algebraic equations for the unknowns $y_2, ..., y_N$, that can be solved numerically with any of the methods presented in Chapter 4.

Example 9-3 shows the solution of a linear second-order BVP using the finite difference method. In this example, the temperature distribution in a pin fin is calculated for the case where only convection is included in the analysis.

Example 9-3: Temperature distribution in a pin fin. Solving a second-order linear ODE (BVP) using the finite difference method.

When only convection is included in the analysis, the steady state temperature distribution, $T(x)$, along a pin fin can be obtained from the solution of the equation:

$$\frac{d^2T}{dx^2} - \frac{h_cP}{kA_c}(T - T_S) = 0, \quad 0 \le x \le L \tag{9.26}$$

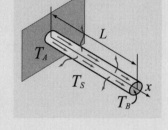

with the boundary conditions: $T(0) = T_A$ and $T(L) = T_B$.

In Eq. (9.26), h_c is the convective heat transfer coefficient, P is the perimeter bounding the cross section of the fin, k is the thermal conductivity of the fin material, A_c is the cross-sectional area of the fin, and T_S is the temperature of the surrounding air.

Determine the temperature distribution if $L = 0.1$ m, $T(0) = 473$ K, $T(0.1) = 293$ K, and $T_S = 293$ K. Use the following values for the parameters in Eq. (9.26): $h_c = 40$ W/m^2/K, $P = 0.016$ m^2, $k = 240$ W/m/K, and $A_c = 1.6 \times 10^{-5}$ m^2.

Solve the ODE using the finite difference method. Divide the domain of the solution into five equally spaced subintervals.

SOLUTION

Equation (9.26) is a second-order linear ODE. Using the finite difference method, the second derivative $\frac{d^2T}{dx^2}$ is approximated by the central difference formula, Eq. (9.23):

$$\frac{T_{i-1} - 2T_i + T_{i+1}}{h^2} - \beta(T_i - T_S) = 0 \tag{9.27}$$

where $\beta = \dfrac{h_c P}{k A_c}$. Equation (9.27) can be written as:

$$T_{i-1} - (2 + h^2\beta)T_i + T_{i+1} = -h^2\beta T_S \tag{9.28}$$

Next, the domain of the solution is divided into five equally spaced subintervals (defined by six points), as shown in the figure.

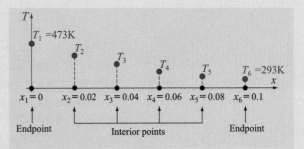

Next, Eq. (9.28) is written for each of the interior points (i.e., $i = 2, 3, 4, 5$):

for $i = 2$ $T_1 - (2 + h^2\beta)T_2 + T_3 = -h^2\beta T_S$ or $-(2 + h^2\beta)T_2 + T_3 = -(h^2\beta T_S + T_1)$ (9.29)

Recall that T_1 is known.

$$\text{for}\quad i = 3\quad T_2 - (2 + h^2\beta)T_3 + T_4 = -h^2\beta T_S \tag{9.30}$$

$$\text{for}\quad i = 4\quad T_3 - (2 + h^2\beta)T_4 + T_5 = -h^2\beta T_S \tag{9.31}$$

for $i = 5$ $T_4 - (2 + h^2\beta)T_5 + T_6 = -h^2\beta T_S$ or $T_4 - (2 + h^2\beta)T_5 = -(h^2\beta T_S + T_6)$ (9.32)

Recall that T_6 is known from the boundary condition $T(0.1) = 293\,\text{K}$.

Equations (9.29)–(9.32) are a system of four linear algebraic equations for the four unknowns T_2, T_3, T_4, and T_5. In matrix form, $[a][T] = [c]$, the system can be written as:

$$\begin{bmatrix} -(2 + h^2\beta) & 1 & 0 & 0 \\ 1 & -(2 + h^2\beta) & 1 & 0 \\ 0 & 1 & -(2 + h^2\beta) & 1 \\ 0 & 0 & 1 & -(2 + h^2\beta) \end{bmatrix} \begin{bmatrix} T_2 \\ T_3 \\ T_4 \\ T_5 \end{bmatrix} = \begin{bmatrix} -(h^2\beta T_S + T_1) \\ -h^2\beta T_S \\ -h^2\beta T_S \\ -(h^2\beta T_S + T_6) \end{bmatrix} \tag{9.33}$$

The system of equations in Eq. (9.33) can be solved with any method described in Chapter 4. The answer is the solution of the ODE in Eq. (9.26) at the interior points.

The following MATLAB program in a script solves Eq. (9.33) and plots of the results.

Program 9-3: Script file. Solving a linear second-order ODE using the finite difference method.

```
% Solution of Example 9-3
clear all
hc = 40; P = 0.016; k = 240; Ac = 1.6E-5;
h = 0.02; Ts = 293;
x = 0:0.02:0.1;                              Create a vector for the x coordinate of the mesh points.
beta = hc*P/(k*Ac);
aDia = -(2 + h^2*beta);
cele = -h^2*beta*Ts;
T(1) = 473; T(6) = 293;
```

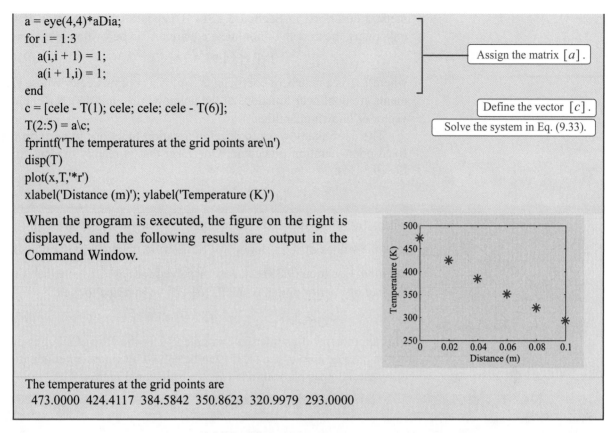

```
a = eye(4,4)*aDia;
for i = 1:3
    a(i,i + 1) = 1;
    a(i + 1,i) = 1;
end
c = [cele - T(1); cele; cele; cele - T(6)];
T(2:5) = a\c;
fprintf('The temperatures at the grid points are\n')
disp(T)
plot(x,T,'*r')
xlabel('Distance (m)'); ylabel('Temperature (K)')
```

Assign the matrix $[a]$.

Define the vector $[c]$.

Solve the system in Eq. (9.33).

When the program is executed, the figure on the right is displayed, and the following results are output in the Command Window.

The temperatures at the grid points are
473.0000 424.4117 384.5842 350.8623 320.9979 293.0000

Additional note

Application of the finite difference method to an ODE does not always result in a tridiagonal system of equations as in the above illustration. The numerical solution in Example 9-3 yielded a tridiagonal system because the ODE is second order and a central difference scheme is used to approximate the second derivative.

Finite difference solution of a nonlinear two-point BVP

The approach for solving a nonlinear ODE with the finite difference method is the same as that used for solving a linear ODE. The only difference is that the resulting system of simultaneous equations is nonlinear. Methods for solving such systems are described in Chapter 3. The task is much more challenging than that of solving a system of linear equations. The most computationally efficient means of solving a system of nonlinear equations is applying some type of iterative scheme. However, as discussed in Chapter 3, iterative methods run the risk of diverging unless the starting or initial values for the iterations are close enough to the final answer.

Application of the finite difference method to a nonlinear ODE results in a system of nonlinear simultaneous equations. One method for solving such nonlinear systems is a variant of the fixed-point iteration method described in Section 3.2.10. If $[y]$ is a column vector of the unknowns, the system of nonlinear equations can be written in the form:

$$[a][y] + [\Phi] = [b] \tag{9.34}$$

where $[a]$ is a matrix of coefficients, $[\Phi]$ is a column vector whose elements are nonlinear functions of the unknowns y_i, and $[b]$ is a column vector of known quantities.

There are many ways to develop iteration functions to perform the fixed point iteration procedure. One form that is readily evident from Eq. (9.34) is:

$$[a][y]^{k+1} = [b] - [\Phi]^k \tag{9.35}$$

where $[y]^{k+1}$ is the vector of unknowns, and $[\Phi]^k$ is known since it uses the values of the solution $[y]^k$ that were obtained in the previous iteration. Equation (9.35) can now be solved for $[y]^{k+1}$. Note that if the number of interior points is small, $[a]$ can be inverted to yield:

$$[y]^{k+1} = [a]^{-1}([b] - [\Phi]^k) \tag{9.36}$$

This approach is illustrated in Example 9-4. In the case of a large number of interior points, Eq. (9.35) can be solved by Gaussian elimination or by the Thomas algorithm if $[a]$ is a tridiagonal matrix.

Example 9-4: Temperature distribution in a pin fin. Solving a second-order nonlinear ODE (BVP) using the finite difference method.

When convection and radiation are included in the analysis, the steady-state temperature distribution, $T(x)$, along a pin fin can be calculated from the solution of the equation:

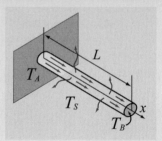

$$\frac{d^2 T}{dx^2} - \frac{h_c P}{k A_c}(T - T_S) - \frac{\varepsilon \sigma_{SB} P}{k A_c}(T^4 - T_S^4) = 0, \quad 0 \le x \le L \tag{9.37}$$

with the boundary conditions: $T(0) = T_A$ and $T(L) = T_B$.
The definition and values of all the constants in Eq. (9.37) are given in Example 9-2.
Determine the temperature distribution if $L = 0.1\,\text{m}$, $T(0) = 473\,\text{K}$, $T(0.1) = 293\,\text{K}$, and $T_S = 293\,\text{K}$.
Solve the ODE using the finite difference method. Divide the domain of the solution into five equally spaced subintervals.

SOLUTION

Equation (9.37) is a second-order nonlinear ODE. Using the finite difference method, the second derivative $\dfrac{d^2 T}{dx^2}$ is approximated by the central difference formula, Eq. (9.23):

$$\frac{T_{i-1} - 2T_i + T_{i+1}}{h^2} - \beta_A(T_i - T_S) - \beta_B(T_i^4 - T_S^4) = 0 \tag{9.38}$$

where $\beta_A = \dfrac{h_c P}{k A_c}$ and $\beta_B = \dfrac{\varepsilon \sigma_{SB} P}{k A_c}$

Equation (9.38) can be written as:

$$T_{i-1} - (2 + h^2 \beta_A) T_i - h^2 q \beta_B T_i^4 + T_{i+1} = -h^2 (\beta_A T_S + \beta_B T_S^4) \tag{9.39}$$

Next, the domain of the solution is divided into five equally spaced subintervals (defined by six points), as shown in the figure.

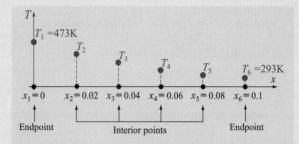

Next, Eq. (9.39) is written for each of the interior points (i.e., $i = 2, 3, 4, 5$):

for $\quad i = 2 \quad T_1 - (2 + h^2 \beta_A) T_2 - h^2 \beta_B T_2^4 + T_3 = -h^2 (\beta_A T_S + \beta_B T_S^4)$

or $\quad -(2 + h^2 \beta_A) T_2 - h^2 \beta_B T_2^4 + T_3 = -h^2 (\beta_A T_S + \beta_B T_S^4) - T_1 \tag{9.40}$

Recall that T_1 is known.

for $\quad i = 3 \quad T_2 - (2 + h^2 \beta_A) T_3 - h^2 \beta_B T_3^4 + T_4 = -h^2 (\beta_A T_S + \beta_B T_S^4) \tag{9.41}$

for $\quad i = 4 \quad T_3 - (2 + h^2 \beta_A) T_4 - h^2 \beta_B T_4^4 + T_5 = -h^2 (\beta_A T_S + \beta_B T_S^4) \tag{9.42}$

for $\quad i = 5 \quad T_4 - (2 + h^2 \beta_A) T_5 - h^2 \beta_B T_5^4 + T_6 = -h^2 (\beta_A T_S + \beta_B T_S^4)$

or $\quad T_4 - (2 + h^2 \beta_A) T_5 - h^2 \beta_B T_5^4 = -h^2 (\beta_A T_S + \beta_B T_S^4) - T_6 \tag{9.43}$

Recall that T_6 is known from the boundary condition, $T(0.1) = 293$ K.

Equations (9.40)–(9.43) are a system of four nonlinear algebraic equations for the four unknowns T_2, T_3, T_4, and T_5. Following Eq. (9.34), the system can be written in the form $[a][T] + [\Phi] = [b]$:

$$\begin{bmatrix} -(2 + h^2 \beta_A) & 1 & 0 & 0 \\ 1 & -(2 + h^2 \beta_A) & 1 & 0 \\ 0 & 1 & -(2 + h^2 \beta_A) & 1 \\ 0 & 0 & 1 & -(2 + h^2 \beta_A) \end{bmatrix} \begin{bmatrix} T_2 \\ T_3 \\ T_4 \\ T_5 \end{bmatrix} + \begin{bmatrix} -h^2 \beta_B T_2^4 \\ -h^2 \beta_B T_3^4 \\ -h^2 \beta_B T_4^4 \\ -h^2 \beta_B T_5^4 \end{bmatrix} = \begin{bmatrix} -h^2 (\beta_A T_S + \beta_B T_S^4) - T_1 \\ -h^2 (\beta_A T_S + \beta_B T_S^4) \\ -h^2 (\beta_A T_S + \beta_B T_S^4) \\ -h^2 (\beta_A T_S + \beta_B T_S^4) - T_6 \end{bmatrix} \tag{9.44}$$

The system of equations in Eq. (9.44) can be solved by using the fixed-point iteration method (see Chapter 3). An iteration formula is obtained by solving Eq. (9.44) for $[T]$:

$$[T]^{k+1} = [a]^{-1} ([b] - [\Phi]^k) \tag{9.45}$$

The iterative solution starts by guessing the values of T_2, T_3, T_4, and T_5, calculating the vector $[b]$ and using Eq. (9.45) to calculate new values of $[T]^{k+1}$. The new values are substituted back in Eq. (9.45), and so on. While various iteration formulas can be written from Eq. (9.44), not all will converge. The equations in Eq. (9.44) are for the case $N = 5$. The system of equations, however, is tridiagonal and can be easily extended if the domain is divided into a larger number of subintervals.

The system can be solved with any of the methods for solving a system of nonlinear equations that have been presented in Chapter 3. A MATLAB program (script file) that numerically solves the problem is listed next. The program is written in terms of the number of subintervals N. The program uses the fixed-point iteration method and executes four iterations. The program displays the temperature at the interior points after each iteration and plots the results from the last iteration.

Program 9-4: Script file. Solving a nonlinear second-order ODE using the finite difference method.

```
% Solution of Chapter 9 Example 4
clear all
hc = 40; P = 0.016; k = 240; Ac = 1.6E-5; epsln = 0.4; seg = 5.67E-8;
betaA = hc*P/(k*Ac); betaB = epsln*seg*P/(k*Ac); Ts = 293;
N = 5; h = 0.1/N;
x = 0:h:0.1;                            Define a vector for the x coordinate of the mesh points.
aDia = -(2 + h^2*betaA);
bele = -h^2*(betaA*Ts + betaB*Ts^4);
h2betaB = h^2*betaB;
Ti(1) = 473; Ti(N + 1) = 293;
Tnext(1) = Ti(1); Tnext(N + 1) = Ti(N + 1);
a = eye(N - 1,N - 1)*aDia;
for i = 1:N - 2
   a(i,i + 1) = 1;                      Assign the matrix [a].
   a(i + 1,i) = 1;
end
ainv = inv(a);
b(1) = bele - Ti(1); b(N-1) = bele - Ti(N+1);
c(2:N - 2) = bele;                      Define the vector [c].
Ti(2:N) = 400;                          Initial guess for the interior points is 400K.
for i = 1:4                             Solve the system in Eq. (9.44).
   phi = -h2betaB*Ti(2:N).^4';
   Tnext(2:N) = ainv*(b' - phi);
   Ti = Tnext;
   fprintf('After iteration number%2.0f, the temperatures at mesh points are:\n',i )
   disp(Tnext)
end
plot(x,Tnext,'*r')
xlabel('Distance (m)'); ylabel('Temperature (K)')
```

When the program is executed, the figure on the right is produced, and the following results are displayed in the Command Window.

In addition, the program was executed with $N = 20$. The results from this execution are shown in the lower figure.

After iteration number 1, the temperatures at mesh points are:
473.0000 423.2293 382.8297 349.1078 319.8155 293.0000
After iteration number 2, the temperatures at mesh points are:
473.0000 423.3492 383.3225 349.8507 320.4519 293.0000
After iteration number 3, the temperatures at mesh points are:
473.0000 423.3440 383.3132 349.8409 320.4456 293.0000
After iteration number 4, the temperatures at mesh points are:
473.0000 423.3441 383.3134 349.8410 320.4457 293.0000

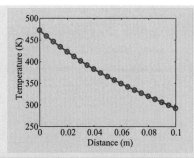

The results agree with the solution that was obtained using the shooting method in Example 9-2.

Additional notes

(1) In this example an iterative solution of the nonlinear system was easily enabled by MATLAB's built-in function for calculating the inverse of a matrix. (It was done for convenience because in this problem the matrix is small.) As stated in Chapter 4, it is in general not a good idea to invert a matrix since it is computationally inefficient. It is better to solve the system in Eq. (9.44) by Gaussian elimination or by the Thomas algorithm.

(2) If a different iteration formula is used (e.g. $[y]_{i+1} = [y]_i - \{[a][y]_i + [\Phi]_i - [b]\}$), the iterations will not converge regardless of how close the initial guesses are to the true solution.

Finite difference solution of a linear BVP with mixed boundary conditions

The finite difference method can also be applied for a BVP with mixed boundary conditions. In this case, a constraint that involves the derivative is prescribed at one or both of the endpoints of the solution domain. In these problems the finite difference method is used to discretize the ODE at the interior points (as in two-point BVPs). However, the system of algebraic equations that is obtained cannot be solved since the solution at the endpoints is not given (there are more equations than unknowns). The additional equations needed for solving the problem are obtained by discretizing the boundary conditions using finite differences, and incorporating the resulting equations into the algebraic equations for the interior points. The solution of a second-order BVP with mixed boundary conditions is illustrated in Example 9-5.

Final note

The foregoing discussion and examples show that neither the finite difference method nor the shooting method has a clear advantage when solving a higher-order nonlinear ODE. The finite difference method requires the solution of a nonlinear system of equations, while the shooting method requires information regarding the higher derivatives of the dependent variable at the leftmost boundary. The choice of which method to use is therefore problem-dependent, depending either on how easily initial guesses can be generated for the derivatives of the dependent variable at a boundary (shooting method) or on how well a particular fixed-point iteration scheme converges (finite difference method).

Example 9-5: Solving a BVP with mixed boundary conditions.

Use the finite difference method to solve the following mixed boundary condition BVP.

$$-2\frac{d^2y}{dx^2} + y = e^{-0.2x}, \quad \text{for} \quad 0 \le x \le 1 \tag{9.46}$$

with the boundary conditions: $y(0) = 1$ and $\left.\frac{dy}{dx}\right|_{x=1} = -y$.

Divide the solution domain into eight subintervals, and use the central difference approximation for all derivatives. Compare the numerical solution with the exact solution:

$$y = -0.2108e^{x/(\sqrt{2})} + 0.1238e^{-x/(\sqrt{2})} + \frac{e^{-0.2x}}{0.92} \tag{9.47}$$

SOLUTION

To use the finite difference method, the second derivative $\frac{d^2y}{dx^2}$ in Eq. (9.46) is approximated with the three-point second-order accurate central difference formula, Eq. (9.23):

$$-2\left(\frac{y_{i-1} - 2y_i + y_{i+1}}{h^2}\right) + y_i = e^{-0.2x_i} \tag{9.48}$$

By combining like terms and multiplying through by h^2, Eq. (9.48) can be written as:

$$-2y_{i-1} + (4 + h^2)y_i - 2y_{i+1} = h^2 e^{-0.2x_i} \tag{9.49}$$

Next, Eq. (9.49) is written for each of the interior points (i.e. $i = 2, 3, 4, 5, 6, 7, 8$):

$$\text{for} \quad i = 2 \quad -2y_1 + (4 + h^2)y_2 - 2y_3 = h^2 e^{-0.2x_2}$$

$$\text{or} \quad (4 + h^2)y_2 - 2y_3 = 2 + h^2 e^{-0.2x_2} \tag{9.50}$$

since $y_1 = y(0) = 1$.

$$\text{for} \quad i = 3: \quad -2y_2 + (4 + h^2)y_3 - 2y_4 = h^2 e^{-0.2x_3} \tag{9.51}$$

$$\text{for} \quad i = 4: \quad -2y_3 + (4 + h^2)y_4 - 2y_5 = h^2 e^{-0.2x_4} \tag{9.52}$$

$$\text{for} \quad i = 5: \quad -2y_4 + (4 + h^2)y_5 - 2y_6 = h^2 e^{-0.2x_5} \tag{9.53}$$

$$\text{for} \quad i = 6: \quad -2y_5 + (4 + h^2)y_6 - 2y_7 = h^2 e^{-0.2x_6} \tag{9.54}$$

$$\text{for} \quad i = 7: \quad -2y_6 + (4 + h^2)y_7 - 2y_8 = h^2 e^{-0.2x_7} \tag{9.55}$$

$$\text{for} \quad i = 8: \quad -2y_7 + (4 + h^2)y_8 - 2y_9 = h^2 e^{-0.2x_8} \tag{9.56}$$

Equations (9.50) through (9.56) are a system of seven linear equations with eight unknowns, y_2, y_3, y_4, y_5, y_6, y_7, y_8, and y_9. In this problem, unlike the two-point BVP, the value of the solution at the right endpoint (i.e., y_9) is not known.

An additional equation can be derived by considering the discretized boundary condition at $x = 1$:

$$\left.\frac{dy}{dx}\right|_{x=1} = -y \tag{9.57}$$

The derivative in Eq. (9.57) is approximated by a finite difference formula. As described in Chapter 6, several formulas can be used for this purpose. Since the derivative condition is given at the last point, it makes sense to use a one-sided backward formula that uses values at the previous points. In addition, since the second derivative in Eq. (9.46) is approximated with a second-order accurate formula, it makes sense to use a three-point backward difference formula for the first derivative, which is also a second-order accurate formula. The general form of the three-point backward difference formula is:

$$\frac{dy}{dx} = \frac{y_{i-2} - 4y_{i-1} + 3y_i}{2h} \tag{9.58}$$

Using Eq. (9.58) to approximate the derivative at the last point (x_9, y_9) in the boundary condition at $x = 1$ gives:

$$\frac{y_7 - 4y_8 + 3y_9}{2h} = -y_9 \tag{9.59}$$

Solving Eq. (9.59) for y_9 yields:

$$y_9 = \frac{-1}{3 + 2h} y_7 + \frac{4}{3 + 2h} y_8 \tag{9.60}$$

Equation (9.60) provides the additional relationship needed for solving all the unknowns. Substituting Eq. (9.60) in Eq. (9.56) gives:

$$\left(\frac{2}{3 + 2h} - 2\right) y_7 + \left(4 + h^2 - \frac{8}{3 + 2h}\right) y_8 = h^2 e^{-0.2x_8} \tag{9.61}$$

Equation (9.61) together with Eqs. (9.50)–(9.55) form a system of seven linear equations with seven unknowns. In matrix form, $[a][y] = [c]$, the system can be expressed by:

$$
\begin{bmatrix}
(4+h^2) & -2 & 0 & 0 & 0 & 0 & 0 \\
-2 & (4+h^2) & -2 & 0 & 0 & 0 & 0 \\
0 & -2 & (4+h^2) & -2 & 0 & 0 & 0 \\
0 & 0 & -2 & (4+h^2) & -2 & 0 & 0 \\
 & & & -2 & (4+h^2) & -2 & 0 \\
 & & & & -2 & (4+h^2) & -2 \\
 & & & & & \left(\frac{2}{3+2h} - 2\right) & \left(4+h^2 - \frac{8}{3+2h}\right)
\end{bmatrix}
\begin{bmatrix}
y_2 \\ y_3 \\ y_4 \\ y_5 \\ y_6 \\ y_7 \\ y_8
\end{bmatrix}
=
\begin{bmatrix}
2 + h^2 e^{-0.2x_2} \\
h^2 e^{-0.2x_3} \\
h^2 e^{-0.2x_4} \\
h^2 e^{-0.2x_5} \\
h^2 e^{-0.2x_6} \\
h^2 e^{-0.2x_7} \\
h^2 e^{-0.2x_8}
\end{bmatrix}
\tag{9.62}
$$

Once Eqs. (9.62) are solved, the value of y_9 can be calculated with Eq. (9.60). Notice that the system of equations in Eq. (9.62) is tridiagonal and can be easily extended if the domain is divided into a larger number of subintervals. The system can be solved with any of the methods for solving a system of linear equations that have been presented in Chapter 4, especially the Thomas algorithm.

The following MATLAB program in a script file presents the solution. The program uses the user-defined function named `Tridiagonal`, which was written in Example 4-9, for solving a tridiagonal system of equations. The program is written in terms of the number of subintervals, N, such that it be easily executed with different values for N.

Program 9-5: Script file. Solving BVP with mixed boundary conditions using finite difference method.

```
% Solution of Chapter 9 Example 5
clear all
a = 0; b = 1;                                        Endpoints of the solution domain.
N = 8; h = (b - a)/N;
x = a:h:b                          Define a vector for the x coordinate of the interior points.
hDenom = 3 + 2*h;
aDia = (4 + h^2);
y(1) = 1;
a = eye(N - 2,N - 2)*aDia;
a(N - 1,N - 1) = aDia - 8/hDenom;
for i = 1:N - 2
    a(i,i + 1) = -2;
    a(i + 1,i) = -2;                                 Assign the matrix [a].
end
a(N - 1,N - 2) = 2/hDenom - 2;
c(1) = 2 + h^2*exp(-0.2*x(2));
c(2:N - 1) = h^2*exp(-0.2*x(3:N));                   Create the vector [c].
y(2:N) = Tridiagonal(a,c);                           Solve the system in Eq. (9.62).
y(N + 1) = -1*y(N - 1)/hDenom + 4*y(N)/hDenom        Solution at the last point using Eq. (9.60).
yExact = -0.2108.*exp(x./sqrt(2)) + 0.1238.*exp(-x./sqrt(2)) + exp(-0.2.*x)./0.92
```

When the program is executed with $N = 8$, the solution displayed in the Command Window is:

```
x =
     0    0.1250    0.2500    0.3750    0.5000    0.6250    0.7500    0.8750    1.0000
y =
   1.0000    0.9432    0.8861    0.8284    0.7701    0.7106    0.6499    0.5874    0.5230
yExact =
   1.0000    0.9432    0.8861    0.8286    0.7702    0.7109    0.6501    0.5878    0.5234
```

The results show a close agreement between the numerical and exact solutions.

9.4 USE OF MATLAB BUILT-IN FUNCTIONS FOR SOLVING BOUNDARY VALUE PROBLEMS

Boundary value problems involve ODEs of second-order or higher with boundary conditions specified at both endpoints of the interval. As was shown in Section 8.9, ODEs of second-order or higher can be transformed into a system of first-order ODEs. MATLAB solves BVPs by solving a system of the first-order ODEs. The solution is obtained with a MATLAB built-in function named bvp4c. This function uses a finite difference method with the three-stage Lobatto IIIa formula.[1] The function bvp4c has different forms, but in order to simplify the presentation it is introduced here in its simplest form, and is applied for solving a two-point BVP.

Recall that a second-order ODE, with x as the independent variable and y as the dependent variable, has the form:

$$\frac{d^2y}{dx^2} = f\left(x, y, \frac{dy}{dx}\right) \text{ for } a \le x \le b \text{ with } y(a) = Y_a \text{ and } y(b) = Y_b \text{ (9.63)}$$

Transforming the second-order ODE into a system of two first-order ODEs gives:

$$\frac{dy}{dx} = w \quad \text{and} \quad \frac{dw}{dx} = f(x, y, w) \qquad (9.64)$$

where w is an additional (new) dependent variable.

The simple form of MATLAB's built-in function `bvp4c` for solving a first-order ODE is:

> **sol = bvp4c('odefun', 'bcfun', solinit)**

where:

odefun   The name of the user-defined function (function file) that calculates $\frac{dy}{dx} = w$ and $\frac{dw}{dx} = f(x, y, w)$ for given values of x, w and y, odefun, can be typed as a string (i.e., 'odefun') or by using a handle (i.e., @odefun).
The format of the user-defined function odefun is:

> **dydx = odefun(x, yw)**

The input argument x is a scalar, and the input argument yw is a column vector with the values of the dependent variables, $\begin{bmatrix} y \\ w \end{bmatrix}$. The output argument dydx is a column vector with the values: $\begin{bmatrix} w \\ f(x, y, w) \end{bmatrix}$.

bcfun   The name of the user-defined function (function file) that computes the residual in the boundary condition. The residual is the difference between the numerical solution and the prescribed boundary conditions (at the points where the boundary conditions are prescribed). bcfun can be typed as a string (i.e., 'bcfun') or by using a handle (i.e., @bcfun).

---

1. The Lobatto formula is a quadrature formula that, unlike Gauss quadrature (see Chapter 7), uses both endpoints of the interval, $(n-1)$ interior points and weights, to make the integration exact for polynomials of degree $2n-1$. The locations of these $(n-1)$ interior points are the roots of a set of orthogonal polynomials known as Jacobi polynomials. Details of this method are beyond the scope of this book.

Boundary condition:

$$y(a) = Y_a \quad \text{and} \quad \left.\frac{dy}{dx}\right|_{x=b} = D_b$$

vector res is: $\begin{bmatrix} ya(1) - Y_a \\ yb(2) - D_b \end{bmatrix}$

Boundary condition:

$$\left.\frac{dy}{dx}\right|_{x=a} = D_a \quad \text{and} \quad y(b) = Y_b$$

vector res is: $\begin{bmatrix} ya(2) - D_a \\ yb(1) - Y_b \end{bmatrix}$

Boundary condition (general case):

$$c_1 \left.\frac{dy}{dx}\right|_{x=a} + c_2 y(a) = C_a \quad \text{and}$$

$$c_3 \left.\frac{dy}{dx}\right|_{x=b} + c_4 y(b) = C_b$$

vector res is (for $c_1, c_3 \neq 0$):

$$\begin{bmatrix} ya(2) - \dfrac{C_a}{c_1} + \dfrac{c_2}{c_1} ya(1) \\ yb(2) - \dfrac{C_b}{c_3} + \dfrac{c_4}{c_3} yb(1) \end{bmatrix}$$

Figure 9-11: The residuals for mixed boundary conditions.

The format of the user-defined function `bcfun` is:

$$\boxed{\texttt{res = bcfun(ya,yb)}}$$

The input arguments `ya` and `yb` are column vectors corresponding to the numerical solution at $x = a$ and at $x = b$. The first elements, `ya(1)` and `yb(1)`, are the values of y at $x = a$, and $x = b$, respectively. The second elements `ya(2)` and `yb(2)`, are the values of dy/dx at $x = a$, and $x = b$, respectively. The output argument `res` is a column vector with the values of the residuals. The function `bcfun` can be used with any boundary conditions. For example, for Dirichlet boundary conditions (see Eq. (9.3)), where Y_a and Y_b are the prescribed boundary conditions, the column vector `res` is: $\begin{bmatrix} ya(1) - Y_a \\ yb(1) - Y_b \end{bmatrix}$. For Neumann boundary conditions (see Eq. (9.4)), where the derivatives D_a and D_b are the prescribed boundary conditions, the column vector `res` is: $\begin{bmatrix} ya(2) - D_a \\ yb(2) - D_b \end{bmatrix}$. Mixed boundary conditions can have several forms (see Eq. (9.5)). The possible expressions for the column vector `res` are listed in Fig. 9-11.

`solinit`   A structure containing the initial guess for the solution. `solinit` is created by a built-in MATLAB function named `bvpinit`. (`solinit` is the output argument of `bvpinit`.)

The format of the built-in function `bvpinit` is:

$$\boxed{\texttt{solinit = bvpinit(x,yinit)}}$$

The input argument `x` is a vector that specifies the initial interior points. For a BVP with the domain $[a, b]$, the first element of x is a, and the last element is b. Often an initial number of ten points is adequate and can be created by typing `x=linspace(a,b,10)`. The input argument `yinit` is the initial guess for the solution. `yinit` is a predefined vector that has one element for each of the dependent variables in the system of first-order ODEs that is solved. In the case of two equations, Eq. (9.64), the vector `yinit` has two elements. The first element is the initial guess for the value of y, and the second element is the initial guess for the value of w. MATLAB uses these initial guesses for all the interior points. `yinit` can also be entered as a name of a user-defined function (i.e., typed as a string 'yinit' or @yinit). In this case, the function has the form y=guess(x), where x is an interior point and y is the vector with the initial guess for the solution, as was explained earlier.

`sol` A structure containing the solution. Three important fields in `sol` are:

`sol.x` The *x* coordinate of the interior points. The number of interior points is determined during the solution process by MATLAB. It is, in general, not the same as was entered by the user in `bvpinit`.

`sol.y` The numerical solution, $y(x)$, which is the *y* value at the interior points.

`sol.yp` The value of the derivative, $\frac{dy}{dx}$, at the interior points.

Using MATLAB's built-in function `bvp4c` for solving a two-point BVP, is illustrated in detail next in Example 9-6.

Example 9-6: Solving a two-point BVP using MATLAB's built-in function `bvp4c`.

Use MATLAB's built-in function `bvp4c` to solve the following two-point BVP.

$$\frac{d^2y}{dx^2} + 2x\frac{dy}{dx} + 5y - \cos(3x) = 0, \quad \text{for} \quad 0 \le x \le \pi \tag{9.65}$$

with the boundary conditions: $y(0) = 1.5$ and $y(\pi) = 0$.

SOLUTION

To be solved with MATLAB, the equation is rewritten in the form:

$$\frac{d^2y}{dx^2} = -2x\frac{dy}{dx} - 5y + \cos(3x) \tag{9.66}$$

Next, by introducing a new dependent variable $w = \frac{dy}{dx}$, $\frac{dw}{dx} = \frac{d^2y}{dx^2}$, the second-order ODE, Eq. (9.66), is transformed into the following system of two first-order ODEs:

$$\frac{dy}{dx} = w \tag{9.67}$$

$$\frac{dw}{dx} = -2xw - 5y + \cos(3x) \tag{9.68}$$

MATLAB's `bvp4c` function has the form: `sol = bvp4c('odefun','bcfun',solinit)`. Before it can be used, the two user-defined functions `odefun` and `bcfun` have to be written. In the present solution, the user-defined function `odefun` is actually named `odefunExample6`. The listing of the function is:

```
function dydx = odefunExample6(x,yw)
dydx = [yw(2)
  -2*x*yw(2) - 5*yw(1) + cos(3*x)];
```

Right-hand side of Eq. (9.67).
Right-hand side of Eq. (9.68).

> Comments about the user-defined function `odefunExample6`:
> - `yw` is a column vector in which `yw(1)` is the value of y and `yw(2)` is the value of w.
> - `dydx` is a column vector in which `dydx(1)` is the value of the right-hand side of Eq. (9.67) and `dydx(2)` is the value of the right-hand side of Eq. (9.68).

In the present solution, the user-defined function `bcfun` is actually named `bcfunExample6`. The listing of the function is:

```
function res = bcfunExample6(ya,yb)
BCa = 1.5; BCb = 0;
res = [ya(1) - BCa
  yb(1) - BCb];
```

The boundary conditions are assigned to `BCa` and `Bcb`.
The residual at $x = a$.
The residual at $x = b$.

> Comments about the user-defined function `bcfunExample6`:
> - `ya` is a column vector in which `ya(1)` is the numerical (calculated by MATLAB) value of the solution, y, at $x = a$. `yb(1)` is the numerical (calculated by MATLAB) value of the solution, y, at $x = b$.
> - `res` is a column vector in which `res(1)` is the value of the residual at $x = a$, and `res(2)` is the value of the residual at $x = b$.

Once the two user-defined functions, `odefunExample6` and `bcfunExample6`, are written, they are used together with MATLAB's built-in functions `bvpinit` and `bvp4c`, in the following program, written in a script file, to solve the BVP in Eq. (9.65).

```
% Solution of Chapter 9, Example 6
clear all
solinit = bvpinit(linspace(0,pi,20),[0.2, 0.2]);
sol = bvp4c('odefunExample6','bcfunExample6',solinit)
plot(sol.x, sol.y(1,:),'r')
xlabel('x'); ylabel('y')
```

Initial grid.
Initial guess of the solution.
Create the structure `solinit` by using the function `bvpinit`.
Use the function `bvp4c` to solve the ODE.

When the program is executed, the figure on the right, with the plot of the solution, is displayed in the Figure Window. In addition, the fields of the structure `sol` are displayed in the Command Window (since no semicolon is typed at the end of the `sol=....` command).

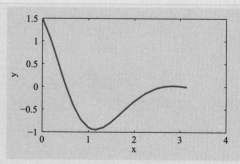

The display in the Command Window is:

```
sol =
     x: [1x24 double]
     y: [2x24 double]
    yp: [2x24 double]
 solver: 'bvp4c'
```

x: is a row vector with the x coordinates of the interior points.

y: is a (2×24) matrix with the values of the solution at the interior points. The first row is the solution for y, and the second row is the solution for w.

yp: is a (2×24) matrix with the values of the derivatives, $\frac{dy}{dx}$ (first row), and $\frac{dw}{dx}$ (second row).

The content of a field can be displayed by typing sol.fieldname. For example, typing sol.x displays the following values of the x coordinates of the interior points:

```
>> sol.x
ans =
  Columns 1 through 12
      0   0.1653   0.2480   0.3307   0.3720   0.4134   0.4960   0.6614   0.8267   0.9921   1.0748   1.1574
  Columns 13 through 24
   1.3228   1.4881   1.6535   1.8188   1.9842   2.2322   2.4802   2.6456   2.8109   2.8936   2.9762   3.1416
```

This shows that the solution was obtained by using 23 subintervals.

Typing sol.y(1,:) displays the first row of y:, which is the solution at the interior points:

```
>> sol.y(1,:)
ans =
  Columns 1 through 12
   1.5000   1.0849   0.8354   0.5728   0.4404   0.3089   0.0544   -0.3909   -0.7116   -0.8908   -0.9303   -0.9404
  Columns 13 through 24
   -0.8901   -0.7750   -0.6271   -0.4711   -0.3242   -0.1438   -0.0276   0.0124   0.0260   0.0247   0.0192        0
```

9.5 ERROR AND STABILITY IN NUMERICAL SOLUTION OF BOUNDARY VALUE PROBLEMS

Numerical error

Numerical error for a boundary value problem depends on the choice of the numerical method used. For the shooting method, the numerical error is the same as for the initial value problem discussed in Section 8.5 and depends on the method used. Obviously, this is because the shooting method solves the BVP by reformulating it as a series of IVPs with guesses for the leftmost boundary condition.

In the case of the finite difference method of solution of the BVP, the error is determined by the order of accuracy of the numerical scheme used. The truncation errors of the different approximations used for the derivatives are discussed in Section 6.3. As discussed in Section 6.9, the total error consists of the truncation error and round-off error. The accuracy of the solution by the finite difference method is therefore determined by the larger of the two truncation errors: that of the difference scheme used for the differential equation or that of the difference scheme used to discretize the boundary conditions. An effort must therefore be made to ensure that the order of the truncation error is the same for the boundary conditions and the differential equation.

Stability

The stability of numerical solutions to IVPs was discussed in Section 8.13. Numerical solutions to BVPs are also susceptible to instability, although the solution can become unstable in different ways. In an IVP, the instability was associated with error that grew as the integration progressed. In contrast, in BVPs, the growth of numerical error as the solution progresses is limited by the boundary conditions. In some cases, there may be valid multiple solutions to the BVP so that when it is solved as an IVP, small changes in the initial constraint (i.e., leftmost boundary condition) can produce one solution or the other for a small change in the leftmost boundary condition.

In the case of the shooting method (Example 9-2), it can be seen that even if the two approximate solutions generated by guesses for the unknown $\left.\dfrac{dT}{dx}\right|_{x=0}$ are laden with error, an accurate numerical solution can still be obtained by trapping the rightmost boundary condition between two approximate solutions generated numerically. Thus, the rightmost constraint keeps the types of errors illustrated in Example 8-13 from growing unboundedly. This does not mean that the solution at the interior points will be error free. The solution at the interior points can still exhibit the types of error propagation shown in Example 8-13, depending on the choice of method and step size.

In some cases, the differential equation itself may be unstable to small perturbations in the boundary conditions, in which case the problem formulation has to be examined. In other cases, multiple, valid solutions to the ODE exist for different rightmost boundary conditions.

Use of the finite difference method for solving a BVP places the burden of stability on the technique used to solve the resulting system of equations simultaneously at all points. Because the solution is determined everywhere simultaneously, the notion of marching forward in time or marching from left to right that is present in IVPs is not relevant for BVPs. Stability of solving a BVP by finite differences therefore rests on stability of the scheme used to solve the resulting set of simultaneous equations. In Example 9-3, the resulting system of equations for the unknown temperatures at the interior points is linear. In this case, all the potential difficulties of solving a linear system of equations discussed in Chapter 4 such as conditioning, apply. In Example 9-4, the system of equations is nonlinear. In this case, stability is determined by the type of method used to solve the system as well as the proximity of the initial guess to the solution. Since fixed-point iteration was used to solve the system of nonlinear equations in Example 9-4, stability of the numerical solution depends on the choice of the iteration function as well as the initial guess used to start the iteration. This instance shows that the rightmost boundary condition is unable to prevent a runaway divergence in the case of some iteration functions.

9.6 PROBLEMS

Problems to be solved by hand
Solve the following problems by hand. When needed, use a calculator, or write a MATLAB script file to carry out the calculations. If using MATLAB, do not use built-in functions for numerical solutions

9.1 Consider the following second-order ODE:

$$\frac{d^2y}{dx^2} = y + x(x-4)$$

(*a*) Using the central difference formula for approximating the second derivative, discretize the ODE (rewrite the equation in a form suitable for solution with the finite difference method).
(*b*) If $h = 1$, what is the value of the diagonal elements in the resulting matrix of coefficients of the system of linear equations that has to be solved?

9.2 Consider the following second-order ODE:

$$\frac{d^2y}{dr^2} + \frac{y}{r} = C$$

where C is a constant.
(*a*) Using the central difference formula for approximating the second derivative, discretize the ODE (rewrite the equation in a form suitable for solution with the finite difference method).
(*b*) If $h = 1$, what is the value of the diagonal elements in the resulting matrix of coefficients of the system of linear equations that has to be solved?

9.3 Consider the following second-order ODE:

$$\frac{d^2y}{dx^2} + x\frac{dy}{dx} + y = 2x \quad \text{for } 0 \le x \le 1 \text{ , with } y(0) = 1 \text{ and } y(1) = 1$$

(*a*) Using the central difference formulas for approximating the derivatives, discretize the ODE (rewrite the equation in a form suitable for solution with the finite difference method).
(*b*) What is the expression for the above-diagonal terms in the resulting matrix of coefficients of the tridiagonal system of linear equations that has to be solved?

9.4 Consider the following boundary value problem.

$$\frac{d^2y}{dx^2} + ay + by^4 = 0 \quad \text{for } 0 \le x \le 1 \text{ , with the boundary conditions: } \left.\frac{dy}{dx}\right|_{x=0} = 0 \text{ and } y(1) = 1$$

where a and b are constants. Discretize the second-order ODE using:
(*a*) Second-order accurate forward difference.
(*b*) Second-order accurate backward difference.
(*c*) Discretize the boundary condition at $x = 0$ using the second-order accurate forward difference.

9.5 Consider the following boundary value problem.

$$\frac{d^3y}{dx^3} + \frac{dy}{dx} - y^4 = 0 \quad \text{for } 0 \le x \le 1 \text{ , with the boundary conditions: } y(0) = 0, \left.\frac{dy}{dx}\right|_{x=0} = 0 \text{ and } y(1) = 10.$$

Discretize the third-order ODE using second-order central differences. When the boundary conditions are discretized, make sure that the order of the truncation error is compatible with that of the ODE.

9.6   Consider the following boundary value problem.

$$\frac{d^2y}{dx^2} + \frac{1}{x}\frac{dy}{dx} = -Q \quad \text{for } 0 \le x \le 1 \text{, with the boundary conditions: } \left.\frac{dy}{dx}\right|_{x=0} = 0 \text{ and } \left.\frac{dy}{dx}\right|_{x=1} = ay(1) + by^4(1) \quad .$$

where a and b are constants. Discretize the ODE using second-order accurate central differences for the derivatives. When the boundary conditions are discretized, make sure that the order of the truncation error is compatible with that of the ODE.

9.7   Consider the following boundary value problem.

$$-\frac{d^2u}{dx^2} + \pi^2 u = 2\pi^2 \cos(\pi x) \text{ for } 0 \le x \le 1 \text{, with the boundary conditions: } u(0) = 1 \text{ and } u(1) = -1$$

What are the diagonal elements of the resulting tridiagonal matrix when the finite difference method with first-order accurate central differences is applied to solve the problem with a step size of 1/8?

9.8   Consider the second-order ODE of the form:

$$\frac{d^2y}{dx^2} + p\frac{dy}{dx} + qy = r(x)$$

where p and q are constants, and $r(x)$ is a given function. Using second-order accurate central differences for the derivatives, discretize the ODE.

Problems to be programmed in MATLAB
Solve the following problems using the MATLAB environment. Do not use MATLAB's built-in functions for solving differential equations.

9.9   Write a user-defined MATLAB function that solves with the shooting method, a second-order boundary value problem of the form:

$$\frac{d^2y}{dx^2} + f(x)\frac{dy}{dx} + g(x)y = h(x) \quad \text{for } a \le x \le b \text{ with } y(a) = Y_a \text{ and } y(b) = Y_b$$

where Y_a and Y_b are constants. The function should first calculate two solutions using two assumed values for the slope at $x = a$, which are specified by the user, and use these solutions for calculating a new initial slope using interpolation (Eq. (9.14)), which is then used for calculating the final solution of the problem. Name the function `[x,y]=BVPShootInt(fOFx,gOFx,hOFx,a,b,n,Ya,Yb,WL,WH)`, where `fOFx`, `gOFx`, and `hOFx` are the names (strings) of the user-defined functions that calculate $f(x)$, $g(x)$ and $h(x)$, respectively, a and b define the domain of the solution, n is the number of subintervals, Ya and Yb are the boundary conditions, and WL and WH are the assumed slopes at $x = a$. Once the first two solutions are calculated, the program should confirm that at $x = b$ the given boundary condition Y_b is between the two solutions, and then calculate the final solution with the interpolated value for the slope. If the boundary condition at $x = b$ is not between the first two solutions, the program should stop and display an error message. Use the user-defined function `Sys2ODEsRK4` that was written in Problem 8.21 for solving the system of the two first-order ODEs within the user-defined function `BVPShootInt`.

Use `BVPShootInt` to solve the boundary value problem in Example 9-6. Use $n = 100$, $W_L = -5$, and $W_H = -1.5$.

9.10 Write a user-defined MATLAB function that solves with the shooting method in conjunction with the bisection method, a second-order boundary value problem of the form:

$$\frac{d^2y}{dx^2} + f(x)\frac{dy}{dx} + g(x)y = h(x) \quad \text{for } a \le x \le b \text{ with } y(a) = Y_a \text{ and } y(b) = Y_b$$

where Y_a and Y_b are constants. For the function name and arguments use `[x,y]=BVPShootBisec(fOFx,gOFx,hOFx,a,b,n,Ya,Yb,WL,WH)`, where `fOFx`, `gOFx`, and `hOFx` are the names (strings) of the user-defined functions that calculate $f(x)$, $g(x)$ and $h(x)$, respectively, a and b define the domain of the solution, n is the number of subintervals, `Ya` and `Yb` are the boundary conditions, and `WL` and `WH` are the assumed slopes at $x = a$ that are used in the first two solutions. Once the first two solutions are calculated, the program should confirm that at $x = b$ the value of the boundary condition Y_b is between the first two solutions. If the boundary condition is not between the first the two solutions, the program should stop and display an error message. Within the user-defined function BVPShootBisec, use the user-defined function `Sys2ODEsRK4` that was written in Problem 8.21 for solving the system of the two first-order ODEs. Stop the iterations when the true relative error at $x = b$ is smaller than 0.001.

Use BVPShootBisec to solve the boundary value problem in Example 9-6. Use $n = 100$, $W_L = -5$, and $W_H = -1.5$.

9.11 Write a user-defined MATLAB function that solves, with the shooting method in conjunction with the secant method, a second-order boundary value problem of the form:

$$\frac{d^2y}{dx^2} + f(x)\frac{dy}{dx} + g(x)y = h(x) \quad \text{for } a \le x \le b \text{ with } y(a) = Y_a \text{ and } y(b) = Y_b$$

where Y_a and Y_b are constants. For the function name and arguments use `[x,y]=BVPShootSecant(fOFx,gOFx,hOFx,a,b,n,Ya,Yb,WL,WH)`, where `fOFx`, `gOFx`, and `hOFx` are the names (strings) of the user-defined functions that calculate $f(x)$, $g(x)$ and $h(x)$, respectively, a and b define the domain of the solution, n is the number of subintervals, `Ya` and `Yb` are the boundary conditions, and `WL` and `WH` are the assumed slopes at $x = a$ that are used in the first two solutions. Within the user-defined function BVPShootSecant, use the user-defined function `Sys2ODEsRK4` that was written in Problem 8.21 for solving the system of the two first-order ODEs. Stop the iterations when the true relative error at $x = b$ is smaller than 0.001.

Use BVPShootSecant to solve the boundary value problem in Example 9-6. Use $n = 100$, $W_L = -5$, and $W_H = -1.5$.

9.12 Write a user-defined MATLAB function that uses the finite difference method to solve a second-order ODE of the form:

$$\frac{d^2y}{dx^2} + p\frac{dy}{dx} + qy = r(x) \quad \text{for } a \le x \le b \text{ with } y(a) = Y_a \text{ and } y(b) = Y_b$$

where p, q, Y_a and Y_b are constants. Discretize the ODE using second-order accurate central differences. Name the function `[x,y]=BVP2ndConst(a,b,n,Ya,Yb,p,q,rOFx)`, where a and b define the domain of the solution, `Ya` and `Yb` are the boundary conditions, n is the number of subintervals, and `rOFx` is the name (string) of the user-defined function that calculates $r(x)$. Within the program, use MATLAB's left division operation to solve the system of linear equations.

Use BVP2ndConst with 50 subintervals to solve the ODE in Problem 9.1 with the boundary conditions $y(0) = 1$, $y(2) = 0$. Plot the solution.

9.13 Write a user-defined MATLAB function that uses the finite difference method to solve a boundary value problem of the form:

$$\frac{d^2y}{dx^2} + f(x)\frac{dy}{dx} + g(x)y = h(x) \quad \text{for } a \le x \le b \quad \text{with} \quad y(a) = Y_a \text{ and } y(b) = Y_b$$

where Y_a and Y_b are constants. Discretize the ODE using second-order accurate central differences. For the function name and arguments use $[\texttt{x,y}]=\texttt{BVP2ndConst(a,b,Ya,Yb,n,pOFx,qOFx,rOFx)}$, where a and b define the domain of the solution, Ya and Yb are the boundary conditions, n is the number of subintervals, pOFx, qOFx, and rOFx are the names (strings) of the user-defined functions that calculate $p(x)$, $q(x)$ and $r(x)$, respectively. Within the program, use MATLAB's left division operation to solve the system of linear equations.

  Use the `BVP2ndConst` with 50 subintervals to solve the boundary value problem in Problem 9.3. Plot the solution.

9.14 Write a user-defined MATLAB function that uses the finite difference method to solve a boundary value problem of the form (see Eq. (9.24)):

$$\frac{d^2y}{dx^2} + f(x)\frac{dy}{dx} + g(x)y = h(x) \quad \text{for } a \le x \le b \quad \text{with} \quad y(a) = Y_a \quad \text{and} \quad \left.\frac{dy}{dx}\right|_{x=b} = D_b$$

where Y_a and D_b are constants. Discretize the ODE using second-order accurate central differences. For function name and arguments use $[\texttt{x,y}]=\texttt{BVP2ndConst(a,b,Ya,Db,n,pOFx,qOFx,rOFx)}$, where a and b define the domain of the solution, Ya and Db are the boundary conditions, n is the number of subintervals, pOFx, qOFx, and rOFx are the names (strings) of the user-defined functions that calculate $p(x)$, $q(x)$ and $r(x)$, respectively. Within the program, use MATLAB's left division operation to solve the system of linear equations.

  Use `BVP2ndConst` with 50 subintervals to solve the following boundary value problem:

$$\frac{d^2y}{dx^2} + \frac{1}{x}\frac{dy}{dx} = -10 \ , \quad y(1) = 1 \quad , \quad \left.\frac{dy}{dx}\right|_{x=3} = -1.2$$

Plot the solution.

Problems in math, science, and engineering
Solve the following problems using the MATLAB environment. As stated, use the MATLAB programs that are presented in the chapter, programs developed in previously solved problems, or MATLAB's built-in functions.

9.15 A flexible cable of uniform density is suspended between two points as shown in the figure. The shape of the cable, $y(x)$, is governed by the differential equation:

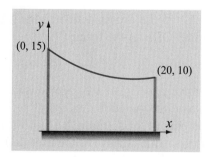

$$\frac{d^2y}{dx^2} = C\sqrt{1 + \left(\frac{dy}{dx}\right)^2}$$

where C is a constant equal to the ratio of the weight per unit length of the cable to the magnitude of the horizontal component of tension in the cable at its lowest point. The cable hangs between two points specified by $y(0) = 15$ m and $y(20) = 10$ m, and $C = 0.041$ m^{-1}. Use MATLAB's built-in functions to determine and plot the shape of the cable between $x = 0$ and $x = 20$ m.

9.16 A simply supported beam of length $L = 4\,\text{m}$ is loaded by a distributed load as shown in the figure. The deflection of the beam, y, is determined from the solution of the following ODE (when the deflections are small):

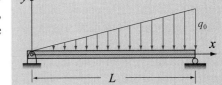

$$EI\frac{d^2y}{dx^2} = \frac{1}{6}q_0\left(Lx - \frac{x^3}{L}\right) \quad y(0) = 0 \text{ and } y(L) = 0$$

where $EI = 1.2 \times 10^7\,\text{N–m}^2$ is the flexural rigidity, and $q_0 = 30 \times 10^3\,\text{N/m}$.

Determine and plot the deflection of the beam as a function of x.

(*a*) Use the user-defined function `BVPShootBisec` that was written in Problem 9.10. Use $n = 100$, $W_L = 0$, and $W_H = -0.005$.

(*b*) Use the user-defined function `BVP2ndConst` with 50 subintervals that was written in Problem 9.12.

(*c*) Use MATLAB built-in functions to solve the ODE.

9.17 A simply supported beam of length $L = 4\,\text{m}$ is loaded by a uniform distributed load as shown in the figure. For large deflections, the deflection of the beam, y, is determined from the solution of the following ODE:

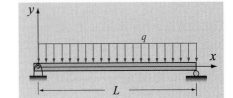

$$EI\frac{d^2y}{dx^2} = \left[1 + \left(\frac{dy}{dx}\right)^2\right]^{3/2}\frac{1}{2}q(Lx - x^2) \quad y(0) = 0 \text{ and } y(L) = 0$$

where $EI = 1.4 \times 10^7\,\text{N–m}^2$ is the flexural rigidity, and $q = 10 \times 10^3\,\text{N/m}$.

Use MATLAB's built-in functions to determine and plot the deflection of the beam as a function of x.

9.18 The temperature distribution in a straight fin, $T(x)$ with a triangular profile is given by the solution of the equation:

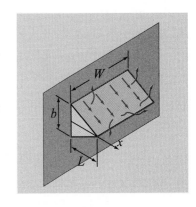

$$\frac{d^2T}{dx^2} - \frac{1}{(L-x)}\frac{dT}{dx} - \frac{2h(LW + bL - bx)(T - T_\infty)}{kbW(L-x)} = 0$$

where $T_\infty = 300\,\text{K}$ is the ambient temperature, x is the coordinate measured along the fin, $k = 237\,\text{W/m/K}$ is the thermal conductivity of aluminum, $h = 15\,\text{W/m}^2/\text{K}$ is the convective heat transfer coefficient, $L = 0.01\,\text{m}$ is the length of the fin, $W = 0.1\,\text{m}$ is its width, and $b = 0.01\,\text{m}$ is the height of the base. The boundary conditions are: $T(x = 0) = 673\,\text{K}$, and $\left.\dfrac{dT}{dx}\right|_{x=L} = 0$.

Find and plot the temperature distribution, $T(x)$, along the fin. Write a program in a script file that solves the problem with the shooting method. Note that a derivative boundary condition is prescribed at $x = L$. Use the three-point backward difference formula to calculate the value of the derivative from the numerical solution at $x = L$. Compare the numerical solution with the prescribed boundary condition and use the bisection method to calculate the new estimate for the slope at $x = 0$. Stop the iterations when the true relative error at $x = L$ is smaller than 0.01. ***Important note:*** The point $x = L$ is a singular point of the ODE. Therefore, the problem cannot be solved as specified. An approximate solution can however be obtained by using $L = 0.00999999\,\text{m}$ for the length of the fin.

9.19 Solve Problem 9.18 using the finite difference method with 100 subintervals. Use second-order accurate central differences for all the derivatives in the ODE, and use appropriate one-sided differences for the boundary condition. As explained in Problem 9.18, use $L = 0.00999999$ m for the length of the fin.

9.20 The fuel rod of a nuclear reactor is a cylindrical structure with the fuel retained inside a cladding as shown in the figure. The fuel causes heat to be generated by nuclear reactions within the cylinder as well as in the cladding. The outer surface of the cladding is cooled by flowing water at $T_\infty = 473$ with a heat transfer coefficient of $h = 10^4$ W/m²/K. The thermal conductivity of the cladding material is $k = 16.75$ W/m/K. The dimensions of the fuel rod are $R = 1.5 \times 10^{-2}$ m, and $w = 3.0 \times 10^{-3}$ m. The temperature distribution in the cladding is determined by the solution of the following boundary value problem:

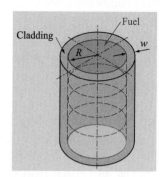

$$\frac{1}{r}\frac{d}{dr}\left(rk\frac{dT}{dr}\right) = -10^8 \frac{e^{-r/R}}{r}, \text{ for } R \le r \le R+w, \text{ with } \left.\frac{dT}{dr}\right|_{r=R} = -\frac{6.32 \times 10^5}{k} \text{ and } \left.\frac{dT}{dr}\right|_{r=R+w} = -\frac{h}{k}(T|_{r=R+w} - T_\infty)$$

Use MATLAB's built-in function `bvp4c` to solve the boundary value problem. Plot the temperature distribution in the cladding as a function of r.

9.21 The axial temperature variation of a current-carrying bare wire is described by:

$$\frac{d^2T}{dx^2} - \frac{4h}{kD}(T - T_\infty) - \frac{4\varepsilon\sigma_{SB}(T^4 - T_\infty^4)}{kD} = -\frac{I^2\rho_e}{k\left(\frac{1}{4}\pi D^2\right)^2}$$

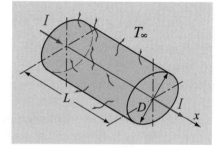

where T is the temperature in K, x is the coordinate along the wire, $k = 72$ W/m/K is the thermal conductivity, $h = 2000$ W/m²/K², is the convective heat coefficient, $\varepsilon = 0.1$ is the radiative emissivity, $\sigma_{SB} = 5.67 \times 10^{-8}$ W/m²/K⁴ is the Stefan-Boltzmann constant, $I = 2$ A is the current, $\rho_e = 32 \times 10^{-8}$ Ωm is the electrical resistivity, $T_\infty = 300$ K is the ambient temperature, $D = 7.62 \times 10^{-5}$ m is the wire diameter, and $L = 4.0 \times 10^{-3}$ is the length of the wire. The boundary conditions are:

$$\text{at } x = 0 \quad T = 300 \text{ K, and at } x = \frac{L}{2} \quad \frac{dT}{dx} = 0$$

Use MATLAB's built-in function `bvp4c` to solve the boundary value problem for $0 \le x \le \frac{L}{2}$, since the temperature distribution is symmetric about $x = \frac{L}{2}$. Plot the temperature distribution along the wire.

Appendix A

Introductory MATLAB

Core Topics

Starting with MATLAB (A.2).

Arrays (A.3).

Mathematical operations with arrays (A.4).

Script files (A.5).

Function files (A.6).

Programming (A.7).

Plotting (A.8).

A.1 BACKGROUND

MATLAB is a powerful language for technical computing. The name MATLAB stands for MATrix LABoratory because its basic data element is a matrix (array). MATLAB can be used for mathematical computations, modeling and simulations, data analysis and processing, visualization and graphics, and algorithm development.

MATLAB is widely used in universities and colleges in introductory and advanced courses in mathematics, science, and especially in engineering. In industry the software is used in research, development, and design. The standard MATLAB program has tools (functions) that can be used to solve common problems. In addition, MATLAB has optional toolboxes that are a collection of specialized programs designed to solve specific types of problems. Examples include toolboxes for signal processing, symbolic calculations, and control systems.

This appendix is a brief introduction to MATLAB. It presents the most basic features and syntax that will enable the reader to follow the use of MATLAB in this book. For a more complete introduction, the reader is referred to MATLAB: An Introduction with Applications, Second Edition, by Amos Gilat, Wiley, 2005.

A.2 STARTING WITH MATLAB

It is assumed that the software is installed on the computer and that the user can start the program. When the program is running, eight windows can be used. A list of the various windows and their purpose is given in Table A-1. Four of the windows—the Command Window, the Figure Window, the Editor Window, and the Help Window—are the most commonly used.

421

Table A-1: MATLAB Windows

Window	Purpose
Command Window	Main window, enters variables, runs programs.
Figure Window	Contains output from graphics commands.
Editor Window	Creates and debugs script and function files.
Help Window	Provides help information.
Launch Pad Window	Provides access to tools, demos, and documentation.
Command History Window	Logs commands entered in the Command Window.
Workspace Window	Provides information about the variables that are used.
Current Directory Window	Shows the files in the current directory.

Command Window: The Command Window is MATLAB's main window and opens when MATLAB is started.

- Commands are typed next to the prompt ($<<$) and are executed when the ***Enter*** key is pressed.

- Once a command is typed and the **Enter** key is pressed, the command is executed. However, only the last command is executed. Everything executed previously is unchanged.

- Output generated by the command is displayed in the Command Window, unless a semicolon (;) is typed at the end.

- When the symbol % (percent symbol) is typed in the beginning of a line, the line is designated as a comment.

- The clc command (type clc and press ***Enter***) clears the Command Window. After working in the Command Window for a while, the display may be very long. Once the clc command is executed, a clear window is displayed. The command does not change anything that was done before. For example, if some variables were defined previously, they still exist in memory and can be used. The up-arrow key (↑) can also be used to recall commands that were typed before.

Figure Window: The Figure Window opens automatically when graphics commands are executed and contains graphs created by these commands.

Editor Window: The Editor Window is used for writing and editing programs. This window is opened from the ***File*** menu in the Command Window. More details on the Editor Window are given in Section A.5 where it is used for creating script files.

Help Window: The Help Window contains help information. This window can be opened from the ***Help*** menu in the toolbar of any MATLAB window. The Help Window is interactive and can be used to obtain information on any feature of MATLAB.

Elementary arithmetic operations with scalars

The simplest way to use MATLAB is as a calculator. With scalars, the symbols of arithmetic operations are:

Operation	Symbol	Example	Operation	Symbol	Example
Addition	+	5 + 3	Right division	/	5 / 3
Subtraction	−	5 − 3	Left division	\	5 \ 3 = 3 / 5
Multiplication	*	5 * 3	Exponentiation	^	5 ^ 3 (means 5^3 = 125)

A mathematical expression can be typed in the Command Window. When the **Enter** key is pressed, MATLAB calculates the expression and responds by displaying `ans =` and the numerical result of the expression in the next line. Examples are:

```
>> 7 + 8/2
ans =
   11
>> (7+8)/2 + 27^(1/3)
ans =
   10.5000
```

Numerical values can also be assigned to variables, which is a name made of a letter or a combination of several letters (and digits). Variable names must begin with a letter. Once a variable is assigned a numerical value, it can be used in mathematical expressions, in functions, and in any MATLAB statements and commands.

```
>> a = 12
a =
   12
>> B = 4;
>> C = (a - B) + 40 - a/B*10
C =
   18
```

Since a semicolon is typed at the end of the command, the value of B is not displayed.

Elementary math built-in functions

In addition to basic arithmetic operations, expressions in MATLAB can include functions. MATLAB has a very large library of built-in functions. A function has a name and an argument in parentheses. For example, the function that calculates the square root of a number is `sqrt(x)`. Its name is `sqrt`, and the argument is x. When the function is used, the argument can be a number, a variable, or a computable expression that can be made up of numbers and/or variables. Functions can also be included in arguments, as well as in expressions. The following shows examples of using the function `sqrt(x)` when MATLAB is used as a calculator with scalars.

```
>> sqrt(64)
ans =
    8
>> sqrt(50 + 14*3)
ans =
    9.5917
>> sqrt(54 + 9*sqrt(100))
ans =
    12
>> (15 + 600/4)/sqrt(121)
ans =
    15
```

| Argument is a number. |
| Argument is an expression. |
| Argument includes a function. |
| Function is included in an expression. |

Lists of some commonly used elementary MATLAB mathematical built-in functions are given in Table A-2. A complete list of functions organized by name of category can be found in the Help Window.

Table A-2: Built-in elementary math functions.

Command	Description	Example

`sqrt(x)`	Square root.	`>> sqrt(81)`
		`ans =`
		    `9`

`exp(x)`	Exponential (e^x).	`>> exp(5)`
		`ans =`
		    `148.4132`

`abs(x)`	Absolute value.	`>> abs(-24)`
		`ans =`
		    `24`

`log(x)`	Natural logarithm.
	Base e logarithm (ln).
		`>> log(1000)`
		`ans =`
		    `6.9078`

`log10(x)`	Base 10 logarithm.	`>> log10(1000)`
		`ans =`
		    `3.0000`

`sin(x)`
 	
`sind(x)`
	Sine of angle x (x in radians).
	
	Sine of angle x (x in degrees).
		`>> sin(pi/6)` `>> sind(30)`
		`ans =` `ans =`
		    `0.5000` `0.5000`

The other trigonometric functions are written in the same way. The inverse trigonometric functions are written by adding the letter "a" in front, for example, `asin(x)`.		

Table A-2: Built-in elementary math functions. (Continued)

Command	Description	Example
round(x)	Round to the nearest integer.	>> round(17/5) ans = 3
fix(x)	Round toward zero.	>> fix(13/5) ans = 2
ceil(x)	Round up toward infinity.	>> ceil(11/5) ans = 3
floor(x)	Round down toward minus infinity.	>> floor(-9/4) ans = -3

Display formats

The format in which MATLAB displays output on the screen can be changed by the user. The default output format is fixed point with four decimal digits (called short). The format can be changed with the format command. Once the format command is entered, all the output that follows is displayed in the specified format. Several of the available formats are listed and described in Table A-3.

Table A-3: Display format

Command	Description	Example
format short	Fixed point with four decimal digits for: $0.001 \leq number \leq 1000$ Otherwise display format short e.	>> 290/7 ans = 41.4286
format long	Fixed point with 14 decimal digits for: $0.001 \leq number \leq 100$ Otherwise display format long e.	>> 290/7 ans = 41.42857142857143
format short e	Scientific notation with four decimal digits.	>> 290/7 ans = 4.1429e+001
format long e	Scientific notation with 15 decimal digits.	>> 290/7 ans = 4.142857142857143e+001
format bank	Two decimal digits.	>> 290/7 ans = 41.43

A.3 ARRAYS

The array is a fundamental form that MATLAB uses to store and manipulate data. An array is a list of numbers arranged in rows and/or columns. The simplest array (one-dimensional) is a row, or a column of numbers, which in science and engineering is commonly called a vector. A more complex array (two-dimensional) is a collection of numbers arranged in rows and columns, which in science and engineering is called a matrix. Each number in a vector or a matrix is called an element. This section shows how to construct vectors and matrices. Section A.4 shows how to carry out mathematical operations with arrays.

Creating a vector

In MATLAB, a vector is created by assigning the elements of the vector to a variable. This can be done in several ways depending on the source of the information that is used for the elements of the vector. When a vector contains specific numbers that are known, the value of each element is entered directly by typing the values of the elements inside square brackets:

> **variable_name = [number number ... number]**

For a row vector, the numbers are typed with a space or a comma between the elements. For a column vector the numbers are typed with a semicolon between them. Each element can also be a mathematical expression that can include predefined variables, numbers, and functions. Often, the elements of a row vector are a series of numbers with constant spacing. In such cases the vector can be created by typing:

> **variable_name = m:q:n**

where m is the first element, q is the spacing, and n is the last element. Another option is to use the linspace command:

> **variable_name = linspace(xi,xf,n)**

Several examples of constructing vectors are:

```
>> yr = [1984  1986  1988  1990  1992  1994  1996]      Row vector by typing elements.
yr =
       1984      1986      1988      1990      1992      1994      1996
>> pnt = [2;  4;  5]                                    Column vector by typing elements.
pnt =
     2
     4
     5
>> x = [1:2:13]                                         Row vector with constant spacing.
x =
     1    3    5    7    9   11   13
>> va = linspace(0,8,6)       Row vector with 6 elements, first element 0, last element 8.
```

va =

    0   1.6000   3.2000   4.8000   6.4000   8.0000

Creating a two-dimensional array (matrix)

A two-dimensional array, also called a matrix, has numbers in rows and columns. A matrix is created by assigning the elements of the matrix to a variable. This is done by typing the elements, row by row, inside square brackets [ ]. In each row the elements are separated with spaces or commas.

> variable_name = [1st row elements; 2nd row ele-
> ments;....; last row elements]

The elements that are entered can be numbers or mathematical expressions that may include numbers, predefined variables, and functions. All the rows must have the same number of elements. If an element is zero, it has to be entered as such. MATLAB displays an error message if an attempt is made to define an incomplete matrix. Examples of matrices created in different ways are:

```
>> a = [5  35  43; 4  76  81; 21  32  40]          Semicolons are typed between rows.
a =
     5    35    43
     4    76    81
    21    32    40
>> cd = 6; e = 3; h = 4;                            Variables are defined.
>> Mat = [e, cd*h, cos(pi/3); h^2, sqrt(h*h/cd), 14]    Elements are entered as mathemat-
Mat =                                                    ical expressions.
    3.0000   24.0000    0.5000
   16.0000    1.6330   14.0000
```

- All variables in MATLAB are arrays. A scalar is an array with one element; a vector is an array with one row, or one column, of elements; and a matrix is an array with elements in rows and columns.

- The variable (scalar, vector, or matrix) is defined by the input when the variable is assigned. There is no need to define the size of the array (single element for a scalar, a row or a column of elements for a vector, or a two-dimensional array of elements for a matrix) before the elements are assigned.

- Once a variable exists as a scalar, a vector, or a matrix, it can be changed to be any other size, or type, of variable. For example, a scalar can be changed to a vector or a matrix, a vector can be changed to a scalar, a vector of different length, or a matrix, and a matrix can be changed to have a different size, or to be reduced to a vector or a scalar. These changes are made by adding or deleting elements.

Array addressing

Elements in an array (either vector or matrix) can be addressed individually or in subgroups. This is useful when there is a need to redefine only some of the elements, or to use specific elements in calculations, or when a subgroup of the elements is used to define a new variable.

The address of an element in a vector is its position in the row (or column). For a vector named ve, ve(k) refers to the element in position k. The first position is 1. For example, if the vector *ve* has nine elements:

ve = 35 46 78 23 5 14 81 3 55

then

$ve(4) = 23$, $ve(7) = 81$, and $ve(1) = 35$.

The address of an element in a matrix is its position, defined by the row number and the column number where it is located. For a matrix assigned to a variable *ma*, *ma*(k,p) refers to the element in row k and column p.

For example, if the matrix is: $ma = \begin{bmatrix} 3 & 11 & 6 & 5 \\ 4 & 7 & 10 & 2 \\ 13 & 9 & 0 & 8 \end{bmatrix}$

then, $ma(1,1) = 3$, and $ma(2,3) = 10$.

It is possible to change the value of one element by reassigning a new value to the specific element. Single elements can also be used like variables in mathematical expressions.

```
>> VCT = [35 46 78 23 5 14 81 3 55]                      Define a vector.
VCT =
   35   46   78   23    5   14   81    3   55
>> VCT(4), VCT(6) = 273             Assign new values to the fourth and sixth elements.
VCT =
   35   46   78   23    5  273   81    3   55
>> VCT(5)^VCT(8) + sqrt(VCT(7))                    Use vector elements in a
ans =                                              mathematical expression.
   134
>> MAT = [3 11 6 5; 4 7 10 2; 13 9 0 8]                   Define a matrix.
MAT =
    3   11    6    5
    4    7   10    2
   13    9    0    8
>> MAT(3,1) = 20                        Assign new values to the (3,1) element.
MAT =
    3   11    6    5
    4    7   10    2
   20    9    0    8
>> MAT(2,4) - MAT(1,2)                              Use matrix elements in a
                                                   mathematical expression.
```

ans =
 -9

Using a colon : in addressing arrays

A colon can be used to address a range of elements in a vector or a matrix. For a vector, *va*(:) refers to all the elements of the vector *va* (either a row or a column vector). *va*(*m:n*) refers to elements *m* through *n* of the vector *va*.

For a matrix, *A*(:,*n*)refers to the elements in all the rows of column *n*. *A*(*n*,:)refers to the elements in all the columns of row *n*. *A*(:,*m:n*) refers to the elements in all the rows between columns *m* and *n*. *A*(*m:n*,:) refers to the elements in all the columns between rows *m* and *n*. A(*m:n,p:q*) refers to the elements in rows *m* through *n* and columns *p* through *q*.

```
>> v = [4  15  8  12  34  2  50  23  11]          Define a vector.
v =
   4   15   8   12   34   2   50   23   11
>> u = v(3:7)          Vector u is created from the elements 3 through 7 of vector v.
u =
   8   12   34   2   50
>> A = [1  3  5  7  9  11; 2  4  6  8  10  12; 3  6  9  12  15  18; 4  8  12  16  20  24; 5  10  15  20  25  30]          Define a matrix.
A =
   1    3    5    7    9   11
   2    4    6    8   10   12
   3    6    9   12   15   18
   4    8   12   16   20   24
   5   10   15   20   25   30
C = A(2,:)          Vector C is created from the second row of matrix A.
C =
   2   4   6   8   10   12
>> F = A(1:3,2:4)          Matrix F is created from the elements in rows 1 through 3 and columns 2 through 4 of matrix A.
F =
   3    5    7
   4    6    8
   6    9   12
```

MATLAB has many built-in functions for managing and handling arrays. Several are listed in Table A-4.

Table A-4: Built-in functions for handling arrays.

Command	Description	Example
length(A)	Returns the number of elements in vector A.	>> A = [5 9 2 4]; >> length(A) ans = 4

Table A-4: Built-in functions for handling arrays. (Continued)

Command	Description	Example
size(A)	Returns a row vector [m,n], where m and n are the size $m \times n$ of the array A. (m is number of rows. n is number of columns.)	>> A = [6 1 4 0 12; 5 19 6 8 2] A = 6 1 4 0 12 5 19 6 8 2 >> size(A) ans = 2 5
zeros(m,n)	Creates a matrix with *m* rows and *n* columns, in which all the elements are the number 0.	>> zr = zeros(3,4) zr = 0 0 0 0 0 0 0 0 0 0 0 0
ones(m,n)	Creates a matrix with *m* rows and *n* columns, in which all the elements are the number 1.	>> ne = ones(4,3) ne = 1 1 1 1 1 1 1 1 1 1 1 1
eye(n)	Creates a square matrix with *n* rows and *n* columns in which the diagonal elements are equal to 1 (identity matrix).	>> idn = eye(5) idn = 1 0 0 0 1 0 0 0 1

Strings

- A string is an array of characters. It is created by typing the characters within single quotes.

- Strings can include letters, digits, other symbols, and spaces.

- Examples of strings: 'ad ef', '3%fr2', '{edcba:21!'.

- When a string is being typed in, the color of the text on the screen changes to maroon when the first single quote is typed. When the single quote at the end of the string is typed the color of the string changes to purple.

   Strings have several different uses in MATLAB. They are used in output commands to display text messages, in formatting commands of plots, and as input arguments of some functions. Strings can also be assigned to variables by simply typing the string on the right side of the assignment operator, as shown in the next example.

>> a = 'FRty 8'

a =
FRty 8
>> B = 'My name is John Smith'
B =
My name is John Smith

A.4 MATHEMATICAL OPERATIONS WITH ARRAYS

Once variables are created in MATLAB, they can be used in a wide variety of mathematical operations. Mathematical operations in MATLAB can be divided into three categories:

1. Operations with scalars ((1 × 1) arrays) and with single elements of arrays.
2. Operations with arrays following the rules of linear algebra.
3. Element-by-element operations with arrays.

Operations with scalars and single elements of arrays are done by using the standard symbols as in a calculator. So far, all the mathematical operations in the appendix have been done in this way.

Addition and subtraction of arrays

With arrays, the addition, subtraction, and multiplication operations follow the rules of linear algebra (see Chapter 2). The operations + (addition) and – (subtraction) can be carried out only with arrays of identical size (the same number of rows and columns). The sum, or the difference of two arrays, is obtained by adding, or subtracting, their corresponding elements.

In general, if A and B are two arrays (for example, (2×3) matrices),

$$A = \begin{bmatrix} A_{11} & A_{12} & A_{13} \\ A_{21} & A_{22} & A_{23} \end{bmatrix} \quad \text{and} \quad B = \begin{bmatrix} B_{11} & B_{12} & B_{13} \\ B_{21} & B_{22} & B_{23} \end{bmatrix}$$

then, the matrix that is obtained by adding A and B is:

$$\begin{bmatrix} (A_{11} + B_{11}) & (A_{12} + B_{12}) & (A_{13} + B_{13}) \\ (A_{21} + B_{21}) & (A_{22} + B_{22}) & (A_{23} + B_{23}) \end{bmatrix}$$

In MATLAB, when a scalar (number) is added to, or subtracted from, an array, the number is added to, or subtracted from, all the elements of the array. (Note that such an operation is not defined in linear algebra.) Examples are:

```
>> VA = [8  5  4]; VB = [10  2  7];        Define two vectors VA and VB.
>> VC = VA + VB                            Define a vector VC that is equal to VA + VB.
VC =
   18    7   11
>> A = [5 -3 8; 9 2 10], B = [10 7 4; -11 15 1]   Define two matrices A and B.
```

```
A =
   5   -3   8
   9    2  10
B =
  10    7   4
 -11   15   1
>> C = A + B
```
Define a matrix C that is equal to A + B.
```
C =
  15    4  12
  -2   17  11
>> C - 8
```
Subtract 8 from the matrix C.

8 is subtracted from each element of C.
```
ans =
   7   -4   4
 -10    9   3
```

Multiplication of arrays

The multiplication operation * is executed by MATLAB according to the rules of linear algebra (see Section 2.4.1). This means that if A and B are two matrices, the operation $A*B$ can be carried out only if the number of columns in matrix A is equal to the number of rows in matrix B. The result is a matrix that has the same number of rows as A and the same number of columns as B. For example, if E is a (3×2) matrix and G is a (2×4) matrix, then the operation C=A*B gives a (3×4) matrix:

```
>> A = [2 -1; 8 3; 6 7], B = [4 9 1 -3; -5 2 4 6]
```
Define two matrices A and B.
```
A =
   2   -1
   8    3
   6    7
B =
   4    9    1   -3
  -5    2    4    6
>> C = A*B
```
Multiply A*B.

C is a (3×4) matrix.
```
C =
  13   16   -2  -12
  17   78   20   -6
 -11   68   34   24
```

Two vectors can be multiplied only if both have the same number of elements, and one is a row vector and the other is a column vector. The multiplication of a row vector times a column vector gives a (1×1) matrix, which is a scalar. This is the dot product of two vectors. (MATLAB also has a built-in function, named dot(a,b), that computes the dot product of two vectors.) When using the dot function, the vectors a and b can each be a row or a column vector. The multiplication of a column vector times a row vector, both with n elements, gives an $(n \times n)$ matrix.

```
>> AV = [2  5  1]                          Define three-element row vector AV.
AV =
    2   5   1
>> BV = [3;  1;  4]                         Define three-element column vector BV.
BV =
    3
    1
    4
>> AV * BV              Multiply AV by BV. The answer is a scalar. (Dot product of two vectors.)
ans =
   15
>> BV * AV                                  Multiply BV by AV. The answer is a (3 × 3)
ans =                                       matrix. (Cross product of two vectors.)
    6   15   3
    2    5   1
    8   20   4
```

Array division

The division operation in MATLAB is associated with the solution of a system of linear equations. MATLAB has two types of array division, which are the left division and the right division. The two operations are explained in Section 4.8.1. Note that division *is not* a defined operation in linear algebra (see Section 2.4.1). The division operation in MAT-LAB performs the equivalent of multiplying a matrix by the inverse of another matrix (or vice versa).

Element-by-element operations

Element-by-element operations are carried out on each element of the array (or arrays). Addition and subtraction are already by definition element-by-element operations because when two arrays are added (or subtracted) the operation is executed with the elements that are in the same position in the arrays. In the same way, multiplication, division, and exponentiation are carried out on each element of the array. When two or more arrays are involved in the same expression, element-by-element operations can only be done with arrays of the same size.

Element-by-element multiplication, division, and exponentiation of two vectors or matrices are entered in MATLAB by typing a period in front of the arithmetic operator.

Symbol	Description	Symbol	Description
.*	Multiplication	./	Right division
.^	Exponentiation	.\	Left Division

If two vectors a and b are $a = \begin{bmatrix} a_1 & a_2 & a_3 & a_4 \end{bmatrix}$ and $b = \begin{bmatrix} b_1 & b_2 & b_3 & b_4 \end{bmatrix}$, then element-by-element multiplication, division, and exponentiation of the two vectors are:

$$a .* b = \begin{bmatrix} a_1b_1 & a_2b_2 & a_3b_3 & a_4b_4 \end{bmatrix}$$

$$a ./ b = \begin{bmatrix} a_1/b_1 & a_2/b_2 & a_3/b_3 & a_4/b_4 \end{bmatrix}$$

$$a .^\wedge b = \begin{bmatrix} (a_1)^{b_1} & (a_2)^{b_2} & (a_3)^{b_3} & (a_4)^{b_4} \end{bmatrix}$$

If two matrices A and B are:

$$A = \begin{bmatrix} A_{11} & A_{12} & A_{13} \\ A_{21} & A_{22} & A_{23} \\ A_{31} & A_{32} & A_{33} \end{bmatrix} \quad \text{and} \quad B = \begin{bmatrix} B_{11} & B_{12} & B_{13} \\ B_{21} & B_{22} & B_{23} \\ B_{31} & B_{32} & B_{33} \end{bmatrix}$$

then element-by-element multiplication and division of the two matrices give:

$$A .* B = \begin{bmatrix} A_{11}B_{11} & A_{12}B_{12} & A_{13}B_{13} \\ A_{21}B_{21} & A_{22}B_{22} & A_{23}B_{23} \\ A_{31}B_{31} & A_{32}B_{32} & A_{33}B_{33} \end{bmatrix} \quad A ./ B = \begin{bmatrix} A_{11}/B_{11} & A_{12}/B_{12} & A_{13}/B_{13} \\ A_{21}/B_{21} & A_{22}/B_{22} & A_{23}/B_{23} \\ A_{31}/B_{31} & A_{32}/B_{32} & A_{33}/B_{33} \end{bmatrix}$$

Examples of element-by-element operations with MATLAB are:

```
>> A = [2  6  3;  5  8  4]              Define a (2 × 3) matrix A.
A =
    2    6    3
    5    8    4
>> B = [1  4  10;  3  2  7]             Define a (2 × 3) matrix B.
B =
    1    4   10
    3    2    7
>> A .* B                              Element-by-element multiplication of arrays A and B.
ans =
    2   24   30
   15   16   28
>> C = A ./ B                          Element-by-element division of array A by array B.
C =
   2.0000   1.5000   0.3000
   1.6667   4.0000   0.5714
>> B .^ 3                              Element-by-element exponentiation of array B.
ans =
      1     64   1000
     27      8    343
```

Element-by-element calculations are very useful for calculating the value of a function at many values of its argument. This is done by first defining a vector that contains values of the independent variable and then by using this vector in element-by-element computations to create a vector in which each element is the corresponding value of the function. For example, calculating $y = \dfrac{z^3 + 5z}{4z^2 - 10}$ for eight values of z, $z = 1, 3, 5, \ldots, 15$, is accomplished as follows:

```
>> z = [1:2:15]
z =
    1    3    5    7    9    11    13    15
>> y = (z.^3 + 5*z)./(4*z.^2 - 10)]
y =
   -1.0000   1.6154   1.6667   2.0323   2.4650   2.9241   3.3964   3.8764
```

Define a vector z with eight elements.

Vector z is used in element-by-element calculation of the elements of vector y.

In the last example element-by-element operations are used three times; to calculate z^3 and z^2 and to divide the numerator by the denominator.

MATLAB has many built-in functions for operations with arrays. Several of these functions are listed in Table A-5.

Table A-5: Built-in functions for handling arrays.

Command	Description	Example
mean(A)	If A is a vector, the function returns the mean value of the elements of the vector.	>> A = [5 9 2 4]; >> mean(A) ans = 5
sum(A)	If A is a vector, the function returns the sum of the elements of the vector.	>> A = [5 9 2 4]; >> sum(A) ans = 20
sort(A)	If A is a vector, the function arranges the elements of the vector in ascending order.	>> A = [5 9 2 4]; >> sort(A) ans = 2 4 5 9
det(A)	The function returns the determinant of a square matrix A.	>> A = [2 4; 3 5]; >> det(A) ans = -2

A.5 SCRIPT FILES

A script file is a file that contains a sequence of MATLAB commands, which is also called a program. When a script file is run, MATLAB executes the commands in the order they are written just as if they were typed in the Command Window. When a command generates output (e.g., assignment of a value to a variable without semicolon at the end), the output is displayed in the Command Window. Using a script file is convenient because it can be stored, edited later (corrected and/or changed), and executed many times. Script files can be typed and edited in any text editor and then pasted into the MATLAB editor. Script files are also called M-files because the extension .m is used when they are saved.

Creating and saving a script file

Script files are created and edited in the Editor/Debugger Window. This window is opened from the Command Window. In the *File* menu, select *New* and then select *M-file*. Once the window is open, the commands of the script file are typed line by line. MATLAB automatically numbers a new line every time the *Enter* key is pressed. The commands can also be typed in any text editor or word processor program and then copied and pasted in the Editor/Debugger Window.

Before a script file can be executed, it has to be saved. This is done by choosing *Save As...* from the *File* menu, selecting a location (folder), and entering a name for the file. The rules for naming a script file follow the rules of naming a variable (must begin with a letter, can include digits and underscore, and be up to 63 characters long). The names of user-defined variables, predefined variables, MATLAB commands, or functions should not be used to name script files.

A script file can be executed either by typing its name in the Command Window and then pressing the *Enter* key, or directly from the Editor Window by clicking on the *Run* icon. Before this can be done, however, the user has to make sure that MATLAB can find the file (i.e., that MATLAB knows where the file is saved). In order to be able to run a file, the file must be in either the current directory or the search path.

The current directory is shown in the "Current Directory" field in the desktop toolbar of the Command Window. The current directory can be changed in the Current Directory Window.

When MATLAB is asked to run a script file or to execute a function, it searches for the file in directories listed in the search path. The directories included in the search path are displayed in the Set Path Window that can be opened by selecting *Set Path* in the *File* menu. Once the Set Path Window is open, new folders can be added to, or removed from, the search path.

Global variables

Global variables are variables that, once created in one part of MATLAB, are recognized in other parts of MATLAB. This is the case for variables in the Command Window and script files since both operate on variables in the workspace. When a variable is defined in the Command Window, it is also recognized and can be used in a script file. In the same way, if a variable is defined in a script file, it is also recognized and can be used in the Command Window. In other words, once the variable is created, it exists, can be used, and can be reassigned a new value in both the Command Window and a script file. (There are different types of files in MATLAB, called function files, that normally do not share their variables with other parts of the program. This is explained in Section A.6.)

Input to a script file

When a script file is executed, the variables used in the calculations within the file must have assigned values. The assignment of a value to a variable can be done in three ways, depending on where and how the variable is defined. One option is to define the variable and assign it a value in the script file. In this case the assignment of value to the variable is part of the script file. If the user wants to run the file with a different variable value, the file must be edited and the assignment of the variable changed. Then, after the file is saved, it can be executed again.

A second option is to define the variable and assign it a value in the Command Window. In this case, if the user wants to run the script file with a different value for the variable, the new value is assigned in the Command Window and the file is executed again.

The third option is to define the variable in the script file but assign a specific value in the Command Window when the script file is executed. This is done by using the `input` command to create the variable.

Output from a script file

As discussed earlier, MATLAB automatically generates a display when some commands are executed. For example, when a variable is assigned a value, or the name of a previously assigned variable is typed and the **Enter** key is pressed, MATLAB displays the variable and its value. In addition, MATLAB has several commands that can be used to generate displays. The displays can be messages that provide information, numerical data, and plots. Two commands frequently used to generate output are `disp` and `fprintf`. The `disp` command displays the output on the screen, while the `fprintf` command can be used to display the output on the screen or to save the output to a file.

The `disp` command is used to display the elements of a variable without displaying the name of the variable and to display text. The format of the `disp` command is:

```
disp(name of a variable) or disp('text as string')
```

Every time the `disp` command is executed, the display it generates appears in a new line.

The `fprintf` command can be used to display output (text and data) on the screen or to save it to a file. With this command the output can be formatted. For example, text and numerical values of variables can be intermixed and displayed in the same line. In addition, the format of the numbers can be controlled. To display a mix of text and a number (value of a variable), the `fprintf` command has the form:

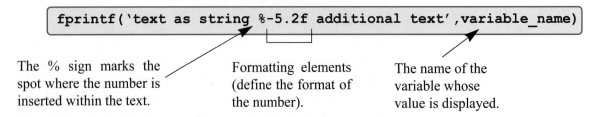

The % sign marks the spot where the number is inserted within the text.

Formatting elements (define the format of the number).

The name of the variable whose value is displayed.

A.6 FUNCTION FILES

A function file is a MATLAB program that is used as a function. It has input and output arguments that can be one or several variables, and each can be a scalar, vector, or an array of any size. Schematically, a function file can be illustrated by:

It can be used for a math function and as subprograms in large programs. In this way large computer programs can be made up of smaller "building blocks" that can be tested independently. MATLAB has built-in functions for all the standard math functions and for many mathematical operations (e.g., solving a nonlinear equation, curve fitting). The user can create user-defined functions that can be used for any purpose.

Function files are created and edited, like script files, in the Editor/ Debugger Window. The first executable line in a function file must be the function definition line that has the form:

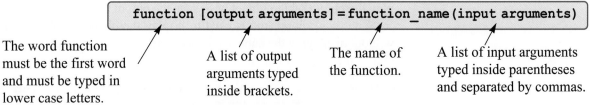

The word function must be the first word and must be typed in lower case letters.

A list of output arguments typed inside brackets.

The name of the function.

A list of input arguments typed inside parentheses and separated by commas.

The word "function", typed in lower case letters, must be the first word in the function definition line. The input and output arguments are used to transfer data into and out of the function. The input arguments are listed inside parentheses following the function name. Usually, there is at least one input argument. If there are more than one, the input arguments are separated by commas. The computer code that performs the calculations within the function file is written in terms of the input arguments and assumes that the arguments have assigned numerical values.

The output arguments, which are listed inside brackets on the left side of the assignment operator in the function definition line, transfer the output from the function file. Function files can have one, several, or no output arguments. If there are more than one, the output arguments are separated with commas. If there is only one output argument, it can be typed without brackets. In order for the function file to work, the output arguments must be assigned values in the computer program that is in the function body.

Following the function definition line, there are usually several lines of comments. They are optional but frequently used to provide information about the function. Next, the function contains the computer program (code) that actually performs the computations. The code can use all MATLAB programming features, including calculations, assignments, any built-in or user-defined functions, and flow control (conditional statements and loops; see Section A.7).

All the variables in a function file are local. This means that the variables are defined and recognized only inside the function file. When a function file is executed, MATLAB uses an area of memory that is separate from the workspace (the memory space of the Command Window and the script files). In a function file, the input variables are assigned values each time the function is called. These variables are then used in the calculations within the function file. When the function file finishes its execution, the values of the output arguments are transferred to the variables that were used when the function was called. Thus, a function file can have variables with the same name as variables in the Command Window or in script files. The function file does not recognize variables with the same name that have been assigned values outside the function. The assignment of values to these variables in the function file will not change their assignment elsewhere.

A simple function, named `loan`, that calculates the monthly and total pay for a loan for a given loan amount, interest rate, and duration is listed next.

```
function [mpay,tpay] = loan(amount,rate,years)        Function definition line.
%loan calculates monthly and total payment of loan.
%Input arguments:
%amount:  loan amount in $.
%rate:   annual interest rate in percent.
%years:   number of years.
%Output arguments:
%mpay:   monthly payment.
%tpay:   total payment.

format bank
ratem = rate*0.01/12;
a = 1 + ratem;
```

```
b = (a^(years*12) - 1)/ratem;
mpay = amount*a^(years*12)/(a*b);
tpay = mpay*years*12;
```

Assign values to the output arguments.

The function `loan` is next used in the Command Window for calculating the monthly and total pay of a four-year loan of $25,000 with interest rate of 4%:

```
>> [month total] = loan(25000,7.5,4)
month =
     600.72
total =
    28834.47
```

A.7 PROGRAMMING IN MATLAB

A computer program is a sequence of computer commands. In a simple program the commands are executed one after the other in the order that they are typed. Many situations, however, require more sophisticated programs in which different commands (or groups of commands) are executed when the program runs with different input variables, In other situations there might be a need to repeat a sequence of commands several times within a program. For example, programs that solve equations numerically repeat a sequence of calculations until the error in the answer is smaller than some measure.

MATLAB provides several tools that can be used to control the flow of a program. Conditional statements make it possible to skip commands or to execute specific groups of commands in different situations. `For` loops and `while` loops make it possible to repeat a sequence of commands several times.

Changing the flow of a program requires some kind of decision-making process within the program. The computer must decide whether to execute the next command or to skip one or more commands and continue at a different line in the program. The program makes these decisions by comparing values of variables. This is done by using relational and logical operators.

A.7.1 Relational and Logical Operators

Relational and logical operators are used in combination with other commands in order to make decisions that control the flow of a computer program. A relational operator compares two numbers by determining whether a comparison statement (e.g., 5 < 8) is true or false. If the statement is true, it is assigned a value of 1. If the statement if false, it is assigned a value of 0. Relational operators in MATLAB are given in the table that follows.

Relational Operator	Description	Relational Operator	Description
<	Less than	>=	Greater than or equal to
>	Greater than	= =	Equal to 5
<=	Less than or equal to	~=	Not equal to

Note that the "equal to" relational operator consists of two = signs (with no space between them), since one = sign is the assignment operator. Also, in other relational operators that consist of two characters there is no space between the characters (<=, >=, ~=). Two examples are:

```
>> 5 > 8
ans =
    0
>> 4 = = 6
ans =
    0
```

A logical operator examines true/false statements and produces a result that is true (1) or false (0) according to the specific operator. Logical operators in MATLAB are:

Logical operator	Name	Description
& Example: A&B	AND	Operates on two operands (A and B). If both are true, the result is true (1); otherwise the result is false (0).
| Example: A|B	OR	Operates on two operands (A and B). If either one, or both are true, the result is true (1); otherwise (both are false) the result is false (0).
~ Example: ~A	NOT	Operates on one operand (A). Gives the opposite of the operand. True (1) if the operand is false, and false (0) if the operand is true.

Logical operators can have numbers as operands. A nonzero number is true, and a zero is false. Several examples are:

```
>> 3&7
ans =
    1
>> a = 5|0
a =
    1
>> ~25
ans =
    0
```

3 and 7 are both true (nonzero), so the outcome is 1.

1 is assigned to a since at least one number is true (nonzero).

A.7.2 Conditional Statements, if-else Structures

A conditional statement is a command that allows MATLAB to make a decision of whether to execute a group of commands that follow the conditional statement, or to skip these commands. In a conditional statement, a conditional expression is stated. If the expression is true, a group of commands that follow the statement is executed. If the expression is false, the computer skips the group.

The if-end structure

The simplest form of a conditional statement is the `if-end` structure, which is shown schematically in Fig. A-1. The figure shows how the

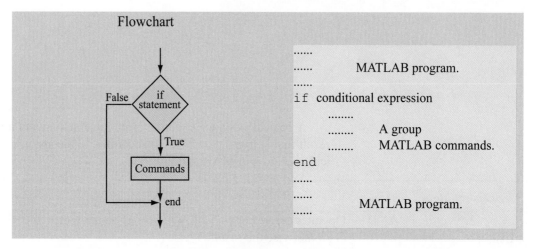

Figure A-1: The structure of the `if-end` conditional statement.

commands are typed in the program and presents a flowchart that symbolically shows the flow, or the sequence, in which the commands are executed. As the program executes, it reaches the `if` statement. If the conditional expression in the `if` statement is true (1), the program continues to execute the commands that follow the `if` statement all the way down to the `end` statement. If the conditional expression is false (0), the program skips the group of commands between the `if` and the `end`, and continues with the commands that follow the `end` statement.

The if-else-end structure

The `if-else-end` structure provides a means for choosing one group of commands, out of a possible two groups, for execution (see Fig. A-2). The first line is an `if` statement with a conditional expression. If the conditional expression is true, the program executes the group 1 commands between the `if` and the `else` statements, and then skips to the `end`. If the conditional expression is false, the program skips to the `else` statement and executes the group 2 commands between the `else` and the `end` statements.

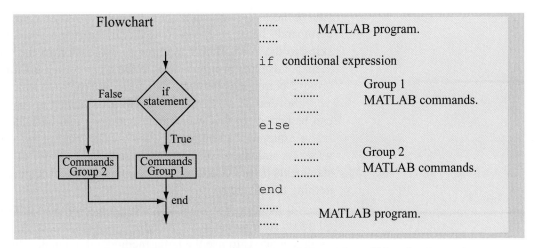

Figure A-2: The structure of the `if-else-end` conditional statement.

The if-elseif-else-end structure

The `if-elseif-else-end` structure is shown in Fig. A-3. This structure includes two conditional statements (`if` and `elseif`) that make it possible to select one out of three groups of commands for execution. The first line is an `if` statement with a conditional expression. If the conditional expression is true, the program executes the group 1 commands between the `if` and the `elseif` statements and then skips to the `end`. If the conditional expression in the `if` statement is false, the program skips to the `elseif` statement. If the conditional expres-

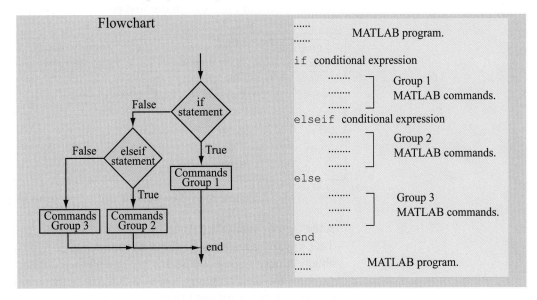

Figure A-3: The structure of the `if-elseif-else-end` conditional statement.

sion in the elseif statement is true, the program executes the group 2 commands between the elseif and the else statements and then skips to the end. If the conditional expression in the elseif statement is false, the program skips to the else statement and executes the group 3 commands between the else and the end statements.

Several elseif statements and associated groups of commands can be added. In this way more conditions can be included. Also, the else statement is optional. This means that in the case of several elseif statements and no else statement, if any of the conditional statements is true, the associated commands are executed, but otherwise nothing is executed.

In general, the same task can be accomplished by using several elseif statements or if-else-end structures. A better programming practice is to use the latter method, which makes the program easier to understand, modify, and debug.

A.7.3 Loops

A loop is another means to alter the flow of a computer program. In a loop, the execution of a command, or a group of commands, is repeated several times consecutively. Each round of execution is called a pass. In each pass at least one variable (but usually more than one) that is defined within the loop is assigned a new value.

for-end loops

In for-end loops, the execution of a command or a group of commands is repeated a predetermined number of times. The form of the loop is shown in Fig. A-4. In the first pass k = f, and the computer exe-

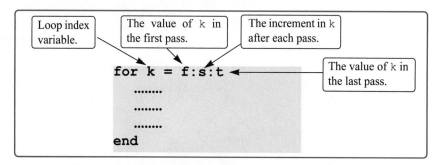

Figure A-4: The structure of a for-end loop.

cutes the commands between the for and the end commands. Then, the program goes back to the for command for the second pass. k obtains a new value equal to k = f + s, and the commands between the for and the end commands are executed with the new value of k. The process repeats itself until the last pass where k = t. Then the program does not go back to the for, but continues with the commands that follow the end command. For example, if k = 1:2:9, then there are five loops, and the value of k in the passes is 1, 3, 5, 7, and 9. If the increment value s is omitted, its value is 1 (default) (i.e., k = 3:7 produces five passes with k = 3, 4, 5, 6, 7).

A program that illustrates the use of conditional statements and loops is shown next (script file). The program changes the elements of a given vector such that elements that are positive and are divisible by 3 and/or 5 are doubled. Elements that are negative but greater than –5 are raised to the power of 3, and all the other elements are unchanged.

```
V = [5, 17, -3, 8, 0, -7, 12, 15, 20 -6, 6, 4, -2, 16];
n = length(V);
for k = 1:n          In the kth pass of the loop the kth element is checked and changed, if needed.
    if  V(k) > 0 & (rem(V(k),3) == 0 | rem(V(k),5) == 0)
        V(k) = 2*V(k);
    elseif  V(k) < 0 & V(k) > -5
        V(k) = V(k)^3;
    end
end
V
```

When the program is executed, the following new vector V is displayed in the Command Window:

```
V =
   10   17  -27   8   0  -7   24   30   40  -6   12   4  -8   16
```

A.8 PLOTTING

MATLAB has many commands that can be used for creating different types of plots. These include standard plots with linear axes, plots with logarithmic axes, bar and stairs plots, polar plots, and many more. The plots can be formatted to have a desired appearance.

Two-dimensional plots can be created with the plot command. The simplest form of the command is:

$$\boxed{\texttt{plot(x,y)}}$$

The arguments x and y are each a vector (one-dimensional array). Both vectors must have the same number of elements. When the plot command is executed, a figure is appears in the Figure Window, which opens automatically. The figure has a single curve with the x values on the abscissa (horizontal axis) and the y values on the ordinate (vertical axis). The curve is constructed of straight line segments that connect the points whose coordinates are defined by the elements of the vectors x and y. The vectors, of course, can have any name. The vector that is typed first in the plot command is used for the horizontal axis, and the vector that is typed second is used for the vertical axis. The figure that is displayed has axes with linear scale and default range. For example, if a vector x has the elements 1, 2, 3, 5, 7, 7.5, 8, 10, and a vector y has the elements 2, 6.5, 7, 7, 5.5, 4, 6, 8, a simple plot of y versus x can be produced by typing the following in the Command Window:

```
>> x = [1  2  3  5  7  7.5  8  10];
>> y = [2  6.5  7  7  5.5  4  6  8];
>> plot(x,y)
```

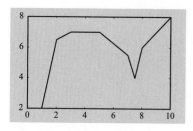

Figure A-5: A plot of data points.

Once the `plot` command is executed, the plot that is shown in Fig. A-5 is displayed in the Figure Window.

The `plot` command has additional optional arguments that can be used to specify the color and style of the line and the color and type of markers, if any are desired. With these options the command has the form:

$$\boxed{\texttt{plot(x,y, 'line specifiers',)}}$$

Line specifiers can be used to define the style and color of the line and the type of markers (if markers are desired). The line style specifiers are:

Line Style	Specifier	Line Style	Specifier
solid (default)	-	dotted	:
dashed	--	dash-dot	-.

The line color specifiers are:

Line Color	Specifier	Line Style	Specifier	Line Color	Specifier	Line Color	Specifier
red	r	blue	b	magenta	m	black	k
green	g	cyan	c	yellow	y	white	w

The marker type specifiers are:

Marker	Specifier	Marker	Specifier	Marker	Specifier
plus sign	+	asterisk	*	square	s
circle	o	point	.	diamond	d

The specifiers are typed inside the `plot` command as strings. Within the string the specifiers can be typed in any order.

The `plot` command creates bare plots. The plot can be modified to include axis labels, a title, and other features. Plots can be formatted by using MATLAB commands that follow the `plot` command. The

Figure A-6: Formatted plot.

formatting commands for adding axis labels and a title are:

```
xlabel('text as string')
ylabel('text as string')
title('text as string')
```

For example, the program listed below produces the plot that is displayed in Fig. A-6. The data is plotted with a dashed red line and asterisk markers, and the figure includes axis labels and a title.

```
>> yr = [1988:1:1994];
>> sle = [8  12  20  22  18  24  27];
>> plot(yr,sle,'--r*','linewidth',2,'markersize',12)
>> xlabel('YEAR')
>> ylabel('SALES (Millions)')
>> title('Sales Records')
```

Formatting of plots can also be done in the Figure Window using the Plot Editor.

A.9 PROBLEMS

A.1 Define the variables a, b, c, and d as $a = 14.75$, $b = -5.92$, $c = 61.4$, and $d = 0.6(ab - c)$. Evaluate:

(a) $\quad a + \dfrac{ab(a+d)^2}{c}\dfrac{}{\sqrt{|ab|}}$

(b) $\quad de^{\left(\frac{d}{2}\right)} + \dfrac{(ad + cd)/\left(\dfrac{25}{a} + \dfrac{35}{b}\right)}{(a + b + c + d)}$

A.2 Define two variables: $alpha = 5\pi/9$, $beta = \pi/7$. Using these variables, show that the following trigonometric identity is correct by calculating the value of the left and right sides of the equation.

$$\sin\alpha\cos\beta = \frac{1}{2}[\sin(\alpha - \beta) + \sin(\alpha + \beta)]$$

A.3 Create a row vector with 16 equally spaced elements in which the first element is 4 and the last element is 61.

A.4 Create a column vector in which the first element is 31, the elements decrease with increments of –4, and the last element is –9. (A column vector can be created by the transpose of a row vector.)

A.5 Create the matrix shown by using the vector notation for creating vectors with constant spacing when entering the rows (i.e., do not type individual elements).

```
A =
       0    1.0000    2.0000    3.0000    4.0000    5.0000    6.0000
  3.0000    9.1667   15.3333   21.5000   27.6667   33.8333   40.0000
 28.0000   27.7500   27.5000   27.2500   27.0000   26.7500   26.5000
  6.0000    5.0000    4.0000    3.0000    2.0000    1.0000         0
```

A.6   Create the matrix A in Problem A.5, and then use colons to address a range of elements to create the following vectors:

(*a*) Create a four-element row vector named va that contains the second through fifth elements of the third row of *A*.

(*b*) Create a four-element column vector named vb that contains the elements of the sixth column of *A*.

A.7   Create the matrix A in Problem A.5, and then use colons to address a range of elements to create the following matrices:

(*a*) Create a 3 × 4 matrix *B* from the first, second, and fourth rows, and the first, second, fourth, and seventh columns of the matrix *A*.

(*b*) Create a 2 × 3 matrix *C* from the second and fourth rows, and the second, fifth, and sixth columns of the matrix *A*.

A.8   For the function $y = \dfrac{(x^3 + 1)^2}{x^2 + 2}$, calculate the value of *y* for the following values of *x*: $-1.6\ -1.2\ -0.8$
$-0.4\ 0\ 0.4\ 0.8\ 1.2$. Solve the problem by first creating a vector *x*, and then creating a vector *y*, using element-by-element calculations. Make a plot of the points with asterisk marker for the points and black line connecting the points. Label the axes.

A.9   Define *k* as scalar $k = 0.8$ and *x* as the vector $x = -3, -2.8, -2.6, \ldots, 2.6, 2.8, 3$. Then use these variables to calculate *y* by: $y = \dfrac{8a^2}{x^2 + 4a^2}$. Plot *y* versus *x*.

A.10 Use MATLAB to show that the sum of the infinite series $\displaystyle\sum_{n=0}^{\infty}(-1)^n\dfrac{1}{(2n+1)}$ converges to $\pi/4$. Do it by computing the sum for:

(*a*) $n = 100$

(*b*) $n = 1{,}000$

(*c*) $n = 5{,}000$

   In each part create a vector n in which the first element is 0, the increment is 1 and the last term is 100, 1,000, or 5,000. Then, use element-by-element calculation to create a vector in which the elements are $(-1)^n\dfrac{1}{(2n+1)}$. Finally, use the function sum to add the terms of the series. Compare the values obtained in parts *a*, *b*, and *c* with the value of $\dfrac{\pi}{4}$. (Do not forget to type semicolons at the end of commands that otherwise will display large vectors.)

A.11 The Gateway Arch in St. Louis is shaped according to the equation:

$$y = 693.8 - 68.8\cosh\left(\dfrac{x}{99.7}\right) \text{ ft.}$$

Make a plot of the Arch for $-299.25 \le x \le 299.95$ ft.

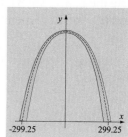

A.12 A simply supported beam is pinned at one end and is supported by a roller at the other end. The deflection y at point x of a beam loaded with the distributed load shown is given by the equation:

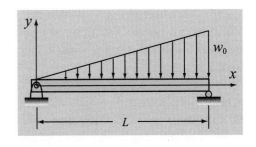

$$y = \frac{-w_0 x}{360 EIL}(3x^3 - 10L^2 x^2 + 7L^4)$$

where E is the elastic modulus, I is the moment of inertia, and L is the length of the beam. For the beam shown in the figure $L = 6$ m, $E = 70 \times 10^9$ Pa (aluminum), $I = 9.19 \times 10^{-6}$ m⁴, and $w_0 = 800$ N/m.

Plot the deflection of the beam y as a function of x.

A.13 Write a user-defined MATLAB function that converts the mass and height of a person from SI units (kg and cm) to weight and height in standard units (lb and in.). For the function name and arguments use [in, lb] = STtoSI(cm, kg). The input arguments are the height in centimeters and mass in kg, and the output arguments are the height in inches and weight in pounds. Use the function in the Command Window to determine in standard units the height and mass of a 85 kg person who is 178 cm tall.

A.14 Write a user-defined MATLAB function that determines the cross-sectional area, A, and moments of inertia I_{xx} and I_{yy} of an "I" beam. For the function name and arguments use [A,Ixx,Iyy] = MofIner(b,h,tf,tw). The input arguments are the width, b, height, h, flange thickness, t_f, and web thickness, t_w, of the beam, in millimeter, as shown in the figure. The output arguments are the cross-sectional area (in mm²), and the moments of inertia (in mm⁴). Use the function in the Command Window to determine the cross-sectional area and moments of inertia and of a beam with $b = 140$ mm, $h = 400$ mm, $t_f = 8.8$ mm, and $t_w = 6.4$ mm.

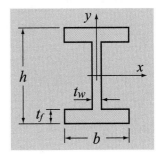

A.15 Write a user-defined MATLAB function that calculates the equivalent resistance, R_{eq} of n resistors $R_1, R_2, ..., R_n$ connected in parallel. For function name and arguments use Req = EqResistance(R). The input argument is a vector whose elements are the values of the resistors. The output argument is the value of the equivalent resistance. The function should work for any number of resistors. Use the function in the Command Window to determine the equivalent resistance of the following five resistors that are connected in parallel $R_1 = 200\Omega$, $R_2 = 600\Omega$, $R_3 = 1000\Omega$, $R_4 = 100\Omega$, and $R_5 = 500\Omega$.

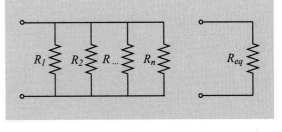

A.16 A vector is given by: $x = [15\ \ 85\ 72\ 59\ 100\ 80\ 44\ 60\ 91\ 38]$. Using conditional statements and loops write a program that determines the average of the elements of the vector that are larger than 59.

A.17 Write a user-defined function that creates a vector whose elements are the primary numbers between two numbers. For the function name and arguments use pr = Primary(a,b). The input to the function are two numbers (integers) a and b (such that $a < b$), and the output pr is a vector in which the elements are the primary between a and b.

A.18 Write a user-defined function that sorts the elements of a vector (of any length) from the largest to the smallest. For the function name and arguments use y = downsort(x). The input to the function is a vector x of any length, and the output y is a vector in which the elements of x are arranged in descending order. Do not use the MATLAB sort function. Test your function by using it in the Command Window to rearrange the elements of the following vector: [-2, 8, 29, 0, 3, -17, -1, 54, 15, -10, 32].

A.19 A cylindrical, vertical fuel tank has hemispheric end cap at the bottom and a conic end cap at the top as shown. The radius of the cylinder and the hemispheric end cap is $r = 60$ cm.

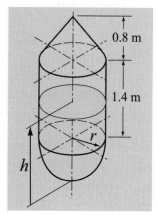

Write a user-defined function (for the function name and arguments use V = Vfuel(h)) that gives the volume of the fuel in the tank as a function of the height h. Use the function to make a plot of the volume as a function of h for $0 \le h \le 2.8$ m.

Appendix B

MATLAB Programs

The following table is a summary of the major MATLAB programs (script files and user-defined functions) that are listed in the book.

Program Number	Program Name or Title	File Type	Description	Page
3-1	Bisection method	Script file	Solves a nonlinear equation using the bisection method.	58
3-2	NewtonRoot	User-defined function	Finds the root of a nonlinear equation using Newton's method.	64
3-3	SecantRoot	User-defined function	Finds the root of a nonlinear equation using the secant method.	69
4-1	Gauss	User-defined function	Solves a system of linear equations using the Gauss elimination method.	103
4-2	GaussPivot	User-defined function	Solves a system of linear equations using Gauss elimination method with pivoting.	107
4-3	LUdecompCrout	User-defined function	Decomposes a matrix into lower triangular and upper triangular matrices using Crout's method.	120
4-4	ForwardSub	User-defined function	Applies forward substitution.	120
4-5	BackwardSub	User-defined function	Applies backward substitution.	121
4-6	Crout's LU decomposition	Script file	Solves a system of linear equations using Crout's LU decomposition method.	121
4-7	InverseLU	User-defined function	Calculates the inverse of a matrix using Crout's LU decomposition method.	124

Program Number	Program Name or Title	File Type	Description	Page
4-8	Gauss Seidel iteration	Script file	Solves a system of four linear equations using the Gauss–Seidel iteration method.	129
4-9	Tridiagonal	User-defined function	Solves a tridiagonal system of linear equations using the Thomas algorithm.	138
4-10	QRFactorization	User-defined function	Factors a matrix into an orthogonal matrix and an upper-triangular matrix.	166
5-1	LinearRegression	User-defined function	Curve fitting with linear function using the least squares method.	185
5-2	Curve fitting with non-linear functions	Script file	Curve fitting with nonlinear functions using the least squares method.	191
5-3	Curve fitting using polynomial regression	Script file	Curve fitting of a fourth-order polynomial.	195
5-4	LagrangeINT	User-defined function	Interpolation using Lagrange polynomial.	201
5-5	NewtonINT	User-defined function	Interpolation using Newton's polynomial.	208
5-6	LinearSpline	User-defined function	Interpolation using linear splines.	211
6-1	derivative	User-defined function	Calculates the derivative of a function that is given by a set of points.	238
7-1	trapezoidal	User-defined function	Numerical integration using the composite trapezoidal method.	275
7-2	Romberg	User-defined function	Numerical integration using the Romberg integration method.	297
8-1	odeEuler	User-defined function	Solves a first-order ODE using Euler's explicit method.	315
8-2	Solving a first-order ODE	Script file	Solves a first-order initial value problem using Euler's implicit method.	322

Program Number	Program Name or Title	File Type	Description	Page
8-3	`odeModEuler`	User-defined function	Solving first-order ODE using the modified Euler method.	325
8-4	`odeRK4`	User-defined function	Solves a first-order ODE using the fourth-order Runge-Kutta method.	338
8-5	`Sys2ODEsRK2`	User-defined function	Solves a system of two first-order ODEs using the second-order Runge-Kutta method.	351
8-6	`Sys2ODEsRK4`	User-defined function	Solves a system of two first-order ODEs using the fourth-order Runge-Kutta method.	356
9-1	Solving a second-order ODE (BVP)	Script file	Solving second-order ODE (BVP) using the shooting method.	394
9-2	Solving a second-order ODE (BVP)	Script file	Solving second-order ODE (BVP) using the shooting method in conjunction with the bisection method.	396
9-3	Solving a linear second-order ODE (BVP)	Script file	Solving a linear second-order ODE using the finite difference method.	400
9-4	Solving a nonlinear second-order ODE (BVP)	Script file	Solving a nonlinear second-order ODE using the finite difference method.	404
9-5	Solving a linear second-order ODE (BVP with mixed boundary conditions)	Script file	Solving a linear second-order ODE with mixed boundary conditions using the finite difference method.	407

Index